Rajendra Vasant Gurjar, Kayo Masuda, Masayoshi Miyanishi
Affine Space Fibrations

De Gruyter Studies in Mathematics

Volume 79

Rajendra Vasant Gurjar, Kayo Masuda,
Masayoshi Miyanishi

Affine Space Fibrations

DE GRUYTER

Mathematics Subject Classification 2010
Primary: 14R25; Secondary: 14R10

Authors

Prof. Dr. Rajendra Vasant Gurjar
Indian Institute of Technology Mumbai
Department of Mathematics
Mumbai 400076
India
gurjar@math.iitb.ac.in

Prof. Dr. Kayo Masuda
Kwansei Gakuin University
Department of Mathematical Sciences
2-1 Gakuen
Sanda 669-1337
Japan
kayo@kwansei.ac.jp

Prof. Dr. Masayoshi Miyanishi
Kwansei Gakuin University
Research Center for Mathematical Sciences
2-1 Gakuen
Sanda 669-1337
Japan
miyanisi@kwansei.ac.jp

ISBN 978-3-11-057736-5
e-ISBN (PDF) 978-3-11-057756-3
e-ISBN (EPUB) 978-3-11-057742-6
ISSN 0179-0986

Library of Congress Control Number: 2021933810

Bibliographic information published by the Deutsche Nationalbibliothek
The Deutsche Nationalbibliothek lists this publication in the Deutsche Nationalbibliografie; detailed bibliographic data are available on the Internet at http://dnb.dnb.de.

Typesetting: VTeX UAB, Lithuania
Printing and binding: CPI books GmbH, Leck

www.degruyter.com

Preface

Abhyankar–Moh–Suzuki theorem (abbreviated as *AMS theorem*, see Theorem 1.3.14) asserts that an algebraic curve C on the affine plane $\mathbb{A}^2$ which is isomorphic to $\mathbb{A}^1$ is a fiber of an $\mathbb{A}^1$-bundle morphism $f : \mathbb{A}^2 \to \mathbb{A}^1$ which is a trivial bundle. There are now more than a dozen of proofs of this theorem. In principle, the following geometric interpretation will explain the difficulties lying in the proofs of the theorem. Embed $\mathbb{A}^2$ into the projective plane $\mathbb{P}^2$ as the complement of a line at infinity ℓ_∞, i. e., $\mathbb{P}^2 = \mathbb{A}^2 \cup \ell_\infty$. Let $\overline{C}$ be the closure of C in $\mathbb{P}^2$. Then $\overline{C}$ has only one point P_∞ outside C and $\overline{C}$ meets ℓ_∞ at the point P_∞. Let $d = \deg \overline{C}$ and e the multiplicity of $\overline{C}$ at P_∞. Then $\overline{C}$ is linearly equivalent to $d\ell_\infty$. Namely, for a polynomial $f(x,y)$ such that C is defined by $f(x,y) = 0$, the polynomial $f(x,y)$ considered as a rational function on $\mathbb{P}^2$ has C as the zero and $d\ell_\infty$ as the pole. Hence we can consider a linear pencil Λ generated by $\overline{C}$ and $d\ell_\infty$ which has a base point P_∞ and whose general member is the zero of a rational function $f(x,y) + c$ with $c \in \mathbb{C}$. We eliminate the base points of Λ by blowing up base points and taking the proper transforms of Λ. We can decompose this process as finitely many blowing-ups iterated. Let

$$f : V_n \xrightarrow{\sigma_n} V_{n-1} \xrightarrow{\sigma_{n-1}} V_{n-2} \longrightarrow \cdots \longrightarrow V_1 \xrightarrow{\sigma_1} V_0 := \mathbb{P}^2$$

be the necessary blowing-ups. Let $f_i = \sigma_1 \cdots \sigma_i : V_i \to V_0$ and E_i be the exceptional curve arising from σ_i. Then the proof consists of showing:

(1) The proper transform $\Lambda_i := f_i'(\Lambda)$ has only one base point $P_\infty^{(i)}$ lying over P_∞ if $i < n$.
(2) The exceptional curve E_i appears with positive coefficient in the member of Λ_i mapping to $d\ell_\infty$ of Λ if $i < n$.
(3) The last exceptional curve E_n is a cross-section of the linear pencil Λ_n which is free from base points. Further, the morphism $\Phi_{\Lambda_n} : V_n \to \mathbb{P}^1$ defines a $\mathbb{P}^1$-fibration and has only one singular fiber which contains the proper transform of ℓ_∞ as a component.

Then $\mathbb{A}^2$ is contained in V_n as an open set whose complement is the union of the proper transform of ℓ_∞ and all exceptional curves. Hence $\mathbb{A}^2$ as an open set of V_n is *ruled* by curves which are all isomorphic to $\mathbb{A}^1$.

These arguments lead us to introduce the following two notions:

(i) $\mathbb{A}^1$-fibrations. Namely, a dominant morphism $f : X \to Y$ of algebraic varieties is an $\mathbb{A}^1$-fibration if general (closed) fibers X_y with $y \in Y$ are isomorphic to $\mathbb{A}^1$.
(ii) A given curve C on an algebraic variety X is a fiber of an $\mathbb{A}^1$-fibration $f : X \to Y$. Then we say that *the curve C moves in an $\mathbb{A}^1$-fibration*.

We summarize various properties of $\mathbb{A}^1$-fibrations in Section 2.1, Chapter 2. Among others, there exists a Zariski open set U of Y such that the restriction $f|_{f^{-1}(U)} : f^{-1}(U) \to U$ is a locally trivial $\mathbb{A}^1$-bundle in the Zariski topology. An $\mathbb{A}^1$-fibration $f : X \to Y$ has

https://doi.org/10.1515/9783110577563-201

different properties depending on whether the base variety Y is complete or not. We say that f is *of complete type* if Y is complete, and *of affine type* if Y is affine. If an $\mathbb{A}^1$-fibration $f : X \to Y$ is of affine type with an affine variety X, it is shown that f is the quotient morphism of an action of the additive group G_a on X under a mild condition. Namely, the coordinate ring of Y is the G_a-invariant subring A^{G_a} of the coordinate ring A, and the morphism f is determined by the canonical inclusion $A^{G_a} \hookrightarrow A$. Hence it is important to consider G_a-actions on affine varieties and the associated quotient morphisms. One can say that G_a-actions give means to study $\mathbb{A}^1$-fibrations on an affine variety *from the interior*.

There are also means to study $\mathbb{A}^1$-fibrations *from the exterior*. We say that a smooth affine surface is a *Gizatullin surface* if there exists a smooth projective surface V and an effective divisor $D = D_1 + \cdots + D_r$ such that $X = V \setminus D$ and D is a *zigzag*, i. e., D is a divisor with simple normal crossings consisting of smooth rational curves and the dual graph of D is a linear chain. Then any curve C on X moves in an $\mathbb{A}^1$-fibration on X if C is isomorphic to $\mathbb{A}^1$ (see Theorems 2.6.2 and 2.6.12). In fact, a Gizatullin surface X has at least two independent G_a-actions (hence $\mathbb{A}^1$-fibrations) and $\Gamma(X, \mathcal{O}_X)^* = \mathbb{C}^*$, i. e., no nonconstant invertible functions on X, while X is not necessarily factorial. Here we say that X is factorial if the coordinate ring is a unique factorization domain. The existence of two independent G_a-actions is shown by a deft change of the divisor D by blowing-ups and blowing-downs. A Gizatullin surface X is isomorphic to $\mathbb{A}^2$ if and only if X is factorial. Thus we know that AMS theorem holds for affine surfaces belonging to a broader category of Gizatullin surfaces.

C. P. Ramanujam [192] defined the *fundamental group at infinity* $\pi_1^\infty(X)$ for a complex smooth affine surface X, and characterized the affine plane $\mathbb{A}^2$ as X which is topologically contractible and has $\pi_1^\infty(X) = (1)$. The group $\pi_1^\infty(X)$ is defined as $\pi_1(\partial U)$, where ∂U is the boundary of a *tubular neighborhood* U of the boundary divisor $D := V \setminus X$ in the metric topology, when X is embedded as an open set into a smooth projective surface V so that D is a divisor with simple normal crossings. It is curious enough that the finiteness of $\pi_1^\infty(X)$ implies the existence of an $\mathbb{A}^1$-fibration if X is smooth and D is connected (see Theorem 1.3.19). The situation is still quite subtle as shown by the result that if X is normal but singular, $X \cong \mathbb{A}^2/G$ if X is topologically contractible and $\pi_1^\infty(X)$ is finite for a finite subgroup $G \subset \mathrm{GL}(2, \mathbb{C})$, where X has no $\mathbb{A}^1$-fibrations unless G is cyclic (see Theorem 1.3.20).

Iitaka [95] introduced the *logarithmic Kodaira dimension* $\overline{\kappa}(X)$ for $X = V \setminus D$, where V is a smooth projective variety and D is a divisor with simple normal crossings. It is defined by the divisor $K_V + D$ in the same way as the Kodaira dimension $\kappa(V)$ by the canonical divisor K_V. The dimension $\overline{\kappa}(X)$ is independent of the choice of a pair (V, D). At the beginning of this theory, there was an optimistic forecast that the Enriques–Kodaira classification for projective algebraic surfaces can be achieved more or less in the same ways for *open (incomplete) algebraic surfaces*. In fact, this forecast was correct if modified in details. In the case $\overline{\kappa}(X) = -\infty$ and X is affine, for example, X has an $\mathbb{A}^1$-fibration if X is smooth and X contains $\mathbb{A}^2/G$ as an open set if X is singular.

A connection between $\pi_1^\infty(X)$ and $\overline{\kappa}(X)$ is given by a result that $|\pi_1^\infty(X)| < \infty$ implies $\overline{\kappa}(X) = -\infty$.

It is our purpose in writing the present book to explain in details the above linkage of different concepts from the viewpoint of $\mathbb{A}^1$-fibrations, or more generally the *affine space fibrations* of higher relative dimension. In Chapter 1, we give an overview in the case of surfaces and the readers are referred to various papers when they are interested. The details will be developed in Chapter 2 with emphasis on $\mathbb{A}^1$-fibrations induced by G_a-actions. Chapter 3 is a collection of our trials to extend the results in the case of surfaces to the case of higher dimensional affine varieties. We, the authors, firmly believe that there should exist exciting results in the higher-dimensional case, but it is also certain that we need new angles to observe various phenomena in incomplete worlds.

During the preparation of the manuscript, we were given research and financial supports from the Mathematisches Forschungsinstitut Oberwolfach (Rip 2018). The first author was an INSA Scientist at the Indian Institute of Technology Bombay, and the second and third authors were supported by JSPS research grants (Grant-in-Aid for Scientific Research (C) 15K04831 and 16K05115). For all of assistance, we would like to express our deepest gratitude. We would like to express our deep gratitude to P.K. Russell and late M. Koras for the collaboration we had for many years and at different occasions. A considerable part of the results contained in this volume is based on this collaboration.

R. V. Gurjar, K. Masuda, M. Miyanishi

Contents

1 Preliminaries

The research field which we will describe in this book is called *Affine Algebraic Geometry*. Though this name bears an impression that research deals with only affine algebraic varieties, it is not always so because structures and properties of affine varieties depend on how they behave at infinity. Since such varieties are not compact, we have to add some conditions which control behaviors at infinity. One way is to consider a pair (V, D) of a complete algebraic variety V and an effective divisor D with simple normal crossings such that $V \setminus D$ is a given affine variety X. This is a viewpoint of what is called *Logarithmic Geometry*, and research from this direction is fast growing in algebraic geometry. We employ this direction by using *logarithmic Kodaira dimension* introduced first by Iitaka [95]. Important is that this dimension does not depend on the choice of a pair (V, D) if $X = V \setminus D$ is the same. By taking this viewpoint, we can use abundant tools like divisors, linear systems, morphisms defined by linear systems, etc., which have been used to treat mainly complete algebraic varieties.

If algebraic varieties are defined over the complex number field $\mathbb{C}$, we can consider the complex metric and topology on algebraic varieties. This enables us to consider good neighborhoods $\{U_n\}$ of the boundary at infinity. The fundamental group of the boundary ∂U_n, which is defined in the interior part X and hence denoted by $\pi_1^\infty(X)$, is a topological tool to judge the behavior of X near the boundary at infinity. There are pioneering works of Zariski about the abelian property of the fundamental group of $\mathbb{P}^2 \setminus C$, where C is a plane curve with at most nodal singular points. We are very much influenced by the works of Mumford [171] and Ramanujam [192] which enable us to write the fundamental group by means of generators and relations.

Affine algebraic varieties, or more generally incomplete algebraic varieties which we call also *open algebraic varieties,* have more room for topological arguments or tools like fundamental groups, singular homology or cohomology groups, cohomology groups with compact support, Poincaré and Lefschetz dualities. It is one of the features particular in affine algebraic geometry that more topological arguments are used in research.

We have to add the importance of algebraic group actions in the research. In particular, unipotent group actions examplified by the additive group G_a play significant roles in research. The additive group action on an affine algebraic variety $X = \operatorname{Spec} A$ corresponds to a locally nilpotent derivation δ on the coordinate ring A. Commutative algebra is then available by this correspondence. There are invariants like the *Makar-Limanov invariant* or the *Derksen invariant*, which are defined by the additive group actions on X. The existence of an additive group action allows us to single out locally one variable from the coordinate ring. These invariants measure how many variables can be singled out.

Affine space fibration, which is the main subject of the present volume, is a dominant morphism $f : X \to Y$ whose general fibers are isomorphic to $\mathbb{A}^n$. It is equally sig-

https://doi.org/10.1515/9783110577563-001

nificant in affine algebraic geometry as a projective space fibration in projective algebraic geometry. This subject will be explained in details in Chapters 2 and 3. Since the arguments involve the above different directions, logarithmic geometry, topology and transformation groups, we think it is better for the readers that they understand what kind of results and arguments leading to such results are used. So, in Chapter 1, together with many problems attached at the end of the chapter, we give some overview of results obtained for affine algebraic surfaces with the emphasis on topological approaches and introduction to the theory of open algebraic surfaces.

Some definitions and results will be discussed again in the subsequent chapters and developed in details.

General references for basic definitions and results are [90, 158, 19, 213, 225].

By abuse of notation, the set-theoretic complement of a closed subset Y of a topological space X is written as $X - Y$ or $X \setminus Y$. We assume throughout Chapter 1 that the ground field k is the complex number field $\mathbb{C}$. The affine space of dimension n is denoted by $\mathbb{A}^n$, and the affine line $\mathbb{A}^1$ minus one point by $\mathbb{A}^1_*$, which is the underlying space of the multiplicative group G_m.

1.1 Topological preliminaries

A main reference to topological arguments given below is Spanier's book [213].

1.1.1 Local fundamental groups and homology groups

We will discuss several useful results about topology of complex algebraic (or analytic) varieties. These will be used in many proofs later in this book. For a complex algebraic variety X, we can consider a complex analytic variety corresponding to X by considering the defining polynomial equations as convergent power series equations. We denote the complex analytic variety corresponding to X by the same symbol X unless confusion is feared.

Let X be a complex analytic variety. An important property of X is the following result.

Lemma 1.1.1. *Given a closed analytic subvariety $Y \subset X$, there is a triangulation of X such that Y is a subcomplex in this triangulation.*

This implies that every point $P \in X$ has a fundamental system of neighborhoods U_n, each of which strongly deforms to P. More generally, Y has a fundamental system of neighborhoods U_n of which Y is a strong deformation retract. Further, if $X - Y$ is a complex manifold then U_n can be so chosen that ∂U_n is a C^∞ manifold. The inclusion $U_n \subset U_m$ can be chosen so that for $n > m$ the neighborhood U_m strongly deforms to U_n.

If further X is a normal complex space and $P \in Y = \mathrm{Sing}(X)$, then all $U_n - Y$ are diffeomorphic so that for $n > m$ the punctured complement $U_n - Y$ is a strong deformation retract of $U_m - Y$.

In view of these observations, we define $\pi_1^P(X)$ as the fundamental group $\pi_1(U_n - \mathrm{Sing}(U_n))$. Similarly, we define $H_i^P(X;\mathbb{Z})$ to be $H_i(U_n - \mathrm{Sing}\, U_n;\mathbb{Z})$ for $i \geq 0$. We call $\pi_1^P(X)$ (resp. $H_i^P(X;\mathbb{Z})$) the *local fundamental group* of X at P (resp. the *local* ith *integral homology group* of X at P). Often we will use $\mathbb{Q}$ or $\mathbb{C}$ as the coefficient group instead of $\mathbb{Z}$.

Given a complex analytic space X and a closed complex subspace Y, we can use the standard machinery of algebraic topology like Mayer–Vietoris sequence, relative homology or cohomology sequence for the pair (X, Y) with $\mathbb{Z}$, $\mathbb{Q}$ or $\mathbb{C}$ coefficients.

Lemma 1.1.2. *Every complex algebraic variety has finitely generated homology and cohomology groups with $\mathbb{Z}$-coefficients.*

Proof. If X is complete there are only finitely many simplices in the triangulation. Hence homology groups are finitely generated. If X is smooth, we embed X as an open set of a complete variety V, and let $D := V - X$. Then $\dim D < \dim X$. Use the long exact cohomology sequence of the pair (V, D) to compute $H^i(V, D;\mathbb{Z})$. Since $H^*(V;\mathbb{Z})$ is finitely generated and so is $H^*(D;\mathbb{Z})$ by induction on $\dim X$, $H^*(V, D;\mathbb{Z})$ is finitely generated. Then by the duality $H^i(V, D;\mathbb{Z}) \cong H_{2n-i}(X;\mathbb{Z})$, $H_*(X;\mathbb{Z})$ is finitely generated. The universal coefficient theorem implies that $H^*(X;\mathbb{Z})$ is finitely generated. If X is singular, we consider X and its singular locus S. By the smooth case and by induction, $X - S$ and S have finitely generated homology groups. So, by using duality for cohomology with compact support (see Section 1.1.2 below), X has finitely generated homology groups. The details are left to the readers. □

1.1.2 Cohomology with compact supports

Let X be a complex analytic variety. We can find a sequence of compact subsets $K_1 \subset K_2 \subset \cdots$ with union X. If $n > m$, the inclusion of pairs $(X, X \setminus K_n) \hookrightarrow (X, X \setminus K_m)$ induces a homomorphism

$$H^i(X, X \setminus K_m) \to H^i(X, X \setminus K_n)$$

with the coefficient group $\mathbb{Z}$, $\mathbb{Q}$, and $\mathbb{C}$. The limit $\varinjlim_n H^i(X, X \setminus K_n)$ as $n \to \infty$ is called the *cohomology group with compact support* of X, denoted by $H_c^i(X)$. If X is compact then we just have an isomorphism $H_c^i(X) \cong H^i(X)$. For the definition of relative cohomology with compact support $H_c^i(X, Y)$, see Spanier [213].

Let (X, Y) be a pair of complex analytic variety and a closed analytic subspace. Then there is a long exact sequence

$$\begin{aligned} \cdots &\to H_c^i(X, Y) \to H_c^i(X) \to H_c^i(Y) \to H_c^{i+1}(X, Y) \\ &\to H_c^{i+1}(X) \to H_c^{i+1}(Y) \to \cdots \end{aligned}$$

The fact that ordinary and compact support cohomologies for any compact space are equal follows immediately from the definition of the latter. Similarly for relative cohomology sequences. Duality for compact manifolds with or without boundaries is very important. So, for noncompact manifolds compact support cohomology is equally important.

There is a definition of pseudomanifolds (and oriented pseudomanifolds) in exercises of Spanier [213]. Roughly speaking, a simplicial complex of dimension m is a pseudomanifold if any $m-1$ simplex is a face of exactly two m simplexes. A triangulable manifold is a pseudomanifold, but not conversely. A nice example is a singular complex analytic variety V. First of all, it is triangulable such that its singular locus is a subcomplex. The smooth locus is a manifold, hence a pseudomanifold. But $\dim(\operatorname{Sing} V)$ is at most $n-1$, so real dimension of $\operatorname{Sing} V$ is at most $2n-2$. From this it follows that any $2n-1$ dimensional simplex is a face of exactly two $2n$ dimensional simplexes (most of the portion of the $2n-1$ simplex lies in the smooth locus of V which is a pseudomanifold). This observation is not so well known. For exactly same reason, even a singular complex analytic variety is an orientable simplicial complex.

1.1.3 Duality for singular varieties

Let X be an irreducible normal complex analytic variety of dimension n. Let R be either $\mathbb{Z}$ or $\mathbb{Q}$ as the coefficient group. Assume that X has only isolated singular points. Further, assume that for any point $P \in X$ there exists a neighborhood U_P of P for which $H_i(U_P - \{P\}; R) \cong H_i(S^{2n-1}; R)$ for $i \geq 0$, where S^r is the r-dimensional sphere. Then, for any closed complex subspace Y containing the singular locus of X, there is an isomorphism

$$H_c^i(X, Y; R) \cong H_{2n-i}(X - Y; R), \quad i \geq 0.$$

If $X - Y$ is orientable, this duality holds without the above assumptions. Hence it holds for a pair (X, Y) of a complex algebraic variety X and its closed set Y containing the singular locus of X. This result has the following useful consequence.

Lemma 1.1.3. *For any irreducible noncompact complex algebraic variety X of dimension n, we have $H_c^0(X; \mathbb{Z}) = 0$ and $H_c^{2n}(X; \mathbb{Z}) \cong \mathbb{Z}$.*

Proof. Let $S = \operatorname{Sing} X$. If X is smooth then $H_c^0(X) \cong H_{2n}(X)$, which is zero since X is noncompact. If X is singular then $\dim S \leq n-1$. The long exact sequence of cohomology groups with compact support for (X, S) gives an exact sequence

$$\cdots \to H_c^{2n-1}(S) \to H_c^{2n}(X, S) \to H_c^{2n}(X) \to H_c^{2n}(S) \to \cdots$$

where $H_c^{2n}(S) = H_c^{2n-1}(S) = 0$. Hence we have

$$H_c^{2n}(X) \cong H_c^{2n}(X, S) \cong H_0(X - S),$$

which is isomorphic to $\mathbb{Z}$ because $X - S$ is connected. In order to show that $H_c^0(X) = 0$, we use induction on n. We assume the case $\dim X = 1$ for the moment. Now choose a noncompact hypersurface H of X containing S. Then $X - H$ is smooth. Use the exact sequence

$$0 \to H_c^0(X, H) \to H_c^0(X) \to H_c^0(H) \to \cdots.$$

By duality, $H_c^0(X, H) \cong H_{2n}(X - H) = 0$ by the smooth case. By induction, $H_c^0(H) = 0$, whence $H_c^0(X) = 0$.

We show that $H_c^0(X) = 0$ if $\dim X = 1$. Write $X = \bigcup_i K_i$, where K_i is a suitable increasing sequence of compact sets in X. Then $X - K_i$ splits into a disjoint union of punctured neighborhoods of the places at infinity for X. We have the long exact sequence of cohomology groups for the pair $(X, X - K_i)$,

$$0 \to H^0(X, X - K_i) \to H^0(X) \to H^0(X - K_i) \to \cdots.$$

Since the natural map $H_0(X - K_i) \to H_0(X)$ is a surjection, the induced map on cohomology $H^0(X) \to H^0(X - K_i)$ is an injection. Hence $H^0(X, X - K_i) = 0$. This shows that $H_c^0(X) = 0$. □

1.1.4 Theorem of Kaup–Narasimhan–Hamm

For Stein manifolds (or spaces), we refer to a book by Grauert and Remmert [62]. A complex affine variety is a Stein space, but the converse is not true (see [222] for counterexamples). The following theorem was first shown by A. Andreotti and T. Frankel [4] for a Stein manifold of dimension n (smooth case) and generalized by R. Narasimhan [178] and L. Kaup [112] for an arbitrary Stein space. The relative case was proved by H. A. Hamm [88].

Theorem 1.1.4. *Let X be an irreducible Stein space of dimension n. Then $H_i(X; \mathbb{Z}) = 0$ for $i > n$ and $H_n(X; \mathbb{Z})$ is torsion free. If further X is an affine variety then $H_i(X; \mathbb{Z})$ are finitely generated for $i \geq 0$. If A is a closed subspace of X, then $H_n(X, A; \mathbb{Z})$ is a free abelian group.*

1.1.5 Ehresmann's fibration theorem

The following result is due to C. Ehresmann [48].

Theorem 1.1.5. *Let $f : X \to Y$ be a proper surjective differentiable map, where X, Y are C^∞ manifolds. If the induced tangent map*

$$f_* : T_{X,x} \to T_{Y,f(x)}$$

is surjective for all $x \in X$ then f is a C^∞ locally trivial fiber bundle.

More generally, let $Z \subset X$ be a proper closed C^∞ submanifold such that for any $z \in Z$ the induced map

$$f_* : T_{Z,z} \to T_{Y,f(z)}$$

is a surjection then $f|_{X-Z} : X - Z \to Y$ is a C^∞ locally trivial fiber bundle.

We will use this result in the following situation:

Let X, Y be smooth, irreducible complex algebraic varieties. Let $f : X \to Y$ be a surjective morphism. Then there is a proper closed subvariety $S \subset Y$ such that $f : X - f^{-1}(S) \to Y - S$ is a C^∞ locally trivial fiber bundle (see [182, Lemma 1.5]).

1.1.6 Results of M. Nori and Xiao Gang

First of all, we state the following result [182].

Lemma 1.1.6. *Let X, Y be smooth irreducible complex algebraic varieties, and let $f : X \to Y$ be a surjective morphism with an irreducible general fiber F. Assume that there is a proper closed subvariety $S \subset Y$ of codimension greater than one such that for any $y \in Y - S$, the fiber $f^*(y)$*[1] *is reduced at some point. Then the natural sequence of fundamental groups*

$$\pi_1(F) \to \pi_1(X) \to \pi_1(Y) \to (1)$$

is exact.

Example 1.1.7. Let $\mathrm{SL}(2, \mathbb{C})$ be the special linear group defined by 2×2 matrices $\left(\begin{smallmatrix} a & b \\ c & d \end{smallmatrix}\right)$ with $ad - bc = 1$. Then $\mathrm{SL}(2, \mathbb{C})$ is the closed hypersurface in $\mathbb{A}^4$ defined by $z_1 z_4 - z_2 z_3 = 1$. For any $z_1 = c \in \mathbb{C}$, the surface $\{c z_4 - z_2 z_3 = 1\}$ is irreducible and reduced, and for $c \neq 0$ it is isomorphic to $\mathbb{A}^2$. Applying Nori's exact sequence to the projection $(z_1, z_2, z_3, z_4) \mapsto z_1$ we deduce that $\mathrm{SL}(2, \mathbb{C})$ is simply-connected. □

The next result is important for the *Ramified Covering Trick* below (see Bundagaard-Nielsen [22], Fox [54] and Chau [26]).

Lemma 1.1.8 (Fenchel's conjecture). *Let C be a smooth irreducible quasi-projective curve of genus g. Let $P_1, \ldots, P_r$ be distinct points in C and $m_1, \ldots, m_r$ be arbitrary integers larger than one, where we assume that if $C \cong \mathbb{P}^1$ then $r \geq 2$ and if further $r = 2$ then $m_1 = m_2$. Then there is a finite Galois covering $\pi : D \to C$ which is branched precisely over $P_1, \ldots, P_r$ and the ramification index over P_i is m_i for each i.*

1 It is called the *schematic fiber* and defined as $X \times_Y \operatorname{Spec} k(y)$, where $k(y)$ is the residue field of Y at $y \in Y$.

Lemma 1.1.9 (Ramified covering trick). *Let $f : X \to C$ be a surjective morphism with an irreducible general fiber F, where X is a smooth irreducible algebraic surface and C is a smooth irreducible quasiprojective curve. Let $P_1, \dots, P_r$ be the points in C such that $f^*(P_i)$ has multiplicity $m_i > 1$ and for any $P \in C$ other than $P_1, \dots, P_r$ the multiplicity of the fiber $f^*(P)$ is 1. We assume that if $C \cong \mathbb{P}^1$ then $r \geq 2$ and in case $r = 2$ we have $m_1 = m_2$. Let $\pi : D \to C$ be as in Fenchel's conjecture. Then the normalized fiber product $\overline{X \times_C D}$, which is the normalization of the fiber product $X \times_C D$, is a finite Galois étale covering of X.*

For a fiber $F = \sum_{i=1}^r m_i F_i$ with irreducible components, the multiplicity m is defined as $\gcd(m_1, \dots, m_r)$. If $m > 1$, the fiber F is called a multiple fiber. Hence it is possible that a multiple fiber F is reducible. For $m > 1$, the fact that the normalized fiber product $\overline{X \times_C D}$ is étale over X is proved by using a local analysis. This goes back to Kodaira's paper. The splitting of F in the normalized fiber product depends on the topology of F, multiplicities of its components, etc.

The next result is a generalization of the uniformization theorem where ramification is allowed. We state it for simplicity when C is a smooth, irreducible projective curve.

Lemma 1.1.10. *Let C be a smooth, irreducible projective curve of genus g. Let $P_1, \dots, P_r$ be distinct points in C and $m_1, \dots, m_r$ arbitrary integers greater than one. In case $C \cong \mathbb{P}^1$ we assume that $r \geq 2$, and if $r = 2$ then $m_1 = m_2$. Then there is a Galois ramified covering (possibly of infinite degree) $\pi : \widetilde{C} \to C$ such that $\widetilde{C}$ is simply-connected, and π is ramified precisely over $P_1, \dots, P_r$ with ramification index over P_i equal to m_i.*

Further, there is a group Γ of complex analytic isomorphisms of $\widetilde{C}$ defined by generators $\alpha_1, \beta_1, \dots, \alpha_g, \beta_g, c_1, \dots, c_r$ with relations

$$c_1^{m_1} = c_2^{m_2} = \cdots = c_r^{m_r} = 1 = [\alpha_1, \beta_1] \cdots [\alpha_g, \beta_g] c_1 c_2 \cdots c_r,$$

and $\widetilde{C}/\Gamma$ is isomorphic to C.

Using this result we obtain a useful generalization of Nori's result. See also [229, 230]. We state it for simplicity in the case when C is a smooth, irreducible, projective curve.

Lemma 1.1.11. *Let X, C be a smooth projective irreducible surface and a smooth projective curve, respectively. Let $f : X \to C$ be a projective morphism with an irreducible general fiber F. Let $P_1, \dots, P_r \in C$ be the distinct points such that $f^*(P_i)$ has multiplicity $m_i > 1$ for $i = 1, \dots, r$ and for any P not equal to any P_i the multiplicity of $f^*(P)$ is 1. If $C \cong \mathbb{P}^1$ then we assume that $r \geq 2$, and if $r = 2$ then $m_1 = m_2$. Then there is a short exact sequence*

$$\pi_1(F) \to \pi_1(X) \to \Gamma \to (1),$$

where Γ is defined as in Lemma 1.1.10.

Proof. The Riemann surface in Lemma 1.1.10 is simply-connected. By Lemma 1.1.6, there is a surjection

$$\pi_1(F) \longrightarrow \pi_1(\overline{X \times_C \widetilde{C}}),$$

where $\overline{X \times_C \widetilde{C}}$ is the normalized fiber product of X and $\widetilde{C}$ over C, which is a topological covering of X with Galois group Γ. Hence there is an exact sequence

$$(1) \to \pi_1(\overline{X \times_C \widetilde{C}}) \to \pi_1(X) \to \Gamma \to (1).$$

Combining these two sequences, we obtain the desired result. □

1.1.7 Some more results on the topology of an algebraic variety

We have the following result.

Lemma 1.1.12. *The following assertions hold:*

(1) *Let X be an irreducible normal complex analytic variety. Then for any proper closed subvariety $Z \subset X$ the natural homomorphism $\pi_1(X - Z) \to \pi_1(X)$ is a surjection. In particular, $H_1(X-Z; R) \to H_1(X; R)$ is a surjection for any abelian coefficient group R.*

(2) *Let X, Y be irreducible, normal complex analytic varieties and $f : X \to Y$ a surjective proper holomorphic map with finite fibers. Then for every $i \geq 0$ there is a homomorphism* (transfer) *$H_i(Y; \mathbb{Z}) \to H_i(X; \mathbb{Z})$ such that the composite*

$$H_i(Y; \mathbb{Z}) \to H_i(X; \mathbb{Z}) \to H_i(Y; \mathbb{Z})$$

is the multiplication by d, where $d = \deg f$.

This is proved in [58] by triangulating X, Y so that f is a simplicial map.

Corollary 1.1.13. *With X, Y as above in Lemma* 1.1.12*, the natural homomorphisms $H_i(X; \mathbb{Q}) \to H_i(Y; \mathbb{Q})$ are surjections for each i.*

Regarding the induced morphism on π_1 in assertion (2) of Lemma 1.1.12, we have the following useful result.

Lemma 1.1.14. *With the above notations, the image of the natural homomorphism $\pi_1(X) \to \pi_1(Y)$ has finite index which divides* $\deg f$*. If further X, Y are algebraic varieties and f is a (not necessarily finite) dominant morphism then the result is still valid for the statement about π_1.*

The proof involves the observation that there is a proper closed subvariety $S \subset Y$ such that $X - f^{-1}(S) \to Y - S$ is a finite unramified covering and the use of the assertion (1) of Lemma 1.1.12.

A slightly more general result is often useful.

Lemma 1.1.15. *Let $f : X \to Y$ be a dominant morphism with X, Y irreducible, normal algebraic varieties. Then the image of the natural homomorphism $\pi_1(X) \to \pi_1(Y)$ is of finite index, say r. If a general fiber of f has d irreducible components then $r \mid d$. In particular, if $d = 1$ then $\pi_1(X) \to \pi_1(Y)$ is a surjection.*

Proof. We observe that there is a proper closed subvariety $S \subset Y$ such that $X - f^{-1}(S) \to Y - S$ is a C^∞ locally trivial fiber bundle, and use the long exact homotopy sequence of a fiber bundle. □

Let X be a connected complex manifold and $D = D_1 + D_2 + \cdots + D_r$ a divisor in X with irreducible components D_i. For each D_i, we can find a small loop $\gamma_i \subset X - D$ such that γ_i is the boundary of a small (complex) 1-dimensional disc Δ_i in X transverse to D_i at one point in D_i. Then the natural homomorphism $\pi_1(X - D) \to \pi_1(X)$ is a surjection with kernel the normal subgroup generated by $\gamma_1, \dots, \gamma_r$ (with a suitable base point).

Consider for simplicity the case where D is smooth and irreducible. We can cover X by open sets of $X - D$ and a suitable tubular neighborhood N of D. Then, by Van Kampen's theorem, $\pi_1(X)$ is the push-out of the following square:

$$\begin{array}{ccc} \pi_1(\partial N) \cong \pi_1((X - D) \cap N) & \longrightarrow & \pi_1(N) \\ \downarrow & & \downarrow \\ \pi_1(X - D) & \longrightarrow & \pi_1(X), \end{array}$$

where $(X - D) \cap N$ has ∂N as a strong deformation retract, and D is also a strong deformation retract of $N - \partial N$. Hence $\pi_1((X - D) \cap N) \cong \pi_1(\partial N)$. Meanwhile, since ∂N is an S^1-bundle over D, we have an exact sequence

$$(1) \to (\gamma) \to \pi_1(\partial N) \to \pi_1(D) \to (1)$$

and γ is central in $\pi_1(\partial N)$ since N is a complex manifold. Hence we have an exact sequence

$$(1) \to (\gamma) \to \pi_1(X - D) \to \pi_1(X) \to (1).$$

The following result of F. Serrano [208] is the homological analogue of the above result.

Lemma 1.1.16. *Let $f : X \to C$ be a proper surjective holomorphic map from a connected complex manifold X onto a smooth quasiprojective curve C such that a general fiber F of f is irreducible. Let $P_1, \dots, P_r$ be all the points in C for which the fibers $f^*(P_i)$ have multiplicity $m_i > 1$. Then the following natural sequence is exact:*

$$H_1(F; \mathbb{Z}) \to H_1(X; \mathbb{Z}) \to H_1(C; \mathbb{Z}) \times G(f) \to 0,$$

where $G(f)$ is the cokernel of the homomorphism $\mathbb{Z} \to \bigoplus_{i=1}^r \mathbb{Z}/(m_i)$ sending $1 \mapsto (\bar{1}, \dots, \bar{1})$.

1.1.8 Suzuki–Zaidenberg formula for the topological Euler–Poincaré characteristic

For any simplicial complex K, let $e(K)$ denote the topological Euler–Poincaré characteristic of K. The next result is very useful in the study of smooth affine algebraic surfaces.

Theorem 1.1.17. *Let $f : X \to C$ be an affine surjective morphism from a smooth irreducible surface X onto a smooth algebraic curve C such that a general fiber F of f is irreducible. Let $F_1, \dots, F_r$ be all the singular fibers of f (this is the smallest set of fibers of f outside which f is a C^∞ locally trivial fiber bundle). Then we have*

$$e(X) = e(C) \cdot e(F) + \sum_{i=1}^{r} (e(F_i) - e(F)).$$

Further, $e(F_i) \geq e(F)$ for all i and if equality holds for some i then F is either $\mathbb{A}^1$ or $\mathbb{A}^1_*$ and F_i is isomorphic to F, if taken with the reduced structure.

The proof in [218] and [232] use pluri-subharmonic functions. A more algebro-geometric proof is given in [69].

1.1.9 Fixed point locus of a G_a-action

Let the additive group G_a act regularly on an irreducible algebraic variety. Generalities on G_a-actions on algebraic varieties are found in Section 2.2 of Chapter 2. The fixed point set X^{G_a} is a closed subvariety of X. The following result is proved in Białynicki-Birula [17].

Lemma 1.1.18. *The following assertions hold:*
1. *$H^i_c(X) \cong H^i_c(X^{G_a})$ for $i = 0, 1$, where the cohomologies are with $\mathbb{Z}$-coefficients.*
2. *If X is incomplete of dimension > 0 then X^{G_a} cannot have an isolated fixed point.*
3. *If X is complete then X^{G_a} is connected.*

Proof. The second assertion follows from the observation that $H^0_c(X) = 0$ since X is incomplete but for an isolated fixed point $P \in X$ we have $H^0_c(P) \neq 0$.

For assertion (3), since $H^0_c(X) \cong H^0(X) \cong \mathbb{Z}$ as X is connected, we get $H^0_c(X^{G_a}) \cong H^0(X^{G_a}) \cong \mathbb{Z}$. Hence X^{G_a} is connected. □

1.1.10 An application of topological method

Let V be a normal projective surface with a $\mathbb{P}^1$-fibration $f : V \to C$ onto a smooth projective curve C. A fiber f^*P of f is called a *singular fiber* if f^*P is not isomorphic to a general fiber. Let $F_0 = (f^*P)_{\mathrm{red}}$, the underlying topological space.

Lemma 1.1.19. *F_0 is simply-connected.*

Proof. We can find a small Euclidean neighborhood U of P in C such that F_0 is a strong deformation retract of $f^{-1}(U)$. By using the proof of the generalization of Nori's exact sequence (see Lemma 1.1.11), we get an exact sequence

$$\pi_1(\mathbb{P}^1) \to \pi_1(f^{-1}(U)) \to \mathbb{Z}/(m) \to (1),$$

where m is the multiplicity of f^*P. This implies that $\pi_1(f^{-1}(U))$ (and hence $\pi_1(F_0)$) is finite. Hence $H_1(F_0; \mathbb{Q}) = 0$. This implies that the components of F_0 are rational curves with at worst unibranch singular points and the dual graph of F_0 has no loops. An easy repeated application of Van Kampen theorem implies that F_0 is simply-connected. If an irreducible component of F_0 is not smooth then the arithmetic genus $p_a(F_0) > 0$, which contradicts the invariance of arithmetic genus under flat and proper morphism. □

As a corollary we see that the dual graph of F_0 has no loops. We apply this to give a topological proof of the following result in [159, Chapter 3, Lemma 1.4.2].

Theorem 1.1.20. *Let X be an irreducible normal affine surface with an $\mathbb{A}^1$-fibration $\varphi : X \to C$ over a normal algebraic curve C. Let $F_0 = \varphi^*P$ be a singular fiber of φ. Then F_0 is a disjoint union of irreducible curves each isomorphic to $\mathbb{A}^1$. Further, X has at most cyclic quotient singularities.*

Proof. We can embed $X \subset V$, where V is a normal projective surface such that φ extends to a $\mathbb{P}^1$-fibration $\Phi : V \to \overline{C}$. Let $\overline{F}_0$ be the fiber of Φ containing F_0. By Lemma 1.1.19, $\overline{F}_0$ is simply-connected. Since X is affine, the closure of each irreducible component Δ of F_0 meets $\overline{F}_0 - F_0$. Since $\overline{F}_0$ is simply-connected, we deduce easily that any two irreducible components Δ_1, Δ_2 of F_0 are disjoint and contractible.

Let Δ contain a singular point Q of X and let m be the multiplicity of Δ in φ^*P. We use the ramified covering trick to construct a finite cyclic ramified cover of X, say $Y \to X$ such that Y has an $\mathbb{A}^1$-fibration $\psi : Y \to D$ and the inverse image of Δ occurs as a reduced curve in the fibers of ψ. Then Y is smooth at any point mapping to Q. This implies, by Mumford's result [171], that X has a cyclic quotient singular point. In fact, note that the morphism $Y \to X$ is a finite cyclic morphism. For any point $Q' \in Y$ mapping to Q in X, the inclusion of the analytic local ring $\mathcal{O}^h_{X,Q}$ into $\mathcal{O}^h_{Y,Q'}$ is étale and finite. Since Q' is smooth, by the covering space theory, the local fundamental group at Q is finite cyclic. Hence Q is a cyclic singularity. □

1.2 Logarithmic Kodaira dimension and characterizations of affine spaces

Let X be an algebraic variety. By Nagata's complete embedding theorem, we can embed X into a complete variety V as an open set. By the resolution of singularities, we can assume that V is smooth in an open neighborhood of $D := V \setminus X$ and D consists of irreducible subvarieties of codimension one $D = D_1 + \cdots + D_r$, where intersections of the D_i are normal crossings. Suppose that every D_i is smooth. Then normal crossing means that if $P \in D_i \cap D_j$ with $i \neq j$ then $T_{V,P} = T_{D_i,P} + T_{D_j,P}$, where $T_{V,P}$ and $T_{D_i,P}$ are the tangent spaces at P of V and D_i. If $\dim V > 2$, we define $D = \sum_{i=1}^r D_i$ to be a divisor with normal crossings if, for any $1 \le i \le r$, the divisor $\sum_{j \neq i} D_j \cap D_i$ is a divisor with normal crossings on D_i. When every irreducible component D_i is smooth, we say that $D = \sum_{i=1}^r D_i$ a divisor with *simple normal crossings* or an *SNC-divisor*. If $D := V \setminus X$ is a divisor with simple normal crossings, we call V (or (V, D) more precisely) a *normal completion* of X and D the *boundary divisor*. Here the term *normal* refers to the way of intersection in D, not to the normal singularities. We also say a *normal compactification* instead of a normal completion.

Let X be a smooth algebraic variety and let (V, D) be a normal completion of X. If X is quasiprojective, we may assume that V is projective. In fact, embed X into a projective variety V'. Desingularization of V' and making the boundary $V' \setminus X$ an SNC-divisor are achieved by blowing-ups of smooth centers in $V' \setminus X$ which preserve the projectiveness of V'. The *logarithmic Kodaira dimension* of X is defined as the Iitaka dimension $\kappa(V, D + K_V)$. Important fact is that $\kappa(V, D + K_V)$ is independent of the choice of a normal completion (V, D) of X. We denote it by $\overline{\kappa}(X)$. We refer to [159] for a detailed account of the theory. Logarithmic Kodaira dimension is an adaptation of Kodaira dimension $\kappa(V)$ for a complete algebraic variety V. If X is complete, $\overline{\kappa}(X) = \kappa(X)$. The introduction of logarithmic Kodaira dimension by Iitaka, together with the minimal model theory by Kawamata, made it possible to develop a sufficient theory of classification of open algebraic surfaces. For details of this theory we can refer to [159], and it is enough for the purposes of this book to pick up the properties which will be used later.

Lemma 1.2.1. *The following assertions hold:*

(1) *Let U be a nonempty open set of a smooth algebraic variety X. Then $\overline{\kappa}(X) \le \overline{\kappa}(U)$. For two smooth algebraic varieties X_1, X_2, we have $\overline{\kappa}(X_1 \times X_2) = \overline{\kappa}(X_1) + \overline{\kappa}(X_2)$.*

(2) (Easy addition formula) *Let $f : X \to Y$ be a fibration with X smooth and let X_y be a general closed fiber. Then we have an inequality*

$$\overline{\kappa}(X) \le \overline{\kappa}(X_y) + \dim Y.$$

(3) (Addition formula) *Let $f : X \to Y$ be a fibration with X and Y smooth and of relative dimension ≤ 1. Then, with a general closed fiber X_y of f, we have an inequality*

$$\overline{\kappa}(X) \ge \overline{\kappa}(X_y) + \overline{\kappa}(Y).$$

(4) *Let X be a smooth algebraic variety with $\overline{\kappa}(X) \ge 0$. Then there exist a proper birational morphism $\mu : X^{\#} \to X$ and a fibration $f : X^{\#} \to Y$ (called an* Iitaka fibration *of X) such that*

(i) $X^{\#}$ *is smooth,*
(ii) $\dim Y = \overline{\kappa}(X)$,
(iii) $\overline{\kappa}(X_y) = 0$ *for a general closed fiber* X_y *of* f.

Lemma 1.2.2. *Let* $\varphi : V \to W$ *be a dominant morphism of smooth complete algebraic varieties* V, W. *Let* D, B *be SNC-divisors on* V, W, *respectively, such that* $\varphi^{-1}(\operatorname{Supp} B) \subseteq \operatorname{Supp} D$. *Let* $X = V \setminus D$, $Y = W \setminus B$ *and let* $f : X \to Y$ *be the restriction of* φ *onto* X. *Then there exists an effective divisor* $\overline{R}_f$ *on* V *such that*

$$K_V + D \sim \varphi^*(K_W + B) + \overline{R}_f.$$

This relation yields an inequality $\overline{\kappa}(Y) \leq \overline{\kappa}(X)$. *The divisor* $\overline{R}_f$ *is called the* log-ramification divisor *of* f.

The subsequent three theorems are important in the theory of open algebraic surfaces. Let (V, D) be a normal completion of a smooth algebraic variety X. If the divisor D is connected, we say that X is *connected at infinity*. This concept is independent of the choice of a normal completion of X.

Theorem 1.2.3. *The following assertions hold for a smooth algebraic variety* X:
(1) *If* $\dim X = 1$ *and* $\overline{\kappa}(X) = -\infty$ *then* $X \cong \mathbb{A}^1$ *or* $X \cong \mathbb{P}^1$.
(2) *If* X *has an* $\mathbb{A}^1$*-fibration then* $\overline{\kappa}(X) = -\infty$.
(3) *Let* X *be a smooth affine surface (or connected at infinity instead of assuming the affineness). Then* $\overline{\kappa}(X) = -\infty$ *if and only if* X *has an* $\mathbb{A}^1$*-fibration.*

Proof. (1) Suppose X is not complete. Then X is affine. Let (C, D) be a normal completion of X, where $d = \deg D > 0$ and D is ample. Then $\deg(K_C + D) = 2g - 2 + d < 0$ because $\overline{\kappa}(X) = 1$ if $g \geq 1$ and $\overline{\kappa}(X) \geq 0$ if $g = 0$ and $d \geq 2$. Hence $d = 1$ and X is isomorphic to $\mathbb{P}^1$ minus one point. Hence $X \cong \mathbb{A}^1$.

(2) Let $f : X \to Y$ be an $\mathbb{A}^1$-fibration. It suffices to show that $\overline{\kappa}(f^{-1}(U)) = -\infty$ for an open set U of Y. Hence we may assume also that Y is smooth. By Theorem 2.1.9 below, there exists a dominant morphism of finite degree $Y' \to Y$ such that $X' := X \times_Y Y'$ contains an $\mathbb{A}^1$*-cylinder* $U' \times \mathbb{A}^1$. By Lemma 1.2.2, we have $\overline{\kappa}(X) \leq \overline{\kappa}(X')$. Since $\overline{\kappa}(X') \leq \overline{\kappa}(U' \times \mathbb{A}^1) = \overline{\kappa}(U') + \overline{\kappa}(\mathbb{A}^1) = -\infty$ by Lemma 1.2.1, it follows that $\overline{\kappa}(X) = -\infty$. There is another argument. By shrinking Y if necessary, we can assume that there exist a proper morphism $\overline{f} : \overline{X} \to Y$ with $\overline{X}$ smooth and an open immersion $\iota : X \hookrightarrow \overline{X}$ such that $f = \overline{f} \cdot \iota$. Then there exists an irreducible divisor D_1 of $\overline{X} \setminus X$ which lies horizontally along $\overline{f}$ and meets general fibers in single points transversally. This implies that D_1 is relatively ample. Hence f is an affine morphism. Since we may also assume that f is faithfully flat, Lemma 2.1.3 implies that the $\mathbb{A}^1$-fibration $f : X \to Y$ is generically trivial. Hence X contains an $\mathbb{A}^1$-cylinder and $\overline{\kappa}(X) = -\infty$.

(3) The *if* part follows from the assertion (2). The *only if* part is involved and we will be referred to [159, Chapter 2, §2]. □

Let G be a finite group acting on $\mathbb{A}^2$ faithfully. Then the G-action is linearizable, i. e., after a suitable change of coordinates of $\mathbb{A}^2$ the G-action is given via the general linear group $\mathrm{GL}(2,k)$. We assume that the G-action is linear. Then the G-action commutes with the natural G_m-action $(t,(x,y)) \mapsto (tx,ty)$ for $t \in G_m$. The algebraic quotient $\mathbb{A}^2/\!/G$ is defined[2] as $\operatorname{Spec} k[x,y]^G$, where $k[x,y]^G$ is the invariant subring. Let $X = (\mathbb{A}^2/\!/G) \setminus \{\overline{O}\}$ be the smooth locus of $\mathbb{A}^2/\!/G$, where $\overline{O}$ is the image of the point of origin and is a unique (quotient) singular point. Since the G_m-action on $\mathbb{A}^2$ descends down to $\mathbb{A}^2/\!/G$ and has $\overline{O}$ as a unique fixed point if G is *small* (see explanations before Theorem 1.2.12), so, the G_m-action gives rise to an $\mathbb{A}^1_*$-fibration $q : X \to \mathbb{P}^1$. If G is not cyclic, with this $\mathbb{A}^1_*$-fibration, X is called a *Platonic* $\mathbb{A}^1_*$*-fiber space*. For the $\mathbb{A}^1_*$-fibration $q : X \to \mathbb{P}^1$, every fiber is isomorphic to $\mathbb{A}^1_*$ except for exactly three irreducible multiple fibers m_iF_i for $i = 1,2,3$ and $F_i \cong \mathbb{A}^1_*$. A triplet $\{m_1, m_2, m_3\}$ is one of the *Platonic triplets* $\{2,2,n\}$ $(n \geq 2), \{2,3,3\}, \{2,3,4\}, \{2,3,5\}$ if we assume that $m_1 \leq m_2 \leq m_3$. The Platonic $\mathbb{A}^1_*$-fiber space X has $\mathbb{A}^2 - \{O\}$ as the universal covering space and the quotient morphism $\pi : \mathbb{A}^2 - \{O\} \to X$ is the universal covering. Hence $\pi_1(X) \cong G$. For the details, see [159, Chapter 3, §2.5].

A crux of the theory of open algebraic surfaces of $\overline{\kappa}(X) = -\infty$ is the following result [159, Chapter 3, Theorem 2.5.4].

Theorem 1.2.4. *Let X be a smooth algebraic surface which is not necessarily connected at infinity. Then $\overline{\kappa}(X) = -\infty$ implies that either X has an $\mathbb{A}^1$-fibration or X contains an open set U which is isomorphic to a Platonic $\mathbb{A}^1_*$-fiber space.*

Apart from $\mathbb{A}^1$-fibrations, frequently used are $\mathbb{A}^1_*$-fibrations. The following theorem was proved by Kawamata [115].

Theorem 1.2.5. *Let X be a smooth affine surface. If $\overline{\kappa}(X) = 1$, then X has an $\mathbb{A}^1_*$-fibration. The existence of $\mathbb{A}^1_*$-fibration on X implies that $\overline{\kappa}(X) \leq 1$.*

One more result from the theory of open algebraic surfaces is an inequality between logarithmic Chern numbers of a smooth algebraic surface analogous to the well-known Miyaoka–Yau inequality (see [159, Chapter 2, Theorem 6.6.2]). It was proved by R. Kobayashi, S. Nakamura, and F. Sakai [121]. In this volume we need the following consequence of this result.

Theorem 1.2.6. *Let X be a smooth affine surface with $e(X) \leq 0$. Then $\overline{\kappa}(X) \leq 1$.*

The following results are about the characterizations of the affine spaces and affine-space-like varieties. The one for $\mathbb{A}^2$ is definitive, while those on $\mathbb{A}^3$ are not and have still room for improvement. References for these results are [148, 155, 156].

2 This quotient is called the *algebraic quotient* and denoted by $\mathbb{A}^2/\!/G$ to distinguish it from other geometric quotients which are usually denoted by $\mathbb{A}^2/G$. But, in this case, both quotients are isomorphic, i. e., $\mathbb{A}^2/\!/G \cong \mathbb{A}^2/G$.

Theorem 1.2.7. *Let X be a normal affine surface. Then $X \cong \mathbb{A}^2$ if and only if the following three conditions are satisfied:*
(1) *X is factorial.*
(2) *$k[X]^* = k^*$, where $k[X]$ is the coordinate ring of X.*
(3) *X has an $\mathbb{A}^1$-fibration.*

The following characterizations, slightly different from each other, are given in [155] and [156].

Theorem 1.2.8. *Let $X = \operatorname{Spec} A$ be a smooth affine threefold. Then X is isomorphic to $\mathbb{A}^3$ if and only if*
(1) *$A^* = k^*$.*
(2) *A is factorial.*
(3) *$H_3(X; \mathbb{Z}) = 0$.*
(4) *X contains an open set of the form $U \times \mathbb{A}^2$ such that complement $X \setminus (U \times \mathbb{A}^2)$ consists of smooth irreducible components.*

Theorem 1.2.9. *Let $X = \operatorname{Spec} A$ be an affine algebraic variety of dimension 3. Then X is isomorphic to $\mathbb{A}^3$ if and only if the following conditions are satisfied:*
(1) *$A^* = k^*$.*
(2) *A is factorial.*
(3) *X has topological Euler number 1.*
(4) *X contains a* cylinder-like open set *$U \times \mathbb{A}^2$ such that the complement $X \setminus (U \times \mathbb{A}^2)$ consists of irreducible components which are factorial.*

The following is an improvement of Theorem 1.2.8 by [98].

Theorem 1.2.10. *Let $X = \operatorname{Spec} A$ be a smooth affine algebraic variety of dimension 3. Then X is isomorphic to $\mathbb{A}^3$ if and only if the following conditions are satisfied:*
(1) *$A^* = k^*$.*
(2) *A is factorial.*
(3) *$H_3(X; \mathbb{Z}) = 0$.*
(4) *X contains a cylinder-like open set $U \times \mathbb{A}^2$ such that each irreducible component of the complement $X \setminus (U \times \mathbb{A}^2)$ has at most isolated singularities.*

In the above three characterizations of $\mathbb{A}^3$, one can notice that a topological condition appears in terms of homology groups. It seems that topological conditions are necessary to set the framework (perhaps the underlying topological space) in dealing with affine algebraic varieties. Note that the affine space $\mathbb{A}^n$ is contractible to a single point. Hence $H_i(X; \mathbb{Z}) = 0$ for all $i > 0$ and $\pi_1(X) = 1$, where $X = \mathbb{A}^n$.

Conversely, we can ask what kind of algebraic structures can be put into an algebraic variety X satisfying the condition $H_i(X; \mathbb{Z}) = 0$ for all $i > 0$. If X is further smooth, we call X a *homology n-space*,[3] where $n = \dim X$. If $n = 2$, X is called a *homology plane*. These objects will be observed after studying algebraic properties of X which can be derived from topological restrictions.

The affine space $\mathbb{A}^n$ is the most basic and elementary material in building an algebraic variety, but its properties and behaviors had not been fully explored for a long time. Perhaps the following topological characterization due to Ramanujam [192] was the first result in this direction. In order to state the result, we introduce the following group.

Let X be a smooth algebraic surface defined over $\mathbb{C}$ and let (V, D) be a normal completion. Let $D = D_1 + \cdots + D_r$. View V as a complex manifold with the Euclidean metric topology. Then we can take a small neighborhood $\overline{U}_\varepsilon$ of D such that the distance of a point P in $\overline{U}_\varepsilon$ to the nearest component of D is bounded by a small number ε. Let $\partial\overline{U}_\varepsilon$ be the boundary of $\overline{U}_\varepsilon$. The fundamental group $\pi_1(\partial\overline{U}_\varepsilon)$ is independent of ε as long as ε is sufficiently small and the embedding of X into V. So, this group is called the *fundamental group at infinity* and denoted by $\pi_1^\infty(X)$. Here is a famous theorem of Ramanujam [192].

Theorem 1.2.11. *Let X be a complex smooth algebraic surface defined over $\mathbb{C}$. Then X is isomorphic to $\mathbb{A}^2$ if and only if X is contractible*[4] *and $\pi_1^\infty(X) = 1$.*

We revisit this result in Section 1.3.10 below. Ramanujam's proof shows that contractibility can be replaced by a very weak condition, *viz.*, affiness of X. Fujita proved in [56] that a smooth $\mathbb{Z}$-homology plane is affine.

A more algebraic characterization of $\mathbb{A}^2$ is given by the following theorem [67, 151].

Theorem 1.2.12. *Let X be a smooth affine algebraic surface such that there is a finite morphism $\pi : \mathbb{A}^2 \to X$. Then X is isomorphic to $\mathbb{A}^2$.*

On the other hand, the structure of an algebraic variety can be clarified by algebraic actions of algebraic groups. One of the main subjects of the book is to do such a clarification via actions of the additive group G_a. But the first thing to do is perhaps with finite groups. Namely, given an algebraic variety X, we can consider the following problems:

(1) What are finite subgroups G contained in $\mathrm{Aut}(X)$, where $\mathrm{Aut}(X)$ is the group of all automorphisms of X and a subgroup G of $\mathrm{Aut}(X)$ gives a G-action on X?
(2) For a finite group G acting on an affine algebraic variety $X = \operatorname{Spec} A$, we can consider the *algebraic quotient* X/G as $\operatorname{Spec} A^G$, where A^G is the ring of G-invariant

3 In [84], it is originally called a *homology n-fold*. But, in the analogy between homology plane and affine plane, we rename it as *homology n-space* because it is a homological analogue of the affine n-space $\mathbb{A}^n$.

4 By a theorem of J. H. C. Whitehead, this is equivalent to the condition that X is a homology plane with $\pi_1(X) = 1$.

elements of A which becomes an affine domain over k. Clarify the structure of X/G.

Consider these problems for $X = \mathbb{A}^2$. As already remarked before Theorem 1.2.4, a finite group G of $\mathrm{Aut}(\mathbb{A}^2)$ is conjugate to a subgroup of $\mathrm{GL}(2,k)$, i. e., any finite group action on $\mathbb{A}^2$ is *linearizable*. Hence we can consider only finite linear subgroups. An element $g \in \mathrm{GL}(2,k)$ is called a *pseudoreflection* if the eigenvalues of g, counted with multiplicity, are 1 except for one value. Given a finite subgroup G of $\mathrm{GL}(2,k)$, the subgroup N of G generated by all pseudoreflections of G is a normal subgroup, and the quotient $\mathbb{A}^2/N$ is isomorphic to $\mathbb{A}^2$ by a theorem of Chevalley–Shephard–Todd [27, 212]. Hence $\mathbb{A}^2/G$ is isomorphic to $\mathbb{A}^2/\overline{G}$, where $\overline{G} = G/N$. So, we may assume that G is a *small* subgroup, i. e., G contains no pseudoreflections.

A small finite subgroup G of $\mathrm{GL}(2,k)$ is classified up to conjugation in Brieskorn [20, Satz 2.9]. The quotient $\mathbb{A}^2/G$ is a normal affine surface with a unique singular point $\overline{O}$, which is the image of the point of origin O of $\mathbb{A}^2$ and called a *quotient singular point*. The minimal resolution of singularity of $\overline{O}$ is described in terms of the weighted graphs. The details will be given in [159, Chapter 2, §2]. The importance of $\mathbb{A}^2/G$ is exhibited by the following two results in [154] and [76]. These results are revisited in Section 1.3.

Theorem 1.2.13. *Let X be a normal affine surface such that X is the image of a finite morphism $\pi : \mathbb{A}^2 \to X$. Then X is isomorphic to $\mathbb{A}^2/G$ for a small finite subgroup of* $\mathrm{GL}(2,k)$.

Theorem 1.2.14. *Let X be a normal affine surface defined over $\mathbb{C}$ such that X is topologically contractible and $\pi_1^\infty(X)$ is a finite group. Then X is isomorphic to $\mathbb{A}^2/G$.*

By the explanation before Theorem 1.2.4, $\mathbb{A}^2/G$ is related to the structure of algebraic surfaces with $\overline{\kappa} = -\infty$.

1.3 Topological approaches to affine surfaces

We will see that various topological conditions on the boundary at infinity of an affine surface are effective to determine the surface. The ground field k is an algebraically closed field of characteristic zero. Whenever we employ topological arguments, we tacitly assume that $k = \mathbb{C}$.

1.3.1 Derivations and vector fields

We will revisit the subject in more specified ways in Chapter 2, Section 2.2.

Let X be an affine irreducible variety with coordinate ring A. Then the module of k-derivations of A is by definition

$$\mathrm{Der}_k A := \mathrm{Hom}_A(\Omega^1_{A/k}, A).$$

If A is normal then $\mathrm{Der}_k A$ is a finitely generated, reflexive A-module whose rank as an A-module is the same as $\dim A$.

Let $A = k[x_1, \ldots, x_n]/\mathfrak{P}$. Then any $\delta \in \mathrm{Der}_k A$ is of the form

$$\sum_i \overline{a}_i \frac{\partial}{\partial x_i},$$

where $a_i \in k[x_1, \ldots, x_n]$ and $\overline{a}_i$ is its image in A. It holds that for any $f \in \mathfrak{P}$, $\sum_i a_i \frac{\partial f}{\partial x_i} \in \mathfrak{P}$.

Let $P \in X$ be a smooth point. A tangent vector to X at P is of the form

$$\sum_{i=1}^n \lambda_i \frac{\partial}{\partial x_i},$$

where $\lambda_i \in k$ and $\sum_i \lambda_i \frac{\partial f}{\partial x_i}(P) = 0$ for any $f \in \mathfrak{P}$. It follows that if $\delta = \sum_i \overline{a}_i \frac{\partial}{\partial x_i}$ is a k-derivation of A then $\sum_i \overline{a}_i(P) \frac{\partial}{\partial x_i}$ is a tangent vector to X at P. Thus, a k-derivation of A gives an *algebraic* vector field on $X - \mathrm{Sing}\, X$. Conversely, an algebraic vector field $\mathcal{V}$ on $X - \mathrm{Sing}\, X$ is given by

$$\sum_{i=1}^n \overline{a}_i \frac{\partial}{\partial x_i},$$

where $a_i \in A$ and $\sum_i \overline{a}_i \frac{\partial f}{\partial x_i} \in \mathfrak{P}$ for all $f \in \mathfrak{P}$.

Let X be an irreducible algebraic variety. Since $\Omega_{X/k}$ is a coherent sheaf on X, so is its dual $\mathcal{H}om_{\mathcal{O}_X}(\Omega^1_{X/k}, \mathcal{O}_X)$. It is known that any reflexive coherent sheaf $\mathcal{F}$ on a normal irreducible variety X is equal to $i_*(\mathcal{F}|_{X - \mathrm{Sing}\, X})$ for the inclusion $i : X - \mathrm{Sing}\, X \hookrightarrow X$, i. e., $\mathcal{F}$ is determined by its restriction to $X - \mathrm{Sing}\, X$.

1.3.2 G_a-actions on $\mathbb{A}^n$

Let G_a act regularly on $\mathbb{A}^n$. In general, the ring of invariants $A := k[x_1, \ldots, x_n]^{G_a}$ is not finitely generated as a k-algebra. If the G_a-action is linear then A is finitely generated over k by Weitzenböck's theorem [224] (see [209] for a modern treatment). Let X be the corresponding affine variety if A is finitely generated over k. Even in this case the quotient morphism $q : \mathbb{A}^n \to X$ is in general not a surjection. Instead we have the following result.

Lemma 1.3.1. *$X - q(\mathbb{A}^n)$ has codimension ≥ 2 in X.*

Proof. By Lemma 2.2.1, A is a UFD. Hence a closed subvariety W of X of codimension one is principal. Namely, there exists a prime element $p \in A$ such that pA is the defining ideal of W. By Lemma 2.2.1, $\widetilde{W} := q^{-1}(W)$ is an irreducible and reduced subvariety of $\mathbb{A}^n$ such that $q(\widetilde{W})$ dominates W. □

We will prove now the following important results (see Bonnet [18]).

Theorem 1.3.2. *Let $n \leq 3$. Then A is finitely generated over k. Further, $X \cong \mathbb{A}^{n-1}$ and $q : \mathbb{A}^n \to X$ is a surjection.*

Proof. We note first that we may assume the base field k to be the complex number field $\mathbb{C}$ by the Lefschetz principle (see the footnote in the proof of Lemma 2.2.17). We will first consider the case $n = 2$. We have seen that in this case A is finitely generated over k by Zariski [234] (see also Theorem 2.2.4). Now A is factorial, i. e., a UFD, and $A^* = k^*$. Hence $A \cong k[t]$. Indeed, $X = \operatorname{Spec} A$ is a smooth affine algebraic curve which has trivial Picard group. Hence X is rational, and isomorphic to $\mathbb{P}^1$ with m points removed. If $m = 1$ then $X \cong \mathbb{A}^1$. If $m \geq 2$ then $A^* \cong k^* \times \mathbb{Z}^{\times(m-1)}$. Hence $m = 1$. This implies that q is surjective. A general fiber of $q : \mathbb{A}^2 \to X$ is isomorphic to $\mathbb{A}^1$ (see Lemma 2.2.1). By the AMS theorem, which is discussed later in this chapter, every fiber of q is isomorphic to $\mathbb{A}^1$.

Now consider the case $n = 3$. We will indicate the outline of a proof and leave the details to [154]. By Theorem 2.2.4, A is finitely generated over k, A is factorial, and $A^* = k^*$. We claim that the following assertions hold.

Claim.
(1) $\pi_1(X - \operatorname{Sing} X) = (1)$.
(2) *The logarithmic Kodaira dimension* $\overline{\kappa}(X - \operatorname{Sing} X) = -\infty$.

We first prove assertion (1). Let $X^\circ = X - \operatorname{Sing} X$ and let $Y^\circ = q^{-1}(X^\circ)$. Since $\operatorname{Sing} X$ is a finite set and $q : \mathbb{A}^3 \to X$ has no fibers of dimension two, $\mathbb{A}^3 \setminus Y^\circ$ is a closed set of codimension ≥ 2. Hence $\pi_1(Y^\circ) = \pi_1(\mathbb{A}^3) = (1)$. Namely, Y° is simply-connected. Let $f : U \to X^\circ$ be a topological covering. Then the natural projection $Y^\circ \times_{X^\circ} U \to Y^\circ$ is a topological covering. Hence it decomposes to a direct sum of copies of Y°. This implies that the morphism $q^\circ := q|_{Y^\circ}$ splits as

$$q^\circ : Y^\circ \longrightarrow U \xrightarrow{f} X^\circ.$$

Hence f is a finite morphism from an algebraic surface U to X°. If $\deg f > 1$, then a general fiber of q°, hence of q is not connected. This contradicts the fact that a general fiber of the quotient morphism $q : \mathbb{A}^3 \to X$ is isomorphic to $\mathbb{A}^1$. This implies that $\pi_1(X^\circ) = (1)$.

For the second assertion, we show the existence of a linear hyperplane H with respect to a fixed system of coordinates on $\mathbb{A}^3$ such that H meets in finitely many points the fibers of q lying over points of $\operatorname{Sing} X$. Since $\operatorname{Sing} X$ is a finite set, the restriction of q onto $H' = H - S$ is a dominating morphism onto X°, where $S = H \cap q^{-1}(\operatorname{Sing} X)$ (see [154, Lemma 4]). Since $H \cong \mathbb{A}^2$, H' contains a *cylinder-like open set* $\Delta \times \mathbb{A}^1$, where Δ is an affine curve. By [159, Chapter 2, Section 2.1], $\overline{\kappa}(H') = -\infty$. By Lemma 1.2.2, we have $\overline{\kappa}(H') \geq \overline{\kappa}(X^\circ)$. This implies that $\overline{\kappa}(X^\circ) = -\infty$ and completes the proof of Claim. □

If X contains a cylinder-like open set, $X \cong \mathbb{A}^2$ by the following result which is equivalent to Theorem 1.2.7 (see [159, Chapter 3, Theorem 2.2.1]).

Lemma 1.3.3. *Let $X = \operatorname{Spec} A$ be an affine normal surface defined over k. Then X is isomorphic to $\mathbb{A}^2$ if and only if the following conditions are satisfied:*
(1) *A is factorial.*
(2) *$A^* = k^*$.*
(3) *X contains a cylinder-like open set $\Delta \times \mathbb{A}^1$, where Δ is an affine curve.*

If X does not contain a cylinder-like open set and $\overline{\kappa}(X^\circ) = -\infty$, Theorem 1.2.4 is available. We restate it as follows.

Lemma 1.3.4. *Let $X = \operatorname{Spec} A$ be a normal affine surface satisfying the following conditions:*
(1) *The divisor class group $\mathrm{C}\ell(X)$ has $\mathbb{Q}$-rank zero.*
(2) *$A^* = k^*$.*
(3) *$\overline{\kappa}(X^\circ) = -\infty$ but X° contains no cylinder-like open sets, where $X^\circ = X - \operatorname{Sing} X$.*

Then X is a Platonic $\mathbb{A}^1_*$-fiber space.

For the ring $A = k[x_1, x_2, x_3]^{G_a}$, A is factorial. Then the above condition (1) is satisfied. The projection formula for divisors shows that condition (2) is satisfied since there exists a quasifinite moephism $f : U := \mathbb{A}^2 \setminus S \to X^\circ$. Condition (3) holds by the assumption. Hence X is a Platonic $\mathbb{A}^1_*$-fiber space and $\pi_1(X^\circ) \cong G \neq (1)$. This contradicts the first assertion of Claim in the proof of Theorem 1.3.2. Thus we conclude that $\mathbb{A}^3 /\!/ G_a \cong \mathbb{A}^2$.

Let $U = q(\mathbb{A}^3)$. Then $X - U$ is a finite set of points in X by Lemma 1.3.1. Bonnet [18] proved that $H_3(\mathbb{A}^3; \mathbb{Z}) \to H_3(U; \mathbb{Z})$ is a surjection. But if $U \neq X$ then $H_3(U; \mathbb{Z})$ is a free abelian group of rank $\#(X - U)$. In fact, since $\mathbb{A}^2 - \{O\}$ is homotopic to the 3-sphere S^3, we have $H_i(\mathbb{A}^2 - \{O\}) = 0$ for $i \neq 0, 3$ and $\cong \mathbb{Z}$ for $i = 0, 3$. Let $X - U = \{P_1, \ldots, P_r\}$, $U' = X - \{P_1\}$ and $U'' = X - \{P_2, \ldots, P_r\}$. We have $U = U' \cap U''$ and $X = U' \cup U''$. The Mayer–Vietoris exact sequence of homology groups

$$H_{i+1}(X) \to H_i(U) \to H_i(U') \oplus H_i(U'') \to H_i(X)$$

shows that $H_i(U) \cong H_i(U') \oplus H_i(U'')$ for $i > 1$. By induction on r, we have $H_3(U; \mathbb{Z}) \cong \mathbb{Z}^{\oplus r}$. This is a contradiction because $H_3(\mathbb{A}^3; \mathbb{Z}) = 0$. Thus q is a surjection. □

Theorem 1.3.2 no longer holds if $n = 4$ as shown by the following examples of Bonnet [18] and Winkelmann [226].

Example 1.3.5. (1) (Bonnet) Let x, y, u, v denote the coordinates in $\mathbb{A}^4$, let t be a parameter of the additive group G_a, and consider a G_a-action σ on $\mathbb{A}^4$ defined by

$$\sigma(t; x, y, u, v) = (x, y, u - ty, v + tx).$$

Then $k[x, y, u, v]^{G_a} = k[x, y, xu+yv]$, whence $V := \mathbb{A}^4//G_a \cong \mathbb{A}^3$. The quotient morphism $q : \mathbb{A}^4 \to V$ is given by

$$(x, y, u, v) \mapsto (x, y, xu + yv).$$

Then q is not surjective because

$$q(\mathbb{A}^4) = \mathbb{A}^3 \setminus \{(x_1, x_2, x_3) \mid x_1 = x_2 = 0, x_3 \neq 0\}.$$

(2) (Winkelmann) Let $X = \mathbb{A}^4 = \operatorname{Spec} k[x_1, x_2, x_3, x_4]$ equipped with a G_a-action defined by a locally nilpotent derivation

$$\delta = x_1 \frac{\partial}{\partial x_2} + x_2 \frac{\partial}{\partial x_3} + (x_2^2 - 2x_1x_3 - 1)\frac{\partial}{\partial x_4}.$$

Thus the G_a-action is given by

$$\sigma(t; x_1, x_2, x_3, x_4) = \left(x_1, x_2 + tx_1, x_3 + tx_2 + \frac{1}{2}t^2x_1, x_4 + t(x_2^2 - 2x_1x_3 - 1)\right).$$

Then $k[x_1, x_2, x_3, x_4]^{G_a} = \operatorname{Ker} \delta = k[\xi_1, \xi_2, \xi_3, \xi_4]$, where

$$\begin{aligned}
\xi_1 &= x_1,\\
\xi_2 &= x_2^2 - 2x_1x_3,\\
\xi_3 &= x_1x_4 - x_2(x_2^2 - 2x_1x_3 - 1),\\
\xi_4 &= x_1x_4^2 - 2x_2x_4(x_2^2 - 2x_1x_3 - 1) + 2x_3(x_2^2 - 2x_1x_3 - 1)^2,
\end{aligned}$$

and $Y := X//G_a$ is a hypersurface $\xi_1\xi_4 = \xi_3^2 - \xi_2(\xi_2 - 1)^2$ in $\mathbb{A}^4 = \operatorname{Spec} k[\xi_1, \xi_2, \xi_3, \xi_4]$. Then Y has a unique singular point $(\xi_1, \xi_2, \xi_3, \xi_4) = (0, 1, 0, 0)$, and the fiber $q^{-1}(\alpha_1, \alpha_2, \alpha_3, \alpha_4)$ is given as follows:

$$\begin{cases}
\mathbb{A}^1 & \text{if } \alpha_1 \neq 0,\\
\mathbb{A}^1 + \mathbb{A}^1 & \text{if } \alpha_1 = 0,\ \alpha_2 \neq 0, 1,\\
2\mathbb{A}^1 & \text{if } \alpha_1 = \alpha_2 = 0,\\
\emptyset & \text{if } \alpha_1 = 0,\ \alpha_2 = 1,\ \alpha_3 = 0,\ \alpha_4 \neq 0,\\
\mathbb{A}^2 + \mathbb{A}^2 & \text{if } (\alpha_1, \alpha_2, \alpha_3, \alpha_4) = (0, 1, 0, 0).
\end{cases}$$ □

1.3.3 The Makar-Limanov invariant and affine lines on ML$_i$ surfaces

The subject will be treated in details in Sections 2.6, 2.7, and 2.8. So, this section is only an introduction to results to be treated later.

Let X be a normal affine variety with coordinate ring A. We define the subring $A_0 := \bigcap A^{G_a}$, where the intersection is taken over all possible G_a-actions on X. Following [87], we say that X is an ML$_i$ variety if $\operatorname{tr.deg}_k A_0 = i$. Let $d = \dim X$. Then X is ML$_i$ means that X admits $(d - i)$ independent G_a-actions. Thus X is ML$_d$ just means that there is no nontrivial G_a-action on X.

Remark 1.3.6. It is easy to see that if there is a nontrivial G_a-action on X then $\overline{\kappa}(X) = -\infty$. This is because in this case X contains a cylinder-like open set $U \times \mathbb{A}^1$ (see Lemma 2.2.1). Since $\overline{\kappa}(U \times \mathbb{A}^1) = -\infty$, we have $\overline{\kappa}(X) = -\infty$. It was proved by Fujita [57] and Miyanishi–Sugie [164] that for a smooth affine surface X the conditions $\overline{\kappa}(X) = -\infty$ and that X has a cylinder-like open set are equivalent.

For a smooth affine surface X, the condition that $\overline{\kappa}(X) = -\infty$ is strictly weaker than the existence of a nontrivial G_a-action on X. This is because in the latter case $C := X//G_a$ is an affine curve, which means that there is an $\mathbb{A}^1$-fibration $f : X \to C$. On the other hand, $\overline{\kappa}(X) = -\infty$ means that there is an $\mathbb{A}^1$-fibration $\varphi : X \to B$, where B might be a complete curve. For a smooth affine surface X with ML_2 property and at least one $\mathbb{A}^1$-fibration over a complete base curve, see Example 2.7.7. □

Let X be a smooth affine ML_i surface with $i = 0$ or 1. Then there is an $\mathbb{A}^1$-fibration $f : X \to B$, where B is a smooth affine curve. Given a curve $C \subset X$ such that $C \cong \mathbb{A}^1$ we can ask if there is an $\mathbb{A}^1$-fibration $\varphi : X \to \Gamma$ such that C is contained in a fiber of φ. This question was considered in [87] and will be considered in details in Chapter 2 (see Theorem 2.6.12 for the ML_0 case and Theorem 2.7.12 for the ML_1 case). Here we present a rough story as a motivation. The following result was proved.

Theorem 1.3.7. *With the above notation, assume that X is an ML_0 surface. Suppose that $\mathrm{Pic}\, X$ is a torsion group (hence a finite group). Then any $C \subset X$ which is isomorphic to $\mathbb{A}^1$ is contained in a fiber of an $\mathbb{A}^1$-fibration $\varphi : X \to \mathbb{A}^1$.*

If $\mathrm{Pic}\, X$ is not torsion and X is an ML_0 surface then there are examples for which this is not true (see Proposition 2.6.19). If X is an ML_1 surface this is not the case even if $\mathrm{Pic}\, X$ is torsion (see Proposition 2.7.9).

These topics appear in Chapter 2 again with more details. A key idea is to use topological and algebro-geometric approaches combined together. For the proof of the case when the Picard number $\rho(X)$ is zero, i. e., $\mathrm{Pic}(X)$ is finite, we consider $e(X - C)$. Since X is ML_0 and $\rho(X) = 0$, we deduce that $e(X) = 1$, so that $e(X - C) = 0$. Since $\mathrm{Pic}\, X$ is torsion, there exists $f \in \Gamma(X, \mathcal{O}_X)$ such that the principal divisor (f) is supported by C on X. Then the fibration $f : X \to \mathbb{A}^1$ defined by the function f has the property that $f^*(0) = mC$ for some $m \geq 1$. From $e(X-C) = 0$, we infer by using the Suzuki–Zaidenberg formula that f is an $\mathbb{A}^1$-fibration. Thus C is a fiber component of an $\mathbb{A}^1$-fibration on X with base $\mathbb{A}^1$.

1.3.4 Ascent and descent of ML_0 property

The ML_0 property behaves nicely under finite topological coverings of smooth affine surfaces. Let $\varphi : X \to Y$ be a finite morphism between smooth affine surfaces. The question of ML_0 property on X descending to Y was considered in [87, Theorem 4.2] and will be discussed in details in Section 2.8.

Theorem 1.3.8. *With the above notation, assume that X is ML_0. If either φ is étale or Galois then Y is also ML_0. Conversely, if Y is ML_0 and φ is étale then X is also ML_0.*

Remark 1.3.9. Since every smooth affine surface has a Noether normalization $\varphi : X \to \mathbb{A}^2$, we cannot expect the ML_0 property of Y to imply the same for X. □

However, we can show the following result.

Theorem 1.3.10. *Let $\varphi : X \to Y$ be a finite morphism between smooth affine surfaces. If X is ML_0 and $\rho(X) = 0$ then Y is also ML_0 with $\rho(Y) = 0$.*

1.3.5 $\mathbb{A}^1$-fibration over a complete curve

Let X be a smooth affine surface which admits a surjective $\mathbb{A}^1$-fibration $\varphi : X \to B$ with B normal. It can happen that for any $\mathbb{A}^1$-fibration on X the base B is affine. For example, let $X = \mathbb{A}^2$. Then $e(X) = 1$. It follows easily from Suzuki–Zaidenberg formula that for any $\mathbb{A}^1$-fibration $\varphi : \mathbb{A}^2 \to B$ the base $B \cong \mathbb{A}^1$.

The following result is proved in [87, Theorem 5.1].

Theorem 1.3.11. *Let X be a smooth affine ML_0 surface such that $\operatorname{Pic} X$ has rank > 0. Then there exists a surjective $\mathbb{A}^1$-fibration $\varphi : X \to \mathbb{P}^1$.*

The proof of this result uses the *Derksen invariant* as defined below.

Definiton 1.3.12. *Let X be a smooth affine surface. We consider the subalgebra of the coordinate ring R of X generated by all rings of invariants R^{G_a} for all possible G_a-actions on X. This subalgebra is called the* Derksen invariant *of X.*

1.3.6 The AMS theorem

In this subsection we will state the following Theorem 1.3.14, called the *AMS theorem* first proved by S. S. Abhyankar–T. T. Moh [2] and independently by M. Suzuki [217]. There are many different proofs [149, 197, 194, 179, 110, 8, 3, 73, 68, 236, 124, 139, 100] with the references chronologically aligned. The proof given in Section 1.4 below is based on [68].

The affine plane $\mathbb{A}^2 = \operatorname{Spec} k[x,y]$ is embedded into $\mathbb{P}^2$ in such a way that for a system of homogeneous coordinates (X,Y,Z) of $\mathbb{P}^2$, the embedded $\mathbb{A}^2$ is an open set which is the complement of the line $\{Z = 0\}$ and $x = X/Z, y = Y/Z$. A geometric local ring R over k is *analytically irreducible* if the completion $\widehat{R}$ is an integral domain. Suppose that R is the local ring of an irreducible curve C at a closed point P. Then we say that P is a *one-place point* of C if the function field $k(C)$ has a unique discrete valuation ring $\mathcal{V}$ which dominates the local ring $R = \mathcal{O}_{C,P}$. The following three conditions are equivalent:

(1) $R = \mathcal{O}_{C,P}$ is analytically irreducible.
(2) Let $\nu : \widetilde{C} \to C$ be the normalization of C in $k(C)$. Then $\nu^{-1}(P)$ consists of a single point.
(3) P is a one-place point of C.

Given a curve C on $\mathbb{A}^2$, let $\overline{C}$ be its closure in $\mathbb{P}^2$ with the above embedding $\mathbb{A}^2 \hookrightarrow \mathbb{P}^2$. We say that C has a one-place point *at infinity* if $\overline{C} - C$ consists of one point which is a one-place point of $\overline{C}$. The following lemma is a key in the proof of the AMS theorem.

Lemma 1.3.13. *Let $f(x,y)$ be an irreducible polynomial in $k[x,y]$ such that the curve $\{f = 0\}$ has only one place at infinity. Then for any $\lambda \in k$ the polynomial $f - \lambda$ is irreducible and the curve $\{f = \lambda\}$ has only one place at infinity.*

The AMS theorem is stated as follows.

Theorem 1.3.14. *With the above notation, suppose that the curve $\{f = 0\}$ is isomorphic to $\mathbb{A}^1$. Then there is an automorphism of $\mathbb{A}^2$ which maps the curve $\{f = 0\}$ to the line $\{x = 0\}$.*

The proof of the theorem shows that there is an open embedding $\mathbb{A}^2 \subset V$ into a smooth projective surface V such that $f : \mathbb{A}^2 \to \mathbb{A}^1$ extends to a $\mathbb{P}^1$-fibration $\Phi : V \to \mathbb{P}^1$, and $V - \mathbb{A}^2$ consists of a smooth irreducible cross-section of Φ, say S, and all irreducible components of one reducible fiber. Further, the closure $\overline{\{f = \lambda\}}$ meets S transversally in one point and $\overline{\{f = \lambda\}}$ is a full fiber of Φ. From this it is easy to deduce that each curve $\{f - \lambda = 0\}$ is isomorphic to $\mathbb{A}^1$.

1.3.7 The Lin–Zaidenberg theorem

V. Y. Lin and M. G. Zaidenberg [233] proved the following result by making use of deep results from Teichmüller theory.

Theorem 1.3.15. *Let $C \subset \mathbb{A}^2$ be an irreducible algebraic curve which is topologically contractible. Then there exist affine coordinates x, y on $\mathbb{A}^2$ such that in terms of these coordinates C is defined by the equation $\{x^m = y^n\}$, where* $\gcd(m,n) = 1$.

Besides the original proof [233], there are several new proofs [73, 124, 186]. In [73], the proof uses the fact that $e(\mathbb{A}^2 - C) = 0$ and hence $\overline{\kappa}(\mathbb{A}^2 - C) \le 1$. By using the detailed knowledge of smooth affine surfaces with $\overline{\kappa} \le 1$, we deduce the result. We produce this proof in Section 1.4.

1.3.8 Smooth affine surfaces with an $\mathbb{A}^1_*$-fibration

Let X be a smooth affine surface with an $\mathbb{A}^1_*$-fibration $f : X \to B$ onto a smooth curve B. We will describe all possible singular fibers that can occur for f.

Let $X \subset V$ be an open embedding into a smooth surface V such that f extends to a $\mathbb{P}^1$-fibration $\Phi : V \to \overline{B}$, where $B \subset \overline{B}$ is the completion. Then $V - X$ contains either two cross-sections of Φ (the *untwisted* case) or a single irreducible smooth 2-section of Φ (the *twisted* case). Each fiber of f is contained in a fiber of Φ as an open subset. We know a complete description of singular fibers of a $\mathbb{P}^1$-fibration on a smooth projective surface $\Phi : V \to \overline{B}$. Every fiber of Φ is a tree of smooth rational curves. Let F_0 be a fiber of f which is scheme-theoretically not isomorphic to $\mathbb{A}^1_*$. Every irreducible component of F_0 meets $V-X$. Since the fiber of Φ containing F_0 has no nontrivial loops, we deduce that F_0 is written as follows: $(F_0)_{\mathrm{red}} = \Gamma + \Delta$, where $\Gamma = \emptyset, \mathbb{A}^1_*$ or $\mathbb{A}^1 + \mathbb{A}^1$, $\Delta = \emptyset$ or a disjoint union of the $\mathbb{A}^1$s and $\Gamma \cap \Delta = \emptyset$. In the case $\Gamma = \mathbb{A}^1 + \mathbb{A}^1$, we can say the following.

Corollary 1.3.16. *Suppose the scheme-theoretic fiber F_0 contains a connected component $m_1C_1 + m_2C_2$ where $C_1 \cong C_2 \cong \mathbb{A}^1$ and $C_1 \cap C_2 \neq \emptyset$. Then* $\gcd(m_1, m_2) = 1$.

Proof. Suppose $m = \gcd(m_1, m_2) > 1$. Using the ramified covering trick, we get a smooth irreducible affine surface $\widetilde{X}$ with an $\mathbb{A}^1_*$-fibration $\varphi : \widetilde{X} \to \widetilde{B}$ such that a singular fiber of φ has m connected components which look like $C_1 \cup C_2$. This is not possible by the description of all possible singular fibers of an $\mathbb{A}^1_*$-fibration. □

1.3.9 Theorem of Mumford

We will discuss an important result of D. Mumford [171] which is very useful to study topology of singularities and affine surfaces.

Let X be a 2-dimensional complex manifold and $E := E_1 \cup E_2 \cup \cdots \cup E_r$ a connected complete divisor with simple normal crossings on X. Let g_i be the genus of E_i. Assume that *the dual graph of E has no nontrivial loops* (This is a very crucial assumption). We can find a fundamental system of neighborhoods $\{U_n \mid U_n \supset E\}$ satisfying the following properties:

(1) E is a strong deformation retract of U_n, and ∂U_n is a compact orientable C^∞ manifold.
(2) ∂U_n is a strong deformation retract of $U_n - E$.

It is well-known that the diffeomorphism type of ∂U_n is independent of the given system $\{U_n\}$. We define the fundamental group of ∂U_n as the *local fundamental group* of E in X. Two special cases are of interest to us:

(a) E is the exceptional divisor of a resolution of a normal surface singularity (S, Q). The local fundamental group of E in X is equal to the local fundamental group $\pi_1^Q(S) = \pi_1(\partial B)$, where B is a small ball around Q in S. We denote $\pi_1^Q(S)$ also by $\pi_1(S; Q)$.
(b) E supports a divisor Δ with $\Delta^2 > 0$.

The case (b) happens, for example, if E is the divisor at infinity in a compactification of a normal affine surface $X \subset V$, where V is a normal projective surface which is smooth outside X.

Now we describe the presentation of $\pi_1(\partial U_n)$ by generators and relations as given by Mumford [171]. Let $e_1, \ldots, e_r$ be generators, e_i corresponding to E_i, and a_{ij}, b_{ij} generators for each i and for $1 \le j \le g_i$. Then $\pi_1(\partial U_n)$ is generated by

$$e_1, \ldots, e_r, a_{11}, b_{11}, \ldots, a_{1g_1}, b_{1g_1}, \ldots\ldots, a_{r1}, b_{r1}, \ldots, a_{rg_r}, b_{rg_r},$$

having the following three relations with $e_{ij} := (E_i \cdot E_j)$:

$$\begin{aligned} &(1) \quad \prod_{k=1}^{g_i} [a_{ik}, b_{ik}] \cdot \prod_{j=1}^{r} e_j^{e_{ij}} = 1; \\ &(2) \quad [a_{ik}, e_i] = [b_{ik}, e_i] = 1; \\ &(3) \quad e_i \cdot e_j = e_j \cdot e_i \ \text{ if } E_i \cap E_j \neq \emptyset. \end{aligned}$$

The proof of this assertion uses Van Kampen's theorem. We will always assume that if E_i is a (-1) curve then E_i meets at least three other E_js.

An important special case that is useful in affine algebraic geometry is when $g_i = 0$ for all i. Further, if $\pi_1(\partial U_n)$ is finite then the dual graph of E is very special as explained below.

To wit, assume that the intersection matrix $((E_i \cdot E_j))$ is negative definite. Then E is the exceptional divisor of a resolution of normal surface singularity. In this case, either the dual graph is a linear chain, or the dual graph of E is star-shaped with a unique branch point and three linear chains $\Gamma_1, \Gamma_2, \Gamma_3$ meeting the branch point at one of their end vertices. Further, if d_i is the absolute value of the determinant of the intersection form of curves in Γ_i, then

$$\frac{1}{d_1} + \frac{1}{d_2} + \frac{1}{d_3} > 1.$$

This implies that $\{d_1, d_2, d_3\}$ is, up to permutations, one of the Platonic triplets $\{2, 2, m\}$, $\{2, 3, 3\}$, $\{2, 3, 4\}$, $\{2, 3, 5\}$. Here m is an arbitrary integer ≥ 2.

The fundamental result of Mumford can now be stated as follows.

Theorem 1.3.17 (Mumford's theorem). *If $\pi_1(\partial U_n) = (1)$ then E_i is a (-1) curve for some i, and E_i meets at most two other E_js.*

In the case that E is the exceptional divisor of a resolution of normal surface singularity (S, Q), this theorem yields a famous criterion of simplicity, saying that Q is a smooth point of S if and only if the local fundamental group $\pi_1(S; Q) = (1)$.

We will briefly indicate Mumford's argument. We assume that every (-1) curve in E meets three other curves in E. The aim of the proof is to show that this leads to a contradiction. Now, by using the group-theoretic lemma below, Mumford shows that if E

has a (−1) curve E_i, which is a branch point of the dual graph of E, then E_i meets another (−1) curve E_j. This contradicts the negative definiteness of the intersection form.

Lemma 1.3.18 (Mumford's lemma). *Let $G_1, \dots, G_r$ be nontrivial groups and $g_i \in G_i$ an arbitrary element for $i = 1, \dots, r$. If $r \geq 3$ then the free product $G_1 * \cdots * G_r$ modulo the normal subgroup generated by $g_1 g_2 \cdots g_r$ is a nontrivial group.*

Mumford [171] used non-Euclidean geometry to prove this. A purely group-theoretic proof was given in [77]. In fact, it was proved therein that if $\pi_1(\partial U_n)$ is a finite nontrivial group then either the dual graph of E is a linear chain, or $r = 3$ and G_i is a cyclic group generated by g_i for $i = 1, 2, 3$. Further, the orders of G_1, G_2, G_3 form, up to permutations, a Platonic triplet $\{2, 2, m\}, \{2, 3, 3\}, \{2, 3, 4\}, \{2, 3, 5\}$.

This lemma is used as follows. Let E be as above. For E_i, let $\Gamma_1, \dots, \Gamma_\ell$ be the connected components of the union of the E_js for $j \neq i$; $\Gamma_1, \dots, \Gamma_\ell$ are called *branches* of E at E_i. Let $E_{11}, \dots, E_{\ell 1}$ be the irreducible curves in $\Gamma_1, \dots, \Gamma_\ell$ respectively which intersect E_i. Since the dual graph of E has no nontrivial loops, E_{j1} is well-defined for each j. Each Γ_j corresponds to a group G_j given by Mumford's presentation. It is easy to see from Mumford's presentation that $\pi_1(\partial U_n)/\langle e_i \rangle$ is the group

$$\frac{G_1 * G_2 * \cdots * G_\ell}{\langle e_{11} e_{21} \cdots e_{\ell 1} \rangle},$$

where e_{j1} is the generator corresponding to E_{j1} in Mumford's presentation.

If $\pi_1(\partial U_n)$ is finite then it is easy to see that each $E_i \cong \mathbb{P}^1$ and the dual graph of E is a tree. Suppose that $\pi_1(\partial U_n)$ is trivial. With the above notation, using Mumford's lemma, it follows that if $\ell \geq 3$ at least one G_j is trivial. Since the intersection form on the components of Γ_j is negative definite, by using induction, we see that the curve E_{j1} meeting E_i is a (−1) curve. Arguing this way, Mumford showed that there are two (−1) curves in E which intersect. But the intersection form on the union of these two curves is not negative definite. This contradiction proves Mumford's theorem.

A close look at the proof shows that Mumford's presentation is not needed for the proof. Only Van Kampen's theorem is enough. But Mumford's presentation had many other important applications for finding dual graphs, calculating local first homology group of the singular point, etc.

1.3.10 Theorem of Ramanujam

We use the ideas in Mumford's theorem to study affine surfaces. Let X be a smooth affine surface. Embed $X \subset V$, where V is a smooth projective surface and $D := V - X$ is a divisor with simple normal crossings. Suppose that the fundamental group at infinity $\pi_1^\infty(X)$ is trivial, which is defined before Theorem 1.2.11. Hence, for a suitable neighborhood U_n of D, we have $\pi_1(U_n - D) = (1)$. We then say that X is *simply-connected at infinity*. It then follows that D is a connected union of smooth rational curves such that the dual graph of D has no nontrivial loops. So we can study $\pi_1(U_n - D)$ using the Mumford's presentation.

C. P. Ramanujam [192] proved the following theorem.

Theorem 1.3.19. *Let X be a complex smooth affine surface. Then X is isomorphic to $\mathbb{A}^2$ if and only if X is topologically contractible and $\pi_1^\infty(X) = (1)$.*

Now D supports an ample divisor. Hence the intersection form on a connected union of curves in D (a partial sum) is negative definite, has 0 as an eigenvalue, or has a positive eigenvalue. Because of this C. P. Ramanujam's proof of his characterization of $\mathbb{A}^2$ is more involved.

Recently, R. V. Gurjar and S. R. Gurjar [70] has obtained a simplified proof of this result using some basic results from the theory of open algebraic surfaces.

In the same paper [192], Ramanujam proved that the boundary divisor of a minimal normal completion[5] of the affine plane has a linear dual graph of smooth rational curves. Based on this result, Morrow [170] gave a complete classification of such dual graphs. There is a new proof due to Kishimoto [111].

1.3.11 Some more applications of Mumford's ideas

Let $E = \sum_{i=1}^r E_i$ be an SNC connected divisor on X such that $E_i \cong \mathbb{P}^1$ for all the components E_i of E, and the dual graph of E is a tree. Using the Mumford's presentation for $\pi_1(\partial U_n)$, we see that $H_1(\partial U_n; \mathbb{Z})$ is a finite abelian group of order $\|(E_i \cdot E_j)\|$, the absolute value of the determinant of the intersection matrix $(E_i \cdot E_j)$, provided this determinant is nonzero. If the determinant is zero, then $H_1(\partial U_n; \mathbb{Z})$ is an infinite group.

In case $(E_i \cdot E_j)$ is negative definite, the determinant is nonzero. This happens, in particular, if E contracts to a quotient singular point P on a normal surface Y. Mumford proved that the divisor class group of the analytic local ring $\mathcal{O}_{Y,P}$ is isomorphic to $H_1(\partial U_n; \mathbb{Z})$.

An important corollary of Mumford's work is that an analytic surface singularity (Y, P) is rational if and only if the divisor class group of the complete (or analytic) local ring $\widehat{\mathcal{O}}_{Y,P}$ (or $\mathcal{O}_{Y,P}$) is finite.

H. Flenner [49] generalized this result to higher dimension. Let (Y, P) be a germ of a rational singularity of arbitrary dimension. Then the divisor class group of the analytic local ring $\mathcal{O}_{Y,P}$ is isomorphic to $H^2(U \setminus \operatorname{Sing} Y; \mathbb{Z})$, where U is a suitable Stein neighborhood of P in Y.

A closely related result was proved in [70].

Theorem 1.3.20. *Let Y be a normal algebraic surface with a singular point P. Assume that the exceptional divisor E of a resolution of singularity at P has unimodular inter-*

5 For the definition, see Section 2.6.1.

section form and the dual graph of E is a tree of $\mathbb{P}^1$s. Then the local ring $\mathcal{O}_{Y,P}$ is factorial.

Proof. We will give a sketch. Given any line bundle L on X, where $X \to Y$ is the resolution of singularity at P, we can find a divisor Δ supported on E such that $L|_{E_i} \cong \Delta|_{E_i}$ for each i. Here we are using the assumption that E_i is a smooth rational curve and the intersection matrix $(E_i \cdot E_j)$ is unimodular. Then $(L-\Delta)|_{E_i}$ is trivial for each i. Since the dual graph of E is a tree, by a suitable induction on the number of components of E, we can show that $(L - \Delta)|_E$ is trivial. From this we can deduce that $L - \Delta$ is linearly equivalent to a divisor Δ' which is disjoint from E. This implies that $\mathcal{O}_{Y,P}$ is factorial. □

1.3.12 Surfaces with finite fundamental group at infinity

Let V be a smooth projective surface, E a connected SNC divisor on V such that E supports a divisor Δ with $\Delta^2 > 0$. Assume that there is a fundamental system of neighborhoods U_n of E such that $\pi_1(\partial U_n)$ is finite. In [211], there is a complete description of the weighted dual graph of E. The weighted dual graph of E is one of the following after finitely many blowing-ups and contractions of (–1) curves. All the components of E are smooth rational curves.

(i)

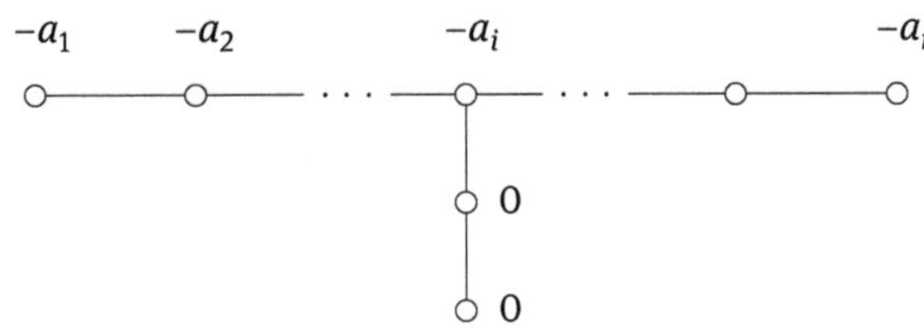

where $a_i \geq 2$ for all i and $1 \leq i \leq r$.

(ii)

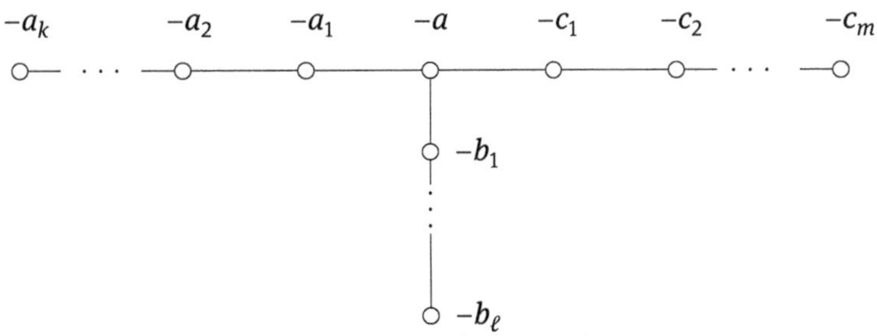

where a_i, b_j, c_p are all $\geq$ 2 and $a \leq 1$. The determinants d_i of the three linear branches satisfy the inequality

$$\frac{1}{d_1} + \frac{1}{d_2} + \frac{1}{d_3} > 1.$$

(iii) The weighted dual graph obtained by attaching a partial chain

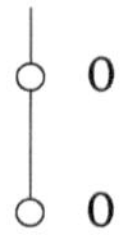

to the weighted dual graph of a minimal resolution of a quotient singular point at any of its vertices. See [159, p. 55] for all possible weighted graphs. The two curves C_1, C_2 corresponding to the two vertices of the chain to be attached are isomorphic to $\mathbb{P}^1$.

The proof of this result in [211] uses ideas from 3-dimensional topology like Dehn's lemma, Loop theorem, etc.

With the previous notation, let $X := V - E$. We then say that X is *connected at infinity* as E is connected. This situation occurs, for example, if X is affine. In [71] the following result was proved.

Theorem 1.3.21. *With the above notation assume that X is connected at infinity and the fundamental group of X at infinity is a finite group. Then we have $\overline{\kappa}(X) = -\infty$.*

Proof. We give a sketch. If the dual graph is obtained by attaching a partial chain 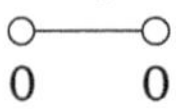, let C_1, C_2 be the curves with self intersection zero. If the partial chain is attached at C_1, then the linear pencil $|C_2|$ gives a $\mathbb{P}^1$-fibration f on V with C_1 a cross-section and all other components of E contained in a singular fiber of f. Clearly, f induces an $\mathbb{A}^1$-fibration on $X := V - E$. Hence $\overline{\kappa}(X) = -\infty$.

If E is of type (ii), we arrive at a contradiction by assuming $\overline{\kappa}(X) \neq -\infty$. Let $K_V + E \approx P + N$ be the Zariski–Fujita decomposition. We refer the readers to [159] for the relevant results.

Case 1. Every component of N is a component of a maximal twig of E. Then Lemma 6.17 in [56] implies that the dual graph of E is a rational fork (abnormal rational rod in the terminology of [56]). In this case, the intersection form on E is negative definite. (Alternatively, we can verify that $P^2 < 0$.)

Case 2. N is not $\mathrm{Bk}^*(\Gamma)$ (see [56] for the notation), where Γ is the dual graph of E. By Lemma 6.20 of [56], there exists a (-1) curve L on V such that either $L \cap E = \emptyset$, or $L \cdot E = 1$ and L meets a component of a twig of E. If $L \cap E = \emptyset$, we contract L without affecting E. Then a suitable induction completes the proof.

If L meets a twig of E then we blow down L and any (-1) curves which arise in the image of E. In finitely many steps we reach a stage when there is no (-1) curve L meeting a bark of the image of E transversally in one point. Then we are reduced to Case 1. This completes the proof of the theorem. □

1.3.13 A topological characterization of $\mathbb{A}^2/G$

As a generalization of C. P. Ramanujam's characterization of $\mathbb{A}^2$, the following result was proved in [76].

Theorem 1.3.22. *Let X be an affine normal surface. Then X is isomorphic to $\mathbb{A}^2/G$ for a finite group G of automorphisms of $\mathbb{A}^2$ if and only if X is topologically contractible and has a finite fundamental group at infinity.*

As explained before Theorem 1.2.4, the minimal resolution of singularity of $X \cong \mathbb{A}^2/G$ with the locus E of exceptional curves is embedded as an open set into a smooth projective surface V with a $\mathbb{P}^1$-fibration $p : V \to \mathbb{P}^1$ such that the $\mathbb{A}^1_*$-fibration on X is the restriction of p onto X. In fact, the $\mathbb{P}^1$-fibration p is taken in a minimal way so that there are two singular fibers (case G is cyclic) or three singular fibers (case G is not cyclic) and the singular fibers respectively contain single (-1) curves which cut out multiple fibers of $p|_X$. The exceptional locus E then contains an irreducible component E_0 which is a cross-section of p and other components of E are fiber components. Then the divisor $D = V - X$ contains an irreducible component D_0 and other components are fiber components of p. Then the dual graph and the intersection matrix of D is uniquely determined by those of E. It is then easily ascertained that the smooth locus $X - \operatorname{Sing} X$ deforms to a punctured neighborhood of the singular point, as well as to $U - D$ for a tubular neighborhood U of D. Hence the local fundamental group of the singular point, which is isomorphic to G, is isomorphic to the fundamental group at infinity $\pi_1(\partial U)$.

Proof. If $X \cong \mathbb{A}^2/G$, then contractibility of X is well-known (see [131, Theorem B]). Since the quotient morphism $q : \mathbb{A}^2 \to \mathbb{A}^2/G$ is finite, we see using Lemma 1.1.14 that X has a finite fundamental group at infinity.

Conversely, suppose that X is contractible and has a finite fundamental group at infinity. By the proof of Theorem 1.3.20 which uses the description of the divisor at infinity of X, we know that $\overline{\kappa}(X - \operatorname{Sing} X) = -\infty$.

Case 1. $X - \operatorname{Sing} X$ has an $\mathbb{A}^1$-fibration f. Then f extends to an $\mathbb{A}^1$-fibration $f : X \to B$, where $B \cong \mathbb{A}^1$. In fact, by Lemma 1.1.14, $\pi_1(B)$ is a finite group since X is contractible. Hence $B \cong \mathbb{A}^1$. Further, X has at most cyclic quotient singular points by [153]. By a Lefschetz theorem due to Nori [182, Corollary 2.3], $\pi_1^\infty(X)$ surjects onto $\pi_1(X - \operatorname{Sing} X)$. Hence $\pi_1(X - \operatorname{Sing} X)$ is finite. Now it follows that every fiber of f is irreducible and at most one fiber is $m\mathbb{A}^1$ for an integer $m \geq 1$. Every reduced fiber of f does not contain any singular point of X. It is easy to see that $X \cong \mathbb{A}^2/(\mathbb{Z}/m\mathbb{Z})$.

Case 2. $X - \operatorname{Sing} X$ does not have an $\mathbb{A}^1$-fibration. By a result of [168] (see also [159, Chapter 3, Theorem 2.5.4]), $X \cong \mathbb{A}^2/G$ for a noncyclic small finite subgroup $G \subset GL(2, \mathbb{C})$. □

1.3.14 Cofinite subalgebras of $\mathbb{C}[x, y]$

The following result was proved in [154, 76].

Theorem 1.3.23. *Let A be a $\mathbb{C}$-subalgebra of a polynomial ring $\mathbb{C}[x, y]$, where $\mathbb{C}[x, y]$ is integral over A and A is integrally closed. Then $A \cong \mathbb{C}[x, y]^G$, where G is a small finite subgroup in $\mathrm{GL}(2, \mathbb{C})$. In particular, if A is regular, A is a polynomial ring in two variables.*

Proof. We will indicate a topological proof using arguments from [67, 76]. The assumption implies that A is a normal algebra of dimension 2 which is finitely generated over $\mathbb{C}$. Let X be the affine normal surface defined by A and let $f : \mathbb{A}^2 \to X$ the induced morphism. Then f is finite. It was proved in [67] that X is contractible and has a finite fundamental group at infinity. Then by Theorem 1.3.22, $X \cong \mathbb{A}^2/G$ as desired. □

The following result [85] is a result in the same trend of Mumford's theorem in which the triviality of the local fundamental group shows the absence of a singular point. Meanwhile, here the triviality of the fundamental group of the smooth part will imply the smoothness of an algebraic surface.

Theorem 1.3.24. *Let X be a complex normal affine surface. Suppose that X is topologically contractible and the smooth part $X^\circ := X - \mathrm{Sing}\, X$ is simply connected. Then X is a smooth surface.*

1.4 Proofs of AMS and Lin–Zaidenberg theorems

Concerning the topologically contractible curves embedded in the affine plane $\mathbb{A}^2$, we present proofs to the theorems of Abhyankar–Moh–Suzuki and Lin–Zaidenberg which are based on the theory of open algebraic surfaces.

By the Lefschetz principle, we may and shall assume that all algebraic varieties considered in this section are defined over the complex number field $\mathbb{C}$. A smooth projective rational curve with self-intersection number $-n$ on a smooth algebraic surface is called a $(-n)$-curve. We assume that a fibration $f : X \to B$ from a surface X to a smooth curve B is always surjective.

1.4.1 Necessary results from theory of open algebraic surfaces

We will implicitly use the following easy results about the logarithmic Kodaira dimension:

(1) A smooth irreducible affine curve C has $\overline{\kappa}(C) = -\infty$ if and only if it is isomorphic to $\mathbb{A}^1$, and $\overline{\kappa}(C) = 0$ if and only if C is isomorphic to $\mathbb{A}^1_*$.
(2) Let $f : X \to Y$ be a dominant morphism with X and Y smooth algebraic varieties. Then $\overline{\kappa}(X) \geq \overline{\kappa}(Y)$. If further f is a proper birational morphism, then the equality holds. If X is a Zariski open set of Y, this implies $\overline{\kappa}(X) \geq \overline{\kappa}(Y)$.

Proof. (1) Let $\overline{C}$ be a smooth completion of C. Then $D := \overline{C} - C$ is a nonzero effective divisor. Then the assertion is easily verified. We leave the proof to the readers as an exercise. See also Theorem 1.2.3.

(2) The first assertion follows from Lemma 1.2.2. The second assertion follows from the proper birational invariance of logarithmic Kodaira dimension (see [95]). The last assertion follows from the addition formula in Lemma 1.2.1. □

For any topological space T, $e(T)$ denotes its topological Euler–Poincaré characteristic. In what follows, by a surface we mean an algebraic surface and by a curve we mean an algebraic curve.

Let X be a smooth quasiprojective surface and V a smooth projective compactification of X such that the divisor $D := V - X$ has simple normal crossings. We say that (V, D) is a *minimal normal compactification* of X if any (-1)-curve in D meets at least three other irreducible components of D.

The next result is a main point in the proof of the topological characterization of $\mathbb{A}^2$ by C. P. Ramanujam [192] (see Theorem 1.3.19).

Lemma 1.4.1. *Let (X, D) be a minimal normal compactification of $\mathbb{A}^2$. Then the dual graph of D is a linear chain of $\mathbb{P}^1$'s.*

Let X be a smooth surface with a morphism $f : X \to B$, where B is a smooth curve. For any scheme-theoretic fiber F of f, the greatest common divisor of the multiplicities of irreducible components of F is called the *multiplicity* of F. If the multiplicity of F is larger than one, then we call F a *multiple fiber*.

We use the notion of F-fibration for $F = \mathbb{A}^1, \mathbb{A}^1_*$ or $\mathbb{P}^1$ though the definitions in the general settings are given in the next chapter. A dominant morphism $f : X \to C$ from a smooth algebraic surface X to a smooth algebraic curve C is called an *F-fibration* if general closed fibers are isomorphic to F. A fiber $f^*(P) = (X, f) \times_C \operatorname{Spec} k(P)$ with $P \in C$ is a *singular fiber* if $f^*(P)$ is not isomorphic to F.

We will often use the following elementary result about singular fibers of a $\mathbb{P}^1$-fibration on a smooth projective surface (see [150, Chapter II, Lemma 2.3]).

Lemma 1.4.2. *Let $f : V \to C$ be a $\mathbb{P}^1$-fibration from a smooth projective surface V to a smooth complete curve C and let F be a singular fiber of f. Then the following assertions hold:*

(1) *F is a connected tree of $\mathbb{P}^1$s.*
(2) *F contains a (-1)-curve and any (-1)-curve in F intersects at most two other irreducible components of F.*
(3) *If F_1 is an irreducible component of F occurring with multiplicity 1 in F then F contains a (-1)-curve other than F_1.*

We call Lemma 1.4.2 a *Lemma of Gizatullin*. If $g : X \to B$ is an $\mathbb{A}^1$-fibration from an affine normal surface X to a smooth curve B, then we can find a $\mathbb{P}^1$-fibration $f : V \to C$ as in Lemma 1.4.2 such that X is an open set of V and g is the restriction of f onto X.

Since every fiber of a $\mathbb{P}^1$-fibration is connected and every irreducible component of the $\mathbb{A}^1$-fibration g is an affine curve, we can deduce the following result from Lemma 1.4.2 (see [159, Chapter 3, Lemma 1.4.2]).

Lemma 1.4.3. *Let $g : X \to B$ be an $\mathbb{A}^1$-fibration on an affine normal surface X. Let F be a singular fiber of g. Then F is a disjoint union of $\mathbb{A}^1$s, and singularities of X are at most cyclic quotient singularities. Further, each fiber component has at most one singular point.*

We now need some easy properties of twisted and untwisted $\mathbb{A}^1_*$-fibrations. For a detailed explanation, see Remark 2.1.13 below. Let $f : X \to B$ be an $\mathbb{A}^1_*$-fibration on a smooth quasiprojective surface X. Then there exists a projective embedding $X \subset V$ into a smooth projective surface V such that f extends to a $\mathbb{P}^1$-fibration $\overline{f} : V \to \overline{B}$. Let $D := V - X$. If D has a unique irreducible component which dominates $\overline{B}$, we say that the $\mathbb{A}^1_*$-fibration f is *twisted* and call this irreducible component the *horizontal component* of D. If D has exactly two irreducible components dominating $\overline{B}$, we say that f is *untwisted*. These components are cross-sections of the $\mathbb{P}^1$-fibration $\overline{f} : V \to \overline{B}$ and called the *horizontal components* of D.

The next result describes the nature of a singular fiber of a $\mathbb{P}^1$-fibration arising from an $\mathbb{A}^1_*$-fibration on an affine surface. For the elementary proofs, we refer the readers to a paper of Fujita [56, Lemmas 7.6 and 7.7]. It is, in fact, proved by a repeated use of Lemma 1.4.2 and the assumption that X is affine and hence D is connected.

Lemma 1.4.4. *Let $f : X \to B$ be an untwisted $\mathbb{A}^1_*$-fibration on a smooth affine surface X. Let $X \subset V$ be a smooth compactification such that $D := V - X$ is an SNC divisor and V admits a $\mathbb{P}^1$-fibration $\overline{f} : V \to \overline{B}$ extending f. Denote by D_1, D_2 the horizontal components of D. Let F be a singular fiber of $\overline{f}$. Assume that no irreducible component of D contained in F is a (-1)-curve. Then we have the following assertions:*
(1) *Suppose that $F \cap X$ is irreducible and occurs with multiplicity larger than one in F. If further $D \cap F$ is disconnected, then the dual graph of F is linear and the closure of $F \cap X$ is the unique (-1)-curve in F.*
(2) *Suppose that F is irreducible and $F \cap X \neq \emptyset$. If $D \cap F$ is connected, then F is the closure of $F \cap X$ and $D_1 \cap D_2 \cap F \neq \emptyset$ and $F \cap X \cong \mathbb{A}^1$.*

The next result follows easily from the observation that the irreducible components at infinity in a smooth compactification V for $\mathbb{A}^2$ generate $\operatorname{Pic}(V)$ freely. See the answer of Problem 7 in Section 1.5, where it is shown that $\operatorname{Pic}(V) \cong H^2(V;\mathbb{Z}) \cong H^2(D;\mathbb{Z}) \cong \oplus_{i=1}^r \mathbb{Z}[D_i]$ for the irreducible decomposition $D = \sum_{i=1}^r D_i$.

Lemma 1.4.5. *Let $f : \mathbb{A}^2 \to B$ be an $\mathbb{A}^1_*$-fibration. Then the following assertions hold:*
(1) *The fibration is untwisted, and B is isomorphic to $\mathbb{P}^1$ or $\mathbb{A}^1$.*
(2) *If $B \cong \mathbb{P}^1$, then every fiber of f is irreducible, and there is exactly one fiber isomorphic to $\mathbb{A}^1$ if taken with reduced structure.*
(3) *If $B \cong \mathbb{A}^1$ then exactly one fiber is reducible and it contains two irreducible components, say C_1, C_2. Further, either $C_1 \cong C_2 \cong \mathbb{A}^1$ and C_1, C_2 intersect each other*

transversally in a single point, or $C_1 \cong \mathbb{A}^1, C_2 \cong \mathbb{A}^1_*$ and they are disjoint. All other fibers of f are isomorphic to $\mathbb{A}^1_*$ if taken with reduced structure.

Proof. (1) If f is twisted, there is a 2-torsion element in $\mathrm{Pic}(\mathbb{A}^2)$, which is a contradiction. See Remark 2.1.13. Since f is a dominant morphism, $\Gamma(B, \mathcal{O}_B)$ is a k-subalgebra of $\Gamma(\mathbb{A}^2, \mathcal{O}_{\mathbb{A}^2})$. Since an invertible regular function is a constant on $\mathbb{A}^2$, this implies that there are no non-constant invertible regular functions on B. Further, B is a rational curve. Hence either $B \cong \mathbb{P}^1$ or $B \cong \mathbb{A}^1$. Assertions (2) and (3) are left to the readers as Problem 11 of Section 1.5. □

1.4.2 Proof of AMS theorem

In this subsection we prove Theorem 1.3.14, which is the AMS theorem. Let $C \subset \mathbb{A}^2$ be a closed embedding of the curve C isomorphic to $\mathbb{A}^1$. Denote the complement $\mathbb{A}^2 - C$ by X. Then $e(X) = 0$. By Theorem 1.2.6, $\overline{\kappa}(X) \leq 1$. The next result is the main step in the proof of Theorem 1.3.14.

Lemma 1.4.6. *With the above notations, we have $\overline{\kappa}(X) = -\infty$.*

Proof. Suppose that $\overline{\kappa}(X) = 0$ or 1. We will obtain a contradiction. First consider the case $\overline{\kappa}(X) = 0$. We choose a regular function f on $\mathbb{A}^2$ such that the divisor of f is C, i. e., $C = V(f)$. Consider the morphism $f : \mathbb{A}^2 \to \mathbb{A}^1$ given by $P \mapsto f(P)$. Since f is a prime element in the coordinate ring $k[x, y]$ of $\mathbb{A}^2$, a general fiber of f is irreducible. By the addition formula in Lemma 1.2.1 applied to the morphism $f|_X : X \to \mathbb{A}^1_*$, we get

$$0 = \overline{\kappa}(X) \geq \overline{\kappa}(F) + \overline{\kappa}(\mathbb{A}^1_*) = \overline{\kappa}(F),$$

where F is a general fiber of f. Suppose $\overline{\kappa}(F) = -\infty$. Then $F \cong \mathbb{A}^1$ and f is an $\mathbb{A}^1$-fibration. Any $\mathbb{A}^1$-fibration on a smooth quasiprojective surface is trivial on a Zariski open subset of the base curve (see Lemma 2.1.3 below). Hence X contains a Zariski open subset which is isomorphic to a cylinder-like open set $B \times \mathbb{A}^1$. It follows that $\overline{\kappa}(X) = -\infty$ (see Theorem 1.2.3).

Hence assume that $\overline{\kappa}(F) = 0$ and F is isomorphic to $\mathbb{A}^1_*$. Namely, f is an $\mathbb{A}^1_*$-fibration and C is a full fiber of f. But this contradicts Lemma 1.4.5 because the fiber containing C must be reducible as $C \cong \mathbb{A}^1$.

Assume now that $\overline{\kappa}(X) = 1$. By Lemma 1.2.5, X contains a Zariski open subset U with an $\mathbb{A}^1_*$-fibration g. We need to consider three cases separately:

CASE 1. *g does not extend to a morphism on $\mathbb{A}^2$.*

Then the closures of the fibers of g have a common point, say P, in $\mathbb{A}^2$. This point cannot lie in X. For, otherwise X will contain a family of affine rational curves with one-place at infinity and passing through P. By blowing up successively at P and its infinitely near points, we resolve the base locus and get a smooth surface $\widetilde{X}$ which admits a morphism to a curve with general fiber isomorphic to $\mathbb{A}^1$. But then $\overline{\kappa}(\widetilde{X}) = -\infty$ as remarked above. Since $\widetilde{X}$ is obtained from X by a sequence of blowing-ups, $\overline{\kappa}(\widetilde{X}) = \overline{\kappa}(X)$, which contradicts the assumption that $\overline{\kappa}(X) = 1$. Hence $P \in C$. This gives an $\mathbb{A}^1_*$-fibration $h : \mathbb{A}^2 \setminus \{P\} \to B$ such that $g = h|_X$.

We claim that $B \cong \mathbb{P}^1$. To see this we observe that since the closure of every fiber of h in $\mathbb{A}^2$ passes through P, every fiber of h intersects the boundary of a small neighborhood of P. Hence this boundary, which is a compact set, maps onto B and hence B is compact.

Clearly, the Picard group of $\mathbb{A}^2 \setminus \{P\}$ is trivial and any nonconstant regular function on $\mathbb{A}^2 \setminus \{P\}$ vanishes at some point. Hence the proof of Lemma 1.4.5 applies and shows that all the fibers of h are irreducible. This also implies that g is an untwisted $\mathbb{A}^1_*$-fibration on X. Note that clearly $\mathbb{A}^2 \setminus \{P\}$ is simply-connected. Suppose h has three or more multiple fibers $m_1F_1, \ldots, m_sF_s$ lying over the points $P_1, \ldots P_s$ of $\mathbb{P}^1$. Using Lemma 1.1.9, we can construct a finite Galois covering $\tau : \widetilde{B} \to \mathbb{P}^1$ such that the ramification index at a point over P_i is m_i. Then the normalization of the fiber product $(\mathbb{A}^2 \setminus \{P\}) \times_{\mathbb{P}^1} \widetilde{B}$, which we denote by $\overline{(\mathbb{A}^2 \setminus \{P\}) \times_{\mathbb{P}^1} \widetilde{B}}$, is a finite unramified covering of $\mathbb{A}^2 \setminus \{P\}$, which is impossible. Hence there are at most two multiple fibers for h. If g has only one multiple fiber, then X contains a Zariski open subset U isomorphic to $\mathbb{A}^1_* \times \mathbb{A}^1_*$ and hence $\overline{\kappa}(X) \le \overline{\kappa}(U) = 0$. This contradicts the assumption that $\overline{\kappa}(X) = 1$. The same argument shows that $C \setminus \{P\}$ is a reduced fiber of h, for otherwise g has only one multiple fiber. Hence g has exactly two multiple fibers. Let m_1F_1, m_2F_2 be the two multiple fibers. Then again by Lemma 1.1.9, $\gcd(m_1, m_2) = 1$.

Let V be a smooth projective compactification of $\mathbb{A}^2 \setminus \{P\}$ such that the divisor at infinity D has simple normal crossings and there is a $\mathbb{P}^1$-fibration $\Phi : V \to \mathbb{P}^1$ extending h. The fibration g is an untwisted fibration with one horizontal irreducible component of D lying over P and the other coming from the compactification divisor for $\mathbb{A}^2$. We now apply Lemma 1.4.2. Let $\overline{C}$ be the closure of C in V. Since $\overline{C}$ occurs with multiplicity 1 in the fiber of Φ which contains it, using Lemma 1.4.2 repeatedly, we can assume that V has the following properties:

(1) D has two horizontal components D_1, D_2 such that D_1 lies over P and D_2 is an irreducible component at infinity for $\mathbb{A}^2$.
(2) The closure $\overline{C}$ of C in V is a full fiber of Φ.
(3) $D_1 \cap D_2 \cap \overline{C} = \emptyset$.
(4) The fiber $\widetilde{F}_i$ of Φ containing F_i is a linear chain of smooth rational curves, the closure $\overline{F}_i$ of F_i is the unique (-1)-curve in $\widetilde{F}_i$ and D_1, D_2 intersect the two end irreducible components of $\widetilde{F}_i$ for $i = 1, 2$ by Problem 11 in Section 1.5 (see also [56, Lemmas 7.6, 7.7]).
(5) $\widetilde{F}_1, \widetilde{F}_2$ are the only singular fibers of Φ.

Indeed, as $\overline{C}$ has multiplicity 1 in the fiber, by applying Lemma 1.4.2 repeatedly, we can make $\overline{C}$ a full fiber of Φ by contracting (-1)-curves other than $\overline{C}$ in the fiber, which also lie outside $\mathbb{A}^2 \setminus \{P\}$. In this process, the components D_1, D_2 do not meet on $\overline{C}$. Hence the properties (1), (2), and (3) hold. By a similar argument, (5) holds. For the proof of (4), we use Lemma 1.4.4. For this, let E be any (-1)-curve contained in $D \cap \widetilde{F}_i$. By the remark before Lemma 1.4.4, E meets at most two other irreducible components of $\widetilde{F}_i$, and if it meets two such irreducible components then it occurs with multiplicity ≥ 2 in $\widetilde{F}_i$. In the latter case E cannot meet either of D_1, D_2 as these are cross-sections. Hence we can contract E and get a smaller compactification of $\mathbb{A}^2 \setminus \{P\}$ which satisfies all the properties of X. By this process we reach a situation where $D \cap \widetilde{F}_i$ does not contain any (-1)-curve. The hypothesis of Lemma 1.4.4 is therefore satisfied, and we may apply it. See also Problem 11 in Section 1.5 for the case $F_i \cong \mathbb{A}^1_*$.

Let Δ_1, Δ_2 be the connected components of D containing D_1, D_2, respectively, where Δ_1 contracts to the point P smoothly. Since no irreducible component of Δ_1 other than D_1 is a (-1)-curve, we see easily that D_1 is a (-1)-curve and Δ_1 is a linear chain. As D_1 is a cross-section and $m_i > 1$, $\widetilde{F}_i$ contains at least one irreducible component of Δ_1 for $i = 1, 2$ which intersects D_1. Clearly, V contains a Zariski open subset, say W, which is obtained from $\mathbb{A}^2$ by a finite succession of blowing-ups at the point P such that Δ_1 is the complete exceptional divisor. By the above observation, Δ_1 contains at least three irreducible components and it is easy to see that after successive contractions of (-1)-curves from Δ_1 and its images, the image of C' becomes singular at the point P, where C' is the proper transform of C in W. This contradicts the assumption that C is a smooth curve. Hence Case 1 cannot occur.

Next we consider the case where g extends to a morphism $h : \mathbb{A}^2 \to B$. Then we have two separate cases depending on whether or not C is contained in a fiber of h. If C is contained in a fiber, h is an $\mathbb{A}^1_*$-fibration, and if C is not contained in a fiber, h is an $\mathbb{A}^1$-fibration and C is a cross-section.

Case 2. *g extends to an $\mathbb{A}^1_*$-fibration $h : \mathbb{A}^2 \to B$.*

Then h is untwisted and B is necessarily isomorphic to $\mathbb{A}^1$ or $\mathbb{P}^1$ by Lemma 1.4.5. Clearly, $h(C)$ is a point, say P_0. The proof in this case is somewhat similar to Case 1 above. If $B \cong \mathbb{P}^1$, let $\Phi : V \to B$ be a $\mathbb{P}^1$-fibration on a suitable compactification of $\mathbb{A}^2$. Then as in the proof of Case 1, the closure $\overline{C}$ of C is a reduced fiber of Φ, and h has exactly two multiple fibers m_1F_1, m_2F_2 of relatively prime multiplicities. Minimizing the compactification V of $\mathbb{A}^2$ by contracting (-1) curves contained in D and the fibers of Φ, we may assume that the corresponding fibers of Φ are linear chains of nonsingular rational curves, where $D := V - \mathbb{A}^2$. The horizontal components D_1, D_2 of D meet each other at a point in $\overline{C}$ because $C \cong \mathbb{A}^1$ (see Problem 11 in Section 1.5). Let $\widetilde{F}_i$ be the fiber of Φ containing F_i for $i = 1, 2$. Since $m_i > 1$, $\widetilde{F}_i$ contains at least two extra irreducible components which meet the cross-sections D_1, D_2 and are contained in D. Then D_1 and D_2 are branch points for the dual graph of D, where a branch point of the dual graph signifies a vertex from which sprout three or more edges. In this case (V, D) is a minimal normal compactification of $\mathbb{A}^2$, contradicting Lemma 1.4.1.

Next consider the case where $B \cong \mathbb{A}^1$. By Lemma 1.4.5, the fiber of h containing C is of the form $C \cup C'$, where $C' \cong \mathbb{A}^1$ or $\mathbb{A}^*$. Consider also a suitable compactification

V of $\mathbb{A}^2$ with a $\mathbb{P}^1$-fibration $\Phi : V \to \mathbb{P}^1$. The fiber over the point $P_\infty := \mathbb{P}^1 - B$ is contained in $D := V - \mathbb{A}^2$. As in Case 1, there are exactly two multiple fibers for the morphism $g : X \to \mathbb{A}^1$. Hence the morphism $h|_{X-C'} : X - C' = \mathbb{A}^2 - (C \cup C') \to \mathbb{A}^1_*$ has a multiple fiber, say mF_1 with $m \geq 2$. Let $\rho : \Delta \to \mathbb{P}^1$ be a cyclic covering of degree m totally ramifying over $h(F_1)$ and P_∞. Then the normalization Y of the fiber product $(\Delta - \rho^{-1}(P_\infty)) \times_{\mathbb{A}^1} \mathbb{A}^2$ is an unramified covering of degree m of $\mathbb{A}^2$, a contradiction.

CASE 3. *g extends to an $\mathbb{A}^1$-fibration $h : \mathbb{A}^2 \to B$.*

By Lemma 1.4.5, $B \cong \mathbb{P}^1$ or $\mathbb{A}^1$. As $e(\mathbb{A}^2) = 1$, using Theorem 1.1.17 it is easy to see that $B \cong \mathbb{A}^1$. Also, all the fibers of h are irreducible by the count of the Picard number. They are all reduced, for otherwise we get a finite unramified covering of $\mathbb{A}^2$ by the same argument as in the latter case of Case 2 above. But then every fiber of h is isomorphic to $\mathbb{A}^1$ and h is a trivial $\mathbb{A}^1$-bundle by the results in Section 2.1, especially by Lemmas 2.1.6 and 2.1.7. Hence X is isomorphic to $\mathbb{A}^1 \times \mathbb{A}^1_*$, which has $\bar{\kappa} = -\infty$, contradicting the assumption that $\bar{\kappa}(X) = 1$.

This completes the proof of Lemma 1.4.6.

The rest of the proof of Theorem 1.3.14 is quite well-known. We briefly sketch it. By Theorem 1.2.3, there is an $\mathbb{A}^1$-fibration g on X. The fibers of this morphism are closed in $\mathbb{A}^2$, for otherwise $\mathbb{A}^2$ contains complete curves. Hence this morphism extends to an $\mathbb{A}^1$-fibration $h : \mathbb{A}^2 \to B$ on $\mathbb{A}^2$ with the base curve $B \cong \mathbb{A}^1$. All the fibers of this extended morphism h are reduced and irreducible. As in Case 3 of the proof of Lemma 1.4.6, h is a trivial $\mathbb{A}^1$-bundle and C is a fiber of h. So there exist coordinates x, y on $\mathbb{A}^2$ such that C is defined by $x = 0$. This completes the proof of Theorem 1.3.14. □

1.4.3 Proof of Lin–Zaidenberg theorem

Here we prove Theorem 1.3.15. Let C be a contractible irreducible curve in $\mathbb{A}^2$ and denote by X the complement $\mathbb{A}^2 \setminus C$. Then $e(X) = 0$. By Theorem 1.2.6, $\bar{\kappa}(X) \leq 1$. We consider three cases as in the proof of Theorem 1.3.14.

CASE 1. $\bar{\kappa}(X) = -\infty$.

Then by Theorem 1.2.3, there is an $\mathbb{A}^1$-fibration $g : X \to B$. Since $\mathbb{A}^2$ is affine, it does not contain any complete curves. Hence g extends to an $\mathbb{A}^1$-fibration $h : \mathbb{A}^2 \to B'$, where B' contains B as a Zariski open subset. Next, h extends to a $\mathbb{P}^1$-fibration on a smooth compactification of $\mathbb{A}^2$. Clearly, C is contained in a fiber of h. By Lemma 1.4.2, any fiber of a $\mathbb{P}^1$-fibration on a smooth projective surface is a union of $\mathbb{P}^1$s, hence C is smooth. Therefore $C \cong \mathbb{A}^1$ and the result is already proved in Theorem 1.3.14.

CASE 2. $\bar{\kappa}(X) = 0$.

Let f be a prime element in the coordinate ring of $\mathbb{A}^2$ such that C is defined by $f = 0$. By Lemma 1.2.1, we have

$$\bar{\kappa}(X) \geq \bar{\kappa}(F) + \bar{\kappa}(\mathbb{A}^1_*),$$

where F is a general fiber of the morphism given by f. Hence $\overline{\kappa}(F) \le 0$. But $\overline{\kappa}(F) \ne -\infty$ as in the beginning of the proof of Lemma 1.4.6, for otherwise $\overline{\kappa}(X) = -\infty$. Hence $\overline{\kappa}(F) = 0$, so that $F \cong \mathbb{A}^1_*$. Hence C is a full fiber of an $\mathbb{A}^1_*$-fibration $f : \mathbb{A}^2 \to \mathbb{A}^1$. By Lemma 1.4.5 as before, any irreducible component of an untwisted $\mathbb{A}^1_*$-fibration on a smooth affine surface is smooth. Therefore C is smooth, and we are again done.

Case 3. $\overline{\kappa}(X) = 1$.

Since X is affine, by Theorem 1.2.5 and by the same argument as in the beginning of the proof of Case 1 of Lemma 1.4.6, there is an $\mathbb{A}^1_*$-fibration $g : X \to B$. If this morphism extends to an $\mathbb{A}^1_*$-fibration on $\mathbb{A}^2$ then C is mapped to a point and C is smooth as above. If g extends to an $\mathbb{A}^1$-fibration on $\mathbb{A}^2$, then C is a cross-section and hence smooth and the result is proved by the AMS theorem.

Now assume that C is not smooth and that the rational mapping $g : \mathbb{A}^2 \to \mathbb{P}^1$ given by g on X is not defined at a point P on C.

Claim 1. *The point P is the only singular point of C, and C is analytically irreducible at P.*

Proof. Resolve the indeterminacies of the rational mapping $g : \mathbb{A}^2 \to \mathbb{P}^1$ by a sequence of blowing-ups $\sigma : W \to \mathbb{A}^2$ at the point P and its infinitely near points, where W is a smooth surface. Then the rational mapping g gives rise to an $\mathbb{A}^1$-fibration $p : W \to \mathbb{P}^1$ such that the last exceptional curve of σ is a cross-section of p and the proper transform C' of C is contained in a fiber of p. The $\mathbb{A}^1$-fibration p is extended to a $\mathbb{P}^1$-fibration $\tilde{p} : \widetilde{W} \to \mathbb{P}^1$ on a smooth compactification $\widetilde{W}$ of W. By Lemma 1.4.2, every singular fiber of $\tilde{p}$ is a tree of nonsingular rational curves. Hence C' is smooth. This implies that the point P is the unique singular point of C. The contractibility of C implies that C is homeomorphic to $\mathbb{A}^1$. Hence C is analytically irreducible at P. □

The closure in $\mathbb{A}^2$ of any fiber of $g : X \to B$ passes through the point P. As in the proof of Case 1 of Lemma 1.4.6, the restriction of p gives an untwisted $\mathbb{A}^1_*$-fibration $h : \mathbb{A}^2 \setminus \{P\} \to \mathbb{P}^1$.

Claim 2. *The following assertions hold:*

(1) *Every fiber of h is isomorphic to $\mathbb{A}^1_*$ if taken with reduced structure.*
(2) *There are exactly two multiple fibers m_1F_1, m_2F_2 and h is a trivial $\mathbb{A}^1_*$-fibration outside the union of F_1 and F_2.*
(3) *Let $\overline{F}_i$ be the closure of F_i in $\mathbb{A}^2$ for $i = 1, 2$. Then $\overline{F}_1$ and $\overline{F}_2$ are isomorphic to $\mathbb{A}^1$. Hence $C \setminus \{P\}$ is a reduced fiber of h.*
(4) *The curves $\overline{F}_1$ and $\overline{F}_2$ meet each other transversally at the point P.*

Proof. (1) Let F be a general fiber of h. Then $F \cong \mathbb{A}^1_*$ and $\mathbb{A}^2 \setminus \{P\} - F$ is an affine surface with $e(\mathbb{A}^2 \setminus \{P\} - F) = 0$. Applying Theorem 1.1.17 to the $\mathbb{A}^1_*$-fibration $h|_{\mathbb{A}^2 \setminus \{P\} - F} : \mathbb{A}^2 \setminus \{P\} - F \to \mathbb{A}^1$, we see that every fiber of h other than F is isomorphic to $\mathbb{A}^1_*$ if taken with reduced structure.

(2) If h has three or more multiple fibers, then by Lemma 1.1.9 we can construct a suitable ramified covering $\Delta \to \mathbb{P}^1$ such that the normalized fiber product $\overline{(\mathbb{A}^2 \setminus \{P\}) \times_{\mathbb{P}^1} \Delta}$ is an unramified covering of $\mathbb{A}^2 \setminus \{P\}$. But this is a contradiction because $\mathbb{A}^2 \setminus \{P\}$ is simply-connected. Hence h has at most two multiple fibers and if there are two multiple fibers then their multiplicities are relatively prime. Suppose h has at most one multiple fiber. Then $\overline{\kappa}(\mathbb{A}^2 \setminus C) \le \overline{\kappa}(\mathbb{A}^1_* \times \mathbb{A}^1_*) = 0$, which is a contradiction. Hence h has exactly two multiple fibers m_1F_1, m_2F_2 and h is a trivial $\mathbb{A}^1_*$-fibration outside the union of these two multiple fibers.

(3) Since h is untwisted, $\mathbb{A}^2 - \overline{F}_i$ contains $\mathbb{A}^1_* \times \mathbb{A}^1_*$ as a Zariski open subset for $i = 1, 2$. Hence $\overline{\kappa}(\mathbb{A}^2 - \overline{F}_i) \le 0$. Note that $\overline{F}_i$ is a contractible curve because F_i is isomorphic to $\mathbb{A}^1_*$ and $\overline{F}_i$ is the union of F_i and the point P. Then, by the proofs of Cases 1 and 2 above with C replaced by $\overline{F}_i$, we see that $\overline{F}_i$ is isomorphic to $\mathbb{A}^1$, hence $\overline{F}_i$ is smooth. Since C is a singular curve, C is distinct from $\overline{F}_1$ and $\overline{F}_2$. So, $C \setminus \{P\}$ is a reduced fiber of h.

(4) Note that $\mathbb{A}^2 - (\overline{F}_1 \cup \overline{F}_2)$ is a trivial $\mathbb{A}^1_*$-bundle over $\mathbb{A}^1_*$, hence isomorphic to $\mathbb{A}^1_* \times \mathbb{A}^1_*$. Hence $\overline{\kappa}(\mathbb{A}^2 - \overline{F}_1 \cup \overline{F}_2) = 0$. Then the assertion follows from Lemma 1.4.7 below. □

Lemma 1.4.7. *Let C_1 and C_2 be irreducible curves on $\mathbb{A}^2$ such that C_1 and C_2 are isomorphic to $\mathbb{A}^1$ and that C_1 and C_2 meet in one point P with local intersection multiplicity n. Let $Y = \mathbb{A}^2 - (C_1 \cup C_2)$. Then $\overline{\kappa}(Y) = 0$ or 1, and $\overline{\kappa}(Y) = 0$ if and only if $n = 1$.*

Proof. Since $e(Y) = 0$, $\overline{\kappa}(Y) \le 1$ by Theorem 1.2.6. If $\overline{\kappa}(Y) = -\infty$, then there is an $\mathbb{A}^1$-fibration $g : Y \to B$, which extends to an $\mathbb{A}^1$-fibration $\tilde{g}$ on $\mathbb{A}^2$ so that $C_1 \cup C_2$ is contained in a fiber. This is impossible because any fiber of $\tilde{g}$ is isomorphic to $\mathbb{A}^1$ by the AMS theorem. Hence $\overline{\kappa}(Y) = 0$ or 1.

Suppose $\overline{\kappa}(Y) = 0$. Let $f = 0$ be a defining equation of $C_1 + C_2$ in $\mathbb{A}^2$ and consider the morphism $f : \mathbb{A}^2 \to \mathbb{A}^1$ defined by f. Let F be a general fiber of $f|_Y : Y \to \mathbb{A}^1_*$. Since $\overline{\kappa}(Y) \neq -\infty$, we must have $\overline{\kappa}(F) \neq -\infty$. By Lemma 1.2.1 applied to $f|_Y$, we know that $f|_Y$ is an $\mathbb{A}^1_*$-fibration. Then Lemma 1.4.5 implies that $n = 1$.

Conversely, suppose $n = 1$. Then we may choose affine coordinates x_1, x_2 on $\mathbb{A}^2$ so that C_i is defined by $x_i = 0$. See Problem 12 in Section 1.5 (cf. [150, Theorem 3.2, p. 40]). Hence $Y \cong \mathbb{A}^1_* \times \mathbb{A}^1_*$ and $\overline{\kappa}(Y) = 0$. □

Claim 3. *There exist affine coordinates x_1, x_2 of $\mathbb{A}^2$ such that C is defined by $x_1^{m_1} - x_2^{m_2} = 0$.*

Proof. As in the proof of Lemma 1.4.7 above, we choose affine coordinates x_1, x_2 on $\mathbb{A}^2$ so that $\overline{F}_i$ is defined by $x_i = 0$ for $i = 1, 2$. The fibration $h : \mathbb{A}^2 \setminus \{P\} \to \mathbb{P}^1$ defines a pencil, say Λ, on $\mathbb{A}^2$ of which $m_1\overline{F}_1$ and $m_2\overline{F}_2$ are members. Then we can choose an inhomogeneous coordinate t on $\mathbb{P}^1$ such that $t = x_1^{m_1}/x_2^{m_2}$. Since the given curve C is a member of the pencil Λ, we may assume that C is defined by $t = 1$. So, C is defined by $x_1^{m_1} - x_2^{m_2} = 0$. □

This completes the proof of Lin–Zaidenberg theorem.

1.5 Problems for Chapter 1

1. Let $Y = \mathbb{P}^1 \times \mathbb{P}^1 - \Delta$, where Δ is the diagonal $\{(P,P) \mid P \in \mathbb{P}^1\}$. There is an involution ι on $\mathbb{P}^1 \times \mathbb{P}^1$ defined by $\iota(P,Q) = (Q,P)$ for which Δ is the fixed point locus. Hence the finite group $\mathbb{Z}/2\mathbb{Z}$ acts on Y freely. Let X be the quotient surface. Prove the following assertions (see [76, Example 2] and [157]):
   (1) The cohomology ring $H^*(\mathbb{P}^n;\mathbb{Z}) := \bigoplus_{i=0}^{2n} H^i(\mathbb{P}^n;\mathbb{Z})$ equipped with a ring structure whose multiplication is given by the cup product is isomorphic to $\mathbb{Z}[T]/(T^{n+1})$, where T is a hyperplane generator in $H^2(\mathbb{P}^n;\mathbb{Z})$. In particular, if $n = 2$, $H^i(\mathbb{P}^2;\mathbb{Z}) \cong \mathbb{Z}$ for $i = 0,2,4$ and $H^i(\mathbb{P}^2;\mathbb{Z}) \cong 0$ otherwise.
   (2) Y is isomorphic to a Daniełewski surface $xy = z^2 - 1$ and X is isomorphic to $\mathbb{P}^2 - C$, where C is a smooth conic.
   (3) $e(Y) = 2e(X) = e(\mathbb{P}^1 \times \mathbb{P}^1) - e(\Delta) = 4 - 2 = 2$, whence $e(X) = 1$. Also $\pi_1(X) \cong \mathbb{Z}/2\mathbb{Z}$ since Y is simply-connected. Hence the Betti number $b_1(X) = 0$. Since $H_2(X;\mathbb{Z})$ is torsion-free by Kaup–Narasimhan–Hamm theorem and $b_2(X) = 0$ as $e(X) = 1$, $b_0(X) = 1$ and $b_1(X) = 0$, it follows that $H_2(X;\mathbb{Z}) = 0$.
   (4) The fundamental group at infinity $\pi_1^\infty(X) \cong \mathbb{Z}/4\mathbb{Z}$ because $\pi^{-1}(U_n) \to U_n$ (see Section 1.3 for the notation) is a cyclic étale covering of order 2 and $\pi_1^\infty(Y) \cong \mathbb{Z}/2\mathbb{Z}$ by Mumford's presentation.

   Answer. (1) Standard result. (2) See Section 2.10, Problems 6 and 7. (3) Straightforward. (4) The divisor Δ is the boundary divisor of Y in $\mathbb{P}^1 \times \mathbb{P}^1$. Since $\pi_1^\infty(Y)$ is generated by the class of Δ and $(\Delta^2) = 2$, we have $\pi_1^\infty(Y) \cong \mathbb{Z}/2\mathbb{Z}$ by the Mumford presentation. □

2. Let $\mathbb{F}_0 = \mathbb{P}^1 \times \mathbb{P}^1$. Let $p_1 : \mathbb{F}_0 \to \mathbb{P}_1$ be the projection onto the first factor, ℓ a fiber of p_1 and M a cross-section. Let C be an irreducible curve such that $C \sim 2M + \ell$. Then $p_1|_C : C \to \mathbb{P}^1$ ramifies over two points $P_i \in \mathbb{P}^1$ for $i = 1,2$. Let $F_i = p_1^{-1}(P_i)$. Then $F_i \cap C = \{Q_i\}$. Blow up the point Q_2 and its infinitely near point so that the proper transforms F_2', C' of F_2, C respectively have the following dual graph together with two exceptional curves E_1, E_2:

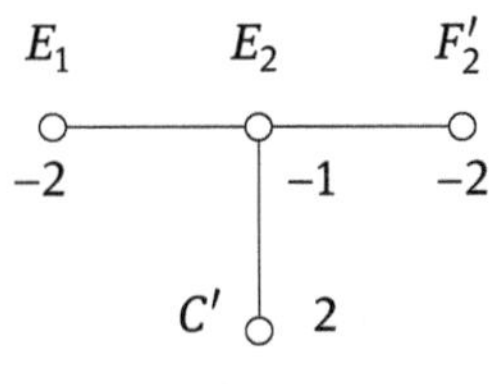

   Contract the curves E_1 and F_2' to cyclic quotient singular points Q_1', Q_2' on a projective normal surface V'. Let $X := V' - (C' + F_1)$, where we confuse the curves with their images on V'. Prove the following assertions (see [76, Example 1]):
   (1) The divisor $C' + F_1$ supports an ample divisor on V'. Hence X is an affine normal surface with two singular points Q_1', Q_2'.

(2) The fundamental group at infinity $\pi_1^\infty(X) \cong (\mathbb{Z}/2\mathbb{Z}) * (\mathbb{Z}/2\mathbb{Z})$ is the free product of two copies of $\mathbb{Z}/2\mathbb{Z}$ which is an infinite group.
(3) X is not topologically contractible.

Answer. (1) Use Nakai–Moishezon criterion (see [89]).
(2) The open set $V' - (C' + F_1)$ deforms to $E_2 - (C' \cap E_2)$, which is topologically a 2-disc, hence contractible to a point. In order to make the divisor $C' + F_1$ an SNC divisor, blow up the point Q_1 and its infinitely near point. We obtain the dual graph similar to the above. So, we use the same notations as above with F_2' replaced by F_1'. Let e_1, e_2, f_1', c' respectively be the classes corresponding to E_1, E_2, F_1', C' in $\pi_1^\infty(V)$ represented *à la* Mumford. The relations are given by

$$e_2 = (f_1')^2, \quad e_2 = e_1^2, \quad e_2 = (c')^0 = 1, \quad e_2 = (f_1')c'e_1,$$

whence we obtain

$$(f_1')^2 = e_1^2 = e_2 = 1, \quad c = (f_1')e_1.$$

This implies the assertion.
(3) Let $\widetilde{X} = \overline{X \times_{\mathbb{P}^1} C}$ be the normalized fiber product. Then the projection $p : \widetilde{X} \to X$ is a finite étale covering of degree 2. □

3. Let C be a connected algebraic curve on $\mathbb{P}^2$ and let $X = \mathbb{P}^2 - C$. Write $C = C_1 + \cdots + C_r$, where C_i is an irreducible component. Prove the following assertions:
(1) $H_1(X;\mathbb{Z}) \cong \mathbb{Z}^{\oplus r}/(d_1, \ldots, d_r) \cong \mathbb{Z}^{\oplus(r-1)} \oplus (\mathbb{Z}/d\mathbb{Z})$, where $d_i = \deg C_i$ and $d = \gcd(d_1, \ldots, d_r)$. Further, $H_2(X;\mathbb{Z}) \cong H^1(C;\mathbb{Z})$.
(2) We say that a complex smooth affine surface X is a *homology plane* or a $\mathbb{Q}$-*homology plane* if $H_i(X;\mathbb{Z}) = 0$ or $H_i(X;\mathbb{Q}) = 0$, respectively, for $i = 1, 2$. Then X as constructed above is a homology plane if and only if $X \cong \mathbb{A}^2$, and X is a $\mathbb{Q}$-homology plane if and only if C is an irreducible rational curve with only one-place singular points, which are called also *cuspidal singular points*.

Answer. (1) We consider the following exact sequence of cohomology groups with coefficients in $\mathbb{Z}$ or $\mathbb{Q}$:

$$\begin{aligned} 0 &\longrightarrow H^0(\mathbb{P}^2, C) \longrightarrow H^0(\mathbb{P}^2) \longrightarrow H^0(C) \\ &\longrightarrow H^1(\mathbb{P}^2, C) \longrightarrow H^1(\mathbb{P}^2) \longrightarrow H^1(C) \\ &\longrightarrow H^2(\mathbb{P}^2, C) \longrightarrow H^2(\mathbb{P}^2) \xrightarrow{h} H^2(C) \\ &\longrightarrow H^3(\mathbb{P}^2, C) \longrightarrow H^3(\mathbb{P}^2) \longrightarrow H^3(C) \\ &\longrightarrow H^4(\mathbb{P}^2, C) \longrightarrow H^4(\mathbb{P}^2) \longrightarrow 0, \end{aligned}$$

where we have:
(i) $H^i(\mathbb{P}^2, C) \cong H_{4-i}(X)$ for $0 \le i \le 4$ by Lefschetz duality and $H^i(\mathbb{P}^2, C) = 0$ for $i = 0, 1$ because X is affine (Kaup–Narasimhan–Hamm theorem).

(ii) $H^2(C) \cong \bigoplus_{i=1}^r H^2(C_i)$ by Mayer–Vietoris exact sequence, and the homomorphism $h : H^2(\mathbb{P}^2) \to H^2(C)$ maps a hyperplane generator $[H]$ of $H^2(\mathbb{P}^2)$ to $\bigoplus_{i=1}^r [C_i \cdot H]$, where $C_i \cdot H$ is the intersection of C_i and H with multiplicity. Hence h is an injection.

(iii) $H^1(\mathbb{P}^2) = H^3(\mathbb{P}^2) = 0$.

These observations show that there are an exact sequence

$$0 \longrightarrow H^2(\mathbb{P}^2) \xrightarrow{h} H^2(C) \longrightarrow H_1(X) \longrightarrow 0$$

and an isomorphism $H^1(C) \cong H_2(X)$. Further we write $d_i = da_i$ for $1 \le i \le r$. Then there exist integers $b_1, \dots, b_r$ such that $a_1b_1 + \cdots + a_rb_r = 1$. By a well-known theorem on unimodular matrices, we can take $(a_1, \dots, a_r)$ as the first row of unimodular matrix A. Let $\{e_1, \dots, e_r\}$ be the standard basis of the free $\mathbb{Z}$-module $\bigoplus_{i=1}^r \mathbb{Z}$ and define a new basis $\{e_1', \dots, e_r'\}$ by

$$(e_1', \dots, e_r') = (e_1, \dots, e_r)^t A.$$

Then e_1' and $e_2', \dots, e_r'$ respectively give the torsion and free parts of $H_1(X; \mathbb{Z})$.

(2) Suppose that X is a homology plane. Then $H_1(X; \mathbb{Z}) = 0$. By (1), it follows that $r = 1$, i. e., C is irreducible, and $d = \deg C = 1$. Hence $X = \mathbb{P}^2 - \ell$, where ℓ is a projective line. Hence $X \cong \mathbb{A}^2$. If $X \cong \mathbb{A}^2$ then it is clear that X is a homology plane. Suppose that X is a $\mathbb{Q}$-homology plane. Then $r = 1$ again and $H_1(X; \mathbb{Z}) \cong \mathbb{Z}/d\mathbb{Z}$ with $d = \deg C$. Note that $H_2(X; \mathbb{Z}) \cong H^1(C; \mathbb{Z}) \cong H_1(C; \mathbb{Z})$, where the second isomorphism follows from the universal coefficient theorem. Suppose that C has a singular point P with more than one analytic branch. Then there is a topological closed loop on C starting the point P via one branch and returning to P via another branch, which is not contracted homotopically to the point P. Hence $H_1(C; \mathbb{Q})$ contains a nonzero element. This implies that all possible singular points on C are one-branch singularities, i. e., one-place points. Similarly, the geometric genus of C must be zero. Hence C is a cuspidal rational curve. Conversely, if C is a cuspidal rational curve, then the long exact sequence in (1) for the pair $(\mathbb{P}^2, C)$ shows that $H_i(X; \mathbb{Q}) = 0$ for $i = 1, 2$. □

4. Let C be a cubic plane curve in $\mathbb{P}^2$ and let $X = \mathbb{P}^2 - C$. Prove the following assertions:
   (1) If C is smooth or nodal, $\overline{\kappa}(X) = 0$. Meanwhile, if C is cuspidal, $\overline{\kappa}(X) = -\infty$.
   (2) If C is cuspidal, let ℓ_0 be a line such that ℓ_0 meets C only at the cusp P of C. Then the linear system $\Lambda = \langle C, 3\ell_0 \rangle$, which is the linear pencil spanned by C and $3\ell_0$, induces an $\mathbb{A}^1$-fibration on X.
   (3) If C is cuspidal, the fundamental group at infinity is isomorphic to the local fundamental group of a surface quotient singularity whose resolution (dual) graph is

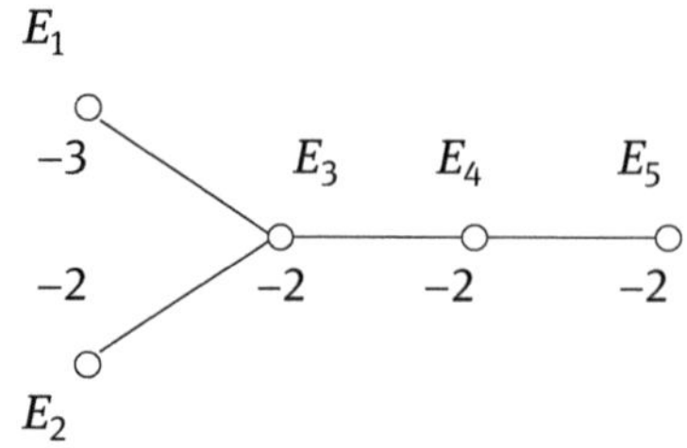

Hence it is a finite group.

(4) We denote by $H_1^\infty(X)$ the homology group $H_1(\partial U_n; \mathbb{Z})$, which we call *the first homology group at infinity* of X. Also $H_1^\infty(X)$ is the abelianization of $\pi_1^\infty(X)$, i. e., the residue group of $\pi_1^\infty(X)$ by its commutator subgroup. It is computed by writing Mumford's presentation additively. In our case, $H_1^\infty(X) \cong \mathbb{Z}/9\mathbb{Z}$.

Answer. (1) For the case where C is smooth, $K_{\mathbb{P}^2} + C \sim 0$. Hence $n(K_{\mathbb{P}^2} + C) \sim 0$, whence $\overline{\kappa}(X) = 0$. For the nodal case, let $f : V \to \mathbb{P}^2$ be the blowing-up of the node P and let E be the exceptional curve. Then X is embedded into V with an SNC boundary divisor $D = C' + E$, where C' is the proper transform of C. Since

$$K_V \sim f^*(K_{\mathbb{P}^2}) + E \sim -f^*(C) + E = -(C' + 2E) + E = -(C' + E),$$

we have $n(K_V + D) \sim 0$. Hence $\overline{\kappa}(X) = 0$. For the cuspidal case, we blow up the cusp P and its infinitely near points until the proper transform C' of C and exceptional curves make an SNC divisor D. A minimal sequence of blowing-ups $f : V \to \mathbb{P}^2$ will produce the following dual graph:

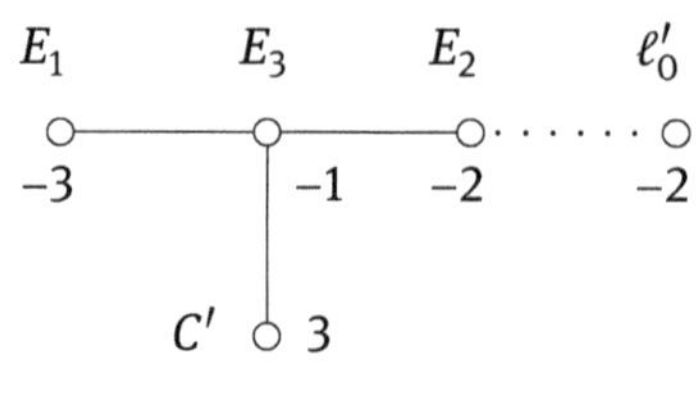

where ℓ_0' is the proper transform of ℓ_0. Then X is embedded into V as an open set with the boundary divisor $D = C' + E_1 + E_2 + E_3$ and $K_V = f^*(-3\ell_0) + E_1 + 2E_2 + 4E_3$. Since $f^*(3\ell_0) \sim f^*(C) = C' + 2E_1 + 3E_2 + 6E_3$, we have

$$\begin{aligned} K_V + D &\sim -(C' + 2E_1 + 3E_2 + 6E_3) + E_1 + 2E_2 + 4E_3 \\ &\quad +(C' + E_1 + E_2 + E_3) \\ &= -E_3. \end{aligned}$$

Hence $\overline{\kappa}(V) = -\infty$.

(2) Since $f^*(C) \sim f^*(3\ell_0)$ and $f^*(\ell_0) = \ell_0' + E_1 + 2E_2 + 3E_3$, we have

$$C' \sim 3\ell_0' + E_1 + 3E_2 + 3E_3.$$

The proper transform $\Lambda' = f'\Lambda$, which is spanned by C' and $3\ell_0' + E_1 + 3E_2 + 3E_3$, still have three base points which consist of $C' \cap E_3$ and its two infinitely near points of the first and second kind. Let $\sigma : W \to V$ be the blowing-ups of these three points and let E_6 be the last exceptional curve. Then Λ'' is spanned by $C'' = \sigma'(C')$ and $3\ell_0' + E_1 + 3E_2 + 3E_3 + 2E_4 + E_5$, where we denote the proper transform of ℓ_0', E_1, E_2, E_3 by σ by the same letters. Now the pencil Λ'' has no base points and has E_6 as a cross-section. Since $X = W - (C + \sum_{i=1}^{6} E_i)$, X has an $\mathbb{A}^1$-fibration for which $3\ell_0''$ is a unique multile fiber.
(3) Instead of the boundary divisor $C' + E_1 + E_2 + E_3$ of X, we consider the boundary divisor $C' + \sum_{i=1}^{6} E_i$ in the proof of assertion (2). This change of the boundary divisors of X is allowable in the computation of the fundamental group at infinity $\pi_1^\infty(X)$ because $\pi_1^\infty(X) = \pi_1(\partial U_n)$ for a good neighborhood U_n of the boundary divisor and ∂U_n is contained in X. Further, the blowing-ups and blowing-downs with centers on the boundary divisor do not affect ∂U_n, hence $\pi_1^\infty(X)$. We blow up one point on C' at a point other than $C' \cap E_6$ and blow down the proper transform of C'. Thus we obtain the dual graph of that in assertion (3) attaching a chain 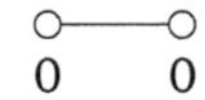.
In the computation of $\pi_1^\infty(X)$, we can consider $({E_6}^2) = ({C'}^2) = 0$. Then, taking the generators corresponding to the boundary components, we have $c^0 e_6 = c e_6^0 e_5 = 1$. Hence $e_6 = 1$ and $c = e_5^{-1}$. Hence $\pi_1^\infty(X)$ is determined by the dual graph in the assertion, which is the resolution graph of a quotient singular point. Hence $\pi_1^\infty(X)$ is a finite group.
(4) We use the boundary divisor in the proof of assertion (1). Writing Mumford's presentation additively, we have

$$e_3 = 3e_1 = 2e_2 = -3c, \quad e_1 - e_3 + c + e_2 = 0.$$

Let $x = e_1 + c$. Then $3x = 0$. We can compute $e_1 = -7c, e_2 = -6c$ and $e_3 = -3c$. Thence we obtain easily $9c = 0$. Hence $H_1^\infty(X) \cong \mathbb{Z}/9\mathbb{Z}$. □

5. Let d be an integer larger than 2 and let C be an irreducible rational curve of degree d with a unique cuspidal singular point P. For example, C is defined by $X_1^{d-1}X_2 = X_0^d$ with respect to a system of homogeneous coordinates (X_0, X_1, X_2). Let $X = \mathbb{P}^2 - C$ and let ℓ_0 be a line meeting C only in the point P. Prove the following assertions:
   (1) The linear pencil $\langle C, d\ell_0 \rangle$ induces an $\mathbb{A}^1$-fibration on X which has a unique multiple fiber $d(\ell_0 \cap X)$. Hence $\overline{\kappa}(X) = -\infty$.
   (2) $H_1^\infty(X) \cong \mathbb{Z}/d^2\mathbb{Z}$.

   Answer. It is just a generalization of the proof of Problem 4. □

6. Let C be an irreducible rational curve in $\mathbb{P}^2$ which is smooth except for a unique cuspidal singular point lying on the line at infinity ℓ_∞. Then show that $\overline{\kappa}(\mathbb{P}^2 - C) = -\infty$.

Answer. Let C_0 be the smooth part of C. Then C_0 is an irreducible curve in $\mathbb{A}^2 := \mathbb{P}^2 - \ell_\infty$ which is isomorphic to $\mathbb{A}^1$. By the AMS theorem, C_0 is a coordinate line $\{x = 0\}$ for a suitable system of coordinates $\{x, y\}$ of $\mathbb{A}^2$. Then $\mathbb{P}^2 - C$ contains a cylinder-like open set $U = \operatorname{Spec} \mathbb{C}[x, x^{-1}, y] \cong \mathbb{A}^1 \times \mathbb{A}^1_*$. Hence $\overline{\kappa}(\mathbb{P}^2 - C) = -\infty$. □

7. Let X be a $\mathbb{Q}$-homology plane.[6] Namely, X is a smooth affine surface with $H_i(X; \mathbb{Q}) = 0$ for every $i > 0$. Embed X into a smooth projective surface V in such a way that $D := V - X$ is a divisor with simple normal crossings. Prove the following assertions:
   (1) The irregularity $q(V) := h^1(V, \mathcal{O}_V) = 0$ and the geometric genus $p_g(V) = h^0(V, K_V) = 0$.
   (2) D is simply-connected. Namely, every component of D is isomorphic to $\mathbb{P}^1$ and the dual graph of D is a tree.
   (3) $\operatorname{Pic}(X)$ is a finite group, and $\Gamma(X, \mathcal{O}_X^*) = \mathbb{C}^*$.
   (4) We have isomorphisms:

$$\operatorname{Pic}(X) \cong H_1(X; \mathbb{Z}) \cong H^2(X; \mathbb{Z}) \cong \operatorname{Coker}(H_2(D; \mathbb{Z}) \to H^2(V; \mathbb{Z})).$$

Answer. (1) We have the following long exact sequence of cohomology groups with coefficients in $\mathbb{Z}$, $\mathbb{Q}$, $\mathbb{R}$ or $\mathbb{C}$ for a pair (V, D):

$$\begin{aligned} 0 &\to H^0(V,D) \to H^0(V) \to H^0(D) \to H^1(V,D) \to H^1(V) \\ &\to H^1(D) \to H^2(V,D) \to H^2(V) \to H^2(D) \to H^3(V,D) \\ &\to H^3(V) \to 0 \to H^4(V,D) \to H^4(V) \to 0, \end{aligned}$$

where

$$H^i(V,D) \cong H_{4-i}(X), \quad H^i(V) \cong H_{4-i}(V)$$

for all i by the Poincaré–Lefschetz duality. Since

$$H^3(V, D; \mathbb{C}) \cong H_1(X; \mathbb{C}) \cong H_1(X; \mathbb{Q}) \otimes_{\mathbb{Q}} \mathbb{C} = 0,$$

we have

$$H^3(V; \mathbb{C}) \cong H_1(V; \mathbb{C}) \cong H^1(V; \mathbb{C}) = 0.$$

By the Hodge decomposition $b_1(V) = h^{1,0} + h^{0,1}$, it follows that $h^{0,1} = \dim H^1(V, \mathcal{O}_V) = q(V) = 0$. By the Hodge theory, it is known (see [13, p. 307]) that the Néron–Severi group $\operatorname{NS}(V)$ is given as

$$\operatorname{NS}(V) = H^{1,1}(V, \mathbb{R}) \cap j^*(H^2(V; \mathbb{Z})),$$

6 It is known by Gurjar–Pradeep–Shastri [78] that a $\mathbb{Q}$-homology plane is rational.

where $\mathrm{NS}(V) := \mathrm{Div}(V)/(\equiv)$ is the divisor class group modulo numerical equivalence and $j^* : H^2(V;\mathbb{Z}) \to H^2(X;\mathbb{C})$ is the natural homomorphism. Since $H^i(V,D;\mathbb{Q}) \cong H_{4-i}(X;\mathbb{Q}) = 0$ for $i = 2,3$, the above exact sequence shows that $H^2(V;\mathbb{Q}) \cong H^2(D;\mathbb{Q})$, where the latter group is a $\mathbb{Q}$-vector space generated by the classes of irreducible components. This implies that $H^{1,1}(V;\mathbb{C}) \cong H^2(V;\mathbb{C})$. Hence $b_2(V) = h^{1,1} = 2p_g(V) + h^{1,1}$, whence $p_g(V) = 0$.
(2) Since $q(V) = 0$, the above exact sequence implies that $H^1(D;\mathbb{C}) = 0$. Let $D = D_1 + \cdots + D_r$ be the irreducible decomposition. We prove the assertion by induction on r. Since D is connected, there exists a component, say D_1, such that $D_1 \cap D_1' \neq \emptyset$, where $D_1' = D_2 + \cdots + D_r$. The Mayer–Vietoris cohomology sequence gives an isomorphism

$$H^1(D;\mathbb{C}) \cong H^1(D_1;\mathbb{C}) \oplus H^1(D_1';\mathbb{C}),$$

whence $H^1(D_i;\mathbb{C}) = 0$ for $1 \le i \le r$ by induction. This implies by the Hodge theory that every D_i is a smooth rational curve. On the other hand, $H^1(D;\mathbb{C}) = 0$ implies that the dual graph of D has no loops.
(3) Consider the standard exact sequence

$$0 \longrightarrow \mathbb{Z} \longrightarrow \mathcal{O}_V \overset{\exp}{\longrightarrow} \mathcal{O}_V^* \longrightarrow 0.$$

This gives an exact sequence

$$\cdots \to H^1(V,\mathcal{O}_V) \to \mathrm{Pic}(V) \to H^2(V;\mathbb{Z}) \to H^2(V,\mathcal{O}_V) \to \cdots,$$

where $H^i(V,\mathcal{O}_V) = 0$ for $i = 1,2$ because $q(V) = p_g(V) = 0$. Hence we have $\mathrm{Pic}(V) \cong H^2(V;\mathbb{Z})$. On the other hand, we have an exact sequence

$$0 \to \bigoplus_{i=1}^{r} \mathbb{Z}[D_i] \to \mathrm{Pic}(V) \to \mathrm{Pic}(X) \to 0,$$

which implies

$$\begin{aligned}\mathrm{Pic}(X) \otimes_{\mathbb{Z}} \mathbb{Q} &\cong (\mathrm{Pic}(V) \otimes_{\mathbb{Z}} \mathbb{Q})/\left(\bigoplus_{i=1}^{r} \mathbb{Q}[D_i]\right) \\ &\cong (\mathrm{Pic}(V) \otimes_{\mathbb{Z}} \mathbb{Q})/H^2(V;\mathbb{Q}) = 0.\end{aligned}$$

Hence $\mathrm{Pic}(X)$ is a finite group.
(4) Consider the following exact sequence of integral homology groups:

$$\begin{aligned}\cdots &\to H_2(D;\mathbb{Z}) \to H_2(V;\mathbb{Z}) \to H_2(V,D;\mathbb{Z}) \\ &\to H_1(D;\mathbb{Z}) \to H_1(V;\mathbb{Z}) \to \cdots,\end{aligned}$$

where $H_1(D;\mathbb{Z}) = 0$ because D is simply-connected. Further, $H_2(V,D;\mathbb{Z}) \cong H^2(X;\mathbb{Z})$ by the duality. Hence the above sequence gives an exact sequence

$$H_2(D;\mathbb{Z}) \to H_2(V;\mathbb{Z}) \to H^2(X;\mathbb{Z}) \to 0,$$

where $H_2(V;\mathbb{Z}) \cong H^2(V;\mathbb{Z}) \cong \mathrm{Pic}(V)$ and $H_2(D;\mathbb{Z}) \cong \bigoplus_{i=1}^r \mathbb{Z}[D_i]$. Hence we have

$$\mathrm{Pic}(X) \cong H^2(X;\mathbb{Z}) \cong \mathrm{Coker}(H_2(D;\mathbb{Z}) \to H^2(V;\mathbb{Z})).$$

On the other hand, by the universal coefficient theorem, we have

$$H^2(X;\mathbb{Z}) \cong \mathrm{Hom}(H_2(X;\mathbb{Z}),\mathbb{Z}) \oplus \mathrm{Ext}^1(H_1(X;\mathbb{Z}),\mathbb{Z}),$$

where $H_2(X;\mathbb{Z}) = 0$ because it is torsion-free by Kaup–Narasimhan–Hamm theorem and $H_2(X;\mathbb{Q}) = 0$, and

$$\mathrm{Ext}^1(H_1(X;\mathbb{Z}),\mathbb{Z}) \cong H_1(X;\mathbb{Z})$$

because $H_1(X;\mathbb{Z})$ is a finite abelian group. □

8. Let $A = k[x,y,z]$ and let $f = f(x,y,z)$ be an irreducible element of A. We say that f is a *coordinate* of A if $A = k[f,g,h]$ for some $g,h \in A$. Denote by $C_{c,d}$ the curve in $\mathbb{A}^3 = \mathrm{Spec}\,A$ defined by $f(x,y,c) = d$ for $c,d \in k$. Suppose the following three conditions:
   (i) The partial derivatives f_x, f_y, f_z generate the unit ideal of A.
   (ii) The hypersurface F_0 defined by $f = 0$ is isomorphic to $\mathbb{A}^2$.
   (iii) The curve $C_{0,0}$ is isomorphic to $\mathbb{A}^1$.
   Prove that f is a coordinate of A by verifying the subsequent results:
   (1) For any $d \in k$ the hypersurface F_d defined by $f = d$ is a smooth surface.
   (2) The curve $C_{c,0}$ is isomorphic to $\mathbb{A}^1$ for $c \in k$, hence irreducible in particular. Further, $C_{c,0} \cap C_{c',0} = \emptyset$ if $c \neq c'$.
   (3) $C_{c,d} \cap C_{c,d'} = \emptyset$ if $d \neq d'$ and $C_{c,d}$ is isomorphic to $\mathbb{A}^1$. In order to prove the second assertion, consider $C_{c,d}$ as a curve on the hyperplane $z = c$ and use the fact that $C_{c,0} \cong \mathbb{A}^1$ in (2) above.
   (4) Let $f_d : F_d \to \mathbb{A}^1 = \mathrm{Spec}\,k[z]$ be the morphism defined by $(x,y,z) \mapsto z$. Then every fiber $C_{c,d} = f_d^{-1}(c)$ being isomorphic to $\mathbb{A}^1$ shows that $F_d \cong \mathbb{A}^2$.
   (5) Let $f : X \to Y$ be the morphism defined by $(x,y,z) \mapsto f(x,y,z)$, where $X = \mathrm{Spec}\,A$ and $Y = \mathrm{Spec}\,k[f]$. Show that f is a trivial $\mathbb{A}^2$-bundle. This shows that f is a coordinate of A.

   Answer. (1) Clear by the Jacobian criterion of singularity.
   (2) Since $F_0 \cong \mathbb{A}^2$, the inclusion $k[z] \subset k[F_0]$ defines an irreducible linear pencil, where $k[F_0]$ is the coordinate ring of F_0. In fact, the normalization of $k[z]$ in $k[F_0]$ gives a smooth affine rational curve with only constant invertible elements. Hence

it is a polynomial ring $k[u]$. If $k[z] \subsetneq k[u]$, then $z = f(u)$ with $\deg f(u) > 1$. Then the curve $C_{0,0}$ defined by $z = 0$ on F_0 is either reducible or nonreduced. Since $C_{0,0} \cong \mathbb{A}^1$, we must have $k[z] = k[u]$. Namely, the family $\{C_{c,0} \mid c \in k\}$ is an irreducible linear pencil. Since $C_{0,0} \cong \mathbb{A}^1$, it follows by AMS theorem that $C_{c,0} \cong \mathbb{A}^1$ for every $c \in k$.
(3) As suggested in the hint, both $C_{c,0}$ and $C_{c,d}$ lie on a hyperplane H_c defined by $z = c$. Since $H_c \cong \mathbb{A}^2$ and $C_{c,0} \cong \mathbb{A}^1$, we deduce as in (2) that $C_{c,d} \cong \mathbb{A}^1$.
(4) The morphism f_d defines an irreducible linear pencil such that every member $C_{c,d} \cong \mathbb{A}^1$ for every $c \in k$. By Theorem 1.2.7, $F_d \cong \mathbb{A}^2$.
(5) Use Kaliman–Zaidenberg theorem (see Theorem 2.1.16) to conclude that $f : X \to Y$ is a trivial $\mathbb{A}^2$-bundle. □

9. This is a famous example of Ramanujam [192]. For the presentation, see Sugie [215]. Let F and G be a smooth conic and a cuspidal cubic in $\mathbb{P}^2$ meeting each other as $F \cdot G = 5R + S$, i. e., F and G meets at R and S with multiplicity 5 and 1, respectively. Blowing up the point S, we remove the proper transforms of F and G. We thus obtain a smooth affine surface X. Hence the exceptional curve E has a nonempty intersection with X. We further blow up the point R and the cusp P so that the proper transforms of F, G, together with the exceptional curves, form an SNC divisor. The obtained projective surface is V and the boundary divisor $D = V - X$. The composite of blowing-ups is a morphism $f : V \to \mathbb{P}^2$. The configuration of F, G on $\mathbb{P}^2$ and the dual graph of $D+E$ are given as below, where $(H_1{}^2) = -3$, $(H_2{}^2) = (E_i{}^2) = -2$ for $1 \le i \le 4$, $(E^2) = (H_3{}^2) = (E_5{}^2) = -1$, $(F'^2) = -2$, and $(G'^2) = -3$.

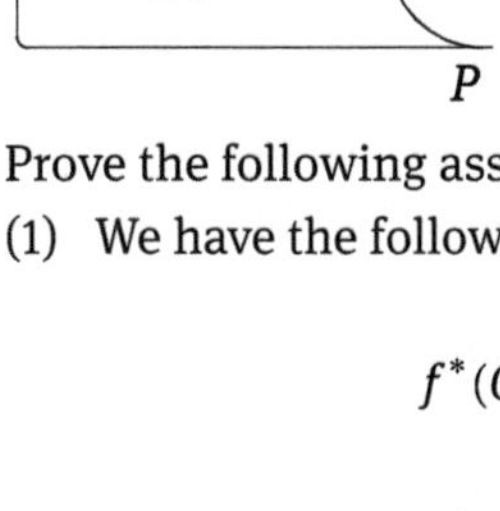

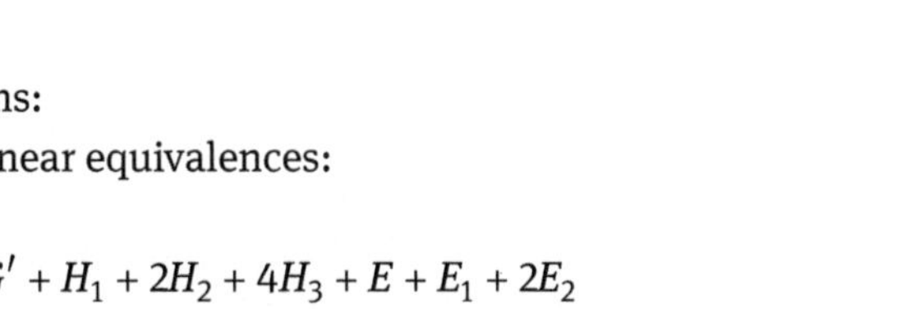

Prove the following assertions:
(1) We have the following linear equivalences:

$$\begin{aligned} f^*(G) &\sim G' + H_1 + 2H_2 + 4H_3 + E + E_1 + 2E_2 \\ &\quad + 3E_3 + 4E_4 + 5E_5, \\ f^*(F) &\sim F' + E + E_1 + 2E_2 + 3E_3 + 4E_4 + 5E_5, \\ K_V &\sim f^*(K_{\mathbb{P}^2}) + E + H_1 + 2H_2 + 4H_3 + E_1 + 2E_2 \\ &\quad + 3E_3 + 4E_4 + 5E_5. \end{aligned}$$

(2) Let

$$D = F' + G' + H_1 + H_2 + H_3 + E_1 + E_2 + E_3 + E_4 + E_5.$$

Then we have

$$D + K_V \sim F' + H_1 + H_2 + H_3 + E_1 + E_2 + E_3 + E_4 + E_5.$$

(3) $\overline{\kappa}(X) = 2$.

Answer. Claims (1) and (2) are straightforward.
(3) We compute

$$\begin{aligned} 5(D + K_V) &\sim 5F' + 5H_1 + 5H_2 + 5H_3 + 5E_1 + 5E_2 \\ &\quad + 5E_3 + 5E_4 + 5E_5 \\ &\sim (4F' + 5H_1 + 5H_2 + 5H_3 + 4E_1 + 3E_2 + 2E_3 + E_4) \\ &\quad + (f^*(F) - E). \end{aligned}$$

Since F is a conic, the linear system $|F-S|$ on $\mathbb{P}^2$, which consists of curves of degree 2 passing through the point S, contains a linear subsystem $|\ell|$, where ℓ is a line. This implies that $\Phi_{|5n(D+K_V)|}(V)$ has dimension 2 if $n \gg 0$. Hence $\overline{\kappa}(X) = 2$. □

10. Let X be a smooth affine surface with an $\mathbb{A}^1$-fibration $f : X \to \mathbb{P}^1$. Suppose that every fiber of f is irreducible, hence isomorphic to $\mathbb{A}^1$ if reduced. Let $m_1F_1, \ldots, m_rF_r$ exhaust all multiple fibers of f lying over points $P_1, \ldots, P_r$ of $\mathbb{P}^1$, where $m_i \geq 2$ and $m_1 = m_2$ if $r = 2$. Then prove the following assertions:
(1) $\pi_1(X)$ is an infinite group if $r \geq 4$.
(2) If $r = 3$ then $\pi_1(X)$ is a finite group if and only if $\sum_{i=1}^{3} \frac{1}{m_i} > 1$. Then $\{m_1, m_2, m_3\}$ is a Platonic triplet.
(3) If $r = 2$ then $\pi_1(X)$ is a finite group.

Answer. (1) By Lemma 1.1.8, there exists a finite Galois covering $p : D \to \mathbb{P}^1$ of degree N which ramifies over all P_i with multiplicity m_i. Consider $\widetilde{X} = \overline{X \times_{\mathbb{P}^1} D}$, the normalized fiber product. For a point P_i there exist s_i points of D which ramify over P_i with ramification index m_i, where $m_is_i = N$. Let t (resp. τ) be a uniformizant of $\mathbb{P}^1$ at P_i (resp. of D at one of the s_i points, say Q_i, over P_i). Then $t \sim \tau^{m_i}$, where $\sim$ means that terms of both sides differ by a unit of $\mathcal{O}_{D,Q_i}$. For any point R_i of F_i, $f^*(t) \sim v^{m_i}$ for an element v of $\mathcal{O}_{X,R_i}$ such that $v = 0$ is a local defining equation of F_i in X. Then $v^{m_i} \sim \tau^{m_i}$ in $\mathcal{O}_{X,R_i}$. By taking the normalization, v/τ is regular on the points of $\widetilde{X}$ lying over the point (R_i, Q_i). Since $(v/\tau)^{m_i} \sim 1$, there are exactly m_i points of $\widetilde{X}$ over (R_i, Q_i). Since there are exactly N/m_i points Q_i over the point P_i, the morphism $\pi : \widetilde{X} \to X$ induced by the projection has all together N distinct points of $\widetilde{X}$ over R_i of X. If $P \neq P_i$ for any i, it is clear that $\widetilde{X}$ is étale over X along the fiber $f^*(P)$. Hence π is étale over X, and $\widetilde{X}$ has an $\mathbb{A}^1$-fibration $\widetilde{f} : \widetilde{X} \to D$. Since

$F_i \cong \mathbb{A}^1, \pi^{-1}(F_i)$ splits into a disjoint union of m_i components which are isomorphic to the affine lines. We know that $\widetilde{f} : \widetilde{X} \to D$ has $\sum_{i=1}^r s_i$ singular fibers, s_i of which consists of m_i mutually disjoint components $\mathbb{A}^1$ with multiplicity 1. Thus we can find an open set U of $\widetilde{X}$ which is an $\mathbb{A}^1$-bundle over D. Since $\pi_1(U)$ surjects onto $\pi_1(\widetilde{X})$ and $\pi_1(U) \cong \pi_1(D)$, we have a surjection $\pi_1(D) \to \pi_1(\widetilde{X})$. On the other hand, any topological covering of D induces a topological covering of $\widetilde{X}$ by taking the fiber product, we know that $\pi_1(\widetilde{X}) \cong \pi_1(D)$.

We show that if $r \geq 4$ then the genus g of D is positive. This will show that $\pi_1(D)$ is an infinite group. Suppose $g = 0$. By the Riemann–Hurwitz formula, we have

$$\begin{aligned} -2 = -2N + \sum_{i=1}^r s_i(m_i - 1) &= -2N + \sum_{i=1}^r \frac{N}{m_i}(m_i - 1) \\ &= N\left(-2 + r - \sum_{i=1}^r \frac{1}{m_i}\right) \geq N\left(-2 + r - \frac{r}{2}\right) \\ &= N\left(-2 + \frac{r}{2}\right). \end{aligned}$$

This is a contradiction since $r \geq 4$.

(2) By the same argument as in (1), we have

$$2g - 2 = N\left(1 - \sum_{i=1}^3 \frac{1}{m_i}\right).$$

This implies that $g = 0$ if and only if $\sum_{i=1}^3 \frac{1}{m_i} > 1$.

(3) Straightforward. In this case, $\pi_1(X) \cong \mathbb{Z}/m\mathbb{Z}$, where $m = m_1 = m_2$. □

11. Let $f : X \to B$ be an $\mathbb{A}^1_*$-fibration from a smooth affine surface to a smooth curve B. Assume that $\text{Pic}(X) = 0$ and $\Gamma(X, \mathcal{O}_X)^* = k^*$. Prove the following assertions:
    (1) The $\mathbb{A}^1_*$-fibration f is untwisted, B is isomorphic to $\mathbb{P}^1$ or $\mathbb{A}^1$. If $B \cong \mathbb{P}^1$ then all fibers of f are irreducible, and if we assume that $b_2(X) = 0$ then there is only one fiber isomorphic to $\mathbb{A}^1$ if taken with reduced structure. If $B \cong \mathbb{A}^1$ then all fibers of f are irreducible except for one reducible fiber which consists of two irreducible components C_1, C_2 such that either $C_1 \cong C_2 \cong \mathbb{A}^1$ and C_1, C_2 intersect each other transversally in a single point, or $C_1 \cong \mathbb{A}^1, C_2 \cong \mathbb{A}^1_*$ and they are disjoint. Further, if $B \cong \mathbb{A}^1$ and $\text{Pic}(X) = 0$ then all irreducible fibers are reduced (and isomorphic to $\mathbb{A}^1_*$ if $b_2(X) = 0$), and the reducible fiber, say $F_1 = \mu_1 C_1 + \mu_2 C_2$, satisfies the condition $\gcd(\mu_1, \mu_2) = 1$.
    (2) Assume that $C \cong \mathbb{P}^1$. Let (V, D) be a smooth normal compactification. We can assume that the $\mathbb{A}^1_*$-fibration f extends to a $\mathbb{P}^1$-fibration $\Phi : V \to B$ in such a way that any (-1)-component of D meets at least three other components of D (hence including the horizontal components) if it is a fiber component. We then say that the compactification (V, D) is *minimal* along the fibration Φ.

Let D_1, D_2 be the horizontal components of D. Let $F = mA$ be a fiber of f with $m > 0$. Let $\widetilde{F}$ be the fiber of Φ which contains the closure $\overline{A}$ of A. Then the dual graph of $\widetilde{F}$ is given as follows:

Case $A \cong \mathbb{A}^1_*$. Then $\widetilde{F}$ is a linear chain and $\overline{A}$ is a unique (-1)-curve in $\widetilde{F}$.

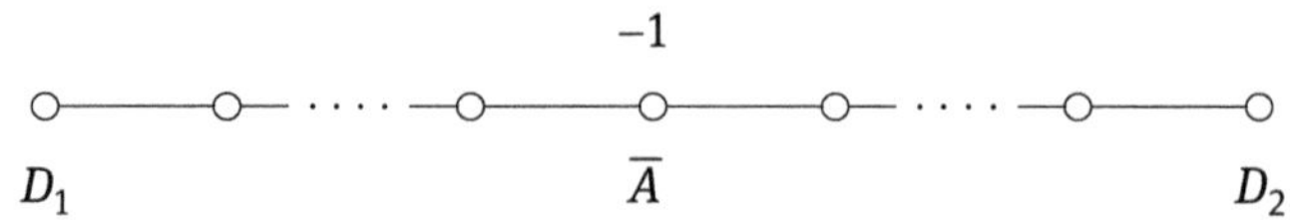

Case $A \cong \mathbb{A}^1$. If $\widetilde{F} \neq \overline{A}$ then $\widetilde{F} - \overline{A}$ is connected.

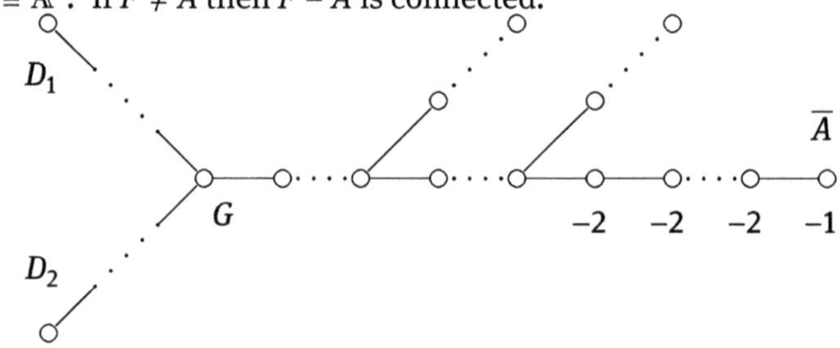

Here all the slanted branches are linear chains. Also $\overline{A}$ is a unique (-1)-component except possibly G when G meets D_1 and D_2. If $\widetilde{F} = \overline{A}$ then D_1, D_2, and $\overline{A}$ meet in one and the same point.

Answer. (1) Since $\mathrm{Pic}(X) = 0$ and $\Gamma(X, \mathcal{O}_X)^* = k^*$, $B \cong \mathbb{P}^1$ or $\mathbb{A}^1$, and the $\mathbb{A}^1_*$-fibration f is untwisted. See the proof of Lemma 1.4.5. Let (V, D) be a smooth normal compactification with a $\mathbb{P}^1$-fibration $\Phi : V \to \overline{B}$ such that $f = \Phi|_X$ and $\overline{B}$ is a smooth compactification of B. Let D_1, D_2 be the horizontal components.
We give the details in the case $B \cong \mathbb{P}^1$. The case $B \cong \mathbb{A}^1$ will be treated similarly. So, assume that $B \cong \mathbb{P}^1$ and let $F_1, \ldots, F_n$ exhaust all singular fibers of f. Write

$$\overline{F}_i = \sum_{j=1}^{r_i} \mu_{ij} C_{ij} + \sum_{j=r_i+1}^{s_i} \delta_{ij} D_{ij}, \tag{1.1}$$

where C_{ij} and D_{ij} are irreducible components such that $C_{ij} \cap X \neq \emptyset$ and $D_{ij} \cap X = \emptyset$. We have a relation

$$D_2 - D_1 \sim \sum_{i,j} \alpha_{ij} C_{ij} + (\text{a linear combination of fiber components of } D). \tag{1.2}$$

Therefore $\mathrm{Pic}(X)$ is an abelian group defined by the following generators and relations

$$\mathrm{Pic}(X) = \left\langle [C_{ij}] \begin{array}{l} i = 1, \ldots, n \\ j = 1, \ldots, r_i \end{array} \middle| \begin{array}{l} \sum_{i,j} \alpha_{ij}[C_{ij}] = 0 \\ \sum_{j=1}^{r_1} \mu_{ij}[C_{ij}] = \cdots = \sum_{j=1}^{r_n} \mu_{nj}[C_{ij}] \end{array} \right\rangle.$$

This implies that

$$\operatorname{rank}(\operatorname{Pic}(X)) \geq \sum_{i=1}^{n} r_i - (n-1) - 1 = \sum_{i=1}^{n} r_i - n.$$

Since $\operatorname{Pic}(X) = 0$ by the hypothesis, we have $r_i = 1$ for all i. Hence each singular fiber $F_i = m_i A_i$ with $A_i \cong \mathbb{A}^1$ or $\mathbb{A}^1_*$.
If $B \cong \mathbb{A}^1$, we conclude that, say $r_1 \leq 2$ and $r_2 = \cdots = r_n = 1$. Suppose that $r_1 = 1$. Since $B \cong \mathbb{A}^1$, D contains an irreducible fiber ℓ_∞. Since the singular fiber $\widetilde{F}_i \sim \ell_\infty$, there exists an integer $N > 0$ such that

$$N(D_2 - D_1) \sim a\ell_\infty + (\text{a linear combination of fiber components of } D).$$

This linear equivalence among the components of D defines a nonconstant invertible function on X. This contradicts the hypothesis that $\Gamma(X, \mathcal{O}_X)^* = k^*$. Thus $r_1 = 2$. Write $F_1 = \mu_1 C_1 + \mu_2 C_2$ and $F_i = m_i A_i$ for $2 \leq i \leq n$. Suppose $\operatorname{Pic}(X) = 0$. Then equations (1.1) and (1.2) imply that the coefficient matrix

$$\begin{pmatrix} \alpha_{11} & \alpha_{12} & \alpha_{21} & \cdots & \alpha_{n1} \\ \mu_1 & \mu_2 & 0 & \cdots & 0 \\ 0 & 0 & m_2 & \cdots & 0 \\ \vdots & \vdots & \vdots & \ddots & \vdots \\ 0 & 0 & 0 & \cdots & m_n \end{pmatrix}$$

is unimodular. Hence $m_2 = \cdots = m_n = 1$ and $\alpha_{11}\mu_2 - \alpha_{12}\mu_1 = 1$. In particular, $\gcd(\mu_1, \mu_2) = 1$. About how the components C_1, C_2 look, see the explanations in Section 1.3.8 and Corollary 1.3.16.
Now suppose that $B \cong \mathbb{P}^1$ and $b_2(X) = 0$. If all fibers are isomorphic to $\mathbb{A}^1_*$ when taken with reduced structures then D is disconnected and consists of two connected components, one containing D_1 and the other D_2. This is a contradiction because the affineness of X implies that D is connected. Let (V, D) be a smooth normal compactification of X. Consider a long exact sequence of cohomologies with $\mathbb{C}$-coefficients

$$\cdots \to H^1(V, D) \to H^1(V) \to H^1(D) \to H^2(V, D) \to \cdots,$$

where $H^1(V, D) \cong H_3(X) = 0$ as X is affine and $H^2(V, D) \cong H_2(X) = 0$ by the assumption $b_2(X) = 0$. On the other hand, V is rational because we may assume that there exists a $\mathbb{P}^1$-fibration $\Phi : V \to \overline{B}$ and $\overline{B} \cong \mathbb{P}^1$. Hence $H^1(V) = 0$. So, we have $H^1(D) = 0$. This implies that D is a tree of rational components. Then D_1, D_2 meet each other in one point and a fiber of Φ passing through the intersection point yields a singular fiber of f which is isomorphic to $\mathbb{A}^1$. Even if $B \cong \mathbb{A}^1$, the assumption $b_2(X) = 0$ implies that D is a tree of smooth rational curves. Since

the fiber at infinity ℓ_∞ is connected to the horizontal sections D_1, D_2, they do not intersect on other fibers. Hence other fibers are isomorphic to $\mathbb{A}^1_*$ because they are irreducible and reduced.
(2) Let F be a singular fiber of f and let $\widetilde{F} = \Phi^{-1}(f(F))$. If $\widetilde{F} \cong \mathbb{P}^1$ then $\widetilde{F} \cap D_1 \cap D_2 \neq \emptyset$ and $F \cong \mathbb{A}^1$. Suppose that $\widetilde{F}$ is reducible. Let G be a (-1)-curve in $\widetilde{F}$ other than $\overline{A}$. Since (V, D) is minimal along the fibers of Φ, the following is the unique possible case

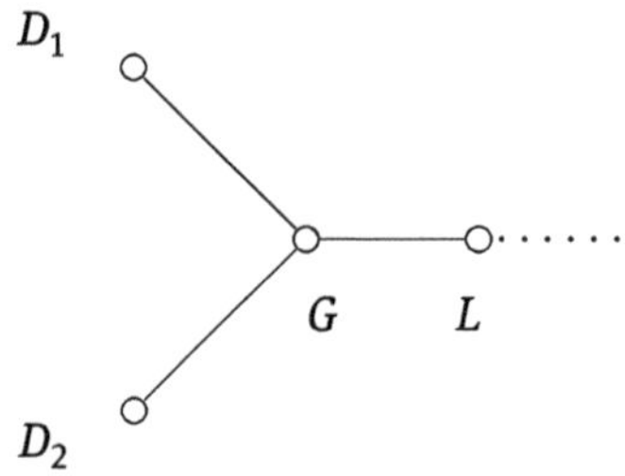

where L is a component of $\widetilde{F}$ because G must meet three other components of D. If $A \cong \mathbb{A}^1_*$ this is impossible for otherwise D is not connected. Hence if $A \cong \mathbb{A}^1_*$, the dual graph of $\widetilde{F}$ is as shown in the problem. If $A \cong \mathbb{A}^1$ and there is a (-1)-curve $G \neq \overline{A}$, we have the situation as shown in the above graph where G occurs with multiplicity 1 in $\widetilde{F}$. Hence there is a (-1)-curve in $\widetilde{F}$ other than G by Lemma 1.4.2. If this (-1)-curve stays outside of X, we have a contradiction to the minimality of (V, D) along the fibers of Φ. Hence it must be equal to $\overline{A}$. Then the dual graph of $\widetilde{F}$ is the one given in the problem. □

12. Let C_1, C_2 be irreducible curves on $\mathbb{A}^2$ such that $C_1 \cong C_2 \cong \mathbb{A}^1$ and that C_1, C_2 meet each other in one point transversally. Prove that there exists a system of coordinates (x, y) of $\mathbb{A}^2$ such that C_1 and C_2 are defined by $y = 0$ and $x = 0$, respectively.

Answer. Let $f_1, f_2 \in k[x, y]$ such that $C_1 = V(f_1)$ and $C_2 = V(f_2)$. For $\alpha, \beta \in k$, let $C_1^\alpha = V(f_1 - \alpha)$ and $C_2^\beta = V(f_2 - \beta)$. We show that for any $\alpha, \beta \in k$, $C_1^\alpha \cong C_2^\beta \cong \mathbb{A}^1$, and C_1^α, C_2^β meet in only one point transversally.
By the AMS theorem, we may assume that C_1 is defined by $y = 0$. Since C_1, C_2 meet in the point $f_2(x, 0) = 0$, we may assume by the assumption that $f_2(x, 0) = x$. Then $f_2(x, 0) - \beta = x - \beta$. Hence the curves C_1, C_2^β meet in the point $(\beta, 0)$ transversally. Now change the role of (C_1, C_2) to (C_2^β, C_1) in the above argument to conclude that C_2^β, C_1^α meet in only one point transversally.
Now we may assume that $f_1 = y$. Then C_1^α and C_2^β meet in the points (y, α), where y is a root of an equation $f_2(x, \alpha) = \beta$. Since α is arbitrary, it follows that $f_2(x, y)$ is linear in x. Namely, we may write $f_2(x, y) = x + g(y)$ with $g(y) \in k[y]$. Then set $x' = x + g(y)$ and $y' = y$. Then (x', y') is a system of coordinates of $\mathbb{A}^2$ such that $C_1 = (y')$ and $C_2 = (x')$. □

13. Let G be a connected algebraic group. Verify the following assertions:
   (1) $\overline{\kappa}(G) = -\infty$ or 0.
   (2) $\overline{\kappa}(G) = 0$ if and only if G is a semiabelian variety, i. e., G is an extension of an abelian variety by an algebraic torus.

Answer. (1) Suppose that G is affine. Let U be the unipotent radical. If $U \neq (1)$ then U contains the additive group G_a as a subgroup. Hence there exists a nontrivial G_a-action on G. So, G contains an $\mathbb{A}^1$-cylinder, and $\overline{\kappa}(G) = -\infty$. Suppose that $U = (1)$. Then G is reductive. Let T be a maximal torus of G and let B be a Borel subgroup containing T. If $T \subsetneq B$ then B contains G_a as a subgroup. Hence $\overline{\kappa}(G) = -\infty$ by the same reason as above. Otherwise, $B = T$ and then $G = T$ is an algebraic torus and $\overline{\kappa}(G) = 0$.
(2) Consider the case where G is not affine. By a theorem of Chevalley, there is an exact sequence

$$(1) \longrightarrow L \longrightarrow G \longrightarrow A \longrightarrow (1),$$

where L is a connected normal linear subgroup of G and A is an abelian variety. By Lemma 1.2.1(2), we have

$$\overline{\kappa}(G) \leq \overline{\kappa}(L) + \dim A.$$

If $\overline{\kappa}(L) = -\infty$ then $\overline{\kappa}(G) = -\infty$. Otherwise, L is an algebraic torus, and hence G is a semiabelian variety.
Conversely, suppose that G is a semiabelian variety, i. e., $L = T$, an algebraic torus in the above exact sequence. In particular, G is commutative. Let $T_1 \subset T$ be a subgroup isomorphic to the multiplicative group G_m, and let $G_1 = G/T_1$. By Lemma 1.2.1(3) applied to the case

$$(1) \longrightarrow G_m \longrightarrow G \longrightarrow G_1 \longrightarrow (1),$$

we have $\overline{\kappa}(G) \geq \overline{\kappa}(G_m) + \overline{\kappa}(G_1)$. Since $\overline{\kappa}(G_m) = 0$, it follows that $\overline{\kappa}(G) \geq \overline{\kappa}(G_1)$. By induction on $\dim T$, we may assume that $\overline{\kappa}(G_1) \geq \overline{\kappa}(A) = 0$. Hence we have $\overline{\kappa}(G) \geq 0$. Now, by Brion–Zhang [21, Corollary 1.2], we have $\overline{\kappa}(G) \leq 0$. Hence $\overline{\kappa}(G) = 0$. □

2 Algebraic surfaces with fibrations

In the present chapter, we develop a theory of affine space fibrations on algebraic varieties with emphasis on $\mathbb{A}^1$-fibrations. Roughly speaking, there are two kinds of affine space fibration depending on either a base variety being affine (*affine type*) or complete (*complete type*). Which type of $\mathbb{A}^1$-fibration a given algebraic variety has is one of geometric properties of the variety. If an $\mathbb{A}^1$-fibration of affine type is given on an affine variety, the $\mathbb{A}^1$-fibration coincides with the algebraic quotient morphism (in a generalized sense) by an action of the additive group scheme G_a. Since a G_a-action on an affine variety is given by a locally nilpotent derivation (abbreviated as *lnd*), this interpretation will provide us with abundant knowledge on $\mathbb{A}^1$-fibrations on affine varieties via numerous studies of lnds for commutative rings. An lnd acting on a commutative ring A is a convenient tool to single out one variable locally from A, viz. to find an element $b \in A$ and to write $A[b^{-1}] = B[b^{-1}][x]$ for a subring B and a variable x. Bad things of the $\mathbb{A}^1$-fibration like singular fibers occurs in the closed set $\{b = 0\}$ of $\operatorname{Spec} A$. We will thus make a detailed study of G_a-actions on algebraic varieties.

A G_a-action has also different natures according to an ambient variety, which is a variety acted on by the group G_a, being affine or complete. We study both affine and complete (in fact, projective) cases. The latter case is dealt with in Section 2.4. A G_a-action on a projective variety is mostly studied via a vector field which is determined by the G_a-action.

An obstacle in a study of G_a-action on an algebraic variety is the existence of quotient variety. Even if G_a acts on an affine variety $\operatorname{Spec} A$, the invariant subring $A_0 := A^{G_a}$, which is to be the coordinate ring of algebraic quotient variety, is not necessarily an affine domain over the ground field k. Namely, counterexamples to the Hilbert's 14th problem matter. Even if A_0 is an affine domain, the quotient morphism $q : \operatorname{Spec} A \to \operatorname{Spec} A_0$ does not necessarily behave nicely. For example, different orbits are not necessarily separated by the morphism q. If we require this separation property, we have to consider the geometric quotient *à la* Mumford's GIT. But the geometric quotient might not exist as an algebraic variety, contrary to the case of reductive group actions. In this respect, Seshadri theorem [210] is decisive, which asserts that it exists as an algebraic space provided the G_a-action is proper. This problem is dealt with in Section 2.3.

Meanwhile, an $\mathbb{A}^1$-fibration with a complete base variety is a more geometric object, and we have insufficient knowledge. It is more or less like a $\mathbb{P}^1$-fibration on a projective variety, and it is a subject of research in future.

An algebraic variety can have two or more independent G_a-actions. To measure how many G_a-actions exist on a given affine variety $X = \operatorname{Spec} A$, the notion of Makar-Limanov invariant $\mathrm{ML}(X)$ is introduced, which is the intersection of invariant subrings A^{G_a} when G_a-actions move over all possible actions on X. We call X an ML_i*-variety* if

https://doi.org/10.1515/9783110577563-002

the quotient field of ML(X) has transcendence degree i over k. If $\dim X = n$, we have $0 \le i \le n$, and the gap $n - i$ implies how many independent G_a-actions exist on X.

We analyze the surface case, i. e., the case $n = 2$. An ML_0-surface is called a *Gizatullin surface* if X is smooth. Such a surface has a complete embedding $X \hookrightarrow V$ into a smooth projective surface V such that the divisor $D = V - X$, called *the boundary divisor*, is a linear chain of smooth rational curves having transversal intersections. This surface is also characterized as a *quasihomogeneous* affine surface, which has by definition an open orbit U of the automorphism group Aut(X) such that $X - U$ is a finite set. A Gizatullin surface has some similarity to the affine plane though the Picard rank of X can be arbitrarily big. Namely, a curve on X isomorphic to $\mathbb{A}^1$ is contained in an $\mathbb{A}^1$-fibration. This is an analogue of AMS theorem on the affine plane. These properties are discussed in Sections 2.6, 2.7, and 2.8.

One can say that geometry called *affine algebraic geometry* was originated from endeavors of understanding geometry of the affine spaces. Extraction of geometric features from the affine spaces involves construction of algebraic varieties which share properties with the affine spaces. An ML_0-surface has similarities with the affine plane $\mathbb{A}^2$. One of them is that any affine line moves in an $\mathbb{A}^1$-fibration as mentioned above (see Theorem 2.6.12), and another is that both an ML_0-surface and $\mathbb{A}^2$ have minimal normal completions whose boundary divisors are linear chains of $\mathbb{P}^1$s, hence the fundamental group π_1^∞ is finite (see Lemma 2.6.3). A difference between both surfaces appears with the Picard number $\rho(X)$ being zero or positive. An ML_0-surface X can take an arbitrary positive integer as $\rho(X)$ (see Proposition 2.6.19). If $\rho(X) = 0$, any $\mathbb{A}^1$-fibration on X has at most one multiple irreducible fiber (see Lemma 2.6.6), and its multiplicity is reflected on Pic(X) which is then a finite cyclic group of order equal to the multiplicity. The vanishing of Pic(X) is equivalent to the condition that the coordinate ring of X is a UFD. Then X is isomorphic to $\mathbb{A}^2$. If $\rho(X) > 0$ then there is an $\mathbb{A}^1$-fibration on X parametrized by the projective line $\mathbb{P}^1$ (see Theorem 2.8.14), though a smooth affine surface with $\rho(X) = 0$ has no $\mathbb{A}^1$-fibrations parametrized by $\mathbb{P}^1$. In particular, a Danielewski surface $xy = p(z)$ with $\deg p(z) \ge 2$, which is a hypersurface in $\mathbb{A}^3$ with $p(z)$ having only simple roots, has a surjective morphism onto $\mathbb{P}^1$ whose general fibers are isomorphic to $\mathbb{A}^1$.

An ML_0-surface with $\rho(X) = 0$ belongs to a class of affine surfaces called *affine pseudoplane*, which is by definition an affine surface X having an $\mathbb{A}^1$-fibration $f : X \to \mathbb{A}^1$ such that there is at most one irreducible multiple fiber (see [144]). There are two kinds of affine pseudoplane, one is ML_0 and the other is ML_1. Furthermore, these surfaces are contained in a larger class of affine surfaces called $\mathbb{Q}$-*homology planes*. There are many results obtained about $\mathbb{Q}$-homology planes, but we do not step far into these areas.

Section 2.9 is devoted to a logarithmic deformation (due to Kawamata) of an affine surface with $\mathbb{A}^1$-fibration. It is shown that the type (affine or complete) of the given $\mathbb{A}^1$-fibration is invariant under such deformations.

Throughout the present chapter, the ground field k is an algebraically closed field of characteristic zero. As in the previous chapter, we assume that k is the complex number field whenever we use topological arguments.

2.1 $\mathbb{A}^1$-fibrations and $\mathbb{P}^1$-fibrations

A morphism of algebraic varieties $f : X \to Y$ is called *dominant* if the image of f contains a nonempty open set of Y. If f is dominant, it induces an injection $f^* : k(Y) \hookrightarrow k(X)$ of the function fields. A dominant morphism $f : X \to Y$ is a *fibration* if the extension $k(Y) \subset k(X)$ is a regular extension, i. e., $k(Y)$ is algebraically closed in $k(X)$. This implies that the *generic fiber* $X_\eta := X \times_Y \operatorname{Spec} k(Y)$ is an algebraic variety defined over $k(Y)$, i. e., a geometrically integral algebraic scheme over $k(Y)$. For a point $y \in Y$, the fiber over y (denoted by $X_y, f^*(y)$ or $f^{-1}(y)$) is the fiber product $X \times_Y \operatorname{Spec} k(y)$, where $k(y)$ is the residue field of $\mathcal{O}_{Y,y}$. Then the condition that $k(X)$ is a regular extension of $k(Y)$ is equivalent to the condition that there exists an open set U of Y such that $f^*(y)$ is an algebraic variety over $k(y)$ for $y \in U$. If y is a closed point, i. e., $k(y) = k$, we say that $f^*(y)$ is a *closed fiber* or simply a *general* fiber. When we say about general fibers, we allow replacing U by a smaller nonempty open set $U' \subset U$ if necessary. If $\dim Y = 1$, the fibration defines a *pencil*. If $Y \cong \mathbb{P}^1$, it is a *linear pencil*; and if Y is irrational, the pencil is *irrational*.

Definition 2.1.1. Let F be an algebraic variety defined over k. With the above notations, we say that $f : X \to Y$ is an *F-fibration* if general closed fibers X_y of f are isomorphic to F over $k(y)$. If F is isomorphic to the affine space $\mathbb{A}^n$ of dimension n, then an F-fibration is called an *$\mathbb{A}^n$-fibration*. Whenever we do not have to specify the relative dimension n, we say that f is an *affine space fibration*. If F is smooth, any fiber X_y is called *singular* if X_y is not isomorphic to F. The set of points y such that X_y is singular is called the *singular locus* and denoted by $\operatorname{Sing}(f)$.

We can similarly define an *$\mathbb{A}^1_*$-fibration* or a *$\mathbb{P}^1$-fibration* by the condition that general fibers of f are isomorphic to the punctured affine line $\mathbb{A}^1 \setminus \{0\}$ or the projective line $\mathbb{P}^1$. One of the most basic problems is formulated by *Dolgachev–Weisfeiler* and asks the following.

Problem 2.1. Let $f : X \to Y$ be an F-fibration. Does there exist an open set U of Y such that $X \times_Y U \cong U \times F$ if $F \cong \mathbb{A}^n$?

So, the above problem asks the *generic (local) triviality* of f in the Zariski topology on an open set of Y. If, for every closed point $y \in Y$, there exists an open neighborhood U_y of Y such that $f^{-1}(U_y) \cong F \times U_y$, we say that $f : X \to Y$ is an F-bundle in the Zariski topology. If f is an F-bundle, the morphism f is surjective. For the case $F = \mathbb{A}^n$, we only know the answer of Problem 2.1 for $n = 1$ and 2, for which the answer is positive. The case $n = 1$ was first treated by [107] and elaborated in [109, 46], and the case $n = 2$ was investigated by [105, 81]. An F-fibration $f : X \to Y$ is called *generically isotrivial* if there exist an open set U of Y and an étale finite covering $U' \to U$ such that $X \times_Y U' \cong U' \times F$.

We state some of related results.

Lemma 2.1.2. *The following assertions hold true:*
(1) *Let $f : X \to Y$ and $g : Y \to Z$ be fibrations. Then $g \cdot f : X \to Z$ is a fibration.*
(2) *Let $f : X \to Y$ be a fibration. Then the base change $f' : X \times_Y Y' \to Y'$ is a fibration for any algebraic variety Y' dominating Y.*
(3) *Let $f : X \to Y$ be a dominant morphism of algebraic varieties. Then $f : X \to Y$ is a generically trivial F-fibration for an algebraic variety F if and only if the generic fiber $X_\eta = X \times_Y \operatorname{Spec} k(Y)$ is $k(Y)$-isomorphic to $F \otimes_k k(Y)$.*

Proof. (1) Maps f and g induce regular extensions of the function fields $f^* : k(Y) \hookrightarrow k(X)$ and $g^* : k(Z) \hookrightarrow k(Y)$. Then $(g \cdot f)^* = f^* \cdot g^* : k(Z) \hookrightarrow k(X)$ is a regular extension. Hence $g \cdot f : X \to Z$ is a fibration.

(2) The extension $f^* : k(Y) \hookrightarrow k(X)$ is regular. Hence $f^* \otimes_{k(Y)} k(Y') : k(Y') \hookrightarrow k(X) \otimes_{k(Y)} k(Y')$ is a regular extension, and $f' : X \times_Y Y' \to Y'$ is a fibration.

(3) The “only if” part is clear. So, we show the “if” part. Let X_η be the generic fiber of f. Then X_η is $k(Y)$-isomorphic to $F \times_Y \operatorname{Spec} k(Y)$. Hence there exist birational mappings $\varphi : X \to F \times Y$ and $\psi : F \times Y \to X$ over Y such that $\psi \cdot \varphi = \mathrm{id}_X$ and $\varphi \cdot \psi = \mathrm{id}_{F \times Y}$. Since φ and ψ are defined as regular mappings over the generic point of Y, there exists an open set U of Y such that the restrictions of φ and ψ over U are morphisms and that $\varphi_U \cdot \psi_U = \mathrm{id}_{F \times U}$ and $\psi_U \cdot \varphi_U = \mathrm{id}_{f^{-1}(U)}$. Hence $f^{-1}(U)$ is isomorphic to $F \times U$ over U. □

The main results of [107] are stated in the following two assertions. When applying assertion (1) to various concrete problems, significant is the assumption that only general *closed* fibers of f are isomorphic to $\mathbb{A}^1$. If we assume that f is a fibration, it is easily realized by restricting Y to a smaller open set of Y that every fiber of f is geometrically integral. The assumption in assertion (2) that the generic fiber is isomorphic to $\mathbb{A}^1$ follows automatically from assertion (1) if Y is an algebraic variety over k. In fact, we may assume by restricting Y to a smaller open set since Y is smooth, and hence deduce that the generic fiber is isomorphic to $\mathbb{A}^1$ if the neighboring closed fibers are isomorphic to $\mathbb{A}^1$.

Lemma 2.1.3. *The following assertions hold:*
(1) *Let Y be a smooth variety and let $f : X \to Y$ be an affine, faithfully flat morphism of algebraic varieties. Assume that each fiber of f is geometrically integral and the general (closed) fibers are isomorphic to $\mathbb{A}^1$. Then X is an $\mathbb{A}^1$-bundle over Y.*
(2) *Let $f : X \to Y$ be an affine, faithfully flat morphism of schemes of finite type. Assume that Y is locally noetherian, locally factorial and integral scheme, and that the generic fiber of f is isomorphic to $\mathbb{A}^1$ and all other fibers are geometrically integral. Then X is an $\mathbb{A}^1$-bundle over Y.*

The next result is a generalization due to Kambayashi–Wright [109] of assertion (2) in Lemma 2.1.3.

Lemma 2.1.4. *Let $f : X \to Y$ be a faithfully flat morphism of finite type such that for each point $y \in Y$ the fiber X_y is isomorphic to $\mathbb{A}^1_{k(y)}$. If the base scheme Y is a noetherian normal scheme, then X is an $\mathbb{A}^l$-bundle over Y.*

In this result, we note that the fiber X_y is isomorphic to $\mathbb{A}^1_{k(y)}$ for every point y of the scheme Y which include not only closed points but also nonclosed points. Nonclosed points correspond to the generic points of irreducible subschemes of Y. For practical use of this result, we state it in the case X and Y are algebraic varieties over k.

Corollary 2.1.5. *Let $f : X \to Y$ be a fibration of algebraic varieties over k. Assume that f is faithfully flat, Y is normal, and the fiber X_y is isomorphic to $\mathbb{A}^1$ for every closed point $y \in Y$. Then f is an $\mathbb{A}^1$-bundle.*

Proof. It suffices to show that $X_y \cong \mathbb{A}^1_{k(y)}$ for every nonclosed point $y \in Y$. Let $d = \dim \overline{\{y\}}$, where $\overline{\{y\}}$ is the closure of y in Y. We proceed by induction on d. Let $Z = \overline{\{y\}}$ and let $\widetilde{Z}$ be the normalization of Z. Let $\widetilde{f} : \widetilde{X} \to \widetilde{Z}$ be the base change of f, where $\widetilde{X} := X \times_Y \widetilde{Z}$. Then $\widetilde{f}$ is faithfully flat and $\widetilde{X}_{\widetilde{z}}$ is isomorphic to $\mathbb{A}^1$ for every closed point $\widetilde{z} \in \widetilde{Z}$. In fact, $\widetilde{X}_{\widetilde{z}} \cong X_z \cong \mathbb{A}^1$, where z is the image of $\widetilde{z}$ in Z. Suppose that $d = 1$. Then $\widetilde{Z}$ is smooth. Hence the generic fiber of $\widetilde{f}$ is isomorphic to $\mathbb{A}^1$ by assertion (1) of Lemma 2.1.3. Since $k(\widetilde{Z}) = k(Z)$, the generic fiber of $\widetilde{f}$ is isomorphic to X_y, whence $X_y \cong \mathbb{A}^1$. Consider the case $d > 1$. By induction, we may assume that the fibers of $\widetilde{f}$ over every point of $\widetilde{Z}$ except for the generic point of $\widetilde{Z}$ is isomorphic to $\mathbb{A}^1$. By a remark before Lemma 2.1.3 about the generic fiber X_η, it follows that the generic fiber of $\widetilde{f}$ is also isomorphic to $\mathbb{A}^1$. Hence $\widetilde{f} : \widetilde{X} \to \widetilde{Z}$ is an $\mathbb{A}^1$-bundle by Lemma 2.1.4. Since $k(\widetilde{Z}) = k(Z)$, we have $\widetilde{X} \times_{\widetilde{Z}} \operatorname{Spec} k(\widetilde{Z}) \cong X_y \cong \mathbb{A}^1$. Hence the fiber X_y is isomorphic to $\mathbb{A}^1$ for every point $y \in Y$. Then we can apply Lemma 2.1.4. □

Dutta's elaboration [46] is about Lemma 2.1.4 and stated as follows. By Dutta's result, it suffices to show that $X_y \cong \mathbb{A}^1$, for $y = \eta$ being the generic point of Y and the generic points y of all irreducible divisors of Y.

Lemma 2.1.6. *Let Y be a locally noetherian normal integral scheme and let $f : X \to Y$ be a faithfully flat, affine morphism of finite type such that*

(i) *The fiber of f at the generic point of Y is $\mathbb{A}^1$.*
(ii) *The fiber of f at the generic point of each irreducible reduced closed subscheme of Y of codimension one is geometrically integral.*

Then X is an $\mathbb{A}^1$-bundle over Y. In particular, if Y is an affine scheme then X is a line bundle over Y.

The last assertion about a line bundle follows from a result of Bass–Connell–Wright [14] which is stated as follows.

Lemma 2.1.7. *Let $Y = \operatorname{Spec} B$ be a noetherian affine scheme and let $f : X \to Y$ be an $\mathbb{A}^n$-bundle in the Zariski topology. Then X is a vector bundle. Namely, the coordinate ring A of X is B-isomorphic to the symmetric algebra $S^{\bullet}_B(P)$ of a projective B-module P of rank n.*

An open set of X of the form $U \times \mathbb{A}^n$ is called an *$\mathbb{A}^n$-cylinder* or an *n-cylinder-like open set.*

Lemma 2.1.8. *The following assertions hold:*

(1) *Let $f : X \to Y$ be a generically trivial $\mathbb{A}^n$-fibration. Then X contains an $\mathbb{A}^n$-cylinder.*
(2) *Let X be an affine variety containing an $\mathbb{A}^1$-cylinder $V = U \times \mathbb{A}^1$. Then there exist a surjective proper birational morphism $\sigma : \Gamma \to X$ and a morphism $\rho : \Gamma \to \overline{Y}$ to a complete variety $\overline{Y}$ such that*
   (i) *The variety Γ contains an open set V' such that $\sigma|_{V'} : V' \to X$ induces an isomorphism onto V.*
   (ii) *$\overline{Y}$ contains U as an open set.*
   (ii) *ρ is a generically trivial $\mathbb{A}^1$-fibration such that $\rho^{-1}(U) = V'$ and $\rho|_{V'} : V' \to U$ is identified with the canonical projection $U \times \mathbb{A}^1 \to U$.*

Proof. Assertion (1) is clear by the definition of generic triviality.

(2) Let $\overline{Y}$ be a complete variety containing U as an open set. Let Γ be the closure in $X \times \overline{Y}$ of the graph

$$V' = \{((u,t),u) \mid u \in U, t \in \mathbb{A}^1\}.$$

Then V' is isomorphic to V and $\sigma^{-1}(V) = V'$, where $\sigma : \Gamma \to X$ is the restriction to Γ of the projection $X \times \overline{Y} \to X$. Hence V' is an open set of Γ. Since $\overline{Y}$ is a complete variety, the projection $X \times \overline{Y} \to X$ is a proper surjective morphism, hence so is the restriction σ onto Γ. The restriction ρ to Γ of the canonical projection $X \times \overline{Y} \to \overline{Y}$ induces the projection $U \times \mathbb{A}^1 \to U$ if restricted onto V'. Since X is affine, the fiber $\rho^{-1}(u)$ is a closed set and isomorphic to $\mathbb{A}^1$. Thus $\rho^{-1}(U) \cong V'$, and hence $\rho : \Gamma \to \overline{Y}$ is a generically trivial $\mathbb{A}^1$-fibration. □

Before going further, we state an important result of Kraft–Russell [108], which is called the *generic equivalence theorem*, where the hypothesis on the transcendence degree of the field k is satisfied if k is the complex field $\mathbb{C}$, for example.

Theorem 2.1.9. *Let k be an algebraically closed field of infinite transcendence degree over the prime field. Let $p : S \to Y$ and $q : T \to Y$ be two affine morphisms where S, T, and Y are k-varieties. Assume that for all closed points $y \in Y$ the two (schematic) fibers $S_y := p^{-1}(y)$ and $T_y := q^{-1}(y)$ are isomorphic. Then there are a dominant morphism of finite degree $\varphi : U \to Y$ and an isomorphism $S \times_Y U \cong T \times_Y U$ over U.*

In the theorem, by replacing U by $\varphi^{-1}(U')$ with an open set U' contained in $\varphi(U)$ and by replacing U by a suitable hyperplane section, we may assume that φ induces a surjective quasifinite morphism $\varphi : U \to V(\hookrightarrow Y)$. We may also assume that U and V are smooth. Then the subset W of V consisting of points $y \in V$ such that $\varphi^{-1}(y)$ contains less than d points, where $d = \deg \varphi$, is a closed set of V, and $\varphi : U \setminus \varphi^{-1}(W) \to V \setminus W$ is a finite étale morphism. Hence we may assume that $\varphi : U \to V$ is an étale finite morphism. Hence Theorem 2.1.9 implies the *generic isotriviality* of an F-fibration. Namely we have the following result.

Corollary 2.1.10. *Let k be an algebraically closed field of infinite transcendence degree over the prime field $\mathbb{Q}$. Let $f : X \to Y$ be an affine F-fibration, i. e., f is an affine morphism as well as an F-fibration, whence F is an affine variety. Let X_η be the generic fiber of f. Then there exists a finite field extension K' of $k(Y)$ such that $X_\eta \otimes_{k(Y)} K' \cong F \otimes_k K'$.*

Hence X_η is a $K'/k(Y)$-form of the affine variety F. If X_η is a *trivial* form, i. e., X_η is $k(Y)$-isomorphic to $F \otimes_k k(Y)$, then the fibration f satisfies the generic triviality (cf. Lemma 2.1.2(3)). We simply say that X_η is a $k(Y)$-form of F if X_η is a $K'/k(Y)$-form for a finite field extension K' of $k(Y)$.

Corollary 2.1.11. *Let k be the same as in Corollary 2.1.10. Let $f : X \to Y$ be an $\mathbb{A}^n$-fibration. Then it is generically trivial if $n = 1, 2$.*

Proof. By Corollary 2.1.10, X_η is a $K'/k(Y)$-form of $\mathbb{A}^n$. If $n = 1, 2$ it is well-known that any form of $\mathbb{A}^n$ is trivial (see [106]). Hence f is generically trivial. □

We consider a $\mathbb{P}^1$-fibration with our interest in its generic isotriviality.

Lemma 2.1.12. *Let F be a smooth complete curve defined over k and let $f : X \to Y$ be an F-fibration. Then the following assertions hold:*
1. *By shrinking Y to an open set, we may assume that f is a smooth projective morphism and Y is smooth.*
2. *The generic fiber X_η is a $k(Y)$-form of the curve F.*
3. *Suppose that $F = \mathbb{P}^1$. Then X_η is a conic defined over $k(Y)$. If $\dim Y = 1$, then X_η is $k(Y)$-isomorphic to $\mathbb{P}^1$. If $\dim Y \geq 2$, this is not necessarily the case. There is a degree 2 dominant morphism $\varphi : Y' \to Y$ such that $f' : X \times_Y Y' \to Y'$ is a generically trivial $\mathbb{P}^1$-fibration.*

Proof. (1) Since the generic fiber X_η is a smooth complete curve, it is a projective curve defined over $k(Y)$. Namely, there exists an ample line bundle L_η on X_η such that the associated mapping $X_\eta \to \mathbb{P}(H^0(X_\eta, L_\eta))$ is a closed immersion. By shrinking Y to a smaller open set, there exists a relatively ample line bundle $\mathcal{L}$ on X such that $f_*(\mathcal{L})$ is a free vector bundle of rank $n+1$ over Y and induces a closed immersion $\Phi : X \to \mathbb{P}^n \times Y$ over Y. Hence we may assume that f is a projective morphism.

(2) With the above notations, note that $\Phi_y = \Phi \times_Y \operatorname{Spec} k(y)$ induces an isomorphism between X_y and F for every closed point $y \in Y$. Let D be the zero locus of a nonzero element of $H^0(X, \mathcal{L})$. Then $D_y = D \cap X_y$ is an effective ample divisor of $X_y \cong F$. Let $F^0 = F \setminus D_y$. Then D_y and $D_{y'}$ are isomorphic to each other for closed points $y, y' \in Y$.

Up to this step we can shrink Y to a smaller open set if necessary. Let $X_\eta^0 = X_\eta \setminus D$. By Theorem 2.1.9, there exists an algebraic extension $K/k(Y)$ such that $X_\eta^0 \otimes_{k(Y)} K \cong F^0 \otimes_k K$. Since both $X_\eta \otimes_{k(Y)} K$ and $F \otimes_k K$ are smooth over K, this isomorphism extends to $X_\eta \otimes_{k(Y)} K \cong F \otimes_k K$.

(3) In the proof of assertion (2), we consider $\mathcal{L} = (\Omega_X^n / f^*(\Omega_Y^{n-1}))^{-1}$, the relative anticanonical divisor, where $n = \dim X$. Then it is relatively very ample and gives rise to a projective embedding $Y \hookrightarrow \mathbb{P}^2 \times Y$ over Y whose image is a conic fibration, i. e., every fiber over $y \in Y$ is a conic in $\mathbb{P}^2$ with coefficients in $k(y)$. If $\dim Y = 1$ then $k(Y)$ is a C_1-field.[1] Hence X_η as a conic in $\mathbb{P}^2$ has a $k(Y)$-rational point, whence X_η is $k(Y)$-isomorphic to $\mathbb{P}^1$. Meanwhile, there is an example by Artin–Mumford [10] of a conic over the function field K of dimension two over $\mathbb{C}$ which is not isomorphic to $\mathbb{P}^1$. Since $X_\eta \subset \mathbb{P}^2$ is a conic and meets in two distinct points with a general hyperplane which are defined over a quadratic extension $K/k(Y)$, let Y' be the smooth part of the normalization of Y in K. Then $f' : X \times_Y Y' \to Y'$ has a rational section. Hence it is generically trivial. □

Remark 2.1.13.

(1) Let $f : X \to Y$ be an $\mathbb{A}^1$-fibration. Then there exist a proper morphism $\overline{f} : \overline{X} \to Y$ and an open immersion $\iota : X \hookrightarrow \overline{X}$ such that $f = \overline{f} \cdot \iota$. By shrinking Y to an open set, we may assume that Y and the morphism $\overline{f}$ are smooth, which implies that $\overline{X}$ is smooth as well. Then a general closed fiber $\overline{X}_y$ is a smooth complete curve containing $\mathbb{A}^1$ as an open set, hence it is isomorphic to $\mathbb{P}^1$. Namely, $\overline{f} : \overline{X} \to Y$ is a $\mathbb{P}^1$-fibration. Let D be an irreducible divisor in $\overline{X} \setminus X$ which lies horizontally to the fibration $\overline{f}$. Then the intersection number $D \cdot \overline{X}_y$ for a general closed point y of Y is equal to 1 with intersection point given by the point $\mathbb{P}^1 \setminus \mathbb{A}^1$. Hence D is a cross-section of $\overline{f}$ and meets the generic fiber $\overline{X}_\eta$ transversally in one point. This point is therefore a $k(Y)$-rational point of the conic $\overline{X}_\eta$. So, $\overline{f} : \overline{X} \to Y$ is generically trivial, and $f : X \to Y$ is also generically trivial as an $\mathbb{A}^1$-fibration.

(2) Let $f : X \to Y$ be an $\mathbb{A}^1_*$-fibration. We can take a $\mathbb{P}^1$-fibration $\overline{f} : \overline{X} \to Y$ as above. Let D be the horizontal divisor of $\overline{X} \setminus X$. Then there are two cases.
   (i) Divisor D is irreducible. Then D is a so-called 2-*section* of $\overline{f}$. In this case $\overline{X}_\eta \setminus X_\eta$ consists of two non-$k(Y)$-rational points which are conjugate over $k(Y)$. We then call f a *twisted* $\mathbb{A}^1_*$-fibration. Divisor D gives rise to a nonzero 2-torsion element in $\mathrm{Pic}(X)$ if $\overline{X}_\eta \cong \mathbb{P}^1$.
   (ii) Divisor D is reducible. Then D consists of two irreducible components $D = D_1 + D_2$. Then points $D_i \cap \overline{X}_\eta$ $(i = 1, 2)$ are two distinct $k(Y)$-rational point of $\overline{X}_\eta$, hence $\overline{X}_\eta \cong \mathbb{P}^1$ in particular. The fiber X_η is isomorphic to $\mathbb{A}^1_*$ over $k(Y)$. We the call f an *untwisted* $\mathbb{A}^1_*$-fibration.

1 A field K is called a C_1-*field* if every nonzero homogeneous polynomial P has a nontrivial zero with coordinates in K provided the number of variables is greater than the degree of P. Tsen's theorem states that the algebraic function field in one variable over an algebraically closed field is a C_1-field.

(3) Let Y be the base variety in (2) above. Consider an exact sequence of étale group schemes over Y,

$$0 \longrightarrow G_{m,Y} \longrightarrow \mathrm{GL}(2)_Y \longrightarrow \mathrm{PGL}(1)_Y \longrightarrow 1.$$

Then we have the following exact sequence of étale cohomology groups:

$$\to H^1_{et}(Y, G_m) \to H^1_{et}(Y, \mathrm{GL}(2)) \xrightarrow{\alpha} H^1_{et}(Y, \mathrm{PGL}(1)) \xrightarrow{\beta} H^2_{et}(Y, G_m) \to .$$

An isotrivial $\mathbb{P}^1$-bundle $f : X \to Y$ represents an element θ of $H^1_{et}(Y, \mathrm{PGL}(1))$. Then f is locally trivial in the Zariski cohomology if and only if $\theta \in \mathrm{Im}(\alpha)$. In fact, it is known that $H^1_{et}(Y, \mathrm{GL}(2)) \cong H^1_{Zar}(Y, \mathrm{GL}(2))$ and each element of $H^1_{Zar}(Y, \mathrm{GL}(2))$ represents an $\mathbb{A}^2$-bundle over Y, hence a vector bundle of rank 2 over Y by Lemma 2.1.7. Hence the obstruction for $f : X \to Y$ to be locally trivial in the Zariski topology is the vanishing of the class $\beta(\theta)$ in $H^2_{et}(Y, G_m)$. The group $H^2_{et}(Y, G_m)$ is called the *cohomological Brauer group* of Y.

Let R be a commutative ring. A polynomial ring $R[x_1, \ldots, x_n]$ is often denoted by $R^{[n]}$ if there is no necessity of specifying variables $x_1, \ldots, x_n$. Note that $\mathbb{A}^n$-fibrations have been considered in various situations related to problems of characterizing polynomial rings. Extension of results in the case $n = 1$ to the case $n \geq 2$ will cause many difficult problems. In the case $n = 2$, there are, however, several results. We first state a *Sathaye theorem* [204].

Lemma 2.1.14. *Let R be a rank one discrete valuation ring of equicharacteristic zero. Let the unique maximal ideal of R be tR. Let $K = Q(R)$ the quotient field of R and let $k = R/tR$ the residue field. Let A be an affine domain over R such that $A \otimes_R K \cong K^{[2]}$ and $A \otimes_R k \cong k^{[2]}$. Then $A \cong R^{[2]}$.*

Corollary 2.1.15. *Let Y be a smooth curve defined over an algebraically closed field k of characteristic zero and let $f : X \to Y$ be an $\mathbb{A}^2$-fibration such that the fiber X_y is isomorphic to $\mathbb{A}^2$ for every point y of Y including the generic point. Then f is a locally trivial $\mathbb{A}^2$-bundle in the Zariski topology. Hence X is a vector bundle over Y of rank two.*

Proof. We just apply Lemma 2.1.14 to $f \otimes \mathcal{O} : X \times_Y \operatorname{Spec} \mathcal{O} \to \operatorname{Spec} \mathcal{O}$ for every closed point $y \in Y$, where $\mathcal{O} = \mathcal{O}_{Y,y}$ is a discrete valuation ring. □

In Corollary 2.1.15, the assumption that the generic fiber X_η is isomorphic to $\mathbb{A}^2$ is annoying from the viewpoint of geometric use of the result. Indeed, the assumption that $X_y \cong \mathbb{A}^2$ for every closed point $y \in Y$ implies that $X_\eta \cong \mathbb{A}^2$ over $k(Y)$. This was first observed in [105, Theorem 0.1] and independently in [81, Theorem 3.10].

So, we state the improved result as a theorem.

Theorem 2.1.16. *Let Y be a smooth algebraic curve defined over an algebraically closed field k of characteristic zero. Let $f : X \to Y$ be an $\mathbb{A}^2$-fibration such that $X_y \cong \mathbb{A}^2$ for every* *closed point $y \in Y$. Then f is a locally trivial $\mathbb{A}^2$-bundle in the Zariski topology. If $Y \cong \mathbb{A}^1$ then $X \cong \mathbb{A}^3$.*

Proof. As in Corollary 2.1.15, X is a vector bundle over Y. If $Y \cong \mathbb{A}^1$ then every vector bundle over Y is trivial. Hence $X \cong \mathbb{A}^2 \times \mathbb{A}^1 \cong \mathbb{A}^3$. □

We state the above-mentioned *Kaliman–Zaidenberg theorem* [105, Theorem 0.1] and prove it by making use of the generic equivalence theorem (Theorem 2.1.9).

Theorem 2.1.17. *Let $f : X \to Y$ be an affine morphism of smooth algebraic varieties defined over $\mathbb{C}$. Assume that every closed fiber X_y is isomorphic to $\mathbb{A}^2$. Then f contains an $\mathbb{A}^2$-cylinder. Namely, there exists an open immersion $\iota : Y_0 \times \mathbb{A}^2 \hookrightarrow X$ such that $f \cdot \iota = \iota_0 \cdot p_{Y_0}$, where $\iota_0 : Y_0 \hookrightarrow Y$ is an open immersion and p_{Y_0} is the projection $Y_0 \times \mathbb{A}^2 \to Y_0$.*

Proof. Let $Z = \mathbb{A}^2 \times Y$ and $g : Z \to Y$ be the projection to Y. Since all closed fibers of f and g are isomorphic to $\mathbb{A}^2$, by Theorem 2.1.9, there exists a dominant quasifinite morphism $\varphi : U \to Y$ such that $X \times_Y U \cong \mathbb{A}^2 \times U$ over U. Hence $X_\eta \otimes_{k(Y)} k(U) \cong \mathbb{A}^2_{k(U)}$ for the generic fiber X_η of f. Namely, X_η is a $k(U)/k(Y)$-form of $\mathbb{A}^2$. By [106], it follows that $X_\eta \cong \mathbb{A}^2_{k(Y)}$.

On the other hand, by replacing Y by an affine open set contained in $\varphi(U)$, we may assume that $X = \operatorname{Spec} A$ and $Y = \operatorname{Spec} B$. Then $A \otimes_B Q(B) \cong Q(B)^{[2]}$. Hence we can find elements $t_1, t_2 \in A$ such that the inclusion $B[t_1, t_2] \subseteq A$ becomes isomorphic over $Q(B)$. Since A is finitely generated over B, write $A = B[a_1, \dots, a_n]$. Then there exists an element $b \in B$ such that every a_i belongs to $B[b^{-1}, t_1, t_2]$. Let $Y_0 = D(b)$ in Y. By the above observation, we have $f^{-1}(Y_0) \cong \mathbb{A}^2 \times Y_0$. □

It is important to take note of a remark that a morphism $f : X \to Y$, every (closed) fiber of which is an affine variety, is not necessarily an affine morphism. The following example shows that f with every fiber isomorphic to $\mathbb{A}^2$ is not an affine morphism.

Example 2.1.18. Let V be a smooth affine hypersurface in

$$\mathbb{A}^5 = \operatorname{Spec} \mathbb{C}[z_1, z_2, z_3, z_4, z_5]$$

defined by an equation

$$z_1 z_4 - z_2 z_3 = z_5(z_5 + 1).$$

Let $\pi : V \to \mathbb{A}^2$ be the projection $(z_1, z_2, z_3, z_4, z_5) \mapsto (z_1, z_2)$. Then every fiber of π, except for $\pi^{-1}(0,0)$, is isomorphic to $\mathbb{A}^2$, and $\pi^{-1}(0,0) = F_1 \coprod F_2$, where $F_1 = \{z_1 = z_2 = z_5 = 0\}$, $F_2 = \{z_1 = z_2 = 0, z_5 = -1\}$, and $F_1 \cong F_2 \cong \mathbb{A}^2$. Let $X = V \setminus F_1$ and $f = \pi|_X : X \to Y = \mathbb{A}^2 = \operatorname{Spec} \mathbb{C}[z_1, z_2]$. Then every fiber of f is isomorphic to $\mathbb{A}^2$ over the respective residue field, but f is not affine because X is not affine.

2.2 G_a-actions on affine varieties and locally nilpotent derivations

Given an algebraic variety X, the existence of a fibration $f : X \to Y$ is a great help in studying what kind of structure the variety X has. In fact, one can reduce the study to algebraic varieties Y and F of lower dimension, where F is a general fiber of f. In order to find a fibration on X, a standard method is to find an algebraic group G acting on X as an automorphism group. The quotient variety Y of X by G and the quotient morphism $q : X \to Y$ gives a fibration whose general fiber is a homogeneous space of G. The quotient variety Y is not uniquely determined. For example, it could happen that Y is an algebraic quotient, but not a geometric quotient. Which kind of quotient exists depends on the group G and the variety X and reflects the properties of X. In this section, we consider G_a-actions on affine varieties. Some preliminary results are given in Section 1.3. G_a-actions on projective varieties will be considered in later sections.

2.2.1 G_a-actions, quotient morphisms, and $\mathbb{A}^1$-fibrations

$\mathbb{A}^1$-fibrations on algebraic varieties are closely related to G_a-actions, where G_a denotes the *additive group scheme*, often abbreviated as the additive group. In this section, we recall basic results on the additive group scheme and its action on an *affine* algebraic variety.

Let R be a commutative ring. The *additive group scheme* $G_{a,R}$ over $\operatorname{Spec} R$ is defined on the underlying scheme $\operatorname{Spec} R[t]$ for a variable t by assigning the following group operations:

$$\begin{aligned} &m : G_{a,R} \times G_{a,R} \to G_{a,R} \quad \text{(multiplication)},\\ &i : G_{a,R} \to G_{a,R} \quad \text{(inverse)},\\ &e : \operatorname{Spec} R \to G_{a,R} \quad \text{(neutral element)}, \end{aligned}$$

which satisfy the axioms:

$$\begin{aligned} &m \cdot (m \times \mathrm{id}_{G_{a,R}}) = m \cdot (\mathrm{id}_{G_{a,R}} \times m) \quad \text{(associativity law)},\\ &m \cdot (\mathrm{id}_{G_{a,R}} \times i) \cdot \Delta = m \cdot (i \times \mathrm{id}_{G_{a,R}}) \cdot \Delta = e,\\ &m \cdot (e \times \mathrm{id}_{G_{a,R}}) = m \cdot (\mathrm{id}_{G_{a,R}} \times e) = \mathrm{id}_{G_{a,R}}, \end{aligned}$$

where $\Delta : G_{a,R} \to G_{a,R} \times G_{a,R}$ is the diagonal morphism and $G_{a,R}$ is identified with $\operatorname{Spec} R \times_{\operatorname{Spec} R} G_{a,R}$ and $G_{a,R} \times_{\operatorname{Spec} R} \operatorname{Spec} R$.

In terms of R-algebra homomorphisms, $m = {}^a\mu$, $i = {}^a\iota$, and $e = {}^a\varepsilon$, where $\mu : R[t] \to R[t] \otimes_R R[t]$ is the *comultiplication*, $\iota : R[t] \to R[t]$ is the *coinverse*, and $\varepsilon : R[t] \to R$ is the *augmentation*, which are defined by $\mu(t) = t \otimes 1 + 1 \otimes t$, $\iota(t) = -t$, and $\varepsilon(t) = 0$.

Let A be an R-algebra and let $X = \operatorname{Spec} A$. Then an *action* of $G_{a,R}$ on X is given by a $\operatorname{Spec} R$-morphism $\sigma : G_{a,R} \times_{\operatorname{Spec} R} X \to X$ satisfying the rules $\sigma \cdot (\mathrm{id}_{G_{a,R}} \times \sigma) = \sigma(m \times \mathrm{id}_X)$

as the morphisms $G_{a,R} \times G_{a,R} \times X \to X$ and $\sigma \cdot (e \times \mathrm{id}_X) = \mathrm{id}_X$ as the morphisms $\operatorname{Spec} R \times_{\operatorname{Spec} R} X \to X$. In terms of an R-algebra homomorphism, we have $\sigma = {}^a\varphi$ for $\varphi : A \to R[t] \otimes_R A = A[t]$. The homomorphism φ is written as

$$\varphi(a) = \sum_{i\geq 0} \delta_i(a)t^i, \quad \delta_i(a) \in A.$$

Making a little detour, one can explain the above setting as follows. Let S be an R-algebra. We define $X(S)$ and $G_{a,R}(S)$ by

$$X(S) = \operatorname{Hom}_{\operatorname{Spec} R}(\operatorname{Spec} S, X) \cong \operatorname{Hom}_{R\text{-alg}}(A, S),$$
$$G_{a,R}(S) = \operatorname{Hom}_{\operatorname{Spec} R}(\operatorname{Spec} S, G_{a,R}) \cong \operatorname{Hom}_{R\text{-alg}}(R[t], S) = S^+,$$

where $G_{a,R}(S)$ is endowed with a group structure $\alpha \cdot \beta$ for $\alpha, \beta \in \operatorname{Hom}_{\operatorname{Spec} R}(\operatorname{Spec} S, G_{a,R})$ given by the composition

$$\alpha \cdot \beta : \operatorname{Spec} S \xrightarrow{(\alpha,\beta)} G_{a,R} \times G_{a,R} \xrightarrow{m} G_{a,R}.$$

Then $G_{a,R}(S)$ has the group structure given by the addition $\alpha + \beta$ of elements α, β of S, thus denoted by S^+. The morphism

$$\sigma(S) : G_{a,R}(S) \times X(S) \to X(S),$$

or equivalently

$$S^+ \times \operatorname{Hom}_{R\text{-alg}}(A, S) \to \operatorname{Hom}_{R\text{-alg}}(A, S),$$

is given by

$$(\sigma(S)(s, \theta))(a) = \varphi^\theta(a)|_{t=s} = \sum_{i\geq 0} \theta(\delta_i(a))s^i$$

for $s \in S^+$ and $\theta \in \operatorname{Hom}_{R\text{-alg}}(A, S)$. We denote $\sigma(S)(s, \theta)$ by ${}^s\theta$.

Since $\varphi(a) \in A[t]$, it follows that $\delta_i(a) = 0$ for $i \gg 0$. The rule $\sigma \cdot (\mathrm{id}_X \times e) = \mathrm{id}_X$ gives $\delta_0 = \mathrm{id}_A$. Since φ satisfies $\varphi(ab) = \varphi(a)\varphi(b)$, it follows that $\delta_1(ab) = a\delta_1(b) + b\delta_1(a)$. Namely, δ_1 is an R-derivation of A into A itself. Secondly, the rule $\sigma \cdot (\mathrm{id}_{G_{a,R}} \times \sigma) = \sigma(m \times \mathrm{id}_X)$ yields the relation

$$\begin{aligned}\sum_{i,j\geq 0} \delta_j\delta_i(a)t^j \otimes t^i &= \sum_{i,j\geq 0} \delta_{i+j}(a)(t \otimes 1 + 1 \otimes t)^{i+j} \\ &= \sum_{i,j\geq 0} \binom{i+j}{i} \delta_{i+j}(a)t^j \otimes t^i,\end{aligned}$$

whence we have

$$\binom{i+j}{i} \delta_{i+j} = \delta_j\delta_i, \quad i, j \geq 0.$$

Suppose that R is a k-algebra. Since k has characteristic zero, this relation finally gives a relation that $\delta_i = \frac{1}{i!}\delta_1^i$ for $i \geq 0$, where $\delta_1^0 = \mathrm{id}$. Hence, by setting $\delta = \delta_1$, σ is given by a *locally nilpotent derivation* δ as

$$\varphi(a) = \sum_{i\geq 0} \frac{1}{i!}\delta^i(a)t^i, \tag{2.1}$$

where $\delta|_k = 0$. Then δ is called *k-trivial*. The term *locally nilpotent derivation* is too long to write and often abbreviated as *LND* or *lnd*. By the form of (2.1), the homomorphism $\varphi : A \to A[t]$ is called the *exponential map* associated with δ and denoted by $\exp(t\delta)$.

Conversely, given a k-trivial lnd δ on a k-algebra A, define a k-algebra homomorphism $\varphi : A \to A[t]$ by the formula (2.1). Then ${}^a\varphi : G_a \times X \to X$ defines a G_a-action on $X = \operatorname{Spec} A$. There is, thus, a bijective correspondence between the set of G_a-actions on an affine scheme $X = \operatorname{Spec} A$ defined over k and the set of k-trivial lnds on A. We refer to Freudenburg [55] and Miyanishi [150] for algebraic properties of lnds. Among others, the following results are noteworthy and often used in related problems.

Lemma 2.2.1. *Let k be a field of characteristic zero and let A be an algebra domain.*[2] *Let δ be a k-trivial lnd on A and let $B = \operatorname{Ker} \delta$. Then the following assertions hold:*

(1) *The k-subalgebra B is* factorially closed *in A. Namely, if an element $b \in B \setminus \{0\}$ splits as $b = a_1a_2$ in A, then $a_1, a_2 \in B$. If A is integrally closed in $Q(A)$, so is B in $Q(B)$, where $Q(A)$ and $Q(B)$ respectively denote the quotient fields (the field of fractions, in other terms) of A and B, respectively.*
(2) *The unit group A^* is equal to B^*.*
(3) *If δ is nontrivial, i. e., $B \subsetneq A$, then there exists an element $u \in A$ such that $b := \delta(u) \in B \setminus (0)$ and $A[b^{-1}] = B[b^{-1}][u]$. Such an element u is called a* local slice, *and writing $A[b^{-1}]$ as a polynomial ring over $B[b^{-1}]$ with local slice u as a variable is called the* local slice construction. *If $\delta(u) \in B^*$ then $A = B[u]$. In the last case, u is called a* slice.
(4) *Let $X = \operatorname{Spec} A$ and $Y = \operatorname{Spec} B$. The inclusion $B \hookrightarrow A$ defines a dominant*[3] *morphism $q : X \to Y$. Suppose that both X and Y are algebraic varieties, i. e., both A and B are affine domains. Then q is a generically trivial $\mathbb{A}^1$-fibration.*

Proof. (1) We use the k-algebra homomorphism $\varphi : A \to A[t]$ defined in the equation (2.1). Since $b = \varphi(b) = \varphi(a_1)\varphi(a_2)$ and $A[t]$ is a polynomial ring over an integral domain A, both $\varphi(a_1)$ and $\varphi(a_2)$ are constants as polynomials in t. This implies that $\delta(a_1) = \delta(a_2) = 0$. Hence $a_1, a_2 \in B$. Suppose that an element $\eta \in Q(B)$ is integral over B. Since A is integrally closed in $Q(A)$, η belongs to A. Write $a = \eta$. Let $f(a) = 0$ be

2 It is a k-algebra which is an integral domain as well.

3 This implies that the generic point of Y corresponding to the zero ideal of B is induced by the zero ideal of A.

a monic integral relation of a over B with minimum degree,

$$f(a) = a^n + b_1 a^{n-1} + \cdots + b_{n-1}a + b_n, \quad b_i \in B.$$

Suppose $a \notin B$. Applying δ we then have a monic relation of smaller degree

$$a^{n-1} + \frac{n-1}{n} b_1 a^{n-2} + \cdots + \frac{2}{n} b_{n-2} a + \frac{1}{n} b_{n-1} = 0.$$

This is a contradiction. Hence $a \in B$, and B is integrally closed in $Q(B)$.

(2) Let $a \in A^*$. Then $aa' = 1$ for $a' \in A$. By (1), we have $a, a' \in B$. Hence $a \in B^*$. The inclusion $B^* \subseteq A^*$ is clear.

(3) We define the *δ-degree* of an element $a \in A$ by $\deg_\delta(a) = \max\{n \mid \delta^n(a) \neq 0\}$ if $a \neq 0$. We set $\deg_\delta 0 = -\infty$. Then $a \in B$ if and only if $\deg_\delta(a) \leq 0$. In the localization $A[b^{-1}]$, we can extend δ to an lnd in a natural way so that $\operatorname{Ker}\delta = B[b^{-1}]$ and the element a/b is a slice. Replacing A by $A[b^{-1}]$ and setting $u = a/b$, we may assume that u is a slice such that $\delta(u) = 1$. Let $a \in A$ and let $n = \deg_\delta(a)$. We show by induction on n that $a \in B[u]$. If $n = 0$ then $a \in B$. Note that $\delta^n(a) \in B$ because $\delta^{n+1}(a) = 0$. Then $a - \frac{1}{n!}\delta^n(a)u^n$ has δ-degree less than n. By induction, it is an element of $B[u]$. Hence $a \in B[u]$.

(4) Since $q^{-1}(D(b)) \cong D(b) \times \mathbb{A}^1$, the assertion follows immediately. □

Definition 2.2.2. Let X, Y, and q be the same as in assertion (4) of Lemma 2.2.1. We call Y the *algebraic quotient* of the G_a-action on X if B is an affine domain and call the morphism $q : X \to Y$ the *quotient morphism*. The subalgebra B is called the *invariant subring* of A under the G_a-action and denoted by A^{G_a}. The algebraic quotient Y is denoted by $X/\!/G_a$ to distinguish it from other quotients like the geometric quotient which is often denoted by X/G_a.

The following three conditions are equivalent for a k-algebra domain A with an lnd δ and a k-subalgebra B of A; their proof should be clear:

(i) $B = \operatorname{Ker}\delta$.

(ii) $B = \operatorname{Ker}(\varphi - \mathrm{id})$, where $\varphi : A \to A[t]$ is a k-algebra homomorphism defined by (2.1).

(iii) $B = \{a \in A \mid {}^\alpha a = a, \forall \alpha \in k\}$, where ${}^\alpha a = \varphi(a)|_{t=\alpha}$.

Let (A_i, δ_i) $(i = 1, 2)$ be a pair of k-algebra A_i and an lnd δ_i. A k-algebra homomorphism $h : A_2 \to A_1$ defines a G_a-equivariant morphism if $h\delta_2 = \delta_1 h$. It is then clear that $h(\operatorname{Ker}\delta_2) \subseteq \operatorname{Ker}\delta_1$. In particular, if $\delta_2 = 0$, $h(A_2) \subseteq \operatorname{Ker}\delta_1$. This implies that the quotient morphism $q : X \to X/\!/G_a$ is the categorical quotient in the category of affine k-schemes.[4]

4 Let C be a category. Let X be an object of C with an action of a group object G. Then a morphism $q : X \to Y$ is called a *categorical quotient* of X by G in C if any G-equivariant morphism $f : X \to Z$ with a trivial G-action on Z splits as $f = f' \cdot q$ for a morphism $f' : Y \to Z$. A categorical quotient is determined uniquely up to isomorphisms if it exists.

Even in the case where G_a acts on an affine variety $X = \operatorname{Spec} A$ and $B = A^{G_a}$ is not finitely generated over k, there exists a quasiaffine variety Z and a rational map $\pi : X \to Z$ such that $B = k[Z]$ (see Theorem 2.2.6 below). It is not clear whether Z is realized as an open set of $\operatorname{Spec} B$, or which properties the map $\pi : X \to Z$ satisfies as the quotient map. The construction of Z will be given later in the case where X is smooth and the G_a-action on X is proper.

Let $K = Q(A)$ be the quotient field and let $L = Q(B)$. The lnd δ on A naturally extends to a k-trivial derivation δ_K on K by

$$\delta_K\left(\frac{a_1}{a_2}\right) = \frac{\delta(a_1)a_2 - a_1\delta(a_2)}{a_2^2}.$$

Then $L \subseteq \operatorname{Ker}\delta_K$ and $B \subseteq L \cap A$. Further, we have the following result.

Lemma 2.2.3. *With the above notations, the following assertions hold:*

(1) *Define a k-algebra homomorphism $\Phi : K \to K[[t]]$ by*

$$\Phi(\xi) = \sum_{i\geq 0} \frac{1}{i!}\delta_K^i(\xi)t^i, \quad \xi \in K.$$

Then $\Phi(\xi) = \varphi(a_1)/\varphi(a_2)$ if $\xi = a_1/a_2$ with $a_1, a_2 \in A$.

(2) *Let $n_1 = \deg_\delta(a_1)$ and $n_2 = \deg_\delta(a_2)$. If $\xi = a_1/a_2 \in \operatorname{Ker}\delta_K$ then $n_1 = n_2$ and $\delta^{n_2}(a_2)a_1 = \delta^{n_1}(a_1)a_2$. Hence $\xi \in Q(B)$.*

Proof. (1) For a derivation D on an integral domain R containing $\mathbb{Q}$, we have a formula called the *generalized Leibniz rule*, namely

$$D^i(r_1r_2) = \sum_{j=0}^{i} \binom{i}{j} D^j(r_1)D^{i-j}(r_2), \quad r_1, r_2 \in R.$$

This rule yields an equality $\Phi(r_1r_2) = \Phi(r_1)\Phi(r_2)$, where $\Phi : R \to R[[t]]$ is a mapping defined by $\Phi(r) = \sum_{i\geq 0} \frac{1}{i!}D^i(r)t^i$. This formula applied to $\xi = a_1/a_2 \in K = Q(A)$ and δ_K gives an equality $\varphi(a_2)\Phi(\xi) = \varphi(a_1)$.

(2) If $\xi \in \operatorname{Ker}\delta_K$, then $\Phi(\xi) = \xi$. If we write $\xi = a_1/a_2$ then $\xi\varphi(a_2) = \varphi(a_1)$. This implies assertion (2). □

The 14th problem of Hilbert, slightly modified from the original, is the following question.

The 14th problem of Hilbert. *Let k be a field and let A be an affine domain. Let $K = Q(A)$ and let L be a subfield of K containing k. Is the subdomain $L \cap A$ finitely generated over k?*

There is a detailed account of principal counterexamples to this problem in Freudenburg [55]. Counterexamples with the lowest $\operatorname{tr.deg}_k L$ were given by Kuroda

[134] in the case $\mathrm{tr.deg}_k K = \mathrm{tr.deg}_k L = 3$. Interesting is that most counterexamples are obtained as the invariant subring of G_a acting on A. On the other hand, Zariski [234] contributed to a positive result which asserts the following.

Theorem 2.2.4. *Let k be a field and let A be a normal affine domain over k. Let L be a subfield of $K := Q(A)$ containing k. If $\mathrm{tr.deg}_k L \le 2$ then $B := L \cap A$ is an affine k-domain.*

This theorem is called *Zariski finiteness theorem* by Freudenburg [55]. When we treat the invariant subrings, the finite generation of these rings is always a big obstruction. But an effort to obtain positive results for higher $\mathrm{tr.deg}_k L$ has been made. We refer to one of those results as *Bhatwadekar–Daigle theorem* [16].

Theorem 2.2.5. *Let R be a noetherian domain containing a field k of characteristic zero. If R is a Dedekind domain, i. e., a normal domain of dimension one, then the kernel of any locally nilpotent R-derivation of $R[X, Y, Z]$ is a finitely generated R-algebra. Conversely, if R is neither a field nor a Dedekind domain then there exists a locally nilpotent R-derivation of $R[X, Y, Z]$ whose kernel is not finitely generated over R.*

We observed in Lemma 2.2.1 that a nontrivial G_a-action on an affine variety $X = \operatorname{Spec} A$ has the quotient morphism $q : X \to Y$ which is a generically trivial $\mathbb{A}^1$-fibration provided $B := A^{G_a}$ is finitely generated over k.

There is the following theorem due to J. Winkelmann [227, Theorem 3], which we call *Winkelmann theorem*.

Theorem 2.2.6. *Let k be a field, V an irreducible, reduced, normal k-variety and $G \subset \operatorname{Aut}(V)$. Then there exists a quasiaffine k-variety Z and a rational map $\pi : V \to Z$ such that:*
1. *The rational map π induces an inclusion $\pi^* : k[Z] \subset k[V]$, where $k[Z]$ (resp. $k[V]$) is the ring of k-regular functions on Z (resp. V).*
2. *The image of the pull-back $\pi^*(k[Z])$ coincides with the ring of invariant functions $k[V]^G$.*
3. *For every affine k-variety W and every G-invariant morphism $f : V \to W$, i. e., f is a G-equivariant morphism with G acting trivially on W, there exists a morphism $F : Z \to W$ such that $F \cdot \pi$ is a morphism and $f = F \cdot \pi$.*

We considered G_a-actions and associated quotient morphisms which are generically trivial $\mathbb{A}^1$-fibrations. We consider the converse problem. Namely, starting with $\mathbb{A}^1$-fibrations, we look for G_a-actions such that the associated quotient morphisms give the given $\mathbb{A}^1$-fibrations.

Lemma 2.2.7. *Let $f : X \to Y$ be an $\mathbb{A}^1$-fibration of affine algebraic varieties defined over k, where $X = \operatorname{Spec} A$ and $Y = \operatorname{Spec} B$. Then there exists a G_a-action on X and a factorially closed k-subalgebra $\widetilde{B}$ of A satisfying the following conditions:*

(1) $\widetilde{B} = A^{G_a}$, $B \subseteq \widetilde{B}$ *and* $Q(\widetilde{B}) = Q(B)$.
(2) *The quotient morphism* $q : X \to \widetilde{Y} = \operatorname{Spec} \widetilde{B}$ *induced by* $\widetilde{B} \hookrightarrow A$ *decomposes the* $\mathbb{A}^1$*-fibration* $f : X \to Y$ *as* $f = \nu \cdot q$, *where* $\nu : \widetilde{Y} \to Y$ *is a birational morphism.*
(3) *Such a decomposition* $f = \nu \cdot q$ *is uniquely determined by* f.

Proof. Let V be an open set of Y consisting of points y such that the fiber $X_y = f^{-1}(y)$ is isomorphic to $\mathbb{A}^1$. We may assume that V is smooth. Then $f^{-1}(V)$ is smooth in X. Hence the restriction $f|_{f^{-1}(V)} : f^{-1}(V) \to V$ is faithfully flat. By Corollary 2.1.5, $f|_{f^{-1}(V)} : f^{-1}(V) \to V$ an $\mathbb{A}^1$-bundle. By shrinking V if necessary, we may assume that $f|_{f^{-1}(V)}$ is trivial, i. e., $f^{-1}(V) \cong V \times \mathbb{A}^1$. We can take V to be an open set $D(b) := \operatorname{Spec} B[b^{-1}]$. Hence $A[b^{-1}] = B[b^{-1}][u]$. Consider a partial derivative $\partial/\partial u$ which is a $B[b^{-1}]$-trivial lnd of $A[b^{-1}]$. Since $\partial/\partial u$ sends $A[b^{-1}]$ to itself and $A = B[a_1, \dots, a_n]$ as A is finitely generated over B, choose an integer $m \geq 0$ such that

$$b^m \frac{\partial a_i}{\partial u} \in A, \quad 1 \leq i \leq n.$$

Set $\delta = b^m \frac{\partial}{\partial u}$. Then δ is a B-trivial lnd of A. Hence it defines a G_a-action $\sigma : G_a \times X \to X$ such that σ is a Y-morphism.

Let $\widetilde{B} = \operatorname{Ker} \delta$ which contains B by the construction of δ and $\widetilde{B}$ is factorially closed in A. Since $\widetilde{B} \subseteq B[b^{-1}]$ it follows that $Q(\widetilde{B}) = Q(B)$. Conditions (1) and (2) are satisfied.

We now consider condition (3). Let C be a factorially closed k-subalgebra of A such that $B \subseteq C$ and $Q(C) = Q(B)$. Let $a \in C$. Then $a \in Q(B)$. Thus $b_1 a = b_2$ for elements $b_1, b_2 \in B$. Since $\widetilde{B}$ is factorially closed, $a \in \widetilde{B}$. Hence $C \subseteq \widetilde{B}$. Conversely, a similar argument shows that $\widetilde{B} \subseteq C$. Hence $C = \widetilde{B}$. This proves the uniqueness of the factorization $B \hookrightarrow \widetilde{B} \hookrightarrow A$. □

We call $\widetilde{B}$ the *factorial closure* of B in A. If A is a normal affine k-domain with $\dim A \leq 3$, then $\widetilde{B}$ is finitely generated over k since the 14th problem of Hilbert has a positive answer by Theorem 2.2.4. If $\dim A = 2$ and B is normal then $\widetilde{B} = B$.

Meanwhile, if $\dim A \geq 4$, the factorial closure $\widetilde{B}$ of B in A may not be finitely generated over k even if B and A are finitely generated over k. This is shown by the following example.

Example 2.2.8. Consider the counterexample of the 14th problem of Hilbert due to Roberts [195], where $A = k[x, y, z, S, T, U, V]$ is a polynomial ring in seven variables and δ is a locally nilpotent derivation given by

$$\delta = x^{t+1} \frac{\partial}{\partial S} + y^{t+1} \frac{\partial}{\partial T} + z^{t+1} \frac{\partial}{\partial U} + (xyz)^t \frac{\partial}{\partial V},$$

where $t \geq 2$. Let $\widetilde{B} = \operatorname{Ker} \delta$. Since S/x^{t+1} is a slice in $A[x^{-1}]$, i. e., $\delta(S/x^{t+1}) = 1$, we have $A[x^{-1}] = \widetilde{B}[x^{-1}][S/x^{t+1}]$, whence $\widetilde{B}[x^{-1}]$ is finitely generated over k. Write $\widetilde{B}[x^{-1}] = k[x, x^{-1}, y, z, f_1, \dots, f_\ell]$, where $f_1, \dots, f_\ell \in \widetilde{B}$. Similarly, T/y^{t+1} (resp. U/z^{t+1}) is a slice in $A[y^{-1}]$ (resp. $A[z^{-1}]$). Hence we can write $\widetilde{B}[y^{-1}] = k[x, y, y^{-1}, z, g_1, \dots, g_m]$ (resp.

$\widetilde{B}[z^{-1}] = k[x,y,z,z^{-1},h_1,\ldots,h_n]$), where $g_1,\ldots,g_m,h_1,\ldots,h_n \in \widetilde{B}$. Let B be the normalization of a subring $k[x,y,z,f_1,\ldots,f_\ell,g_1,\ldots,g_m,h_1,\ldots,h_n]$. Let $\widetilde{q} : \operatorname{Spec} A \to \operatorname{Spec} \widetilde{B}$ be the quotient morphism by the G_a-action associated with δ. Then $B \subsetneq \widetilde{B}$, and the inclusion $B \hookrightarrow A$ defines an $\mathbb{A}^1$-fibration $f : \operatorname{Spec} A \to \operatorname{Spec} B$. By the construction of B, it is clear that f coincides with the quotient morphism $\widetilde{q}$ over the open set $D(x)\cup D(y)\cup D(z)$. Then the factorial closure of B in A is $\widetilde{B}$ since $\widetilde{B} = \operatorname{Ker}\delta$. The subalgebra $\widetilde{B}$ is not finitely generated over k by [195]. □

We note here the following result.

Lemma 2.2.9. *Let $p : X \to Y$ be an $\mathbb{A}^1$-fibration of normal affine algebraic k-varieties $X = \operatorname{Spec} A$ and $Y = \operatorname{Spec} B$ such that B is factorially closed in A, whence p is the quotient morphism by a G_a-action. Then $p(X)$ contains all codimension one points of Y. Namely, for any irreducible codimension one subvariety Z of Y, $p(X) \cap Z$ contains a nonempty open set of Z. In particular, if $\dim X = 2$, p is surjective, and if $\dim X = 3$, the image $p(X)$ is an open set of Y and $Y \setminus p(X)$ consists of a finitely many points if it is not the empty set.*

Proof. Suppose that there exists an irreducible codimension one subvariety Z such that $\overline{p(X) \cap Z} \neq Z$. Namely, we suppose that the generic point of Z is not in the image $p(X)$. Let $\mathfrak{p}$ be a prime ideal of B such that $V(\mathfrak{p}) = Z$. Let $\mathcal{O} = B_{\mathfrak{p}}$, which is a discrete valuation ring of the function field $k(Y) = Q(B)$. Let t be a uniformizant of $\mathcal{O}$, which we may assume to be an element of $\mathfrak{p}$. Then the hypothesis implies that $t(A\otimes_B\mathcal{O}) = A\otimes_B\mathcal{O}$. Hence there exist elements $b \in B \setminus \mathfrak{p}$ and $a \in A$ such that $ta = b$. Meanwhile, a is then a G_a-invariant element, and hence $a \in B$. This is a contradiction because $b \notin \mathfrak{p}$. The rest of the assertions is easy to show. □

Lemma 2.2.7 is no longer effective if one replaces Y by a variety which is not necessarily affine. As an example, consider $X = \mathbb{F}_n \setminus H$, where $\mathbb{F}_n$ is a Hirzebruch surface of degree n and H is an ample section. Then the standard $\mathbb{P}^1$-fibration on $\mathbb{F}_n$ induces an $\mathbb{A}^1$-bundle structure $f : X \to Y$, where $Y \cong \mathbb{P}^1$. Here the $\mathbb{A}^1$-fibration f is not the algebraic quotient morphism by G_a because Y is a complete curve. An $\mathbb{A}^1$-fibration $f : X \to Y$ is said to be *of affine type* (*of complete type, resp.*) if f is factored by the morphism q induced by a G_a-action on X as in Lemma 2.2.7 (otherwise, resp.).

2.2.2 More results on group actions

We give first several definitions in general settings. Let S be a scheme and let G be an S-group scheme. Let $\sigma : G \times_S X \to X$ be an action of G on an S-scheme X. Define the *graph morphism* $\Phi : G\times_S X \to X\times_S X$ by $\Phi(T) : (g,x) \mapsto (gx,x)$, where T is an S-scheme, $x \in X(T) = \operatorname{Hom}_S(T,X)$, $g \in G(T)$ and $gx = \sigma(T)(g,x)$.

Definition 2.2.10.

(1) A G-action σ on X is *free* (resp., *proper*) if Φ is a closed immersion (resp., a proper morphism).

(2) Define the *stabilizer group scheme* F_X of $G \times_S X$ as the base change of $\Phi : G \times_S X \to X \times_S X$ by the diagonal $\Delta_{X/S} : X \to X \times_S X$. For any S-morphism $x : T \to X$, we have

$$\mathrm{Hom}_X(T, F_X) = \{(g, x) \mid gx = x,\ g \in \mathrm{Hom}_X(T, G)\}.$$

We say that the G-action is *fixed-point free* if $F_X = X$ the identity X-group scheme.

By the definition, a G-action on X is fixed-point free if it is free. In fact, for any S-scheme, the freeness of the G-action implies that the graph morphism $\Phi(T) : G(T) \times X(T) \to X(T) \times X(T)$ is injective. Namely, for $x \in X(T)$ and $g \in G(T)$, the relation $gx = x$ implies $g = e$ because $ex = x$. Hence $F_X(T) = X(T)$ for every S-scheme T. Hence $F_X = X$.

In our arguments, we set $S = \mathrm{Spec}\, k$ for the algebraically closed ground field k of characteristic zero and consider algebraic group schemes G and algebraic varieties X. Then G is reduced since k has characteristic zero. Hence G is an algebraic group. We need the notion of *geometric quotient*, whose definition is given in Mumford [171]. But we present the definition with a mixture of Mumford [171] and Seshadri [210]. It is clear that they are equivalent.

Definition 2.2.11. Let G be an algebraic group acting on an algebraic variety X. A morphism $q : X \to Y$ from X to an algebraic variety Y is a *geometric quotient* if the following three conditions are satisfied:
(1) q is a surjective, and affine G-invariant morphism.
(2) For closed points x_1, x_2 of X, $q(x_1) = q(x_2)$ if and only if $O(x_1) = O(x_2)$, where $O(x)$ is the G-orbit of x. This is equivalent to saying that every closed fiber of q, i. e., the fiber $q^{-1}(y)$ for a closed point $y \in Y$, is a G-orbit.
(3) $(q_* \mathcal{O}_X)^G = \mathcal{O}_Y$. Namely, for any open set $V \subset Y$, we have $\Gamma(V, \mathcal{O}_Y) = \Gamma(q^{-1}(V), \mathcal{O}_X)^G$.
(4) q is submersive. Namely, for $V \subset Y$, V is open if and only if $q^{-1}(V)$ is open in X.

If a geometric quotient $q : X \to Y$ exists, it is a G-invariant morphism and induces $\Phi_Y = (\sigma, \mathrm{id}_X) : G \times X \to X \times_Y X$. Here Φ_Y is an X-morphism, where $G \times X$ and $X \times_Y X$ are viewed as X-schemes by the second projections. Condition (2) in Definition 2.2.11 of a geometric quotient ensures that the morphism Φ_Y is surjective if $q : X \to Y$ is a geometric quotient. With the notation in Definition 2.2.11, we say that a G-action on an algebraic variety X is *closed* if the orbit $O(x)$ is a closed set in X for any closed point $x \in X$.

A *unipotent algebraic group* (or simply *unipotent group*) is an algebraic group G over k which has a central series of subgroups

$$(e) \lhd G_1 \lhd G_2 \lhd \cdots \lhd G_{n-1} \lhd G_n = G \tag{2.2}$$

such that each subquotient G_i/G_{i-1} is isomorphic to G_a for $1 \le i \le n$.

It is clear that any G_a-action on an affine variety X is closed. In fact, for a closed point $x \in X$, if $O(x)$ is not closed, the closure $\overline{O(x)}$ in X is a complete curve because G_a has only one place at infinity as an algebraic curve. This is a contradiction. It is not known in general whether any unipotent group action on affine variety is closed.

Definition 2.2.12. Let G and X be the same as in Definition 2.2.11. Assume that there exists a geometric quotient $q : X \to Y$. We say that X is a *principal fiber bundle* over Y with group G in the sense of Zariski (étale, flat, resp.) topology if there exist an open covering $\mathcal{V} = \{(V_i, \alpha_i : V_i \to Y)\}_{i \in I}$ of Y in the Zariski (étale, flat, resp.) topology and local sections $\{s_i : V_i \to X\}_{i \in I}$ such that the fiber product of the X-morphism $\Phi_Y : G \times X \to X \times_Y X$ with $s_i : V_i \to X$ is an isomorphism for every $i \in I$. Here a local section $s_i : V_i \to X$ satisfies $q \cdot s_i = \alpha_i$.

A principal fiber bundle X over Y with group G in the Zariski (étale, resp.) topology is also called a locally trivial (locally isotrivial, resp.) *principal homogeneous space* over Y with group G (or simply G-*torsor*). In any topology, if X is a principal fiber bundle over Y with group G, then the V_i-isomorphism $G \times V_i \xrightarrow{\sim} X \times_Y V_i$ gives, by taking the fiber product with $X \times_Y V_i$, an isomorphism $(\Phi_Y)|_{V_i} : (G \times X) \times_Y V_i \xrightarrow{\sim} (X \times_Y X) \times_Y V_i$ for every $i \in I$. Since $\mathcal{V}$ is an open covering of Y in the respective topology, we have an isomorphism

$$\Phi_Y : G \times X \xrightarrow{\sim} X \times_Y X. \tag{2.3}$$

If $q : X \to Y$ is a faithfully flat morphism, the isomorphism (2.3) implies that X is a principal fiber bundle over Y with group G in the flat topology.

Lemma 2.2.13. *Let G be an affine algebraic group acting on an affine variety $X = \operatorname{Spec} A$. Let $B = A^G$. Assume that there exists a geometric quotient $q : X \to Y$. Then the following assertions hold:*
(1) $X \times_Y X = \operatorname{Spec}(A \otimes_B A)$.
(2) *If G is connected, then $A \otimes_B A$ is irreducible, i. e., the nilradical of $A \otimes_B A$ is a prime ideal. If $A \otimes_B A$ is further reduced, it is an affine domain.*
(3) *Suppose that X is normal and B is noetherian. If closed points of Y are separated by elements of B, i. e., for any two closed points y_1, y_2 of Y there exists an element $b \in B$ such that $b(y_1) \neq 0$ and $b(y_2) = 0$, then Y is an open set of* $\operatorname{Spec} B$.

Proof. (1) Let $T = \operatorname{Spec} B$. We first show that $X \times_Y X \cong X \times_T X$. In order to prove this assertion, let $\mathcal{V} = \{V_i\}_{i \in I}$ be an affine open covering of Y and let $U_i = q^{-1}(V_i)$. Then $\mathcal{U} = \{U_i\}_{i \in I}$ is an affine open covering of X because the quotient morphism $q : X \to Y$ is an affine morphism. Note that $q|_{U_i} : U_i \to V_i$ is a geometric quotient of U_i by G for every $i \in I$. Write $A_i = \Gamma(U_i, \mathcal{O}_X)$ and $B_i = \Gamma(V_i, \mathcal{O}_Y)$. Then $(X \times_Y X) \times_Y V_i = \operatorname{Spec}(A_i \otimes_{B_i} A_i)$, where $A_i \otimes_{B_i} A_i$ is the residue ring of $A_i \otimes_k A_i$ by the ideal generated by $\{\xi \otimes 1 - 1 \otimes \xi \mid \xi \in B_i\}$. On the other hand, $X \times_Y X$ is a closed affine subscheme of $X \times_k X$. By the above observation for local pieces $U_i \times_{V_i} U_i$, the coordinate ring of $X \times_Y X$ is the residue ring of $A \otimes_k A$ by the ideal generated by

$$\begin{aligned}&\{\xi \otimes 1 - 1 \otimes \xi \mid \xi \in k(X),\ \xi|_{U_i} \in B_i\ (i \in I)\}\\ &\quad = \{a \otimes 1 - 1 \otimes a \mid a \in A,\ a|_{U_i} \in B_i\ (i \in I)\}\\ &\quad = \{b \otimes 1 - 1 \otimes b \mid b \in B\},\end{aligned}$$

where we note that $\Gamma(Y, \mathcal{O}_Y) = B$. This implies that the coordinate ring of $X \times_Y X$ is $A \otimes_B A$. Hence $X \times_Y X \cong X \times_T X$.

(2) If G is connected, then $G \times X$ is an integral affine scheme. Hence $\Gamma(G) \otimes_k A$ is an integral domain, and $(A \otimes_B A)_{\mathrm{red}} = (A \otimes_B A)/\sqrt{(0)}$ is realized as a k-subalgebra because the graph morphism $\Phi_Y : G \times X \to X \times_Y X$ is surjective. Hence $(A \otimes_B A)_{\mathrm{red}}$ is an integral domain, and $\sqrt{(0)}$ is a prime ideal. Since $A \otimes_B A$ is the residue ring of $A \otimes_k A$, it is finitely generated over k. Hence $A \otimes_B A$ is an affine domain if $\sqrt{(0)} = (0)$.

(3) Since B is the G-invariant subring of an affine normal domain A, B is integrally closed. If B is noetherian, it is normal as well. On the other hand, there is a canonical morphism $\iota : Y \to T = \operatorname{Spec} B$ which assigns to $y \in Y$ the prime ideal $\mathfrak{p}_y = \{b \in B \mid b(y) = 0\}$ of T. Then $\iota \cdot q$ is the morphism $p : X \to T$ induced by the injection $B \hookrightarrow A$. Note that Y is a normal variety. In fact, for an open set $V_i = \operatorname{Spec} B_i$ as above, we have $B_i = (A_i)^G$ with a normal domain A_i. Hence so is B_i. If elements of B separate closed points of Y, then $\iota : Y \to T$ is injective. Hence, by Zariski's main theorem, ι is an open immersion. □

In view of the above lemma, we define a new notion on algebraic quotients.

Definition 2.2.14. Assume that an algebraic group G acts on an affine variety $X = \operatorname{Spec} A$. Let $B = A^G$ and let $q : X \to Y := \operatorname{Spec} B$ be the morphism defined by the inclusion $B \hookrightarrow A$. Here we do not have to assume that B is an affine domain. We say that the G-action on X is *quotient tight*, or simply *q-tight*, if $A \otimes_B A$ is an integral domain.

Lemma 2.2.15. *Let G be a connected algebraic group acting on an affine variety $X = \operatorname{Spec} A$, let $B = A^G$ and let $q : X \to Y = \operatorname{Spec} B$ be the morphism induced by $B \hookrightarrow A$. Then the G-action is q-tight provided q is a flat morphism and $Q(A)^G = Q(B)$ for the quotient fields $Q(A), Q(B)$ of A, B, where G acts naturally on $Q(A)$.*

Proof. We prove first that $Q(A)$ is a regular extension of $Q(B)$. Since k has characteristic zero, it suffices to show that $Q(B)$ is algebraically closed in $Q(A)$. Suppose that $\xi \in Q(A)$ is algebraic over $Q(B)$. Then ξ satisfies an equation

$$b_0\xi^n + b_1\xi^{n-1} + \cdots + b_{n-1}\xi + b_n = 0, \quad b_i \in B. \tag{2.4}$$

The group G is connected and acts algebraically on $Q(A)$ by ${}^g\xi = {}^g a_1/{}^g a_2$, where $g \in G(k)$ and $\xi = a_1/a_2$. Since ${}^g\xi$ also satisfies equation (2.4) and G is connected, it follows that $\xi \in Q(A)^G = Q(B)$. Since A is B-flat and $Q(A)$ is A-flat, we have inclusions

$$A \otimes_B A \overset{\otimes_B A}{\hookrightarrow} Q(A) \otimes_B A \overset{\otimes_A Q(A)}{\hookrightarrow} Q(A) \otimes_B Q(A).$$

Since $Q(A) \otimes_B Q(A) = Q(A) \otimes_{Q(B)} Q(A)$ and $Q(A)$ is a regular extension of $Q(B)$, it follows that $Q(A) \otimes_{Q(B)} Q(A)$ is an integral domain and hence $A \otimes_B A$ is an integral domain. □

2.2.3 Liftability of G_a-actions under finite extensions

A main reference for subsections 2.2.3 and 2.2.4 is [145]. Given a G_a-action on an affine variety $Y = \operatorname{Spec} B$, we have often to consider the liftability of the G_a-action (or the associated $\mathbb{A}^1$-fibration) via a finite covering $X = \operatorname{Spec} A \to Y = \operatorname{Spec} B$. The problem we have in mind is the following:

Let Y be an algebraic variety with a G_a-action. Assume that the topological fundamental group $\pi_1(Y)$ is a finite group. Let $p : X \to Y$ be the universal covering. Then X is an algebraic variety. Does X have a G_a-action compatible with the covering morphism p? Here the G_a-action $\sigma_X : G_a \times X \to X$ and the G_a-action $\sigma_Y : G_a \times Y \to Y$ must satisfy the relation $p \cdot \sigma_X = \sigma_Y \cdot (\mathrm{id}_{G_a} \times p)$ if σ_X really exists.

This is also a special case of the classical problem of lifting a derivation on B via a finite extension of k-algebras $B \hookrightarrow A$. We set up the situation.

Let A be an integral domain over the field k and let B be a k-subalgebra of A such that A is a finite B-module. Given a k-derivation δ on B, δ extends to the quotient field L of B. Since the quotient field K of A is a simple extension of L, K is written as $K = L(\theta)$ for some $\theta \in K$. Let $f(t)$ be the minimal polynomial of θ over L. Then it is well-known that δ lifts uniquely to a k-derivation Δ on K such that $\Delta(\theta) = -f^\delta(\theta)/f'(\theta)$, where $f^\delta(t)$ is the polynomial with all the coefficients of $f(t)$ replaced by their δ-images. The derivation Δ on K does not necessarily restrict to a derivation on A, i. e., $\Delta(A) \not\subset A$ in general. We recall a result of Seidenberg [207].

Lemma 2.2.16. *Let A be a noetherian integral domain containing the field k and let D be a derivation of A. Then D, naturally extended to a derivation of the quotient field $Q(A)$, gives rise to a derivation of the integral closure $\widetilde{A}$ of A in $Q(A)$.*

Proof. Let $\alpha \in Q(A)$. Then $\alpha \in \widetilde{A}$ if and only if there exists a nonzero element $d \in A$ such that $d\alpha^r \in A$ for every $r \geq 0$. In fact, if $\alpha \in \widetilde{A}$ then α satisfies a monic relation

$$\alpha^n + a_1\alpha^{n-1} + \cdots + a_n = 0, \quad a_i \in A.$$

Write $\alpha = a/b$ and $d = b^{n-1}$. Then $d\alpha^r \in A$ for every $r \geq 0$. Conversely, if there exists a nonzero element $d \in A$ such that $d\alpha^r \in A$ for every $r \geq 0$, then the A-algebra $A[\alpha]$ is an A-submodule of a finite A-module $(1/d)A$. Since A is noetherian, α satisfies a monic relation over A.

Let $\alpha \in \widetilde{A}$ and $d \in A$ as above. Let $E := \exp(tD) := \mathrm{id}_A + \sum_{i>0} \frac{1}{i!} t^i D^i$, which is a k-algebra homomorphism $A \to A[[t]]$, where $A[[t]]$ is the formal power series ring over A. The homomorphism E is naturally extended to a k-algebra homomorphism $E : Q(A) \to Q(A[[t]])$. Since $d\alpha^r \in A$, we have

$$E(d\alpha^r) = E(d)E(\alpha)^r \in A[[t]].$$

Hence we have $dE(d)(E(\alpha) - \alpha)^r \in A[[t]]$. Taking the coefficients of the t^r-term, we see that $d^2D(\alpha)^r \in A$ for every $r \geq 0$. This implies that $D(\alpha) \in \widetilde{A}$. □

We assume hereafter that A and B are noetherian normal domains over k. If A is finite over B, A is noetherian if and only if B is noetherian. In fact, if B is noetherian, then A is a noetherian B-module. Hence any ideal of A is finitely generated since it is finitely generated as a B-module. If A is noetherian then B is noetherian by a theorem of Eakin–Nagata [47, 177]. Furthermore, A is an affine domain over k if and only if so is B. In fact, the *"if"* part is clear. To prove the *"only if"* part, write $A = k[a_1, \ldots, a_n]$. Then a_i satisfies a monic relation

$$a_i^{n_i} + b_{i1}a_i^{n_i-1} + \cdots + b_{in_i} = 0, \quad b_{ij} \in B.$$

Let $B' = k[b_{ij} \mid 1 \leq i \leq n, 1 \leq j \leq n_i]$. Then A is a finite B'-module and B is a B'-submodule. Sine B' is noetherian, B is a finite B'-module. This implies that B is finitely generated over k.

We first consider the liftability in the case where A is étale over B. *Hereafter we assume that the given k-derivation δ of B is an lnd.* This will give a positive answer to the problem at the beginning of this subsection when both X and Y are affine.

Lemma 2.2.17. *Assume that A is an affine domain over k and that A is étale and finite over B. Let Δ is the extension of δ to the quotient field $K = Q(A)$. Then $\Delta(A) \subset A$ and Δ is an lnd on A.*

Proof. Since A is étale over B, it follows that $\Omega_{B/k} \otimes_B A \cong \Omega_{A/k}$. Since the given derivation δ on B is locally nilpotent, there exist a nonzero element $b \in B$ and an element y in $B[b^{-1}]$ such that $\delta(y) = 1$. Hence $B[b^{-1}] \cong R[y]$, where R is the kernel of δ extended to $B[b^{-1}]$ and hence an affine domain. The exact sequence of differential modules applied to the inclusions $R[y] \supset R \supset k$ yields a direct sum decomposition

$$\Omega_{R[y]/k} \cong (\Omega_{R/k} \otimes_R R[y]) \oplus R[y]dy.$$

By tensoring it with $B[b^{-1}]$, we obtain a direct sum decomposition

$$\Omega_{R[y]/k} \otimes_{R[y]} B[b^{-1}] \cong (\Omega_{R/k} \otimes_B B[b^{-1}]) \oplus B[b^{-1}]dy.$$

Since $B[b^{-1}] \subset A[b^{-1}]$ is étale, we have

$$\Omega_{A[b^{-1}]/k} \cong \Omega_{B[b^{-1}]/k} \otimes_{B[b^{-1}]} A[b^{-1}] \cong \Omega_{R[y]/k} \otimes_{R[y]} A[b^{-1}].$$

Hence we have a direct sum decomposition

$$\Omega_{A[b^{-1}]/k} \cong (\Omega_{R/k} \otimes_R A[b^{-1}]) \oplus A[b^{-1}]dy.$$

The derivation Δ of the quotient field $K = Q(A)$, which is the extension of δ, is given as an $A[b^{-1}]$-module homomorphism α from $\Omega_{A[b^{-1}]/k}$ to K. Here the restriction of α onto the direct summand $\Omega_{R/k} \otimes_R A[b^{-1}]$ is zero because δ is zero, and $\alpha(dy) = \Delta(y) = \delta(y) = 1$. This implies that $\Delta(A[b^{-1}]) \subset A[b^{-1}]$. In fact, for any $z \in A[b^{-1}]$, we have $\Delta(z) = \alpha(dz)$ and $dz = \omega + fdy$, where $\omega \in \Omega_{R/k} \otimes_R A[b^{-1}]$ and $f \in A[b^{-1}]$. Then $\Delta(z) = \alpha(\omega + fdy) = f\alpha(dy) = f \in A[b^{-1}]$. We shall show that $\Delta(A) \subset A$ and Δ is locally nilpotent. For this end, we may assume that k is the complex number field $\mathbb{C}$. In fact, all objects in the present situation we are concerned with are defined over a subfield of k which is finitely generated over the prime field $\mathbb{Q}$ and which is hence embedded into $\mathbb{C}$. If our assertion is verified over $\mathbb{C}$, it descends to the subfield and then ascends to k.[5]

Let $C = \operatorname{Spec} R$, where R is an affine domain, let $Y_b = \operatorname{Spec} B[b^{-1}]$ and let $X_b = \operatorname{Spec} A[b^{-1}]$. Then $Y_b \cong C \times \mathbb{A}^1$. Hence Y_b is topologically contractible to C. This implies that $\pi_1(Y_b) \cong \pi_1(C)$ and that $X_b \cong \widetilde{C} \times \mathbb{A}^1$, where $\widetilde{C}$ is a finite étale covering of C. Let $\widetilde{R}$ be the coordinate ring of $\widetilde{C}$. Then $A[b^{-1}] \cong \widetilde{R}[y]$, where $\widetilde{R}$ is étale and finite over R. Since δ is trivial on R, it follows that Δ is trivial on $\widetilde{R}$. Thus Δ is locally nilpotent on $A[b^{-1}]$.

Since A is B-flat by definition, we have isomorphisms of A-modules (see [146, Theorem 7.11]):

$$\operatorname{Der}_k(A, A) \cong \operatorname{Hom}_B(\Omega_{B/k}, A) \cong \operatorname{Der}_k(B, B) \otimes_B A.$$

Hence there exists a unique element Δ' of $\operatorname{Der}_k(A, A)$ which corresponds to $\delta \otimes 1_A$ of $\operatorname{Der}_k(B, B) \otimes_B A$. Hence the restriction of Δ' on B is the given derivation δ. By the uniqueness of the extended derivation on K, we conclude that $\Delta' = \Delta$. Thereby, we conclude that $\Delta(A) \subset A$.

It is now easy to see that Δ itself is an lnd since Δ restricted on $A[b^{-1}]$ is locally nilpotent. □

Theorem 2.2.18. *Let $f : X \to Y$ be a finite étale morphism of normal affine varieties. Then the following assertions hold:*

(1) *Let $\sigma_Y : G_a \times Y \to Y$ be a G_a-action. Then there exists a G_a-action $\sigma_X : G_a \times X \to X$ such that $f \cdot \sigma_X = \sigma_Y \cdot (\mathrm{id}_{G_a} \times f)$.*
(2) *Let $q : Y \to T$ be an $\mathbb{A}^1$-fibration of affine type. Then there exists an $\mathbb{A}^1$-fibration of affine type $\widetilde{q} : X \to \widetilde{T}$ such that $q \cdot f = g \cdot \widetilde{q}$, where $g : \widetilde{T} \to T$ is the normalization morphism of T in the function field $k(X)$.*

Proof. (1) Let $X = \operatorname{Spec} A$ and $Y = \operatorname{Spec} B$. Via f^*, B is viewed as a k-subalgebra of A such that A is an étale and finite B-module. The assertion then follows from Lemma 2.2.17.

5 We refer to this argument as the *Lefschetz principle*.

(2) By Lemma 2.2.7, there exists a G_a-action $\sigma_Y : G_a \times Y \to Y$ such that $q = \nu \cdot p$, where $p : Y \to S$ is the morphism induced by $B^{G_a} \hookrightarrow B$ with $T = \operatorname{Spec} R$ and $\nu : S \to T$ is the morphism associated with $R \hookrightarrow B^{G_a}$. By assertion (1), there exists a G_a-action $\sigma_X : G_a \times X \to X$ such that $f \cdot \sigma_X = \sigma_Y \cdot (\mathrm{id}_{G_a} \times f)$. Then we have a commutative diagram

$$\begin{array}{ccccc} \widetilde{R} & \longrightarrow & A^{G_a} & \longrightarrow & A \\ \uparrow & & \uparrow & & \uparrow \\ R & \longrightarrow & B^{G_a} & \longrightarrow & B, \end{array}$$

where $\widetilde{R}$ is the normalization of R in $Q(A)$. Let $\widetilde{T} = \operatorname{Spec} \widetilde{R}$. Then $\widetilde{q} : X \to \widetilde{T}$ associated with the inclusion $\widetilde{R} \hookrightarrow A$ is an $\mathbb{A}^1$-fibration of affine type such that $q \cdot f = g \cdot \widetilde{q}$. □

With the same notations as after Lemma 2.2.16, let $\mathfrak{R} = \sqrt{\operatorname{Ann}(\Omega_{A/B})}$ the radical ideal of A and let $\mathfrak{b} = B \cap \mathfrak{R}$, which we call the *reduced ramification ideal* and the *reduced branch ideal* of A over B, respectively. If A is étale over B, it is clear that $\mathfrak{R} = 0$ and $\mathfrak{b} = 0$. Next we consider the case where the finite extension $A \supset B$ is not necessarily étale.

Lemma 2.2.19. *Assume that A is an affine k-domain and that $\Delta(A) \subset A$ for the extension Δ of δ on $K = Q(A)$. Then Δ is locally nilpotent if and only if there exists a nonzero element $b \in B$ such that $\delta(b) = 0$ and $A[b^{-1}]$ is étale over $B[b^{-1}]$.*

Proof. Assume first that $\delta(b) = 0$ and $A[b^{-1}]$ is étale over $B[b^{-1}]$ for a nonzero element b of B. By Lemma 2.2.17, the derivation δ on $B[b^{-1}]$ lifts to an lnd δ_b on $A[b^{-1}]$. Then δ_b coincides with Δ on $A[b^{-1}]$. For every element a of A, $\Delta^n(a) = \delta_b^n(a)$ which is zero if $n \gg 0$. Hence Δ is an lnd of A.

For the converse, consider the differential module $\Omega_{A/B}$ which is a finite A-module. For a prime ideal $\mathfrak{P}$ of A and its contraction $\mathfrak{p} = \mathfrak{P} \cap B$, the extension $A_{\mathfrak{P}}$ is ramified over $B_{\mathfrak{p}}$ if and only if $\Omega_{A/B} \otimes_A A_{\mathfrak{P}} \neq (0)$. This condition is equivalent to the condition that $\mathfrak{R} \subset \mathfrak{P}$, where $\mathfrak{R}$ is the reduced ramification ideal. Let $\mathfrak{b} = \mathfrak{R} \cap B$ be the reduced branch ideal. Suppose now that Δ is locally nilpotent. Then Δ defines a G_a-action $\sigma_X : G_a \times X \to X$ on $X = \operatorname{Spec} A$ which extends the G_a-action $\sigma_Y : G_a \times Y \to Y$ on $Y = \operatorname{Spec} B$, i. e., $f \cdot \sigma_X = \sigma_Y \cdot (\mathrm{id}_{G_a} \times f)$ holds on $G_a \times X$, where $f : X \to Y$ is the natural finite morphism. For any $\lambda \in k$, denote by ${}^\lambda\mathfrak{P}$ (resp., ${}^\lambda\mathfrak{p}$) the image $\sigma_X(\lambda, \mathfrak{P})$ (resp., $\sigma_Y(\lambda, \mathfrak{p})$). Then ${}^\lambda\mathfrak{p} = {}^\lambda\mathfrak{P} \cap B$, and $A_{{}^\lambda\mathfrak{P}}$ is unramified over $B_{{}^\lambda\mathfrak{p}}$ if and only if so is $A_{\mathfrak{P}}$ over $B_{\mathfrak{p}}$. In other words, ${}^\lambda\mathfrak{P} \supset \mathfrak{R}$ if and only if $\mathfrak{P} \supset \mathfrak{R}$. Hence ${}^\lambda\mathfrak{p} \supset \mathfrak{b}$ if and only if $\mathfrak{p} \supset \mathfrak{b}$. This implies that ${}^\lambda\mathfrak{R} = \mathfrak{R}$ (resp., ${}^\lambda\mathfrak{b} = \mathfrak{b}$) for every $\lambda \in k$. Hence $\mathfrak{R}$ (resp., $\mathfrak{b}$) is a Δ-ideal of A (resp., a δ-ideal of B) in the sense that $\Delta(\mathfrak{R}) \subseteq \mathfrak{R}$ (resp., $\delta(\mathfrak{b}) \subseteq \mathfrak{b}$). If $\mathfrak{b} = B$, then $\mathfrak{R} = A$ and thereby $\Omega_{A/B} = (0)$, that is to say, A is étale over B.[6] In this case, we take $b = 1$.

6 Assume that $\Omega_{A/B} = (0)$. Then, for any maximal ideal $\mathfrak{M}$ of A and $\mathfrak{m} = \mathfrak{M} \cap B$, it holds that $\mathfrak{M}A_{\mathfrak{M}} = \mathfrak{m}A_{\mathfrak{M}}$ and hence the completions $\widehat{A}_{\mathfrak{M}}$ and $\widehat{B}_{\mathfrak{m}}$ coincides with each other. Since $\widehat{A}_{\mathfrak{M}}$ (resp. $\widehat{B}_{\mathfrak{m}}$) is faithfully flat over $A_{\mathfrak{M}}$ (resp. $B_{\mathfrak{m}}$), it follows that $A_{\mathfrak{M}}$ is a flat $B_{\mathfrak{m}}$-module. Hence A is unramified and flat over B. So, B is étale over A.

Suppose that $\mathfrak{b} \neq B$. Then $\mathfrak{b} \neq (0)$ because $\Omega_{A/B} \otimes_A A_{\mathfrak{P}} = (0)$ for a general choice of $\mathfrak{P}$. Since $\mathfrak{b}$ is a δ-ideal, we may choose a nonzero element b of $\mathfrak{b}$ such that $\delta(b) = 0$. Then it is easy to show that $A[b^{-1}]$ is étale and finite over $B[b^{-1}]$. □

In the nonétale case, we can show the following result.

Theorem 2.2.20. *Assume that A and B are normal affine domains over k. Assume further that there exists a nonzero ideal $\mathfrak{b}$ of B satisfying the conditions:*
(1) *The ideal $\mathfrak{b}$ has height ≥ 2 and satisfies one of the following conditions:*
   (i) *$\mathfrak{b}$ is a δ-ideal, i. e., $\delta(\mathfrak{b}) \subset \mathfrak{b}$.*
   (ii) *$\mathfrak{b}$ is generated by finitely many elements $b_1, \ldots, b_m$ such that $\delta(b_i) = 0$ for every $1 \leq i \leq m$.*
(2) *The associated morphism $\operatorname{Spec} A \to \operatorname{Spec} B$ is étale outside $V(\mathfrak{b})$.*

Then the extension Δ of δ on K induces an lnd of A.

Proof. Suppose condition (i) holds. Then $\operatorname{ht}(\mathfrak{b}A) \geq 2$ because $p : X \to Y$ is a finite morphism. Let $\mathcal{D}er_k(X)$ be the coherent $\mathcal{O}_X$-Module $\mathcal{H}om_{\mathcal{O}_X}(\Omega_{X/k}, \mathcal{O}_X)$. Then, by [63, Corollary 5.10.6], the canonical homomorphism

$$\Gamma(X, \mathcal{D}er_k(X)) \to \Gamma(X - V(\mathfrak{b}A), \mathcal{D}er_k(X))$$

is an isomorphism. Since $\Delta_{X-V(\mathfrak{b}A)}$ is the lifting of $\delta_{Y-V(\mathfrak{b})}$, we have $\Delta_{X-V(\mathfrak{b}A)} \in \Gamma(X - V(\mathfrak{b}A), \mathcal{D}er_k(X))$ by Lemma 2.2.17. Hence $\Delta_{X-V(\mathfrak{b}A)}$ extends to a k-derivation $\Delta' \in \Gamma(X, \mathcal{D}er_k(X))$. Since both Δ and Δ' restricted on the function field K is the extension of δ, it follows that $\Delta' = \Delta$. This implies that $\Delta(A) \subset A$. Since $\mathfrak{b}$ is a nonzero δ-ideal, there exists a nonzero element b of $\mathfrak{b}$ such that $\delta(b) = 0$. Then by condition (2), $A[b^{-1}]$ is étale over $B[b^{-1}]$. Hence the restriction $\Delta|_{A[b^{-1}]}$ is locally nilpotent. Since A is a subalgebra of $A[b^{-1}]$, it follows that Δ is locally nilpotent.

Suppose that $\mathfrak{b}$ satisfies the second condition (ii) of (1). Then $\mathfrak{b}$ is a δ-ideal. Hence the assertion follows from the above case (i). However, we can argue in the following fashion. Choose any b_i and denote it by b. Then the open set $D(b) \subset \operatorname{Spec} B \setminus V(\mathfrak{b})$. Hence $A[b^{-1}]$ is finite and étale over $B[b^{-1}]$ by condition (2). Then Δ is an lnd of $A[b^{-1}]$ by Lemma 2.2.17. In particular, $\Delta(A) \subset A[b^{-1}]$. Hence $\Delta(A) \subset \bigcap_{i=1}^m A[b_i^{-1}]$. Meanwhile, let $\mathfrak{p}$ be a prime ideal of B with $\operatorname{ht}(\mathfrak{p}) = 1$. Since $\operatorname{ht}(\mathfrak{b}) \geq 2$ by condition (1), $\mathfrak{p} \not\ni b_i$ for some i. Let $A_\mathfrak{p} := A \otimes_B B_\mathfrak{p}$. Then $A_\mathfrak{p} \supset A[b_i^{-1}]$. Hence $\Delta(A) \subset A_\mathfrak{p}$. Since we can take $\mathfrak{p}$ as an arbitrary prime ideal of B with height one, we have $\Delta(A) \subset \bigcap_{\operatorname{ht}(\mathfrak{p})=1} A_\mathfrak{p} = A$ since A is a finite B-module and B is normal. □

2.2.4 Liftability of derivations for finite extensions and Scheja–Storch theorem

We consider a k-derivation δ on an affine domain B which is not necessarily locally nilpotent and consider when δ is lifted to a k-derivation under a finite extension $B \hookrightarrow$

A of k-affine domains. Since a k-derivation δ is viewed as a vector field on an affine variety Spec B, the following result will generalize the results given in the previous subsection.

We retain the same notations and assumptions as above. In particular, A and B are assumed to be normal domains. Scheja–Storch [205] proved that $\Delta(A) \subseteq A$ if $\delta(\mathfrak{p}) \subseteq \mathfrak{p}$ for every height 1 prime divisor $\mathfrak{P}$ of $\mathfrak{R}$ and $\mathfrak{p} = \mathfrak{P} \cap B$. In particular, if $\Omega_{A/B} = (0)$, then Δ satisfies $\Delta(A) \subseteq A$, i. e., δ lifts to a derivation Δ of A. Vasconcelos [223] proved that if $\Delta(A) \subseteq A$ is satisfied, then Δ is locally nilpotent if so is δ. Following the line of Vasconcelos, we give a proof of a theorem of Scheja–Storch which is stated as follows.

Theorem 2.2.21. *Given a k-derivation $\delta : B \to B$, δ extends to a k-derivation $\Delta : A \to A$ if $\delta(\mathfrak{p}) \subsetneq \mathfrak{p}$ for every height one primed divisor $\mathfrak{p}$ of $\mathfrak{b}$, where $\mathfrak{b}$ is the reduced branch ideal of A over B.*

Our proof consists of several lemmas.

Lemma 2.2.22. *Let $\mathfrak{P}$ be a prime ideal of height one of A and let $\mathfrak{p} = \mathfrak{P} \cap B$. Then the following assertions hold:*
(1) $\Omega_{A/B} \otimes_A A_{\mathfrak{P}} \cong \Omega_{A_{\mathfrak{P}}/B_{\mathfrak{p}}}$.
(2) $\Omega_{A_{\mathfrak{P}}/B_{\mathfrak{p}}} = 0$ *if and only if* $\operatorname{Ann}(\Omega_{A/B}) \not\subset \mathfrak{P}$.
(3) *If* $\Omega_{A_{\mathfrak{P}}/B_{\mathfrak{p}}} = 0$, *then a k-derivation* $\delta : B \to B$ *extends to a k-derivation* $\Delta : A_{\mathfrak{P}} \to A_{\mathfrak{P}}$.

Proof. (1) See [63, 20.5.9].

(2) Since $\Omega_{A/B}$ is a finite A-module, this is a standard fact in commutative algebras.

(3) We have an exact sequence

$$\Omega_{B_{\mathfrak{p}}/k} \otimes_{B_{\mathfrak{p}}} A_{\mathfrak{P}} \to \Omega_{A_{\mathfrak{P}}/k} \to \Omega_{A_{\mathfrak{P}}/B_{\mathfrak{p}}} \to 0,$$

where $\Omega_{A_{\mathfrak{P}}/B_{\mathfrak{p}}} = 0$ by the assumption. Hence, for $a, u \in A$ with $u \notin \mathfrak{P}$, we have

$$d\left(\frac{a}{u}\right) = \sum_i \frac{a_i}{u_i} d\left(\frac{b_i}{v_i}\right),$$

where $a_i, u_i \in A$ with $u_i \notin \mathfrak{P}$ and $b_i, v_i \in B$ with $v_i \notin \mathfrak{p}$. Let $Q(A)$ (resp., $Q(B)$) be the quotient field of A (resp., B). Since the field extension $Q(A)/Q(B)$ is a separable algebraic extension, the derivation δ extends to a k-derivation $\Delta : A_{\mathfrak{P}} \to Q(A)$ which corresponds to a $A_{\mathfrak{P}}$-homomorphism $\varphi : \Omega_{A_{\mathfrak{P}}/k} \to Q(A)$ defined by

$$\Delta\left(\frac{a}{u}\right) = \varphi d\left(\frac{a}{u}\right).$$

Meanwhile, $\Delta(A_{\mathfrak{P}}) \subsetneq A_{\mathfrak{P}}$ because

$$\varphi d\left(\frac{a}{u}\right) = \sum_i \frac{a_i}{u_i} \varphi d\left(\frac{b_i}{v_i}\right) = \sum_i \frac{a_i}{u_i} \delta\left(\frac{b_i}{v_i}\right) \in A_{\mathfrak{P}}.$$ □

Lemma 2.2.23. *Let $\mathfrak{P}$ be as in Lemma 2.2.22. Then $\Omega_{A_{\mathfrak{P}}/B_{\mathfrak{p}}} = A_{\mathfrak{P}}dt$, where t is a uniformizant of a discrete valuation ring $A_{\mathfrak{P}}$.*

Proof. We have a commutative exact diagram:

$$\begin{array}{ccccccc}
0 & & 0 & & 0 & & \\
\uparrow & & \uparrow & & \uparrow & & \\
\mathfrak{P}A_{\mathfrak{P}}/(\mathfrak{p}A_{\mathfrak{P}}+\mathfrak{P}^2A_{\mathfrak{P}}) & \to & \Omega_{A_{\mathfrak{P}}/B_{\mathfrak{p}}} \otimes_{A_{\mathfrak{P}}} \overline{K} & \to & \Omega_{\overline{K}/\overline{L}} & \to & 0 \\
\uparrow & & \uparrow & & \uparrow & & \\
\mathfrak{P}A_{\mathfrak{P}}/\mathfrak{P}^2A_{\mathfrak{P}} & \to & \Omega_{A_{\mathfrak{P}}/k} \otimes_{A_{\mathfrak{P}}} \overline{K} & \to & \Omega_{\overline{K}/k} & \to & 0 \\
\uparrow & & \uparrow & & \uparrow & & \\
\mathfrak{p}B_{\mathfrak{p}}/\mathfrak{p}^2B_{\mathfrak{p}} \otimes_{\overline{L}} \overline{K} & \to & \Omega_{B_{\mathfrak{p}}/k} \otimes_{B_{\mathfrak{p}}} \overline{K} & \to & \Omega_{\overline{L}/k} \otimes_{\overline{L}} \overline{K} & \to & 0
\end{array}$$

where $\overline{K} = Q(A/\mathfrak{P}) = A_{\mathfrak{P}}/\mathfrak{P}A_{\mathfrak{P}}, L = Q(B/\mathfrak{p}) = B_{\mathfrak{p}}/\mathfrak{p}B_{\mathfrak{p}}$, and $\Omega_{\overline{K}/\overline{L}} = 0$. Thence we have a surjection

$$\mathfrak{P}A_{\mathfrak{P}}/(\mathfrak{p}A_{\mathfrak{P}}+\mathfrak{P}^2A_{\mathfrak{P}}) \longrightarrow \Omega_{A_{\mathfrak{P}}/B_{\mathfrak{p}}} \otimes_{A_{\mathfrak{P}}} \overline{K} \longrightarrow 0.$$

Note that $\Omega_{A_{\mathfrak{P}}/B_{\mathfrak{p}}} \cong \Omega_{A/B} \otimes_A A_{\mathfrak{P}}$ is a finite $A_{\mathfrak{P}}$-module and $\mathfrak{P}A_{\mathfrak{P}}/(\mathfrak{p}A_{\mathfrak{P}}+\mathfrak{P}^2A_{\mathfrak{P}}) = t(A_{\mathfrak{P}}/\mathfrak{P}A_{\mathfrak{P}})$. Hence we have $\Omega_{A_{\mathfrak{P}}/B_{\mathfrak{p}}} \cong A_{\mathfrak{P}}dt$ by Nakayama's lemma. □

Corollary 2.2.24. *$\Omega_{A_{\mathfrak{P}}/B_{\mathfrak{p}}}$ is a torsion $A_{\mathfrak{P}}$-module.*

Proof. Since $\mathfrak{P}$ has height one, $A_{\mathfrak{P}}$ and $B_{\mathfrak{p}}$ are discrete valuation rings such that $A_{\mathfrak{P}} \geq B_{\mathfrak{p}}$, where $(\mathcal{O}, \mathfrak{M}) \geq (\mathfrak{o}, \mathfrak{m})$ is the domination of local rings if and only if $\mathcal{O} \supseteq \mathfrak{o}$ and $\mathfrak{M} \cap \mathfrak{o} = \mathfrak{m}$. Let u be a uniformizant of $B_{\mathfrak{p}}$. Then $u = wt^r$ with $w \in (A_{\mathfrak{P}})^*$. Hence we have $rwt^{r-1}dt + t^r dw = 0$ in $\Omega_{A_{\mathfrak{P}}/B_{\mathfrak{p}}}$. By Lemma 2.2.23, $dw = zdt$ with $z \in A_{\mathfrak{P}}$. Hence $t^{r-1}(rw+tz)dt = 0$, where $rw + tz$ is invertible in $A_{\mathfrak{P}}$. So, $t^{r-1} \in \mathrm{Ann}(\Omega_{A_{\mathfrak{P}}/B_{\mathfrak{p}}})$. □

Lemma 2.2.25. *Let $\mathfrak{P}$ be a height one prime ideal of A and let $\mathfrak{p} = B \cap \mathfrak{P}$. Let $\delta : B_{\mathfrak{p}} \to B_{\mathfrak{p}}$ be a k-derivation and let $\Delta : A_{\mathfrak{P}} \to Q(A)$ be the natural extension of δ. Suppose that $\Omega_{A_{\mathfrak{P}}/B_{\mathfrak{p}}} \neq 0$. Then $\Delta : A_{\mathfrak{P}} \to A_{\mathfrak{P}}$ if $\delta(\mathfrak{p}B_{\mathfrak{p}}) \subseteq \mathfrak{p}B_{\mathfrak{p}}$. Furthermore, if $\Delta(A_{\mathfrak{P}}) \subseteq A_{\mathfrak{P}}$ and $\Delta(\mathfrak{P}A_{\mathfrak{P}}) \subseteq \mathfrak{P}A_{\mathfrak{P}}$ then it follows that $\delta(\mathfrak{p}B_{\mathfrak{p}}) \subseteq \mathfrak{p}B_{\mathfrak{p}}$.*

Proof. We prove the first assertion. With the same notations as in Corollary 2.2.24, write $u = wt^r$. If $r = 1$, then $A_{\mathfrak{P}}$ is unramified over $B_{\mathfrak{p}}$, whence $\Omega_{A_{\mathfrak{P}}/B_{\mathfrak{p}}} = 0$ and the assertion follows from Lemma 2.2.22(3). Thus $r \geq 2$. We have

$$du = t^r dw + rwt^{r-1}dt,$$

where $dt, dw \in \Omega_{A_{\mathfrak{P}}/k}$. By Lemma 2.2.23, we may write

$$dw = zdt + \sum_i a_i db_i,$$

with $z, a_i \in A_{\mathfrak{P}}$ and $b_i \in B_{\mathfrak{p}}$. This expression follows from an exact sequence

$$\Omega_{B_{\mathfrak{p}}/k} \otimes_{B_{\mathfrak{p}}} A_{\mathfrak{P}} \to \Omega_{A_{\mathfrak{P}}/k} \to \Omega_{A_{\mathfrak{P}}/B_{\mathfrak{p}}} \to 0.$$

Hence we have

$$du = t^{r-1}(rw + tz)dt + \sum_i a_i t^r db_i.$$

Write $\Delta = \varphi \cdot d$ with an $A_{\mathfrak{P}}$-homomorphism $\varphi : \Omega_{A_{\mathfrak{P}}/k} \to Q(A)$. Then

$$\delta(u) = t^{r-1}(rw + tz)\Delta(t) + \sum_i a_i t^r \delta(b_i).$$

Since $rw + tz \in (A_{\mathfrak{P}})^*$ and $\delta(u) = t^r z'$ with $z' \in A_{\mathfrak{P}}$ by the assumption $\delta(\mathfrak{p}B_{\mathfrak{p}}) \subseteq \mathfrak{p}B_{\mathfrak{p}}$, we have

$$\Delta(t) = t\Big(z' - \sum_i a_i \delta(b_i)\Big)/(rw + tz) \in \mathfrak{P}A_{\mathfrak{P}}.$$

Let $x \in A_{\mathfrak{P}}$. Then we can write

$$dx = a' dt + \sum_i a_i' d(b_i')$$

as above. Then $\Delta(x) = a'\Delta(t) + \sum_i a_i' \delta(b_i') \in A_{\mathfrak{P}}$.

Next, we show the converse. Suppose $\Delta(A_{\mathfrak{P}}) \subseteq A_{\mathfrak{P}}$ and $\Delta(\mathfrak{P}A_{\mathfrak{P}}) \subsetneqq \mathfrak{P}A_{\mathfrak{P}}$. Then $du = t^r dw + rwt^{r-1}dt$ as above with $r \geq 2$. Hence

$$\delta(u) = t^r \Delta(w) + rwt^{r-1}\Delta(t) \in tA_{\mathfrak{P}}.$$

Since $\delta(u) \in B_{\mathfrak{p}}$, we have $\delta(u) \in \mathfrak{P}A_{\mathfrak{P}} \cap B_{\mathfrak{p}} = \mathfrak{p}B_{\mathfrak{p}}$. □

Proof of Theorem 2.2.21. Suppose that $\delta(\mathfrak{p}) \subseteq \mathfrak{p}$ for every height one prime divisor $\mathfrak{p}$ of $\mathfrak{b}$. Let $\mathfrak{P}$ be a height one prime ideal of A. If $B \cap \mathfrak{P} \not\supseteq \mathfrak{b}$, then $\mathfrak{P} \not\supseteq \mathrm{Ann}(\Omega_{A/B})$. Hence $\Omega_{A_{\mathfrak{P}}/B_{\mathfrak{p}}} = 0$ and δ extends to $\Delta : A_{\mathfrak{P}} \to A_{\mathfrak{P}}$ by Lemma 2.2.22. Suppose that $\mathfrak{p} := B \cap \mathfrak{P} \supseteq \mathfrak{b}$. Then $\delta(\mathfrak{p}) \subseteq \mathfrak{p}$ by the assumption. Hence δ extends to $\Delta : A_{\mathfrak{P}} \to A_{\mathfrak{P}}$ by Lemma 2.2.25. Then the natural extension $\Delta : A \to Q(A)$ satisfies $\Delta(A) \subseteq \bigcap_{\mathfrak{P}} A_{\mathfrak{P}} = A$ because A is normal, where $\mathfrak{P}$ ranges over all height one prime ideals of A. □

2.2.5 Locally finite derivations

Let A be an affine domain over k and let D be a k-derivation of A. We call D a *locally finite derivation* (lfd, for short) or an *algebraic derivation* if the k-vector space $\sum_{n\geq 0} k \cdot D^n(a)$ has finite dimension for every element $a \in A$.

Example 2.2.26.

(1) Let $G = \operatorname{Spec} R$ be an affine algebraic group acting on an affine algebraic variety $X = \operatorname{Spec} A$ by a k-morphism $\sigma : G \times X \to X$. Then, for every element a of A, the

k-vector subspace $\sum_{g\in G(k)} k\cdot {}^g a$ in the ring A generated by all translates ${}^g a$ with $g \in G(k)$ has finite dimension, where $G(k)$ is the set of closed points of G. In fact, let $\sigma^* : A \to R\otimes A$ be the coaction. For $a\in A$, write

$$\sigma^*(a) = \sum_{i=1}^{n} f_i \otimes a_i, \quad f_i \in R, \quad a_i \in A.$$

Then ${}^g a = \sum_{i=1}^n \pi_g(f_i)\cdot a_i$, where $\pi_g : R \to k$ is the k-algebra homomorphism corresponding to a point $g\in G(k)$, and $\sum_{g\in G(k)} k\cdot {}^g a$ is a k-subspace of $\sum_{i=1}^n k\cdot a_i$, whence $\sum_{g\in G(k)} k\cdot {}^g a$ has finite dimension.

(2) In particular, let $G = G_a$. Then the coaction $\sigma^* : A \to A[t]$ is given by $\sigma^*(a) = \sum_{n\ge 0} \frac{1}{n!}\delta^n(a)t^n$, where δ is an lnd of A corresponding to σ. For $\lambda \in k = G_a(k)$, ${}^\lambda a = \sum_{n\ge0}\frac{1}{n!}\delta^n(a)\lambda^n$. By making use of the Vandermonde determinant, we have $\sum_{\lambda\in k} k\cdot {}^\lambda a = \sum_{n\ge 0} k\cdot \delta^n(a)$, which has finite dimension because $\delta^n(a) = 0$ for $n \gg 0$. Hence an lnd is an lfd.

(3) Note that if ξ is an element of the quotient field $Q(A)$, the k-vector space $\sum_{n\ge0} k\cdot \delta^n(\xi)$ does not necessarily have finite dimension. An example is an lnd $\delta = \partial/\partial x$ on a polynomial ring $A = k[x]$. If $\xi = 1/x$, it follows that $\sum_{g\in G_a(k)} k\cdot {}^g\xi = \sum_{n\ge0} k\cdot x^{-n}$, which is not of finite dimension.

(4) Let D be a Euler derivation on a polynomial ring $k[x_1,\dots,x_n]$ given by

$$D = \lambda_1 x_1 \frac{\partial}{\partial x_1} + \cdots + \lambda_n x_n \frac{\partial}{\partial x_n}, \quad \lambda_1,\dots,\lambda_n \in k.$$

Then D is an lfd, but not an lnd.

Let D be a nontrivial lfd on an affine domain A. Define a k-algebra homomorphism $\varphi_D : A \to A[[t]]$ by

$$\varphi_D(a) = \sum_{n\ge0}\frac{1}{n!}D^n(a)t^n, \quad a\in A.$$

For $\lambda\in k$, let $\Delta_\lambda = D - \lambda\cdot 1_A$, where 1_A is the identity morphism of A. We often write $\Delta_\lambda = D-\lambda$. Note that Δ_λ is not a derivation, but we can define φ_{Δ_λ} as Δ_λ is a k-endomorphism of A.

Lemma 2.2.27. *For every $\lambda\in k$, we have $\varphi_D = e^{\lambda t}\varphi_{\Delta_\lambda}$.*

Proof. Set $\Delta = \Delta_\lambda$ for simplicity. Then $D = \Delta + \lambda\cdot 1_A$. For $a \in A$, we compute as follows:

$$\begin{aligned}\varphi_D(a) &= \sum_{n\ge0}\frac{1}{n!}(\Delta+\lambda)^n(a)\,t^n \\ &= a\left(1+\lambda t + \frac{1}{2!}\lambda^2t^2 + \cdots + \frac{1}{n!}\lambda^n t^n + \cdots\right) \\ &\quad + \Delta(a)t\left(1+\lambda t+\cdots+\frac{1}{(n-1)!}\lambda^{n-1}t^{n-1}+\cdots\right) + \cdots\end{aligned}$$

$$
\begin{aligned}
&+\frac{1}{m!}\Delta^m(a)t^m\left(1+\lambda t+\cdots+\frac{1}{(n-m)!}\lambda^{n-m}t^{n-m}+\cdots\right)+\cdots\\
&=e^{\lambda t}\left\{a+\Delta(a)t+\frac{1}{2!}\Delta^2(a)t^2+\cdots+\frac{1}{m!}\Delta^m(a)t^m+\cdots\right\}\\
&=e^{\lambda t}\varphi_\Delta(a).
\end{aligned}
$$

□

For $\lambda \in k$, set

$$A_\lambda = \{a \in A \mid (D-\lambda)^n(a) = 0, \quad n \gg 0\}.$$

Then it is clear that $a \in A_\lambda$ if and only if $\varphi_{\Delta_\lambda}(a) \in A[t]$. In particular,

$$A_0 = \{a \in A \mid \varphi_D(a) \in A[t]\}.$$

By making an essential use of Lemma 2.2.27, we prove the following lemmas.

Lemma 2.2.28. *We have the following assertions:*
(1) *A_0 is a k-subalgebra of A, and D is an lnd of A_0.*
(2) *For every $\lambda \in k$, A_λ is an A_0-module.*

Proof. (1) The assertion follows from the fact that φ_D is a k-algebra homomorphism.
(2) Let $a_0 \in A_0$ and $a \in A_\lambda$. Then, by Lemma 2.2.27, we have

$$\varphi_D(a_0a) = \varphi_D(a_0)\varphi_D(a) = e^{\lambda t}\varphi_D(a_0)\varphi_{\Delta_\lambda}(a).$$

On the other hand, we have $\varphi_D(a_0a) = e^{\lambda t}\varphi_{\Delta_\lambda}(a_0a)$. Hence it follows that $\varphi_{\Delta_\lambda}(a_0a) = \varphi_D(a_0)\varphi_{\Delta_\lambda}(a) \in A[t]$, which implies that $\Delta_\lambda^n(a_0a) = 0$ for some $n > 0$. □

Lemma 2.2.29. *For $\lambda, \mu \in k$, $A_\lambda A_\mu \subseteq A_{\lambda+\mu}$.*

Proof. Let $a \in A_\lambda$ and $b \in A_\mu$. Then we have

$$\varphi_D(ab) = \varphi_D(a)\varphi_D(b) = e^{(\lambda+\mu)t}\varphi_{\Delta_\lambda}(a)\varphi_{\Delta_\mu}(b)$$

and

$$\varphi_D(ab) = e^{(\lambda+\mu)t}\varphi_{\Delta_{\lambda+\mu}}(ab).$$

Hence $\varphi_{\Delta_{\lambda+\mu}}(ab) = \varphi_{\Delta_\lambda}(a)\varphi_{\Delta_\mu}(b) \in A[t]$. This implies that $(D-\lambda-\mu)^n(ab) = 0$ for $n \gg 0$, and hence $ab \in A_{\lambda+\mu}$. □

Lemma 2.2.30. *If $\lambda \neq \mu$, then $A_\lambda \cap A_\mu = 0$.*

Proof. Suppose that k is a subfield of the complex number field $\mathbb{C}$. Let $a \in A_\lambda \cap A_\mu$. Then $\varphi_D(a) = e^{\lambda t}\varphi_{\Delta_\lambda}(a) = e^{\mu t}\varphi_{\Delta_\mu}(a)$. Hence we have

$$e^{(\lambda-\mu)t} = \frac{\varphi_{\Delta_\mu}(a)}{\varphi_{\Delta_\lambda}(a)},$$

where $\varphi_{\Delta_\lambda}(a), \varphi_{\Delta_\mu}(a) \in A[t]$. As explained below, the k-subalgebra generated by coefficients of these polynomials are embedded into $\mathbb{C}$. Hence the function in t on the right-hand side is a rational function on the complex domain $\mathbb{C}$. Meanwhile, the function on the left-hand side is an entire function which does not have zeros and poles. Hence it follows that $\varphi_{\Delta_\mu}(a)$ and $\varphi_{\Delta_\lambda}(a)$ are constants, i. e., $(D-\lambda)(a) = 0$ and $(D-\mu)(a) = 0$. So, $D(a) = \lambda a = \mu a$, which implies $a = 0$.

We may assume that k is a subfield of $\mathbb{C}$. Since A is finitely generated over k, A is isomorphic to the residue ring $k[x_1,\dots,x_n]/(f_1,\dots,f_m)$. Let $\alpha_1,\dots,\alpha_r$ be the coefficients appearing in polynomials $f_1,\dots,f_m \in k[x_1,\dots,x_n]$, and let k_0 be the field $\mathbb{Q}(\alpha_1,\dots,\alpha_r)$. Let

$$B = k_0[x_1,\dots,x_n]/(f_1,\dots,f_m).$$

Then $A = B \otimes_{k_0} k$. Adjoining to k_0 the elements λ, μ and the coefficients of a and $D(\overline{x}_i)$ for $1 \le i \le n$ when they are expressed in the forms of polynomials in $\overline{x}_1,\dots,\overline{x}_n$, where $\overline{x}_i$ is the residue class of x_i in B, we may assume that $\varphi_{\Delta_\lambda}(a)$ and $\varphi_{\Delta_\mu}(a)$ are polynomials in $\overline{x}_1,\dots,\overline{x}_n, t$ with coefficients in k_0. Since k_0 is finitely generated over $\mathbb{Q}$, we may embed k_0 into $\mathbb{C}$. Now the field $K = k_0(\overline{x}_1,\dots,\overline{x}_m)$ is finitely generated over k_0. Hence $K = k_0(\xi_1,\dots,\xi_r,\theta)$, where $\{\xi_1,\dots,\xi_r\}$ is a transcendece basis of the extension K/k_0, and K is a simple extension of $k_0(\xi_1,\dots,\xi_r)$ adjoined with a root θ of an irreducible equation $F(X) = 0$ with $F(X) \in k_0(\xi_1,\dots,\xi_r)[X]$. Then we can embed $k_0(\xi_1,\dots,\xi_r)$ into $\mathbb{C}$ and then embed $Q(B)$ into $\mathbb{C}$ by mapping θ to a suitable root of $F(X) = 0$ in $\mathbb{C}$ whose coefficients are considered as elements of $\mathbb{C}$. □

For an element $a \in A_\lambda$, we define the *λ-height* of a as the nonnegative integer r such that $(D-\lambda)^r(a) \neq 0$ and $(D-\lambda)^{r+1}(a) = 0$. Note that λ-height is defined only for elements of A_λ. The first structure theorem on an affine domain with an lfd is stated as follows.

Theorem 2.2.31. *Let D be an lfd on an affine k-domain A. Then $A = \bigoplus_{\lambda \in k} A_\lambda$, which is a graded ring over A_0. Furthermore, $D(A_\lambda) \subseteq A_\lambda$.*

Proof. Let $a \in A$. Then the vector space $V = \sum_{n \ge 0} kD^n(a)$ has finite dimension as D is an lfd. We may choose an integer $n > 0$ such that $\{a, D(a), D^2(a), \dots, D^{n-1}(a)\}$ is a k-basis of V and $D^m(a)$ is expressed as a linear combination of $a, D(a), \dots, D^{n-1}(a)$ for every $m \ge n$. Such a basis of V exists. In fact, suppose that $a, D(a), \dots, D^{m-1}(a)$ are linearly independent over k and $a, D(a), \dots, D^{m-1}(a)$, $D^m(a)$ are linearly dependent. Then $D^m(a)$ is a k-linear combination of $a, D(a), \dots, D^{m-1}(a)$. Hence every $D^r(a)$ with $r \ge m$ is a k-linear combination of $a, D(a), \dots, D^{m-1}(a)$.

Then D is a k-linear endomorphism on V and V decomposes into a direct sum $V = \bigoplus_\lambda V_\lambda$ where $V_\lambda = \{v \in V \mid (D-\lambda)^n(v) = 0, n \gg 0\}$ (Jordan decomposition). Since $V_\lambda \subseteq A_\lambda$, it follows that $a \in \bigoplus_\lambda A_\lambda$. The previous lemmas show that $A = \bigoplus_\lambda A_\lambda$ is a graded ring over A_0. As for the second assertion, let $a \in A_\lambda$. Let r be the λ-height of a. If $r = 0$, then $D(a) = \lambda a \in A_\lambda$. Suppose that $r > 0$. Since $(D-\lambda)^{r+1}(a) = 0$, the element

$(D - \lambda)(a) \in A_\lambda$ and the λ-height of $(D - \lambda)(a)$ is $r - 1$. By induction, we may assume $(D - \lambda)(a) \in A_\lambda$. Then $D(a) = \lambda a + (D - \lambda)(a) \in A_\lambda$. So, $D(A_\lambda) \subseteq A_\lambda$. □

We look into the properties of the subring A_0.

Lemma 2.2.32. *Let D be an lfd on a normal affine domain A. Then A_0 is integrally closed.*

Proof. Suppose that $\xi \in Q(A_0)$ is integral over A_0. Then it follows that $\xi \in A$ since A is normal. The element $\xi \in A$ satisfies a monic equation

$$\xi^n + \alpha_1 \xi^{n-1} + \cdots + \alpha_n = 0, \quad \alpha_i \in A_0.$$

Hence it follows that

$$\varphi_D(\xi)^n + \varphi_D(\alpha_1)\varphi_D(\xi)^{n-1} + \cdots + \varphi_D(\alpha_n) = 0.$$

So, $\varphi_D(\xi)$ is integral over $k[\varphi_D(\alpha_1), \ldots, \varphi_D(\alpha_n)] \subset A[t]$. Since $A[t]$ is integrally closed and $\varphi_D(\xi) \in Q(A[t])$, it follows that $\varphi_D(\xi) \in A[t]$. Hence $\xi \in A_0$. □

We observe the following two examples. One deals with the Euler derivation and the other does a composite of the Euler derivation and a locally nilpotent derivation.

Example 2.2.33.

(1) Let $A = k[x_1, \ldots, x_n]$ be a polynomial ring in n variables and let $D = \sum_{i=1}^n x_i \frac{\partial}{\partial x_i}$ be the Euler derivation on A. Then D is an lfd and $A = \bigoplus_{d \geq 0} A_d$, where A_d is the k-vector space spanned by monomials $x_1^{m_1} \cdots x_n^{m_n}$ of total degree d. In particular, $A_0 = k$. Meanwhile, $\operatorname{Ker} \widetilde{D}$ is $k(\frac{x_2}{x_1}, \ldots, \frac{x_n}{x_1})$ which has transcendence degree $n - 1$ over k, where $\widetilde{D}$ is the natural extension of D onto $Q(A)$.

(2) Let $A = k[x, y]$ and let $D = x\frac{\partial}{\partial x} + \frac{\partial}{\partial y}$. For nonnegative integers m and n, we have

$$D(x^n y^m) = n x^n y^m + m x^n y^{m-1}. \tag{2.5}$$

Hence for any nonnegative integer r, we have $D^r(x^n y^m) \in k \cdot x^n y^m + k \cdot x^n y^{m-1} + \cdots + k x^n$. Thus it follows that the derivation D on A is an lfd. By (2.5), we have $(D - n)(x^n y^m) = m x^n y^{m-1}$ and $(D - n)^{m+1}(x^n y^m) = 0$. Hence if $f(y) \in k[y]$ with $\deg f(y) = m$, then $(D - n)^{m+1}(x^n f(y)) = 0$ and $x^n f(y) \in A_n$. Since any element $a \in A$ is written as $a = f_0(y) + x f_1(y) + \cdots + x^n f_n(y)$ where $f_i(y) \in k[y]$, it follows that $A = \bigoplus_{n \in \mathbb{Z}_{\geq 0}} A_n$ and $A_n = x^n k[y]$. The ring A is graded by the monoid $\mathbb{Z}_{\geq 0}$ of nonnegative integers.

Let D be an lfd on a normal affine domain A over k. Let $\Lambda = \{\lambda \in k \mid A_\lambda \neq 0\}$, which is a submonoid of $(k, +)$ under the addition, and let M be the abelian subgroup of $(k, +)$ generated by Λ. We call Λ (resp., M) the *monoid* (resp., *abelian group*) associated to D.

Theorem 2.2.34. *With the notations and assumptions as above, assume that M is a totally ordered abelian group with ordering $<$ in the sense that $\lambda < \mu$ implies $\lambda + \nu < \mu + \nu$ for all $\nu \in M$. Then the following assertions hold:*

(1) $Q(A_0) \cap A = A_0$.
(2) *$Q(A_0)$ is algebraically closed in $Q(A)$. If $A_0 \subsetneq A$, then we have* $\mathrm{tr.deg}_k Q(A_0) < \mathrm{tr.deg}_k Q(A)$.
(3) *A_0 is factorially closed in A. Hence if A is factorial, then so is A_0.*
(4) *Assume that A is a polynomial ring of* $\dim \le 3$. *Then A_0 is either a polynomial ring or k.*

Proof. (1) Suppose that $a \in Q(A_0) \cap A$. Then a is written as $a_0 a = a_1$ with $a_0, a_1 \in A_0$. By Theorem 2.2.31, $a = a_{\lambda_1} + \cdots + a_{\lambda_n}$ with $a_{\lambda_i} \in A_{\lambda_i}$ for $1 \le i \le n$. Since $a_0 a = a_0 a_{\lambda_1} + \cdots + a_0 a_{\lambda_n} = a_1$, it follows that $a_{\lambda_i} = 0$ except for some i and $a = a_{\lambda_i} \in A_0$.

(2) Suppose that $\xi \in Q(A)$ is algebraic over $Q(A_0)$. Then there exists $a_0 \in A_0$ such that $a_0\xi$ is integral over A_0. Since $A_0 \subset A$ and A is integrally closed by assumption, $a_0\xi \in A$. The element $a_0\xi$ satisfies

$$(a_0\xi)^n + c_1(a_0\xi)^{n-1} + \cdots + c_n = 0, \quad c_i \in A_0. \tag{2.6}$$

Write

$$a_0\xi = a_{\lambda_1} + \cdots + a_{\lambda_m}, \quad a_{\lambda_1} \ne 0, \quad a_{\lambda_m} \ne 0, \quad \lambda_1 < \cdots < \lambda_m.$$

If $\lambda_m > 0$, then the term $a_{\lambda_m}^n$, which has the highest order, cannot be canceled by the other terms in equation (2.6). Hence $\lambda_m \le 0$. Similarly, $\lambda_1 \ge 0$. So, $a_0\xi \in A_0$ and $\xi \in Q(A_0)$. Suppose that $A_0 \subsetneq A$. Then there exists an element $a \in A_\lambda$ with $\lambda \ne 0$. Then a is algebraically independent over $Q(A_0)$. For otherwise, a is algebraic over $Q(A_0)$ and hence $a \in Q(A_0)$. Then $a \in Q(A_0) \cap A = A_0$, which contradicts the choice of a.

(3) Suppose that $a_0 = a_1 a_2 \in A_0$ with $a_1, a_2 \in A \setminus \{0\}$. Write $a_1 = a_{\lambda_1} + \cdots + a_{\lambda_r}$ and $a_2 = a_{\mu_1} + \cdots + a_{\mu_s}$, where $a_{\lambda_i} \in A_{\lambda_i}$ and $a_{\mu_j} \in A_{\mu_j}$ with $\lambda_1 < \cdots < \lambda_r$ and $\mu_1 < \cdots < \mu_s$. Then the highest term of $a_1 a_2$ is $a_{\lambda_r} a_{\mu_s}$ and the lowest is $a_{\lambda_1} a_{\mu_1}$. Hence $\lambda_r + \mu_s = \lambda_1 + \mu_1 = 0$. It follows that $a_1, a_2 \in A_0$ and A_0 is factorially closed in A.

(4) We may assume that $A_0 \subsetneq A$ and $A_0 \ne k$. Then $\mathrm{tr.deg}_k Q(A_0) < \mathrm{tr.deg}_k Q(A) \le 3$ by assertion (2). Since $\dim A \le 3$, it follows by assertion (1) and Zariski finiteness theorem (Theorem 2.2.4) that A_0 is an affine domain. Let $X = \operatorname{Spec} A$ and $Y = \operatorname{Spec} A_0$. Then the inclusion $A_0 \hookrightarrow A$ induces a dominant morphism $q : X \to Y$. Since A is factorial and $A^* = k^*$, A_0 is factorial by assertion (3) and $A_0^* = k^*$. If $\dim A_0 = 1$, then A_0 is a polynomial ring. Suppose that $\dim A_0 = 2$. Then q is equidimensional over $q(X)$. In fact, suppose that there is an irreducible fiber component of dimension 2 in q. Then there exists a prime element $p \in A$ such that $pA \cap A_0 = \mathfrak{m}$ is a maximal ideal. Then $pa_1 = a \in A_0$. Since A_0 is factorially closed in A, it follows that $p \in A_0$. Hence $pA \cap A_0 = pA_0$, which is a contradiction. Then it follows that Y is isomorphic to $\mathbb{A}^2$ or a Platonic fiber space $\mathbb{A}^2/G$ (see [154, Theorem 3]). If G is nontrivial, q splits to $X \to \mathbb{A}^2 \to \mathbb{A}^2/G = Y$. In fact, Y has a unique singular point, say P, and the smooth locus Y° has the universal covering $\mathbb{A}^2 \setminus \{O\}$, where O is the point of origin. Since $q^{-1}(P)$

is either empty or of dimension one, $\mathbb{A}^3 \setminus q^{-1}(O)$ is simply connected. Hence the fiber product

$$Z = (\mathbb{A}^3 \setminus q^{-1}(P)) \times_{Y^\circ} (\mathbb{A}^2 \setminus \{O\})$$

splits into a disjoint union of copies of $\mathbb{A}^3 \setminus q^{-1}(P)$. Then the restriction of the second projection $p_2 : Z \to \mathbb{A}^2 \setminus \{O\}$ to a connected component of Z provides the above splitting of q. This contradicts the fact that $Q(A_0)$ is algebraically closed in $Q(A)$. □

The following example shows subtlety of a derivation being an lfd.

Example 2.2.35. Let $A = k[x, y]$ and

$$D = x^{n-1}y^n\left(x\frac{\partial}{\partial x} - y\frac{\partial}{\partial y}\right) \quad (n \geq 1).$$

We have by computation

$$\begin{aligned} D(x) = x^n y^n, \quad D^2(x) = 0, \quad &\text{and} \\ D^k(y) = (-1)^k k! x^{k(n-1)} y^{kn+1} \quad &(k \geq 1). \end{aligned}$$

Thus $\sum_{j=0}^{\infty} k \cdot D^j(y)$ is infinite-dimensional, and D is not an lfd.

Next, compute $D^\ell(x^i y^j)$. For $0 \leq i < j$, we have

$$D^\ell(x^i y^j) = (i - j)(i - j - 1) \cdots (i - j - \ell + 1) x^{i+\ell(n-1)} y^{j+\ell n}.$$

For $i = j$, we have $D(x^i y^i) = 0$. For $i > j$, it follows that $x^i y^j = x^{i-j}(xy)^j \in A_0$. Hence $A_0 = k[x, xy]$. Therefore, we have $Q(A_0) = Q(A)$ and

$$Q(A_0) \cap A = Q(A) \cap A = A \supsetneq A_0.$$

If we write $z = xy$, then $A_0 = k[x, z]$ and $D|_{A_0} = z^n \frac{\partial}{\partial x}$. □

2.3 Proper G_a-actions and Seshadri theorem

Contrary to its actions on affine varieties, G_a-actions on projective varieties present different features, much of which are related to the property of vector fields. As the notation, the fiber product of k-schemes X, Y, the fiber product $X \times_{\operatorname{Spec} k} Y$ (or $X \times_k Y$) is denoted simply by $X \times Y$ without reference to $\operatorname{Spec} k$ or k. Similarly, the tensor product of k-algebras A, B is often denoted by $A \otimes B$ instead of $A \otimes_k B$.

We first consider proper G_a-actions on affine algebraic varieties. As defined in Definition 2.2.10, a G_a-action σ on an affine algebraic variety X is called *proper* (resp., *free*) if the *graph morphism* $\Phi : G_a \times X \to X \times X$ defined by $(t, Q) \mapsto (\sigma(t, Q), Q)$ is a proper (hence finite as Φ is an affine morphism) morphism (resp., a closed immersion). By definition, a free action is a proper action.

We define the *isotropy subgroup scheme* F_X of $G_a \times X$ as the base change of $\Phi : G_a \times X \to X \times X$ by the diagonal morphism $\Delta : X \to X \times X$, i. e., $F_X = (G_a \times X, \Phi) \times_{X \times X} (X, \Delta)$. In Definition 2.2.10, it is called the stabilizer subgroup scheme. If the action is proper, then F_X is a finite group scheme over X because Φ is a finite morphism. Since $G_a \times X$ has no finite subgroup scheme (hence $F_X = X$), it follows that *if the G_a-action on X is proper, then the action is fixed-point free.*

In the case of a G_a-action on an affine variety, it is not straightforward to judge if the given action is proper or free. The following result summarizes the known criteria.

Lemma 2.3.1. *Let $\sigma : G_a \times X \to X$ be a nontrivial G_a-action on an affine variety $X = \operatorname{Spec} A$ and $\Phi^* : A \otimes A \to A[t]$ the algebra homomorphism associated with the graph morphism Φ, where t is a parameter of G_a. Let $R = \operatorname{Im} \Phi^*$ which is a subalgebra of $A[t]$. Then the first four conditions are equivalent:*
(1) *σ is a proper action.*
(2) *$A[t]$ is integral over R.*
(3) *The element t is integral over R.*
(4) *Given a commutative diagram of ring homomorphisms with a discrete valuation ring $\mathcal{O}$ and the quotient field $Q(\mathcal{O})$,*

$$\begin{array}{ccc} A[t] & \xrightarrow[\mu]{} & Q(\mathcal{O}) \\ {\scriptstyle i}\uparrow & & \uparrow{\scriptstyle j} \\ R & \xrightarrow[\nu]{} & \mathcal{O}, \end{array}$$

we have $\mu(t) \in j(\mathcal{O})$.

Furthermore the following two conditions are equivalent:
(5) *$t \in R$.*
(6) *σ is a free action.*

The last two conditions imply the first four conditions.

Proof. (1) ⇒ (2). Since Φ is a proper affine morphism, it is a finite morphism. Hence $A[t]$ is a finite $A \otimes A$-module. This is equivalent to saying that $A[t]$ is a finite R-module. Hence $A[t]$ is integral over R.

(2) ⇒ (3). Note that R contains A since $\Phi^*(1 \otimes a) = a$ for every $a \in A$. Since $A[t]$ is integral over R, so is t over R.

(3) ⇒ (4). Since t is integral over R, there is a monic equation

$$t^n + r_1 t^{n-1} + \cdots + r_i t^{n-i} + \cdots + r_n = 0, \quad r_i \in R. \tag{2.7}$$

We denote the images of r_i and t in $\mathcal{O}$ and $Q(\mathcal{O})$ by the same letters. Let ν be the normalized valuation of $\mathcal{O}$. Suppose $\nu(t) < 0$. Then (2.7) gives $\nu(t^n) \geq \min\{\nu(r_1 t^{n-1}), \ldots, \nu(r_n)\}$, where $\nu(r_i) \geq 0$. But this is impossible. Hence $\nu(t) \geq 0$ and $\mu(t) \in \mathcal{O}$.

(4) ⇒ (1). This is a valuative criterion of properness (see [89]).

The given action is free if and only if $\Phi^* : A \otimes_k A \to A[t]$ is surjective. The rest of the proof is obvious. □

By Definition 2.2.14, a G_a-action on an affine variety $X = \operatorname{Spec} A$ is q-tight if $A \otimes_B A$ is an integral domain, where B is the invariant subring A^{G_a}. Since $A \otimes_B A$ is the residue ring of $A \otimes_k A$ by the ideal generated by $\{b \otimes 1 - 1 \otimes b \mid b \in B\}$, $A \otimes_B A$ is finitely generated over k. If the variety X is defined over a field K which is not necessarily algebraically closed, we say that the G_a-action is *geometrically q-tight* if the induced action $\sigma \otimes_K \overline{K}$ on $\overline{X} = X \otimes_K \overline{K}$ is q-tight, where $\overline{K}$ is an algebraic closure of K. We also call a G_a-action *q-flat* if A is a flat B-module.

Lemma 2.3.2. *Let $X = \operatorname{Spec} A$ be an affine variety defined over K with a G_a-action σ. Then the following conditions are equivalent:*
(1) *σ is q-tight.*
(2) *Let K' be a finite algebraic extension of the field K and let $X' = X \otimes_K K'$. Then the induced G_a-action σ' on X' is q-tight.*
(3) *σ is geometrically q-tight.*

If σ is q-flat then it is q-tight.

Proof. (1) ⇒ (3). Note that $\overline{B} = B \otimes_K \overline{K}$ is the invariant subring of the induced G_a-action on $\overline{A} = A \otimes_K \overline{K}$. In fact, $0 \longrightarrow B \longrightarrow A \overset{\delta}{\longrightarrow} A$ is an exact sequence of K-modules. Since $\overline{K}$ is K-flat, we have an exact sequence

$$0 \longrightarrow B \otimes_K \overline{K} \longrightarrow A \otimes_K \overline{K} \xrightarrow{\delta \otimes_K \overline{K}} A \otimes_K \overline{K},$$

whence follows the assertion. Here we show that $A \otimes_B A$ is realized as a subdomain of $A[t]$. In fact, there exists an element $u \in A$ such that $b = \delta(u)$ is a nonzero element of B. Then we have $A[b^{-1}] = B[b^{-1}][u]$. In terms of K-algebra homomorphism, the graph morphism is given by $\varphi = \Phi^*$, where

$$\varphi : A \otimes_K A \longrightarrow A[t], \quad \varphi(a_1 \otimes a_2) = \sum_{i \geq 0} \frac{1}{i!} \delta^i(a_1) a_2 t^i,$$

which is decomposed by the canonical homomorphism ι^* as follows:

$$A \otimes_K A \overset{\iota^*}{\longrightarrow} A \otimes_B A \overset{\varphi_B}{\longrightarrow} A[t], \quad \varphi_B(a_1 \otimes a_2) = \sum_{i \geq 0} \frac{1}{i!} \delta^i(a_1) a_2 t^i.$$

Note that

$$\begin{aligned}(A \otimes_B A) \otimes_B B[b^{-1}] &= (A \otimes_B B[b^{-1}]) \otimes_{B[b^{-1}]} (A \otimes_B B[b^{-1}]) \\ &= B[b^{-1}][u] \otimes_{B[b^{-1}]} B[b^{-1}][u] \cong B[b^{-1}][u_1, u_2],\end{aligned}$$

where $u_1 = u \otimes 1$ and $u_2 = 1 \otimes u$. Since $\varphi_{B[b^{-1}]}(u_1) = u + bt$ and $\varphi_{B[b^{-1}]}(u_2) = u$, it is clear that $\varphi_{B[b^{-1}]}$ is injective. We have a commutative diagram

$$\begin{array}{ccc} A \otimes_B A & \xrightarrow{\varphi_B} & A[t] \\ \lambda \downarrow & & \downarrow \\ A \otimes_B A \otimes_B B[b^{-1}] & \xrightarrow[\varphi_{B[b^{-1}]}]{} & A[b^{-1}][t], \end{array}$$

where the localization $\lambda : A \otimes_B A \to A \otimes_B A \otimes_B B[b^{-1}]$ is injective since $A \otimes_B A$ is an integral domain by assumption. Hence φ_B is injective. This entails that $\overline{A} \otimes_{\overline{B}} \overline{A} = A \otimes_B A \otimes_K \overline{K}$ is a subring of $\overline{A}[t]$, hence it is an integral domain. This implies that the G_a-action σ is geometrically q-tight if it is q-tight.

(1) ⇒ (2) and (2) ⇒ (3) are similar.

(3) ⇒ (1). Since $A \otimes_B A \hookrightarrow \overline{A} \otimes_{\overline{B}} \overline{A}$ is an inclusion as $\overline{K}$ is K-flat, $A \otimes_B A$ is an integral domain if so is $\overline{A} \otimes_{\overline{B}} \overline{A}$.

The last assertion is proved in Lemma 2.2.15. □

It is shown by Dufresne–Kraft [45] that, for the G_a-action on $\mathbb{A}^7$ in Example 2.2.8, $X \times_T X$ with $X = \operatorname{Spec} A$ and $T = \operatorname{Spec} A^{G_a}$ has two irreducible components. Hence the G_a-action is not q-tight.

Meanwhile, Lemma 2.2.13 says that, for an action $\sigma : G \times X \to X$ of an algebraic group G on an affine variety $X = \operatorname{Spec} A$, $X \times_Y X = \operatorname{Spec}(A \otimes_B A)$ with $B = A^G$ provided the action σ has a geometric quotient $q : X \to Y$, where q factorizes the canonical morphism $p : X \to T$ induced by the inclusion $B \hookrightarrow A$ as

$$p : X \xrightarrow{q} Y \xrightarrow{q'} T.$$

Here we must note that the notations Y and T are used to respectively denote the geometric and algebraic quotients of X by G_a. This use of notations might be slightly confusing with the use in Chapter 2.

The morphism q' is not necessarily an isomorphism. One can expect that q' is an open immersion under suitable conditions on the action σ and the variety X. For this end, we recall *Seshadri theorem* [210, Theorem 6.1 and Remark 6.1], which we state in the case where a connected algebraic group G is replaced by G_a.

Theorem 2.3.3. *Let X be a normal algebraic variety with a proper G_a-action. Then there exists a finite morphism $\nu : Z \to X$ such that*

(i) *Z is a normal variety with G_a-action and ν is a surjective G_a-equivariant morphism.*
(ii) *G_a acts freely on Z and the geometric quotient W of Z by G_a exists as an algebraic variety. Further the quotient morphism $p_Z : Z \to W$ is a locally trivial principal fiber bundle with group G_a.*
(iii) *The extension $k(Z)/k(X)$ of the function fields is a Galois extension with Galois group Γ, and the canonical action of Γ on Z commutes with the G_a-action.*

(iv) *If every finite set of points in W is contained in an affine open set, e.g., if W is quasiprojective, then the geometric quotient Y of X by G_a exists.*

In order to state an application of the above theorem, we need a preparatory result.

Lemma 2.3.4. *Let $X = \operatorname{Spec} A$ be an affine variety defined over a field K of characteristic zero. Assume that X is equipped with a proper and q-tight G_a-action σ. Let δ be the associated lnd of A and let $B = \operatorname{Ker}\delta$. We further assume that B is an affine K-domain. Let $T = \operatorname{Spec} B$ and let $q : X \to T$ be the algebraic quotient morphism. Then the graph morphism Φ splits as*

$$\Phi : G_a \times X \xrightarrow{\Phi_{X/T}} X \times_T X \xrightarrow{\iota} X \times X,$$

where ι is a closed immersion and $\Phi_{X/T}$ is a proper and dominant morphism.

Proof. The homomorphism $\varphi_B : A \otimes_B A \to A[t]$ is injective as proved in the proof of Lemma 2.3.2. Hence $\Phi_{X/T} : G_a \times X \to X \times_T X$ is a dominant morphism since $\Phi_{X/T} = {}^a\varphi_B$. Further, note that the canonical homomorphism $\iota^* : A \otimes_K A \to A \otimes_B A$ is surjective, whence ι is a closed immersion. Since $\Phi = \iota \cdot \Phi_{X/T}$ is a proper morphism, so is the morphism $\Phi_{X/T}$. □

We can now state the following result.

Theorem 2.3.5. *Let $X = \operatorname{Spec} A$ be a normal affine variety with a G_a-action. Assume that the G_a-action on X is proper and q-tight and that the G_a-invariant subring B is finitely generated over k. Then the geometric quotient Y of X by G_a exists. The quotient morphism $p : X \to Y$ is a locally trivial principal fiber bundle with group G_a. Further, the variety Y is realized as an open set of $T = \operatorname{Spec} B$ such that* $\operatorname{codim}_T(T \setminus Y) \geq 2$.

Proof. We use the notations of Theorem 2.3.3. The geometric quotient W has an action of a finite group Γ and Y is isomorphic to the geometric quotient W/Γ if it exists. By condition (iv) of Theorem 2.3.3, for the existence of Y, it suffices to show that W is quasiprojective. Write $Z = \operatorname{Spec}\widetilde{A}$ and $\widetilde{B} = \widetilde{A}^{G_a}$. Then $A = \widetilde{A}^{\Gamma}$ and $B = \widetilde{B}^{\Gamma}$. Then $\widetilde{B}$ is integral over B because $f(x) := \prod_{\gamma\in\Gamma}(x - {}^{\gamma}\widetilde{b}) \in B[x]$ and $f(\widetilde{b}) = 0$ for every $\widetilde{b} \in \widetilde{B}$. Let $\widehat{B}$ be the normalization of B in the function field $k(Z)$. Then $\widehat{B}$ is a finite B-module and contains $\widetilde{B}$. Since B is noetherian, $\widetilde{B}$ is a finite B-module. Hence $\widetilde{B}$ is finitely generated over k. Since $\widetilde{A}$ is normal, $\widetilde{B}$ is normal, too. Let $U = \operatorname{Spec}\widetilde{B}$. Then we have $Z \times_W Z \cong Z \times_U Z$ by Lemma 2.2.13.

We claim that W is an open set of U such that $\operatorname{codim}_U(U \setminus W) \geq 2$. This claim implies that W is quasiaffine, hence quasiprojective. We note that $\widetilde{A} \otimes_{\widetilde{B}} \widetilde{A}$ is a finite $A \otimes_B A$-module. In fact, $\widetilde{A} \otimes_k \widetilde{A}$ is a finite $A \otimes_k A$-module. Hence $\widetilde{A} \otimes_{\widetilde{B}} \widetilde{A}$ as the residue ring of $\widetilde{A} \otimes_k \widetilde{A}$ is a finite $A \otimes_k A$-module. But $\widetilde{A} \otimes_{\widetilde{B}} \widetilde{A}$ is a $A \otimes_k A$-module via the residue ring $A \otimes_B A$. Hence $\widetilde{A} \otimes_{\widetilde{B}} \widetilde{A}$ is a finite $A \otimes_B A$-module. On the other hand, we have the following commutative diagram:

$$\begin{array}{ccc} G_a \times Z \cong Z \times_W Z & \xrightarrow{\Phi_{Z/U}} & Z \times_U Z \\ {\scriptstyle (\mathrm{id}_{G_a}, \nu)} \downarrow & & \downarrow {\scriptstyle (\nu,\nu)} \\ G_a \times X & \xrightarrow[\Phi_{X/T}]{} & X \times_T X, \end{array}$$

where $T = \operatorname{Spec} B$. Here $\Phi_{X/T}$ is a finite surjective morphism because it is proper and dominant by Lemma 2.3.4, and (id_{G_a}, ν) is a finite surjective morphism and (ν, ν) is a finite morphism. If the canonical morphism $\tau : W \to U$ is not quasifinite, i. e., there exists an irreducible curve C in W which is mapped to a point of U, take a local lift (as a section) $\widetilde{C}$ of C in Z after shrinking C to a small open set. Fix a point $\widetilde{P} \in \widetilde{C}$. Then the set $\{(\widetilde{P}, \widetilde{Q}) \mid \widetilde{Q} \in \widetilde{C},\ p_Z(\widetilde{P}) \neq p_Z(\widetilde{Q})\}$ is an infinite set of $(Z \times_U Z) \setminus \Phi_{Z/U}(Z \times_W Z)$ which is not possible because $\Phi_{Z/U}$ is an isomorphism as shown above. Then $\tau : W \to U$ is a birational quasifinite morphism. Since W and U are normal, τ is an open immersion by Zariski's main theorem, and $\operatorname{codim}_U(U \setminus W) \geq 2$ because $\Gamma(U, \mathcal{O}_U) = \overline{B} = \Gamma(W, \mathcal{O}_W)$.

The geometric quotient Y is realized as $Y = W/\Gamma$. Then $p : X \to Y$ is a principal G_a-bundle. In fact, W has an affine open covering $\mathcal{W} = \{W_i \mid i \in I\}$ such that each affine open set W_i is Γ-stable and makes $p_Z^{-1}(W_i)$ trivial, i. e., there is a Γ-isomorphism $p_Z^{-1}(W_i) \cong G_a \times W_i$. Then $\{W_i/\Gamma \mid i \in I\}$ is an affine open covering of Y such that $p_Z^{-1}(W_i)/\Gamma \cong p^{-1}(W_i/\Gamma)$ and $(p_Z^{-1}(W_i) \times_{W_i} p_Z^{-1}(W_i))/\Gamma \cong G_a \times (p_Z^{-1}(W_i)/\Gamma)$. Hence we have $G_a \times X \cong X \times_Y X$. Here the quotients by Γ are all algebraic quotients.

The last assertion can be verified by the same argument as above for W and U. □

Corollary 2.3.6. *Let $X = \operatorname{Spec} A$ be a normal affine variety equipped with a proper and q-tight G_a-action. Assume that the ring B of G_a-invariants is an affine domain. Let $q : X \to T$ be the algebraic quotient morphism, where $T = \operatorname{Spec} B$. Then every non-empty fiber of q is a G_a-orbit. Hence there are no G_a-fixed points on X.*

Proof. By Theorem 2.3.5, there exists a geometric quotient $p : X \to Y$ such that $q = \iota \cdot p$ with an open immersion $\iota : Y \to T$. The first assertion follows from this remark and the definition of geometric quotient. By the upper semicontinuity of fiber dimension of p, there are no G_a-fixed points. □

In algebraic terms, a proper and q-tight G_a-action is characterized in the following way.

Proposition 2.3.7. *Let $X = \operatorname{Spec} A$ be a smooth factorial*[7] *affine variety defined over k equipped with a nontrivial G_a-action σ. Let δ be the associated lnd on A and let $B = \operatorname{Ker} \delta$. Assume that B is an affine k-domain. Then the G_a-action is proper and q-tight if and only*

7 We need the factoriality condition of A only to prove $b_i = \delta(u_i)$ in the "*only if*" part below. Instead of factoriality, we can assume $\delta(u_i) \in B[b_i^{-1}]^*$.

if there exist elements $b_1, \ldots, b_r \in B$ and elements $u_1, \ldots, u_r \in A$ such that the following conditions are satisfied:

(1) *$b_1, \ldots, b_r$ generate a unitary ideal in A, i. e., $(b_1, \ldots, b_r)A = A$.*
(2) *For every i, $b_i = \delta(u_i)$ and $A[b_i^{-1}] = B[b_i^{-1}][u_i]$.*

Proof. Let $q : X \to T$ be the quotient morphism, where $T = \operatorname{Spec} B$. Consider the *if* part. Let $W_i = D(b_i) = \operatorname{Spec} B[b_i^{-1}]$. Then $q^{-1}(W_i) \cong W_i \times \mathbb{A}^1$ and the G_a-action is a translation given by $\sigma^*(u_i) = u_i + b_i t$, where u_i is the fiber coordinate. Let $W = \bigcup_{i=1}^r W_i$. Condition (1) implies that $\bigcup_{i=1}^r q^{-1}(W_i) = X$, and condition (2) implies that q splits as $q : X \xrightarrow{q_W} W \hookrightarrow T$, where q_W is an $\mathbb{A}^1$-bundle with a free G_a-action. Hence $q_W : X \to W$ is a geometric quotient of X by G_a, and $X \times_W X = X \times_T X$ by Lemma 2.2.13. Hence the graph morphism Φ splits as

$$\Phi : G_a \times X \cong X \times_W X = X \times_T X \xrightarrow{\iota} X \times X,$$

where ι is a closed immersion. This implies that the G_a-action σ is proper and q-tight.

Consider next the "*only if*" part. With the same notations as in Theorem 2.3.5, $q : X \to T$ is a flat morphism because $q = \iota \cdot p$ with $p : X \to Y$ an $\mathbb{A}^1$-fiber bundle over Y and an open immersion $\iota : Y \to T$. Hence there exists an affine open covering $\mathcal{W} = \{W_i \mid i \in I\}$ of Y such that $W_i = D(b_i)$ for some element $b_i \in B$ and $p^{-1}(W_i) \cong W_i \times \mathbb{A}^1$. Since $p^{-1}(\mathcal{W}) = \{p^{-1}(W_i) \mid i \in I\}$ is an open covering of X and since X is quasicompact, we can choose the index set I to be a finite set $I = \{1, 2, \ldots, r\}$. Namely, $(b_1, \ldots, b_r)A = A$. We have $A[b_i^{-1}] = B[b_i^{-1}][u_i]$ for some element $u_i \in A$. We have $\delta(u_i) \in B[b_i^{-1}]^*$ because the G_a-action on $\operatorname{Spec} A[b_i^{-1}]$ is free. We now use the factoriality condition. Since $B[b_i^{-1}]^*$ is generated by the unit group B^* and the elements $b_{ij}^{\pm 1}$ with b_{ij} exhausting all prime factors of b_i, we have $\delta(u_i) = c_i / \prod_{j=1}^{r_i} b_{ij}^{s_{ij}}$ with $c_i \in B^*$. To avoid complexity, we consider only the case where b_i is irreducible. So, $\delta(u_i) = c_i / b_i^{s_i}$. If $s_i < 0$, replace u_i by u_i / c_i and b_i by $1/b_i^{s_i}$. Suppose $s_i \geq 0$. Then replace u_i by $u_i b_i^{s_i+1}/c_i$. Then we have $\delta(u_i) = b_i$ and the condition (2) is satisfied. □

As shown in the above proposition, the existence of a proper and q-tight G_a-action on an affine variety X will greatly affect the structure of X. Hence there are not many varieties with proper and q-tight G_a-actions. Before giving various examples, we need a q-tightness criterion. We shall state it after preparing two lemmas.

Let $q : X \to T$ be a morphism of schemes. We say that q is *equidimensional* if all nonempty fibers X_y of q have the same dimension. We use the following result due to Zariski [235, Lemma 4] and which we refer to as *Zariski's lemma*.

Lemma 2.3.8. *Let $(\mathfrak{o}, \mathfrak{m})$ be a complete semilocal ring of characteristic zero (with $\mathfrak{m}$ denoting the Jacobson radical which is the intersection of the maximal ideals of $\mathfrak{o}$) and let D be a derivation of $\mathfrak{o}$ with values in $\mathfrak{o}$. Assume that there exists an element t in $\mathfrak{m}$ of $\mathfrak{o}$ such that Dt is a unit in $\mathfrak{o}$. Then $\mathfrak{o}$ contains a ring $\mathfrak{o}_1$ of representatives of the (complete) local ring $\mathfrak{o}/\mathfrak{o}t$, having the following properties:*

(a) *D is zero on $\mathfrak{o}_1$.*
(b) *t is analytically independent over $\mathfrak{o}_1$.*
(c) *$\mathfrak{o}$ is the power series ring $\mathfrak{o}_1[[t]]$.*

It follows that t is not a zero divisor of $\mathfrak{o}$, and hence $\dim \mathfrak{o}_1 = \dim \mathfrak{o} - 1$.

Let $X = \operatorname{Spec} A$ be a smooth affine variety with a G_a-action and let $x \in X \setminus X^{G_a}$ be a closed point and let $y = q(x)$, where X^{G_a} is the fixed point locus and $q : X \to T = \operatorname{Spec} B$ is the algebraic quotient morphism. Then the lnd δ associated to the G_a-action satisfies $\delta(\mathfrak{M}) \not\subset \mathfrak{M}$, where $\mathfrak{M} = \mathfrak{m}_{X,x}$ is the maximal ideal of the local ring $\mathcal{O}_{X,x}$.[8] By Lemma 2.3.8, the completion $\widehat{\mathcal{O}}_{X,x}$ is written as $\widehat{\mathcal{O}}_{X,x} = \mathfrak{o}'[[t]]$, where $\mathfrak{o}'$ is a regular complete local ring with maximal ideal $\mathfrak{m}'$ and t is an element of $\mathfrak{M}$ such that $\delta(t) \notin \mathfrak{M}$. It is clear that $\widehat{\mathcal{O}}_{T,y} \subseteq \mathfrak{o}'$ for $y = q(x) \in T$. We prove the following result.

Lemma 2.3.9. *With the above notations, assume that $q^{-1}(y)$ has dimension one at x. Then the following assertions hold:*
(1) *If $q^{-1}(x)$ is reduced at x, then $\mathfrak{o}'$ is isomorphic to $\widehat{\mathcal{O}}_{T,y}$.*
(2) *In the nonreduced case, the ring $\mathfrak{o}'$ is integral over $\widehat{\mathcal{O}}_{T,y}$.*

Proof. (1) Let $(\mathfrak{o}, \mathfrak{m})$ (resp., $(\mathcal{O}, \mathcal{M})$) be the local ring $\widehat{\mathcal{O}}_{T,y}$ (resp., $\widehat{\mathcal{O}}_{X,x}$). Then we have inclusions of local rings $\mathfrak{o} \subseteq \mathfrak{o}' \subset \mathcal{O}$. Note that the ideal $\mathfrak{m}\mathcal{O}$ defines the fiber $q^{-1}(y)$ in $\mathcal{O}$. Since $q^{-1}(y)$ is reduced, $\mathcal{O}/\mathfrak{m}\mathcal{O}$ is a one-dimensional reduced local ring and $\mathfrak{o}'/\mathfrak{m}\mathfrak{o}'$ is an Artin local subring. Hence we have $\mathfrak{m}\mathfrak{o}' = \mathfrak{m}'$. Since both $\mathfrak{o}$ and $\mathfrak{o}'$ are complete local rings with the same reidue fields, we have $\mathfrak{o} = \mathfrak{o}'$.

(2) First we claim that $\mathfrak{m}\mathfrak{o}'$ is primary for $\mathfrak{m}'$, where $\mathfrak{m} = \widehat{\mathfrak{m}}_{T,y}$ as above. Indeed, if not, we have $\operatorname{ht}(\mathfrak{m}\mathfrak{o}') < \operatorname{ht}(\mathfrak{m}')$. But we have

$$\dim \mathfrak{o}' = \dim \widehat{\mathcal{O}}_{X,x} - 1 \quad \text{and} \quad \operatorname{ht}(\mathfrak{m}\widehat{\mathcal{O}}_{X,x}) = \dim \widehat{\mathcal{O}}_{X,x} - 1.$$

Since $\widehat{\mathcal{O}}_{X,x}$ is $\mathfrak{o}'$-flat, $\operatorname{ht}(\mathfrak{m}\mathfrak{o}') = \operatorname{ht}(\mathfrak{m}\widehat{\mathcal{O}}_{X,x})$. Hence we obtain

$$\operatorname{ht}(\mathfrak{m}\mathfrak{o}') = \operatorname{ht}(\mathfrak{m}\widehat{\mathcal{O}}_{X,x}) < \operatorname{ht}(\mathfrak{m}'\widehat{\mathcal{O}}_{X,x}) < \operatorname{ht}(\mathfrak{M}) = \dim \widehat{\mathcal{O}}_{X,x}.$$

This is a contradiction. Next we use the following result from [175].

Let (R, M) be a complete local ring and let N be an R-module which satisfies $\bigcap_{n \geq 0} M^n N = (0)$. If N/MN is a finite R/M-module, then N is a finite R-module.

Apply this result to $R = \widehat{\mathcal{O}}_{T,y}$ and $N = \mathfrak{o}'$. Since $\mathfrak{m}\mathfrak{o}'$ is primary for $\mathfrak{m}'$, $\mathfrak{o}'/\mathfrak{m}\mathfrak{o}'$ is a finite $\widehat{\mathcal{O}}_{T,y}/\mathfrak{m}$-module. Hence $\mathfrak{o}'$ is a finite $\widehat{\mathcal{O}}_{T,y}$-module. Hence $\mathfrak{o}'$ is integral over $\widehat{\mathcal{O}}_{T,y}$. □

8 In fact, let $\mathfrak{m}$ be the maximal ideal of A corresponding to x. Then $\mathfrak{M} = \mathfrak{m}\mathcal{O}_{X,x}$ and $x \notin X^{G_a}$ if and only if $\mathfrak{m} \not\supset \delta(A)$. Note that δ extends naturally to a k-derivation of $\mathcal{O}_{X,x}$ and $\mathcal{O}_{X,x} = k + \mathfrak{M}$. Hence $x \notin X^{G_a}$ if and only if $\delta(\mathfrak{M}) \not\subset \mathfrak{M}$.

We have the following criterion for q-tightness.

Lemma 2.3.10. *Let $X = \operatorname{Spec} A$ be a smooth affine variety with a fixed-point free G_a-action such that the invariant subring $B = A^{G_a}$ is finitely generated over k. Let $q : X \to T = \operatorname{Spec} B$ be the algebraic quotient morphism. We assume that q is equidimensional and all its (schematic) fibers are reduced. Let $U = q(X)$. Then the following assertions hold:*
(1) *T is smooth at every point y of U, U is an open set of T and q is smooth.*
(2) *The G_a-action is q-tight.*

Proof. (1) Let $y \in U$ and let $x \in q^{-1}(y)$. Since x is not a fixed point, we have $\delta(\mathfrak{m}_{X,x}) \not\subset \mathfrak{m}_{X,x}$. Hence, by Lemma 2.3.8, $\widehat{\mathcal{O}}_{X,x} = \mathfrak{o}'[[t]]$, where $\mathfrak{o}'$ is isomorphic to $\widehat{\mathcal{O}}_{T,y}$ by Lemma 2.3.9 because $q^{-1}(y)$ is reduced by the assumption. Since $\widehat{\mathcal{O}}_{X,x}$ is a formal power series ring over k, $\mathfrak{o}'$ is a regular complete local ring with the residue field k by flat descent. Hence T is smooth at y, and q is flat. This implies that U is an open set of T. The morphism q is then smooth because $\widehat{\mathcal{O}}_{X,x}/\mathfrak{m}_{T,y}\widehat{\mathcal{O}}_{X,x} = \mathfrak{o}'[[t]]/\mathfrak{m}'[[t]] \cong k[[t]]$ at every point x of $q^{-1}(y)$.

(2) Since q is flat, the q-tightness of the G_a-action follows by Lemma 2.3.2. The following is a geometric argument. Let $q_X : X \times_T X \to X$ be the base change of $q : X \to T$. Then q_X is smooth since $q : X \to T$ is smooth by (1) above. Hence $X \times_T X$ is smooth since X is smooth by assumption. Then every irreducible component is a connected component. We show that every point (x, x') of $X \times_T X$ is connected to a point (x'', x'') with x'' lying on a smooth fiber of q. This implies that $X \times_T X$ itself is irreducible. Hence $A \otimes_B A$ is an integral domain, and the G_a-action is q-tight.

Let $y = q(x) = q(x')$. Since $q(X)$ is contained in the smooth locus of T, let C be a smooth affine curve such that $y \in C$ and $C \subset q(Y)$. Let $S = q^{-1}(C)$. Since q is smooth, S is a smooth affine surface with the induced $\mathbb{A}^1$-fibration. We may further assume that $q^{-1}(y)$ is the unique reducible fiber in S. Suppose that $x \in L$ and $x' \in L'$, where L and L' are connected components of $q^{-1}(y)$. Let S_1 (resp., S_2) be the surface $S \setminus (q^{-1}(y) \setminus L)$ (resp., $S \setminus (q^{-1}(y) \setminus L')$). Then S_1 and S_2 are $\mathbb{A}^1$-bundles over C, and $S_1 \times_C S_2$ is a locally closed subscheme of $X \times_T X$. Since $S_1 \times_C S_2$ is an $\mathbb{A}^2$-bundle, it is clear that the point (x, x') of $S_1 \times_C S_2$ is connected to a point (x'', x'') in $S_1 \times_C S_2$, hence also in $X \times_T X$. □

The ring $\mathfrak{o}'$ is not necessarily equal to $\widehat{\mathcal{O}}_{Y,y}$ if $q^{-1}(y)$ is not reduced at x as shown by the following example [162].

Example 2.3.11. Take a Danielewski surface $\widetilde{W} \subset \mathbb{A}^3 = \operatorname{Spec} k[x, y, z]$ defined by the equation $z^2 = 2(xy + 1)$. It has an $\mathbb{A}^1$-fibration $\widetilde{\rho} : \widetilde{W} \to \mathbb{A}^1$ defined by $(x, y, z) \mapsto x$. All the fibers of $\widetilde{\rho}$, except for the one over the point $x = 0$, are isomorphic to $\mathbb{A}^1$, while $\widetilde{\rho}^{-1}(0)$ consists of two disjoint lines $\{x = z - \sqrt{2} = 0\}$ and $\{x = z + \sqrt{2} = 0\}$. Also $\widetilde{W}$ has an involution $\sigma : (x, y, z) \mapsto (-x, -y, -z)$. Let $W = \operatorname{Spec} A$ be the quotient surface $\widetilde{W}/\langle\sigma\rangle$, where $A = k[x^2, y^2, xy, xz, yz]$. Note that $z^2 = 2(xy + 1) \in A$. Let $X = x^2, Y = y^2, Z =$

$xy, U = xz$, and $V = yz$. The k-derivation $\widetilde{\delta}$ on $k[x,y,z]$ defined by $\widetilde{\delta}(x) = 0, \widetilde{\delta}(y) = z$, and $\widetilde{\delta}(z) = x$ induces a locally nilpotent k-derivation δ on A such that

$$\delta(X) = 0, \delta(Y) = 2V, \delta(Z) = U, \delta(U) = X, \delta(V) = 2 + 3Z.$$

In fact, W is isomorphic to the complement of a smooth conic in $\mathbb{P}^2$ and the inclusion $k[X] \hookrightarrow A$ defines an $\mathbb{A}^1$-fibration $\rho : W \to \mathbb{A}^1$ for which the fiber over $X = 0$ has the coordinate ring $\overline{A} = A/XA$. Let u, v be the residue classes of U, V modulo XA. Then $\overline{A} = k[u, v]$ with $u^2 = 0$ (see [162, Lemma 2.4] for the computation). The derivation δ on A induces a locally nilpotent derivation $\overline{\delta}$ on $\overline{A}$ given by $\overline{\delta}(u) = 0$ and $\overline{\delta}(v) = 2 + \frac{3}{2}uv$. So, $R := \operatorname{Ker} \overline{\delta} = k[u]$ with $u^2 = 0$. This implies that the fiber over $X = 0$ is a nonreduced G_a-orbit with multiplicity 2.

Let $\mathfrak{M}$ be the maximal ideal (X, Y, Z, U, V) of A and let $\mathfrak{m} = \mathfrak{M} \cap B = (X)$ with $B = k[X]$. Let $\widehat{\mathcal{O}}$ be the $\mathfrak{M}$-adic completion of A and let $\widehat{\delta}$ be the extension of δ to $\widehat{\mathcal{O}}$. It is then shown in [162] that $\operatorname{Ker} \widehat{\delta}$ contains a series

$$F = U - \frac{1}{2}XV + \frac{3}{8}XZV - \frac{15}{16}X^2YV + \frac{45}{32}X^2YZV + \frac{5}{16}X^2V^3 + \cdots,$$

which is not contained in the $\mathfrak{m}$-adic completion $k[[X]]$ of B. □

Now we observe various G_a-actions.

Example 2.3.12.

(1) Note that a fixed-point free G_a-action is not necessarily a proper action as witnessed by a G_a-action on a Danielewski surface $V = \{xy = z^2 - 1\}$ given by $(t, (x, y, z)) \mapsto (x, y + 2zt + xt^2, z + xt)$. In fact, the algebraic quotient is the affine line $\mathbb{A}^1 = \operatorname{Spec} k[x]$ and the quotient morphism $q : V \to \mathbb{A}^1$ has two disjoint fiber components over the point (0). The G_a-action on these components is given by $(0, y, \pm 1) \mapsto (0, y \pm 2t, \pm 1)$. Hence the action of G_a on the fiber $q^{-1}(0)$ is not transitive. In fact, the surface has no proper G_a-actions (see Makar-Limanov [138]). Note that the G_a-action is q-tight by Lemma 2.3.10.

(2) Deveney–Finston [36] gave the following example of a proper G_a-action on $\mathbb{A}^5 = \operatorname{Spec} k[x_1, x_2, y_1, y_2, z]$ whose orbits are not always separable by G_a-invariant functions. The example in terms of an lnd is

$$\delta = x_1 \frac{\partial}{\partial x_2} + y_1 \frac{\partial}{\partial y_2} + (1 + x_1 y_1^2) \frac{\partial}{\partial z}.$$

Then the G_a-action is not q-tight. In fact, if it is q-tight, Theorem 2.3.5 implies that the geometric quotient U of $\mathbb{A}^5$ by G_a exists as an open set of the algebraic quotient $\mathbb{A}^5 /\!/ G_a$ and the orbits are separable by invariant functions. In [36], it is shown that the quotient $\mathbb{A}^5 /\!/ G_a$ is a hypersurface in $\mathbb{A}^5 = \operatorname{Spec} k[u_1, u_2, u_3, u_4, u_5]$ defined by an equation $u_2 u_5 - u_1^2 u_4 - u_3^3 - 3u_1 u_3 = 0$. Its singular locus is the set defined by $u_1 = u_2 = u_3 = u_5 = 0$, and for each point $(0, 0, 0, \alpha, 0)$ of the singular locus, the fiber of $q : \mathbb{A}^5 \to \mathbb{A}^5 /\!/ G_a$ is isomorphic to $\mathbb{A}^2$.

(3) The algebraic group SL(2, k) has a proper G_a-action σ, which is the right action of the upper triangular unipotent group $T \cong G_a$. The coordinate ring A of SL(2, k) is $k[x, y, z, u]/(xu - yz = 1)$ and the G_a action is given by

$$\begin{pmatrix} x & z \\ y & u \end{pmatrix}\begin{pmatrix} 1 & t \\ 0 & 1 \end{pmatrix} = \begin{pmatrix} x & z + xt \\ y & u + yt \end{pmatrix}.$$

So the corresponding lnd δ on A is given by $\delta(x) = \delta(y) = 0, \delta(z) = x$ and $\delta(u) = y$. Now $B = \operatorname{Ker} \delta = k[x, y]$ and the quotient morphism $q : X \to Y$ has the image $U = \mathbb{A}^2 \setminus \{(0, 0)\}$. The open set U is covered as $U = D(x) \cup D(y)$ and $q^{-1}(D(x)) \cong D(x) \times \mathbb{A}^1$ and $q^{-1}(D(y)) \cong D(y) \times \mathbb{A}^1$ because $u = \frac{y}{x}z + \frac{1}{x}$ and $z = \frac{x}{y}u - \frac{1}{y}$. Hence the action σ is proper by Proposition 2.3.7. In fact, we can write $t = (u\sigma^*(z) - uz) - (z\sigma^*(u) - uz)$.

(4) More generally, let G be a connected reductive algebraic group defined over k and let H be a subgroup which is isomorphic to G_a. Let $q : G \to G/H$ be the quotient morphism by the natural right H-action on G. Then G is an H-torsor over G/H. Hence $G \times H \cong G \times_{(G/H)} G$ and the graph morphism $\Phi : H \times G \to G \times G$ splits as

$$H \times G \cong G \times_{(G/H)} G \longrightarrow G \times G,$$

which is a closed immersion. Hence the H-action on G is a proper action. But the quotient G/H is not affine by a theorem of Matsushima [147]. □

2.4 G_a-actions on projective varieties

One cannot expect the existence of affine space fibrations on projective varieties, but can expect dense open sets in projective varieties which have affine space fibrations. In order to find such open sets, it is important to consider algebraic group actions, notably, unipotent group actions on projective varieties. A G_a-action on an affine variety is described in terms of a locally nilpotent derivation. But, a G_a-action on a projective variety or a complete variety has a different feature which is not seen in the case of an affine variety. In this section, we consider actions of G_a or, more generally, of a unipotent group on projective varieties, first by examples and next by introducing a notion of a vector field with locally nilpotent stratification.

2.4.1 Vector fields on projective varieties

In this subsection, we consider global vector fields on $\mathbb{P}^n$ or on the Hirzebruch surface $\mathbb{F}_n$ ($n \geq 0$) in terms of vector fields on the affine space $\mathbb{A}^n$ naturally embedded into $\mathbb{P}^n$ or $\mathbb{A}^2$ into $\mathbb{F}_n$.

First of all, in the case of the natural embedding $\mathbb{A}^n \hookrightarrow \mathbb{P}^n$, we consider a system of homogeneous coordinates $(X_0, X_1, \ldots, X_n)$ on $\mathbb{P}^n$ and set $x_i = X_i/X_0$ for $1 \leq i \leq n$.

Lemma 2.4.1. *Let Δ be a regular vector field on $\mathbb{A}^n$ and write*

$$\Delta = f_1\frac{\partial}{\partial x_1} + \cdots + f_n\frac{\partial}{\partial x_n} \quad \textit{with } f_1,\ldots,f_n \in k[x_1,\ldots,x_n].$$

Then Δ extends to a regular vector field on $\mathbb{P}^n$ if and only if

$$\begin{aligned} f_1 &= a_1x_1^2 + a_2x_1x_2 + \cdots + a_nx_1x_n + \sum_{i=1}^n b_{1i}x_i + c_1,\\ f_2 &= a_1x_1x_2 + a_2x_2^2 + \cdots + a_nx_2x_n + \sum_{i=1}^n b_{2i}x_i + c_2,\\ &\cdots\\ f_n &= a_1x_1x_n + a_2x_2x_n + \cdots + a_nx_n^2 + \sum_{i=1}^n b_{ni}x_i + c_n, \end{aligned}$$

where a_i $(1 \le i \le n)$, b_{ij} $(1 \le i,j \le n)$, and c_i $(1 \le i \le n)$ are elements of k which we can take arbitrarily. Hence $\dim \Gamma(\mathbb{P}^n, \mathcal{T}_{\mathbb{P}^n}) = n(n+2)$.

Proof. To avoid complicated computations, we exhibit the idea in the case $n = 2$. We set $x = X_1/X_0$ and $y = X_2/X_0$. Let $U_i = \{X_i \neq 0\}$ $(i = 0,1,2)$ be the open sets of $\mathbb{P}^2$ isomorphic to $\mathbb{A}^2$. Hence $U_0 = \operatorname{Spec} k[x,y]$. Let $U_1 = \operatorname{Spec} k[u,v]$ with $u = X_0/X_1 = x^{-1}$ and $v = X_2/X_1 = yx^{-1}$. Write $f = f_1$ and $g = f_2$. Assume that Δ is a regular vector field on $\mathbb{P}^2$. Write

$$\Delta = \xi\frac{\partial}{\partial u} + \eta\frac{\partial}{\partial v}, \qquad \xi,\eta \in k[u,v].$$

Since $\xi = \Delta(u) = -u^{2-d}u^d f(\frac{1}{u},\frac{v}{u})$ with $d = \deg_{x,y} f$, we have $d \le 2$. Similarly, on $U_2 = \operatorname{Spec} k[z,w]$, we have $z = X_0/X_2 = y^{-1}$ and $w = X_1/X_2 = xy^{-1}$. Writing $\Delta = \varphi\frac{\partial}{\partial z} + \psi\frac{\partial}{\partial w}$, we have $\varphi = \Delta(z) = -z^{2-e}z^e g(\frac{w}{z},\frac{1}{z})$ with $e = \deg_{x,y} g$. Hence $e \le 2$, and we can write

$$\begin{aligned} f &= a_0x^2 + a_1xy + a_2y^2 + c_0x + c_1y + c_2,\\ g &= b_0x^2 + b_1xy + b_2y^2 + d_0x + d_1y + d_2. \end{aligned}$$

Then we have

$$\begin{aligned} \xi &= -u^2\left(\frac{a_0}{u^2} + a_1\frac{v}{u^2} + a_2\frac{v^2}{u^2} + \frac{c_0}{u} + c_1\frac{v}{u} + c_2\right)\\ &= -(a_0 + a_1v + a_2v^2 + c_0u + c_1uv + c_2u^2),\\ \eta &= -\frac{v}{u}\cdot u^2\left(\frac{a_0}{u^2} + a_1\frac{v}{u^2} + a_2\frac{v^2}{u^2} + \frac{c_0}{u} + c_1\frac{v}{u} + c_2\right)\\ &\quad + \frac{1}{u}\cdot u^2\left(\frac{b_0}{u^2} + b_1\frac{v}{u^2} + b_2\frac{v^2}{u^2} + \frac{d_0}{u} + d_1\frac{v}{u} + d_2\right)\\ &= -a_0\frac{v}{u} - a_1\frac{v^2}{u} - a_2\frac{v^3}{u} - v(c_0 + c_1v + c_2u)\\ &\quad + \frac{b_0}{u} + b_1\frac{v}{u} + b_2\frac{v^2}{u} + d_0 + d_1v + d_2u. \end{aligned}$$

Hence $b_0 = 0, b_1 = a_0, b_2 = a_1$, and $a_2 = 0$. So, f and g are as stated above for $n = 2$ and $x = x_1, y = x_2$. □

Remark 2.4.2. There is an exact sequence of $\mathcal{O}_{\mathbb{P}^n}$-modules (see [89, Example 8.20.1])

$$0 \to \mathcal{O}_{\mathbb{P}^n} \to \mathcal{O}_{\mathbb{P}^n}(1)^{\oplus(n+1)} \to \mathcal{T}_{\mathbb{P}^n} \to 0.$$

Since $\mathrm{H}^1(\mathbb{P}^n, \mathcal{O}_{\mathbb{P}^n}) = 0$, we have $\dim \mathrm{H}^0(\mathbb{P}^n, \mathcal{T}_{\mathbb{P}^n}) = (n+1)^2 - 1 = n(n+2)$. □

Let $V = \mathbb{F}_n$ be the Hirzebruch surface of degree n and let M be a minimal section. The affine plane $\mathbb{A}^2$ can be embedded into V as the complement $V \setminus (M \cup \ell_\infty)$, where ℓ_∞ is the fiber at infinity. We consider a regular vector field Δ on $\mathbb{A}^2$ and look for a condition with which Δ is extendable to a regular vector field on V.

Write $V = \mathrm{Proj}(\mathcal{O}_{\mathbb{P}^1} \oplus \mathcal{O}_{\mathbb{P}^1}(n))$ and let M be defined by the projection $\mathcal{O}_{\mathbb{P}^1} \oplus \mathcal{O}_{\mathbb{P}^1}(n) \to \mathcal{O}_{\mathbb{P}^1}$. To be more precise, let $\mathbb{P}^1 = U_0 \cup U_1$, where $U_0 = \mathrm{Spec}\, k[x]$ and $U_1 = \mathrm{Spec}\, k[x^{-1}]$. Then $\mathcal{O}_{\mathbb{P}^1}(n)|_{U_0} = \mathcal{O}_{U_0} e_1$ and $\mathcal{O}_{\mathbb{P}^1}(n)|_{U_1} = \mathcal{O}_{U_1} e_1'$, where $e_1' = x^n e_1$. We write the direct summand $\mathcal{O}_{\mathbb{P}^1}$ as $\mathcal{O}_{\mathbb{P}^1} e_0$ to give a base e_0. Then V is covered by four open sets $V = V_0 \cup V_1 \cup V_2 \cup V_3$, where

$$\begin{aligned}
V_0 &= \mathrm{Spec}\, k[x,y], \quad y = \frac{e_0}{e_1},\\
V_1 &= \mathrm{Spec}\, k[u,v], \quad u = \frac{1}{x}, \quad v = \frac{e_0}{e_1'} = \frac{y}{x^n},\\
V_2 &= \mathrm{Spec}\, k[x,z], \quad z = \frac{1}{y} = \frac{e_1}{e_0},\\
V_3 &= \mathrm{Spec}\, k[u,t], \quad u = \frac{1}{x}, \quad t = \frac{1}{v} = \frac{e_1'}{e_0} = \frac{x^n}{y}.
\end{aligned}$$

Write a regular vector field Δ on the open set V_0 as

$$\Delta = f\frac{\partial}{\partial x} + g\frac{\partial}{\partial y},$$

with $f, g \in k[x,y]$. Express Δ on the open sets V_1, V_2, V_3 in terms of the above respective coordinate systems and find the condition for Δ to be regular on each of the above open sets. The computations show the following result.

Theorem 2.4.3. *Embed $\mathbb{A}^2$ into $\mathbb{F}_n$ $(n \geq 0)$ as $\mathbb{A}^2 = \mathbb{F}_n \setminus (M \cup \ell_\infty)$. Let $\Delta = f\frac{\partial}{\partial x} + g\frac{\partial}{\partial y}$ be a regular vector field on $\mathbb{A}^2$. Then Δ is extendable to a regular vector field on $\mathbb{F}_n$ if and only if*

(1) $f(x,y) = a_{20}x^2 + a_{10}x + a_{00}$,

(2)

$$g(x,y) = \begin{cases} b_{n0}x^n + \cdots + b_{10}x + b_{00} + b_{01}y + b_{11}xy & (n > 0),\\ b_{02}y^2 + b_{01}y + b_{00} & (n = 0),\end{cases}$$

where $b_{11} = na_{20}$.

Hence $\dim \mathrm{H}^0(\mathbb{F}_n, \mathcal{T}_{\mathbb{F}_n})$ is equal to $n+5$ if $n > 0$ and 6 if $n = 0$.

Remark 2.4.4. Let V be $\mathbb{P}^2$ or $\mathbb{F}_n$. In Lemma 2.4.1 and Theorem 2.4.3, we tacitly used the coincidence of two k-vector spaces $\mathrm{H}^0(V, \mathcal{T}_V)$ and

$$\Gamma = \left\{ \Delta \in \Gamma(\mathbb{A}^2, \mathcal{T}_{\mathbb{A}^2}) \,\middle|\, \begin{array}{l} \text{a regular vector field on } \mathbb{A}^2 \text{ which is} \\ \text{extendable to a regular vector field on } V \end{array} \right\}$$

Since a given vector field on $\mathbb{A}^2$ is uniquely extendable to a rational vector field on V, where only the coefficients are restricted if it is regular on V, there is a natural correspondence θ which assigns Δ to itself

$$\theta : \Gamma \to \mathrm{H}^0(V, \mathcal{T}_V).$$

Then the correspondence is an isomorphism. In fact, an element $\Delta \in \mathrm{H}^0(\mathbb{A}^2, \mathcal{T}_{\mathbb{A}^2})$ is identified with a k-derivation of the function field $k(\mathbb{A}^2)$. So, an element of $\mathrm{H}^0(V, \mathcal{T}_V)$ is a k-derivation of $k(\mathbb{A}^2)$ which is regular on V. The extendability of a given vector field on $\mathbb{A}^2$ onto V depends on the embedding $\mathbb{A}^2 \hookrightarrow V$. The above remark applies if one replaces $\mathbb{A}^2 \hookrightarrow \mathbb{P}^2$ by $\mathbb{A}^n \hookrightarrow \mathbb{P}^n$.

The dimension of $\mathrm{H}^0(\mathbb{F}_n, \mathcal{T}_{\mathbb{F}^n})$ can be computed by the following result.

Proposition 2.4.5. *Let $V = \mathbb{F}_n$ ($n \geq 0$). Then we have an exact sequence*

$$0 \to \mathrm{H}^0(V, \mathcal{O}_V(2M + n\ell)) \to \mathrm{H}^0(V, \mathcal{T}_V) \to \mathrm{H}^0(\mathbb{P}^1, \mathcal{O}_{\mathbb{P}^1}(2)) \to 0,$$

where $h^0(V, \mathcal{O}_V(2M + n\ell)) = n + 2$ if $n > 0$ and $= 3$ if $n = 0$.

Proof. Let $p : V \to \mathbb{P}^1$ be the canonical projection. Then we have an exact sequence

$$0 \to p^*\Omega^1_{\mathbb{P}^1/k} \to \Omega^1_{V/k} \to \Omega^1_{V/\mathbb{P}^1} \to 0 \tag{2.8}$$

where $\Omega^1_{V/\mathbb{P}^1} \cong \mathcal{O}_V(-2M - n\ell)$. In fact, since $\Omega^2_{V/k} \cong \Omega^1_{V/\mathbb{P}^1} \otimes p^*\Omega^1_{\mathbb{P}^1/k}$, we have

$$\Omega^1_{V/\mathbb{P}^1} \cong \mathcal{O}_V(K_V) \otimes \mathcal{O}_V(2\ell) \cong \mathcal{O}_V(-2M - (n+2)\ell + 2\ell) \cong \mathcal{O}_V(-2M - n\ell).$$

Assume that $n > 0$. Taking the dual sequence of (2.8), we have an exact sequence

$$0 \to \mathcal{O}_V(2M + n\ell) \to \mathcal{T}_V \to p^*\mathcal{O}_{\mathbb{P}^1}(2) \to 0.$$

This yields an exact sequence

$$0 \to \mathrm{H}^0(\mathcal{O}_V(2M + n\ell)) \to \mathrm{H}^0(\mathcal{T}_V) \to \mathrm{H}^0(\mathbb{P}^1, \mathcal{O}_{\mathbb{P}^1}(2)) \to \mathrm{H}^1(\mathcal{O}_V(2M + n\ell)),$$

where $h^0(\mathcal{O}_V(2M + n\ell)) = n + 2$, $h^0(\mathbb{P}^1, \mathcal{O}_{\mathbb{P}^1}(2)) = 3$ and $h^1(\mathcal{O}_V(2M + n\ell)) = n - 1$ (see [163, Lemma 1.7]). Since $h^0(V, \mathcal{T}_V) = \dim \Gamma = n + 5$, we have the stated exact sequence.

Assume next that $n = 0$. Then $\mathcal{T}_V = p_1^*\mathcal{T}_{\mathbb{P}^1} \oplus p_2^*\mathcal{T}_{\mathbb{P}^1}$, which gives

$$h^0(\mathcal{T}_V) = 2h^0(\mathcal{T}_{\mathbb{P}^1}) = 6. \qquad \square$$

Determination of regular vector fields on a smooth algebraic surface is not so easy as for $\mathbb{A}^2$. As an example, we determine those for a Danielewski surface $X = \{xy = z^2 - 1\} \subset \mathbb{A}^3$. Let $K = k(X)$ and let Δ be a regular vector field on X. Then, as a derivation of $k(X)/k$, Δ is written as

$$\Delta = g\frac{\partial}{\partial x} + h\frac{\partial}{\partial z},$$

where $g = \Delta(x)$ and $h = \Delta(z)$. Since $y = x^{-1}(z^2 - 1)$, we have

$$f := \Delta(y) = -\frac{1}{x^2} \cdot xyg + \frac{2z}{x}h,$$

whence $xf + yg = 2zh$. Since Δ corresponds to $\delta \in \mathrm{Hom}_R(\Omega^1_{R/k}, R)$ by $\Delta = \delta \cdot d$ with $d : R \to \Omega^1_{R/k}$ being the universal derivation of R, where R is the coordinate ring of X, we have $f = \delta(dy)$, $g = \delta(dx)$, and $h = \delta(dz)$, whence $f, g, h \in R$.

Lemma 2.4.6. *Since $R = k[x,y] + k[x,y]z$ is a free $k[x,y]$-module, write*

$$f = f_0 + f_1 z, \ g = g_0 + g_1 z, \ h = h_0 + h_1 z,$$

where $f_i, g_i, h_i \in k[x,y]$ for $i = 0, 1$. Then we have:
(1) $h_0 = \frac{1}{2}(xf_1 + yg_1)$.
(2) *There exist $L, M, F \in k[x,y]$ such that*

$$f_0 = 2(xy+1)L + yF, \quad g_0 = 2(xy+1)M - xF, \quad h_1 = xL + yM,$$

where L, M, F, f_1, and g_1 are chosen arbitrarily.
(3) *With the choice of these elements, Δ is written as*

$$\Delta = \{2(xy+1)M - xF + g_1 z\}\frac{\partial}{\partial x} + \left\{\left(\frac{1}{2}f_1 + Lz\right)x + \left(\frac{1}{2}g_1 + Mz\right)y\right\}\frac{\partial}{\partial z}.$$

Proof. Since $xf + yg = 2zh$, we have

$$(xf_0 + yg_0) + z(xf_1 + yg_1) = 2h_1(xy+1) + 2h_0 z,$$

whence

$$xf_0 + yg_0 = 2h_1(xy+1) \tag{2.9}$$

and

$$h_0 = \frac{1}{2}(xf_1 + yg_1). \tag{2.10}$$

Let (x,y) be the maximal ideal in $k[x,y]$. Since $xy + 1 \notin (x,y)$, relation (2.9) implies $h_1 \in (x,y)$. Hence we may write

$$h_1 = xL + yM. \tag{2.11}$$

By (2.9), we have an equality in $k[x,y]$,

$$x\{f_0 - 2(xy+1)L\} = y\{2(xy+1)M - g_0\}.$$

Since $\gcd(x,y) = 1$, we have

$$f_0 = 2(xy+1)L + yF, \quad g_0 = 2(xy+1)M - xF$$

for some $F \in k[x,y]$. Tracing the above computations backward, it is clear that the choice of L, M, F, f_1, g_1 in $k[x,y]$ is arbitrary. □

Let $\sigma : W \to V$ be the blowing-up of a smooth algebraic variety V with a smooth center Z. Let Δ be a vector field on V which is regular along Z. The regularity of Δ near the exceptional subvariety $\sigma^{-1}(Z)$ is given by the following.

Lemma 2.4.7. *With the above notations, let P be a point of Z and let $\{x_1, x_2, \ldots, x_n\}$ be a system of local parameters of V at P such that Z is defined by $x_1 = x_2 = \cdots = x_d = 0$, where $d = \operatorname{codim}_V(Z)$. Write Δ near P as*

$$\Delta = f_1 \frac{\partial}{\partial x_1} + f_2 \frac{\partial}{\partial x_2} + \cdots + f_n \frac{\partial}{\partial x_n},$$

where $f_1, \ldots, f_n \in \mathcal{O}_{V,P}$. Then Δ is regular near $\sigma^{-1}(P)$ if and only if $f_1(P) = \cdots = f_d(P) = 0$.

Proof. Since $\sigma^{-1}(Z)$ is a $\mathbb{P}^{d-1}$-bundle over Z, $\sigma^{-1}(P)$ is a projective space $\mathbb{P}^{d-1}$ with a system of homogeneous coordinates $\{X_1, \ldots, X_d\}$. Fix i with $1 \le i \le d$. Then, on the open set $U_i = \{X_i \ne 0\}$, it holds that $X_j/X_i = x_j/x_i$ for $1 \le j \le d$ and $j \ne i$. Set $u_j = x_j/x_i$ if $1 \le j \le d$ and $j \ne i$ and $u_i = x_i$. For any point $Q \in \sigma^{-1}(P)$, the set $\{u_1 - u_1(Q), \ldots, u_d - u_d(Q), x_{d+1}, \ldots, x_n\}$ is a system of local parameters of W at Q. Hence we can write

$$\Delta = \xi_1 \frac{\partial}{\partial u_1} + \cdots + \xi_d \frac{\partial}{\partial u_d} + f_{d+1} \frac{\partial}{\partial x_{d+1}} + \cdots + f_n \frac{\partial}{\partial x_d},$$

where $\xi_i = f_i$. If $1 \le j \le d$ and $j \ne i$, we have

$$\begin{aligned}\xi_j = \Delta(u_j) = \Delta\left(\frac{x_j}{x_i}\right) &= -\frac{x_j}{x_i^2} f_i + \frac{1}{x_i} f_j \\ &= \frac{1}{x_i}\{f_j(u_1 x_i, \ldots, x_i, \ldots, u_d x_i, x_{d+1}, \ldots, x_n) \\ &\quad - u_j f_i(u_1 x_i, \ldots, x_i, \ldots, u_d x_i, x_{d+1}, \ldots, x_n)\}.\end{aligned}$$

If $f_j(P) \ne 0$, then ξ_j has a simple pole along $\sigma^{-1}(Z)$. So, ξ_j is regular only if $f_j(P) = 0$. This implies that Δ is regular along $\sigma^{-1}(Z)$ only if $f_1(P) = \cdots = f_d(P) = 0$. The converse is clear by the above expression of the ξ_j. □

Example 2.4.8. Let $\sigma : \mathbb{F}_1 \to \mathbb{P}^2$ be the blowing-up with center P. By Lemma 2.4.7, $\Gamma(\mathbb{F}_1, \mathcal{T}_{\mathbb{F}^1})$ is identified with

$$\Gamma = \{\Delta \mid \text{a regular vector field on } \mathbb{P}^2 \text{ which vanishes at } P\}.$$

Hence $\dim \Gamma(\mathbb{F}_1, \mathcal{T}_{\mathbb{F}_1}) = \dim \Gamma(\mathbb{P}^2, \mathcal{T}_{\mathbb{P}^2}) - 2 = 8 - 2 = 6$.

Lemma 2.4.7 implies the following result.

Lemma 2.4.9. *Let $\sigma : W \to V$ be the blowing-up of a smooth projective variety V with center P. Assume that $\dim V = n > 1$. Then we have*

$$\dim \Gamma(V, \mathcal{T}_V) \geq \dim \Gamma(W, \mathcal{T}_W) \geq \dim \Gamma(V, \mathcal{T}_V) - n.$$

Proof. Let $\widetilde{\Delta}$ be a regular vector field on W. Let $E = \sigma^{-1}(P)$. Then $\widetilde{\Delta}|_{W \setminus E}$ is a regular vector field on $V \setminus \{P\}$. Hence it extends to a regular vector field Δ on V such that $\Delta = 0$ at P. Indeed, let $\{x_1, \dots, x_n\}$ be a system of local parameters of V at P. As a rational vector field on V, we can write $\widetilde{\Delta} = \sum_{i=1}^n f_i \frac{\partial}{\partial x_i}$. Then $f_1, \dots, f_n$ are elements of $k(V)$ which are regular on an open neighborhood of P punctured the point P. Then $f_1, \dots, f_n$ are regular at P as well because V is smooth at P and $n \geq 2$. Hence Δ is regular at P and $\widetilde{\Delta}$ is the extension of Δ on W. Then $\Delta = 0$ at P by Lemma 2.4.7. The condition that Δ vanishes at P reduces $\dim \Gamma(V, \mathcal{T}_V)$ by at most n. Hence we obtain the stated inequalities. □

Example 2.4.10. Consider $\mathbb{F}_n$ $(n \geq 0)$ as V. With the notations of Theorem 2.4.3, let P be defined by $u = t = 0$ (the point of origin of the open set V_3). Let Δ be a regular vector field in Theorem 2.4.3. Then the condition that Δ vanishes at P imposes the condition of dimension 1 (resp., 2) if $n > 0$ (resp., $n = 0$), i. e., $a_{20} = 0$ (resp., $a_{20} = b_{02} = 0$) if $n > 0$ (resp., $n = 0$). Furthermore, by the computation in Lemma 2.2.7 repeated for the blowing-up at P, we conclude that

$$\dim \Gamma(\mathbb{F}_{n+1}, \mathcal{T}_{\mathbb{F}_{n+1}}) = \begin{cases} \dim \Gamma(\mathbb{F}_n, \mathcal{T}_{\mathbb{F}_n}) - 1 + 2 & (n > 0), \\ \dim \Gamma(\mathbb{F}_n, \mathcal{T}_{\mathbb{F}_n}) - 2 + 2 & (n = 0). \end{cases}$$

Hence $\dim \Gamma(\mathbb{F}_{n+1}, \mathcal{T}_{\mathbb{F}_{n+1}}) = n + 6$. □

We shall give one more result.

Proposition 2.4.11. *Let V be a del Pezzo surface of degree d, where $d = (K_V^2)$. If $d \leq 5$, then there are no regular vector fields on V.*

Proof. The surface V is obtained by blowing up $(9 - d)$ points $P_1, \dots, P_m$ $(m = 9 - d)$ on $\mathbb{P}^2$ in general position, i. e., no three of them lie on a line, no five of them lie on a conic, and no eight of them lie on an irreducible singular cubic with one of them being a singular point. We can choose the line at infinity ℓ_∞ so that none of $P_1, \dots, P_m$ lies on ℓ_∞. Suppose there exists a nonzero regular vector field $\widetilde{\Delta}$ on V. Then, by the proof of Lemma 2.4.9, there exists a regular vector field Δ on $\mathbb{P}^2$ such that Δ vanishes at points $P_1, \ldots, P_m$. We may choose a system of coordinates $\{x, y\}$ on $\mathbb{A}^2 = \mathbb{P}^2 \setminus \ell_\infty$ so that $P_1 = (0,0), P_2 = (1,0)$ and $P_3 = (0,1)$. With the notations in Lemma 2.4.1, it follows that $c_2 = d_2 = 0, d_0 = 0, a_1 + d_1 = 0, a_0 + c_0 = 0$ and $c_1 = 0$. Hence we have $f = a_0(x^2 - x) + a_1xy$ and $g = a_0xy + a_1(y^2 - y)$. Suppose that $P_4 = (\alpha, \beta)$ is involved. Then $\alpha\beta \neq 0$ because no three of P_1, P_2, P_3, P_4 lie on a line. Since $f(\alpha, \beta) = g(\alpha, \beta) = 0$, we have

$$\begin{vmatrix} \alpha^2 - \alpha & \alpha\beta \\ \alpha\beta & \beta^2 - \beta \end{vmatrix} = 0.$$

For otherwise $a_0 = a_1 = 0$ and $\Delta = 0$ everywhere. The above determinant gives $\alpha + \beta = 1$. Then P_2, P_3 and P_4 are colinear, which is a contradiction. So, P_4 cannot be involved, and $9 - d \leq 3$. □

This implies that there is no G_a-action on V if $d \leq 5$. The last result follows from the following two facts:

(i) $\mathrm{H}^0(V, \mathcal{T}_V)$ is the Lie algebra of the algebraic group $\mathrm{Aut}^0(V)$.

(ii) Let φ be an element of $\mathrm{Aut}^0(V)$ which comes from an automorphism of $\mathbb{P}^2$ fixing the points $P_1, \ldots, P_m$. Then $\varphi = \mathrm{id}$ if $m \geq 4$.

2.4.2 Unipotent varieties

Let X be a smooth algebraic variety with an algebraic group G acting on it, whence there is a group homomorphism

$$\sigma : G \to \mathrm{Aut}(X).$$

We call such a variety X a *G-variety*. Taking the Lie algebra homomorphism of σ, we have

$$d\sigma : \mathfrak{g} \to \Gamma(X, \mathcal{T}_X),$$

where $\mathcal{T}_X$ is the tangent bundle of X and $\mathfrak{g}$ is the Lie algebra of G at the neutral element e.

Suppose further that G is connected. If (X, D) is a pair of a G-variety and a G-stable effective divisor D with simple normal crossings, then G stabilizes each of the irreducible components of $D = D_1 + \cdots + D_m$. Let $x \in X$ and let $\{t_1, \ldots, t_n\}$ be a local system of parameters at $x \in X$ such that D is defined by $t_1 \cdots t_r = 0$. Then an infinitesimal automorphism $\exp(\varepsilon\delta)$ with $\delta \in \mathcal{T}_X$ acts on $\widehat{\mathcal{O}}_{X,x}[\varepsilon] = k[[t_1, \ldots, t_n]][\varepsilon]$ as a $k[\varepsilon]$-algebra homomorphism $\widehat{\mathcal{O}}_{X,x}[\varepsilon] \to \widehat{\mathcal{O}}_{X,x}[\varepsilon]$ defined by

$$\exp(\varepsilon\delta)(a) = a + \delta(a)\varepsilon + \frac{1}{2!}\delta^2(a)\varepsilon^2 + \cdots + \frac{1}{j!}\delta^j(a)\varepsilon^j + \cdots,$$

where $a \in \widehat{\mathcal{O}}_{X,x}$, $\delta(\varepsilon) = 0$ and $\varepsilon^N = 0$ for some $N > 0$. Hence that $\exp(\varepsilon\delta)$ preserves each irreducible component D_i means $\delta(t_i a) \in t_i\widehat{\mathcal{O}}_{X,x}$. If we write δ as

$$\delta = c_1\partial_1 + c_2\partial_2 + \cdots + c_i\partial_i + \cdots + c_n\partial_n \quad \text{with } \partial_i = \frac{\partial}{\partial t_i},$$

the above condition is that $t_1 \mid c_1, \ldots, t_r \mid c_r$. Thus, δ belongs to the stalk

$$\mathcal{T}_X(\log D)_x = \mathcal{O}_x(t_1\partial_1) + \cdots + \mathcal{O}_x(t_r\partial_r) + \mathcal{O}_x(\partial_{r+1}) + \cdots + \mathcal{O}_x(\partial_n).$$

Hence the above Lie algebra automorphism $d\sigma$ factors through a homomorphism

$$\mathfrak{g} \to \Gamma(X, \mathcal{T}_X(\log D)) \to \Gamma(X, \mathcal{T}_X).$$

By Definition 2.2.11, a unipotent (algebraic) group G has a central series

$$G_0 = \{e\} \lhd \cdots \lhd G_{i-1} \lhd G_i \lhd \cdots \lhd G_d = G$$

such that $G_i/G_{i-1} \cong G_a$ for $1 \le i \le d$. By definition, a finite unipotent group is a trivial group $\{e\}$ and a one-dimensional unipotent group is the additive group G_a. We can think of an algebraic group G which is generated by one-dimensional unipotent subgroups, called a *quasi-unipotent* group. A quasi-unipotent group is not necessarily a unipotent group. According to Popov [189, Lemma 1.1], a connected linear algebraic group G is quasi-unipotent if and only if G has no multiplicative characters. Hence a semisimple algebraic group like $\mathrm{SL}(n, \mathbb{C})$ is a quasi-unipotent group. We consider mostly unipotent algebraic groups.

A G-variety X is called a *unipotent variety* (resp., *quasi-unipotent variety*) if G is unipotent (resp., quasiunipotent) and X contains an open G-orbit.

Lemma 2.4.12. *Let X be a normal algebraic variety X and let U be an affine open set of X. Then the complement $X \setminus U$ has pure codimension one.*

Proof. Let Y be an irreducible component of $X \setminus U$. We show that Y has codimension one in X. Let η be the generic point of Y and let $R = \mathcal{O}_{X,\eta}$. Let $Z = U \times_X \operatorname{Spec} R$. Since the open immersion $\iota : U \to X$ is an affine morphism, the scheme Z is an affine open set of $\operatorname{Spec} R$. By construction, $Z = \operatorname{Spec} R \setminus \{\mathfrak{m}\}$, where $\mathfrak{m}$ is the maximal ideal of R. Suppose that $\operatorname{codim}_X Y \ge 2$. Then $\dim R \ge 2$. Since R is normal, R is the intersection of the local rings at all codimension one points of Z. Since Z is affine, we obtain that $Z = \operatorname{Spec} R$. This is a contradiction. □

A G-orbit in a G-variety is written as $G \cdot x$ for a closed point x. We call x the *base point* of the orbit. Then $G \cdot x$ is isomorphic to a homogeneous space G/H, where H is the stabilizer group of x. We consider an orbit of a unipotent group U. We prove first the following result which is already known in [187, Corollary, p. 1043] and [214].

Lemma 2.4.13. *Let G be a unipotent group and $X = G/H$ is a homogeneous space. Then the following assertions hold:*

(1) *The underlying scheme of G is the affine space* $\mathbb{A}^d$ *with* $d = \dim G$.
(2) *The homogeneous space X is isomorphic to* $\mathbb{A}^n$ *with* $n = \dim X$.

Proof. (1) We proceed by induction on $d = \dim G$. If $d = 1$, then $G \cong G_a$, whence the underlying scheme of G is the affine line $\mathbb{A}^1$. In general, the center C of G is nontrivial because G is a nilpotent group. Note that a commutative unipotent group is a vector group, i. e., a direct product of G_a. Hence G contains a subgroup G_1 which is isomorphic to G_a. Then the quotient group $\overline{G} := G/G_1$ is a unipotent group of dimension $d-1$. By the induction hypothesis, $\overline{G}$ has the underlying scheme isomorphic to $\mathbb{A}^{d-1}$. On the other hand, G is an $\mathbb{A}^1$-bundle over $\overline{G}$ by [107]. Since $\overline{G}$ is factorial, the bundle structure is trivial, i. e., $G \cong \overline{G} \times \mathbb{A}^1$. Hence $G \cong \mathbb{A}^d$.

(2) We also proceed by double induction on $\dim G$ and $\dim X$. If $\dim X = 1$, then X is a smooth rational affine curve whose invertible regular functions are constants. Hence $X \cong \mathbb{A}^1$. Let G_a be a central subgroup of G. If $G_a \not\subset H$, then the subgroup G_aH contains properly H. Consider the left G_a-action on $X = G/H$, which gives the quotient morphism $q : G/H \to Y := G_a\backslash G/H = G/G_aH$ and $G_aH \supsetneq H$. Hence $Y \cong \mathbb{A}^{n-1}$ by the induction hypothesis on $\dim X$. The morphism q defines an $\mathbb{A}^1$-fibration such that every fiber is isomorphic to $\mathbb{A}^1$ because all the isotropy groups are trivial.[9] Hence q is an $\mathbb{A}^1$-bundle in the Zariski topology [107]. Since Y is factorial, this implies that $G/H \cong \mathbb{A}^1 \times Y \cong \mathbb{A}^n$. If $G_a \subseteq H$, then $G/H \cong (G/G_a)/(H/G_a)$, where G/G_a is a unipotent group and H/G_a is its unipotent subgroup.[10] By the induction hypothesis on $\dim G$, we know that $G/H \cong \mathbb{A}^n$. □

Corollary 2.4.14. *Let G be a unipotent group and let X be a G-variety containing an open G-orbit. Then the following assertions hold:*
(1) *Assume that X is affine. Then X coincides with the open G-orbit and hence isomorphic to the affine space* $\mathbb{A}^n$.
(2) *Let Y be a G-stable open set of X. Then Y contains the open orbit.*

Proof. (1) It is shown in [190] that a generically transitive action of a unipotent group on an affine variety is transitive. Hence the affine variety X coincides with the open orbit. By Lemma 2.4.13, X is hence isomorphic to the affine space $\mathbb{A}^n$.

(2) Let Y be a G-stable open set. Let W be the open G-orbit. Then $Y \cap W$ is a non-empty G-stable open set of X. Since G acts transitively on W, it follows that $W \subseteq Y$. □

Let G (and often denoted by U) denote a unipotent group as above. Let V be a smooth projective variety which is a G-variety and let X be an open G-orbit. Since X is affine by Lemma 2.4.13, the complement $D = V \setminus X$ is a G-stable subvariety of pure

9 Suppose that $\lambda(gH) = (gH)$ with $\lambda \in G_a, \lambda \neq e$ and $g \in G$ Then $\lambda g = gh$ for $h \in H$. Since λ is central in G, it follows that $\lambda = h \in H$. Since λ has infinite order in G, it follows that $G_a \subseteq H$.
10 An algebraic subgroup of a unipotent group is a unipotent group.

codimension one by Lemma 2.4.12. Let $D = D_1 + \cdots + D_r$ be the irreducible decomposition of D. Since G is connected, each irreducible component is G-stable. We observe in concrete examples what takes place for vector fields in the boundary D.

Proposition 2.4.15. *The following assertions hold:*

(1) *Embed* $\mathbb{A}^2 = \operatorname{Spec} k[x,y]$ *into* $\mathbb{P}^2$ *in the standard way* $(x,y) \mapsto (1,x,y)$. *Let* $D = \ell_\infty$ *be the line at infinity. Then* $\Gamma(\mathbb{P}^2, \mathcal{T}_{\mathbb{P}^2}(\log D))$ *is a k-module generated by the elements*

$$\Delta = (c_0x + c_1y + c_2)\frac{\partial}{\partial x} + (d_0x + d_1y + d_2)\frac{\partial}{\partial y},$$

where $c_i, d_j \in k$. *Hence* $\dim \Gamma(\mathbb{P}^2, \mathcal{T}_{\mathbb{P}^2}(\log D)) = 6$. *Let U be the upper triangular unipotent subgroup of* SL(3), *which consists of matrices of the form*

$$\begin{pmatrix} 1 & s_1 & s_2 \\ 0 & 1 & s_3 \\ 0 & 0 & 1 \end{pmatrix}, \quad s_1, s_2, s_3 \in k.$$

Then U acts on $\mathbb{P}^2$ *from the right as* $(X_0, X_1, X_2) \mapsto (X_0, X_1 + s_1X_0, X_2 + s_3X_1 + s_2X_0)$ *with the line at infinity* $\ell_\infty = \{X_0 = 0\}$ *stabilized under this action. With the inhomogeneous coordinates* $x = X_1/X_0, y = X_2/X_0$, *the action is given as* $(x,y) \mapsto (x + s_1, y + s_3x + s_2)$. *The Lie algebra* $\mathfrak{u}$ *of U is generated by the matrices*

$$\delta_1 = \begin{pmatrix} 0 & 1 & 0 \\ 0 & 0 & 0 \\ 0 & 0 & 0 \end{pmatrix}, \quad \delta_2 = \begin{pmatrix} 0 & 0 & 1 \\ 0 & 0 & 0 \\ 0 & 0 & 0 \end{pmatrix}, \quad \delta_3 = \begin{pmatrix} 0 & 0 & 0 \\ 0 & 0 & 1 \\ 0 & 0 & 0 \end{pmatrix}.$$

The Lie algebra homomorphism $d\sigma : \mathfrak{u} \to \Gamma(\mathbb{P}^2, \mathcal{T}_{\mathbb{P}^2}(\log D))$ *is given by*

$$(\delta_1, \delta_2, \delta_3) \mapsto \left(\frac{\partial}{\partial x}, \frac{\partial}{\partial y}, x\frac{\partial}{\partial y}\right).$$

(2) *Let* $O = (1,0,0)$. *Since U acts from the right on* $\mathbb{P}^2$, *the orbit* $O \cdot U$ *is* $\{(1, s_1, s_2) \mid s_1, s_2 \in k\}$ *and the isotropy group at O is*

$$H = \left\{ \begin{pmatrix} 1 & 0 & 0 \\ 0 & 1 & s_3 \\ 0 & 0 & 1 \end{pmatrix} \,\middle|\, s_3 \in k \right\}.$$

The homogeneous space $H \backslash U$ *is the ordinary* (x,y)*-plane. The line at infinity* ℓ_∞ *has a U-action*

$$(0, X_1, X_2) \mapsto (0, X_1, X_2 + s_3X_1).$$

Hence ℓ_∞ *contains an U-orbit* $O_1 \cdot U = \{(0,1,s_3) \mid s_3 \in k\}$, *where* $O_1 = (0,1,0)$ *and the isotropy group is*

$$H_1 = \left\{ \begin{pmatrix} 1 & s_1 & s_2 \\ 0 & 1 & 0 \\ 0 & 0 & 1 \end{pmatrix} \,\middle|\, s_1, s_2 \in k \right\}.$$

The point $O_2 = (0,0,1)$ is a unique U-fixed point. So, there exists a decomposition of $\mathbb{P}^2$ into strata of U-orbits $\mathbb{P}^2 = O \cdot U \cup O_1 \cdot U \cup \{O_2\}$.

(3) *The observation made in (2) above can be easily generalized to the case of $\mathbb{P}^n$ and the group U_n of upper triangular unipotent matrices in $\mathrm{SL}(n+1)$. The decomposition into U_n-orbits*

$$\mathbb{P}^n = O \cdot U_n \cup O_1 \cdot U_n \cup \cdots \cup O_{n-1} \cdot U_n \cup \{O_n\}$$

is also the decomposition into the B_n-orbits, where B_n is the Borel subgroup of $\mathrm{SL}(n+1)$ consisting of upper triangular matrices.

Proof. We prove only assertion (1). The rest are obvious. By Lemma 2.4.1, $\Gamma(\mathbb{P}^2, \mathcal{T}_{\mathbb{P}^2})$ is a k-module generated by the elements

$$\Delta = (a_0x^2 + a_1xy + c_0x + c_1y + c_2)\frac{\partial}{\partial x} + (a_0xy + a_1y^2 + d_0x + d_1y + d_2)\frac{\partial}{\partial y},$$

where $a_i, c_j, d_\ell \in k$. Meanwhile, ℓ_∞ is defined by $x^{-1} = 0$ near the point $(0,1,0)$ and by $y^{-1} = 0$ near the point $(0,0,1)$. Hence $\Delta \in \Gamma(\mathbb{P}^2, \mathcal{T}_{\mathbb{P}^2}(\log D))$ if and only if $\Delta(x^{-1})$ (resp., $\Delta(y^{-1})$) is divisible by x^{-1} (resp., y^{-1}). Hence we obtain the above expression of Δ. As for the correspondence $\mathfrak{u}$ and $\Gamma(\mathbb{P}^2, \mathcal{T}_{\mathbb{P}^2}(\log D))$, we consider infinitesimal automorphisms $\exp(\varepsilon\delta_i)$ and their actions on (X_0, X_1, X_2), where $i = 1,2,3$ and $\varepsilon^2 = 0$. In fact, we have

$$\exp(\varepsilon\delta_1)(X_0,X_1,X_2) = (X_0,X_1,X_2)\begin{pmatrix} 1 & \varepsilon & 0 \\ 0 & 1 & 0 \\ 0 & 0 & 1 \end{pmatrix} = (X_0, X_1 + \varepsilon X_0, X_2),$$

$$\exp(\varepsilon\delta_2)(X_0,X_1,X_2) = (X_0,X_1,X_2)\begin{pmatrix} 1 & 0 & \varepsilon \\ 0 & 1 & 0 \\ 0 & 0 & 1 \end{pmatrix} = (X_0, X_1, X_2 + \varepsilon X_0),$$

$$\exp(\varepsilon\delta_3)(X_0,X_1,X_2) = (X_0,X_1,X_2)\begin{pmatrix} 1 & 0 & 0 \\ 0 & 1 & \varepsilon \\ 0 & 0 & 1 \end{pmatrix} = (X_0, X_1, X_2 + \varepsilon X_1).$$

Hence we have $\delta_1(x) = 1, \delta_1(y) = 0, \delta_2(x) = 0, \delta_2(y) = 1$, and $\delta_3(x) = 0, \delta_3(y) = x$. □

Example 2.4.16.

(1) With the notations before Theorem 2.4.3, identify $\mathbb{A}^2 = \operatorname{Spec} k[x,y]$ with the open set V_0 of $\mathbb{F}_n$. Let $D = \ell_\infty + M$. Then $\Gamma(\mathbb{F}_n, \mathcal{T}_{\mathbb{F}_n}(\log D))$ is a k-module consisting of vector fields

$$\Delta = \begin{cases} (a_{10}x + a_{00})\frac{\partial}{\partial x} + (b_{n0}x^n + \cdots + b_{10}x + b_{00} + b_{01}y)\frac{\partial}{\partial y} & (n > 0), \\ (a_{10}x + a_{00})\frac{\partial}{\partial x} + (b_{01}y + b_{00})\frac{\partial}{\partial y} & (n = 0). \end{cases}$$

Hence $\dim \Gamma(\mathbb{F}_n, \mathcal{T}_{\mathbb{F}_n}(\log D)) = n + 4$ for $n \geq 0$.

By [140], the automorphism group $G := \mathrm{Aut}^0(\mathbb{F}_n)$ satisfies an exact sequence

$$(1) \to H \to \mathrm{Aut}^0(\mathbb{F}_n) \to \mathrm{PGL}(2) \to (1),$$

where H consists of automorphisms

$$(x, y) \mapsto (x, cy + d_0 + d_1x + \cdots + d_nx^n),$$

with $c \in k^*$ and $d_0, d_1, \ldots, d_n \in k$. Hence $H = U_0 \rtimes G_m$, where U_0 is the unipotent subgroup with $c = 1$ in the above expression and $U_0 \cong G_a^{\times(n+1)}$. Let $\mathfrak{u}_0$ be the Lie algebra of U_0. Then the natural G-action σ on $\mathbb{F}_n$ induces a Lie algebra isomorphism

$$d\sigma : \mathfrak{u}_0 \to \left\{\Delta \mid \Delta = (d_0 + d_1x + \cdots + d_nx^n)\frac{\partial}{\partial y},\ d_0, d_1, \ldots, d_n \in k\right\}.$$

Let U be a maximal unipotent subgroup of G containing U_0. Then U/U_0 is a unipotent subgroup of PGL(2). For the Lie algebra $\mathfrak{u}$ of U, $(d\sigma)(\mathfrak{u}) = k\frac{\partial}{\partial x} + (d\sigma)(\mathfrak{u}_0)$. So, $\mathbb{F}_n$ is a G-variety and V_0 is a U-orbit.

(2) The U_0-action on $\ell_\infty \cup M$ is given by

$$\begin{array}{lll} \ell_\infty & : & u \mapsto u, \quad v \mapsto v + d_0u^n + \cdots + d_n, \\ M & : & x \mapsto x, \quad z \mapsto \frac{z}{1+d_0z+d_1xz+\cdots+d_nx^nz}. \end{array}$$

This shows that $\ell_\infty \setminus \{P_\infty\}$ with $P_\infty = (u = 0, t = 0)$ is an U_0-orbit and all points on M are U_0-fixed points. However, since U/U_0 moves the x-coordinate, $M \setminus \{P_\infty\}$ is a U-orbit and P_∞ is the unique U-fixed point. Thus the decomposition into U-orbits is

$$\mathbb{F}_n = V_0 \cup (\ell_\infty \setminus \{P_\infty\}) \cup (M \setminus \{P_\infty\}) \cup \{P_\infty\}.$$

When we deal with a projective variety V, one cannot expect the existence of an affine space fibration $f : V \to W$ because fibers are closed sets in V and the morphism f is a projective morphism. Instead, we can look for affine open sets in V which are isomorphic to the affine space or have affine space fibrations. Such affine open sets are often obtained as homogeneous spaces of the automorphism group $\mathrm{Aut}(V)$. Similar observations are possible for not only projective varieties but also algebraic varieties in general including affine varieties. We consider the case of a Danielewski surface $X = \{xy = z^2 - 1\}$. The first part of the following lemma is due to [138].

Lemma 2.4.17.

(1) *The automorphism group* $\mathrm{Aut}(X)$ *is generated by the following automorphisms:*

$$\begin{array}{lll} H_\lambda & : & H_\lambda(x) = \lambda x, \quad H_\lambda(y) = \lambda^{-1}y, \quad H_\lambda(z) = z, \quad \lambda \in k^*, \\ \Delta_{f(x)} & : & \Delta(x) = x, \quad \Delta(y) = y + 2zf(x) + xf(x)^2, \quad \Delta(z) = z + xf(x), \quad f(x) \in k[x], \\ I & : & I(x) = y, \quad I(y) = x, \quad I(z) = z, \\ P & : & P(x) = x, \quad P(y) = y, \quad P(z) = -z. \end{array}$$

(2) *Let G be the subgroup of* GL(3) *generated by the elements* H_λ, Δ_t, $(I\cdot\Delta_s\cdot I), I, P$ *which are respectively identified with the matrices*

$$\begin{pmatrix} \lambda & 0 & 0 \\ 0 & \lambda^{-1} & 0 \\ 0 & 0 & 1 \end{pmatrix}, \quad \alpha_t = \begin{pmatrix} 1 & t^2 & t \\ 0 & 1 & 0 \\ 0 & 2t & 1 \end{pmatrix}, \quad \beta_s = \begin{pmatrix} 1 & 0 & 0 \\ s^2 & 1 & s \\ 2s & 0 & 1 \end{pmatrix},$$
$$\begin{pmatrix} 0 & 1 & 0 \\ 1 & 0 & 0 \\ 0 & 0 & 1 \end{pmatrix}, \quad \begin{pmatrix} 1 & 0 & 0 \\ 0 & 1 & 0 \\ 0 & 0 & -1 \end{pmatrix},$$

where $\lambda \in k^*$ *and* $s, t \in k$. *Then G acts transitively on X.*

(3) *Let* G_0 *be the subgroup of* GL(3) *generated by* α_t, β_s $(s, t \in k)$. *Then* G_0 *is a non-unipotent, quasi-unipotent group and it acts transitively on X. Hence X is a quasi-unipotent variety.*

(4) *Let* G_1 *be the subgroup of* GL(3) *generated by* $H_\lambda, \alpha_t, \beta_s$ $(\lambda \in k^*, s, t \in k)$. *Then* G_1 *is isomorphic to* PGL(2), *whence* $G_0 = G_1$. *Hence* G_1 *is semi-simple, and X is a homogeneous space* $T\backslash G_1$, *where T is a maximal torus of* G_1.

Proof. (1) See [138].

(2) We note that Δ_t and $I \cdot \Delta_s \cdot I$ respectively define the G_a-actions given by the locally nilpotent derivations δ_1 and δ_2 such that

$$\delta_1(x) = 0, \quad \delta_1(y) = 2z, \quad \delta_1(z) = x,$$
$$\delta_2(x) = 2z, \quad \delta_2(y) = 0, \quad \delta_2(z) = y.$$

As for the G-transitivity of X, let $\alpha_t = \Delta_t$ and $\beta_s = I \cdot \Delta_s \cdot I$. We take the point $O = (0, 0, 1)$ as a base point. If we identify the matrices α_t, β_s with the associated automorphisms of X, then $\alpha_t \cdot \beta_s(0, 0, 1) = (2s, 2st^2 + 2t, 2st + 1)$. In fact, the action is given by the matrix multiplication $(0, 0, 1)\beta_s \cdot \alpha_t$. Let (a, b, c) be an arbitrary point of X, where $ab = c^2 - 1$. If $a \neq 0$, then take $s = \frac{a}{2}$ and $t = \frac{c-1}{a}$. Then $\alpha_t \cdot \beta_s$ sends the point O to (a, b, c). If $a = 0$, then $c = \pm 1$. Take $t = \frac{b}{2}$. Then α_t sends the point O to $(0, b, 1)$. On the other hand, $\beta_s \cdot \alpha_t$ sends the point O to $(2s^2t + 2s, 2t, 2st + 1)$. So, by putting $t = \frac{b}{2}$ and $s = -\frac{2}{b}$ if $b \neq 0$, $\beta_s \cdot \alpha_s$ sends the point O to $(0, b, -1)$. Thus, G acts transitively on X.

(3) Note that the subgroup of G generated by $\{\alpha_t \mid t \in k\}$ is isomorphic to G_a and so is the subgroup generated by $\{\beta_s \mid s \in k\}$. Hence the group G_0 is a quasi-unipotent group. Furthermore, G_0 acts transitively on X by the proof of the assertion (2) and further actions of α_s and β_t. Since a unipotent group orbit of dimension two is isomorphic to $\mathbb{A}^2$ by Lemma 2.4.13, it follows that the group G_0 is not a unipotent group.

(4) Consider the standard right representation of $\mathrm{SL}(2, k)$,

$$(u, v) \mapsto (u, v)\begin{pmatrix} a & b \\ c & d \end{pmatrix}.$$

Let $x = u^2, y = v^2$ and $z = uv$. Then one can show readily that G_1 is the image of $SL(2) \to GL(3)$ induced by the representation of SL(2) of weight 2 on $ku^2 + kv^2 + kuv$. Hence G_1 is isomorphic to $SL(2)/\mathbb{Z}_2$ which is PGL(2). We have already observed that G_0 acts transitively on X. Fix a point $(0,0,1)$. Then, by the action of G_1 on X from the right, the isotropy group of the point $(0,0,1)$ is the maximal torus T generated by H_λ for $\lambda \in k^*$. In order to show that $G_0 = G_1$, note that $\dim G_1 = 3$ and G_0 is an algebraic subgroup of G_1. If $G_0 \subsetneq G_1$ then $\dim G_0 = 2$. Hence G_0 is isomorphic to either $G_a \times G_a$ or $G_a \rtimes G_m$ (including $G_a \times G_m$). In fact, let U be the unipotent radical of G_0 which is a normal subgroup. If $U = G_0$ then G_0 is a commutative unipotent group of dimension two. Hence G_0 is commutative. If $U \neq G_0$ then the canonical surjection $G_0 \to G_0/U \cong G_m$ is a nontrivial multiplicative character of G_0 but G_0 has no multiplicative character according to Popov [189, Lemma 1.1]. Hence $\dim G_0 = 3$ and $G_0 = G_1$.

We prove the last assertion. Write $T\backslash G_1 = T'\backslash SL(2)$, where T' is the maximal torus $\{(\begin{smallmatrix} t & 0 \\ 0 & t^{-1} \end{smallmatrix}) \mid t \in k^*\}$. Let $(\begin{smallmatrix} a & b \\ c & d \end{smallmatrix})$ be an element of SL(2). Then the left action of T' on SL(2) is given by $(a,b,c,d) \mapsto (ta, tb, t^{-1}c, t^{-1}d)$, and hence the T'-invariant subring A of the coordinate ring $B := k[a,b,c,d]/(ad-bc = 1)$ of SL(2) is generated over k by $x = ac$, $y = bd$, $u = ad$ and $z = bc$. The relation among x, y, u and z are $u = z + 1$ and $xy = uz$. Eliminating u, we have $xy = z(z+1)$, which is a defining equation of the Danielewski surface.

The following argument to show that $X \cong T'\backslash SL(2)$ was communicated to us by V. L. Popov. Consider the diagonal action of SL(2) on $\mathbb{P}^1 \times \mathbb{P}^1$, where SL(2) acts on $\mathbb{P}^1$ via the standard right action on the space of row vectors of length 2. Hence the diagonal action induces the right SL(2)-action on X. In fact, it is readily verified that $X \cong (\mathbb{P}^1 \times \mathbb{P}^1) \setminus \Delta$, where Δ is the diagonal. Since the SL(2)-action on $\mathbb{P}^1$ is 2-transitive, SL(2) acts transitively on X. Let $p = [1:0]$ and $q = [0:1]$. Then the stabilizer groups of p and q are respectively the Borel subgroups $B^\wedge$ and B which consist of lower (resp. upper) triangular matrices. Then the stabilizer subgroup of the point (p,q) is the maximal torus $T' = B^\wedge \cap B$. □

In the following result, we treat a Danielewski surface in a little bit more general form.

Lemma 2.4.18. *Let $X := X(P)$ be now a Danielewski surface $\{xy = P(z)\}$, where $P(z) \in k[z]$ such that $\deg P(z) = n \geq 2$ and $\gcd(P(z), P'(z)) = 1$. Then the following assertions hold:*

(1) *For $\lambda \in k^*$, $s, t \in k$, define k-automorphisms $H_\lambda, \alpha_t, \beta_s$ and I of X by*

$$H_\lambda : H_\lambda(x) = \lambda x, \quad H_\lambda(y) = \lambda^{-1} y, \quad H_\lambda(z) = z,$$

$$\alpha_t : \alpha_t(x) = x, \quad \alpha_t(y) = y + \sum_{i=1}^{n} \frac{P^{(i)}(z)}{i!} x^{i-1} t^i, \quad \alpha_t(z) = z + xt,$$

$$\beta_s : \beta_s(x) = x + \sum_{i=1}^{n} \frac{P^{(i)}(z)}{i!} y^{i-1} s^i, \quad \beta_s(y) = y, \quad \beta_s(z) = z + ys,$$

$$I : I(x) = y, \quad I(y) = x, \quad I(z) = z,$$

where $\beta_s = I \cdot \alpha_s \cdot I$. *Then* $\alpha_t = \exp(t\delta_1)$ *and* $\beta_s = \exp(s\delta_2)$, *where* δ_1 *and* δ_2 *are locally nilpotent derivations of the coordinate ring of* X *defined by*

$$\delta_1 : \delta_1(x) = 0, \quad \delta_1(y) = P'(z), \quad \delta_1(z) = x,$$
$$\delta_2 : \delta_2(x) = P'(z), \quad \delta_2(y) = 0, \quad \delta_2(z) = y.$$

(2) *Let* G_0 *be the subgroup of* $\mathrm{Aut}(X)$ *generated by* α_t *and* β_s *for* $s, t \in k$, *and let* G_1 *be the subgroup generated by* G_0 *and* H_λ *for* $\lambda \in k^*$. *Then the group* G_0 *is an algebraic group if and only if* $n = 2$. *If* $n = 2$ *then* G_0 *is isomorphic to* $\mathrm{PGL}(2)$.
(3) *Suppose that* $P'(1) \neq 0$ *and* $P(1) = 0$. *Then* G_0 *acts transitively on* X.

Proof. The verification of (1) is straightforward. We prove assertion (2). If $n = 2$ then, by a linear change of the variable z, we may assume that $P(z) = z^2 - 1$. Then the group G_0 (and hence G_1) is isomorphic to PGL(2) by Lemma 2.4.17. So, it suffices to show that if $n > 2$ then G_0 is not an algebraic group. We owe the subsequent argument to V. L. Popov.

Suppose that G_0 is an algebraic group. Let $\mathfrak{g}_0$ be the Lie algebra of G_0. Since $\alpha_t = \exp(t\delta_1)$ and $\beta_s = \exp(s\delta_2)$, $\mathfrak{g}_0$ contains δ_1, δ_2 and

$$\sigma := [\delta_1, \delta_2] = \delta_1 \circ \delta_2 - \delta_2 \circ \delta_1,$$

which acts on $k[x, y, z]$ as $\sigma(x) = P''(z)x$, $\sigma(y) = -P''(z)y$ and $\sigma(z) = 0$. Consider the k-linear subspace V of the function field $k(X)$ generated by $\{g(x) \mid g \in G_0\}$. It is a G_0-stable vector space of finite dimension since we assume that G_0 is algebraic (see Example 2.2.26). Hence V is also $\mathfrak{g}_0$-stable. Considering $\sigma^m(x)$ and $\sigma^m(y)$ for $m > 0$, it follows that $(P''(z))^m x$ and $(P''(z))^m y$ belong to V for every $m > 0$. Since $\deg P''(z) = n - 2 > 0$, it follows that $\{(P''(z))^m x \mid m > 0\}$ and $\{(P''(z))^m y \mid m > 0\}$ are linearly independent over k. This is a contradiction to $\dim V < \infty$. So, G_0 is not algebraic.

Assertion (3) needs a supplementary explanation. Let $O = (0, 0, 1)$. Let (a, b, c) be a point of X. If $a \neq 0$, then put $s = a/P'(1)$ and $t = (c - 1)/a$. Then $\alpha_t \cdot \beta_s(O) = (a, b, c)$. If $a = 0$ and $b \neq 0$, then put $t = b/P'(1)$ and $s = (c - 1)/b$. Then $\beta_s \cdot \alpha_t(O) = (-P(1)/b, b, c)$. Hence if $P(1) = 0$ then $\beta_s \cdot \alpha_t(O) = (0, b, c)$. Thus G_0 acts transitively on X. □

There are many interesting results on quasi-unipotent varieties due to Popov [187, 189]. We refer the interested readers to these references.

2.4.3 Affine varieties with abundant G_a-actions

Let X be a smooth affine variety of dimension n. Let $\sigma_i : G_a \times X \to X$ be an action of G_a for $1 \leq i \leq m$. We say that the actions σ_i $(1 \leq i \leq m)$ are *independent* if there exists a point P such that the vector fields Δ_i associated with σ_i span a vector subspace of dimension m in the tangent space $\mathcal{T}_{X,P}$. It is clear that we can choose as P any point from an open set U of X. In fact, let Δ_i be the vector field associated with the action σ_i. Then the mapping $P \mapsto (\Delta_1)_P \wedge \cdots \wedge (\Delta_m)_P \in \bigwedge_{i=1}^m \mathcal{T}_{X,P}$ defines a section

$$\Delta_1 \wedge \cdots \wedge \Delta_m : X \to \bigwedge_{i=1}^{m} \mathcal{T}_X,$$

which is nonzero at the point P by the assumption. Then it is nonzero in an open neighborhood of P. We then prove the following result.

Theorem 2.4.19. *Let X be a smooth affine variety of dimension n with n independent G_a-actions σ_i $(1 \le i \le n)$. Then the following assertions hold:*
(1) *The fundamental group of X is a finite group.*
(2) *Suppose that $n \le 3$, X is factorial and the quotient surface of X with respect to one of the actions σ_i, say σ_1, is smooth. Then X is simply connected.*

Proof. (1) Since the arguments are the same for all dimensions, we assume $n = 3$.[11] We denote the group scheme G_a by G_i if G_a acts on X by σ_i. The quotient surface $Y = X/\!/G_1$ is a normal affine surface. Let $q : X \to Y$ be the algebraic quotient morphism. Let $\tau : G_2 \times G_3 \to Y$ be the morphism defined by $(g_2, g_3) \mapsto q(g_3(g_2P))$, where P is a general point of X. Then τ is a dominant morphism. In fact, if ε_i $(i = 2, 3)$ moves a complex number with $|\varepsilon_i|$ small, then $\exp(\varepsilon_3\Delta_3) \cdot \exp(\varepsilon_2\Delta_2)P$ is considered to be a transversal section of a tubular neighborhood of the orbit G_1P. We may assume that $G_1P \cong \mathbb{A}^1$ by replacing P by a general point in an open neighborhood of P. This implies that $\sigma : G_1 \times G_2 \times G_3 \to X$ defined by $(g_1, g_2, g_3) \mapsto g_3(g_2(g_1P))$ is a dominant morphism. In fact, if $|\varepsilon_1|$ is small, $\exp(\varepsilon_1\Delta_1)P$ is a disc neighborhood of P in the orbit G_1P. Hence $\exp(\varepsilon_3\Delta_3)\cdot\exp(\varepsilon_2\Delta_2)\cdot\exp(\varepsilon_1\Delta_1)(P)$ gives a ball-like, analytic, open neighborhood of the point P. Since $G_1 \times G_2 \times G_3$ has $\mathbb{A}^3$ as the underlying space, σ gives a dominant morphism $\sigma : \mathbb{A}^3 \to X$, and τ is also a dominant morphism. Let $\rho : Z \to X$ be a connected topological covering. Then the fiber product $\rho_{\mathbb{A}^3} : (Z, \rho) \times_X (\mathbb{A}^3, \sigma) \to \mathbb{A}^3$ is a topological covering. Since $\pi_1(\mathbb{A}^3) = (1)$, $Z \times_X \mathbb{A}^3$ contains $\mathbb{A}^3$ as a connected component. Then the morphism σ splits as $\sigma : \mathbb{A}^3 \longrightarrow Z \xrightarrow{\rho} X$. This implies that Z is a smooth affine algebraic threefold and $\deg \rho \le \deg \sigma$. Hence $|\pi_1(X)| \le \deg \sigma$.

(2) Consider the case $n = 3$. Assume that X is factorial. Then $Y = X/\!/G_a$ is factorial because $\Gamma(Y, \mathcal{O}_Y)$ is factorially closed in $\Gamma(X, \mathcal{O}_X)$ as the kernel of Δ_1. Since $\tau : \mathbb{A}^2 \to Y$ is dominant, it follows that $\overline{\kappa}(Y) = -\infty$ and there are no nonconstant units in $\Gamma(Y, \mathcal{O}_Y)$. In fact, Y is smooth by the assumption, and hence we have $\overline{\kappa}(Y) \le \overline{\kappa}(\mathbb{A}^2) = -\infty$. Then Y is isomorphic to $\mathbb{A}^2$. Since the general fiber of q is isomorphic to $\mathbb{A}^1$ and the factorial closedness of $\Gamma(Y, \mathcal{O}_Y)$ in $\Gamma(X, \mathcal{O}_X)$ implies that the fibers over codimension one points

11 If $n > 3$, a similar proof works if one can produce a dominant morphism $\sigma : G_1 \times G_2 \times \cdots \times G_n \to X$.

of Y are all geometrically reduced, we can apply Nori's lemma [182, Lemma 1.5] to obtain an exact sequence

$$\pi_1(\mathbb{A}^1) \to \pi_1(X) \to \pi_1(Y) \to (1),$$

if every fiber of q contains a reduced irreducible component. This implies that $\pi_1(X) = (1)$. If there is a point $Q \in Y$ such that the fiber $q^{-1}(Q)$ has all nonreduced irreducible components, let $Y^\circ = Y \setminus S$, where S is the finite set

$$\{Q \in Y \mid \text{every component of } q^{-1}(Q) \text{ is nonreduced}\}$$

and let $X^\circ = q^{-1}(Y^\circ)$. Since $\pi_1(Y^\circ) = \pi_1(Y) = (1)$, we can apply Nori's lemma to $q^\circ : X^\circ \to Y^\circ$ to obtain $\pi_1(X^\circ) = (1)$. Since $\operatorname{codim}_X(X \setminus X^\circ) \geq 2$, we have $\pi_1(X) = \pi_1(X^\circ) = (1)$.

Now assume that $n = 2$. Then $Y \cong \mathbb{A}^1$ and $q : X \to Y$ is an $\mathbb{A}^1$-bundle because $\Gamma(Y, \mathcal{O}_Y)$ is factorially closed in $\Gamma(X, \mathcal{O}_Y)$ and X is factorial. Hence $X \cong Y \times \mathbb{A}^1 \cong \mathbb{A}^2$. In particular, $\pi_1(X) = (1)$. In the case $n = 1$, it is clear that $X \cong \mathbb{A}^1$. So, $\pi_1(X) = (1)$. □

For the affine threefold X in Theorem 2.4.19 with or without assuming that X is factorial, the universal covering space $\widetilde{X}$ of X is a smooth affine threefold such that $\widetilde{X}$ is simply connected and has three independent G_a-actions since the three given G_a-actions on X lift onto $\widetilde{X}$. It is an interesting problem to find out what kind of structure the threefold $\widetilde{X}$ has. An optimistic expectation is that a simply-connected affine smooth variety X with as many independent G_a-actions as $\dim X$ is isomorphic to the affine space. But this is completely irrelevant as shown by the following examples.

Example 2.4.20. Let X be a product of a Danielewski surface $\{xy = z^2 - 1\}$ and the affine line $\mathbb{A}^1$. Then X has three independent G_a-actions because the Danilewski surface has two independent G_a-actions and the direct product factor $\mathbb{A}^1$ has a third G_a-action making the Danielewski surface invariant. Furthermore, X is simply connected because the Danilewski surface is isomorphic to $\mathbb{P}^1 \times \mathbb{P}^1 - \{\text{diagonal}\}$. Meanwhile, the Picard number of X is one because such is the Picard number of the Danielewski surface. A similar example is a hypersurface $X = \{xyz = u^3 - 1\}$ in $\mathbb{A}^4$. Furthermore, X has three G_a-actions which are defined by locally nilpotent derivations δ_i $(i = 1, 2, 3)$, where $\delta_1(x) = \delta_1(y) = 0$, $\delta_1(z) = 3u^2$ and $\delta_1(u) = xy$. The other two lnds δ_2 and δ_3 are defined in a similar fashion by changing the roles of x, y, z. Also X is simply connected, but the Picard number is nonzero. In fact, the projection $(x, y, z, u) \mapsto x$ is a fibration such that a general fiber is isomorphic to a Danielewski surface and the fiber over $x = 0$ is three disjoint copies of $\mathbb{A}^2$.

We consider an additional condition that X is factorial. If $\dim X = 2$, the proof of Theorem 2.4.19 shows that X is isomorphic to $\mathbb{A}^2$ provided X is a factorial smooth affine surface with two independent G_a-actions. But if $\dim X \geq 3$, the situation does not get improved as shown by the following result of V. L. Popov.

Proposition 2.4.21. *Let X be the affine quadric hypersurface $\{xz - yu = 1\}$ in $\mathbb{A}^4 = \operatorname{Spec} k[x,y,z,u]$. Then the following assertions hold:*
(1) *X is the underlying scheme of* SL(2) *and is factorial.*
(2) *X is not isomorphic to $\mathbb{A}^3$.*
(3) *X has three independent G_a-actions.*

Proof. (1) The first assertions is clear, and the second assertion follows from Lemma 2.4.22.

(2) The proof which we give here is unexpectedly difficult. Let δ be an lnd defined by $\delta(x) = \delta(y) = 0, \delta(z) = y$ and $\delta(u) = x$. This is a left action of the group of lower triangular unipotent matrices to SL(2). Then the quotient surface $\mathrm{SL}(2)/G_a \cong \mathbb{A}^2 = \operatorname{Spec} k[x,y]$. But the fiber of the quotient morphism over the point $(x,y) = (0,0)$ is the empty set. Suppose that $\mathrm{SL}(2) \cong \mathbb{A}^3$. Then the algebraic quotient is again $\mathbb{A}^2$ by Theorem 2.4.23 and the quotient morphism is surjective according to Bonnet [18].

(3) Define locally nilpotent derivations δ_i $(1 \le i \le 4)$ of the coordinate ring $R = k[x,y,z,u]/(xz - yu - 1)$ by the following data:

$$\begin{aligned}
&\delta_1(x) = 0 = \delta_1(y), \quad \delta_1(z) = y, \quad \delta_1(u) = x,\\
&\delta_2(x) = y, \quad \delta_2(z) = 0 = \delta_2(y), \quad \delta_2(u) = z,\\
&\delta_3(x) = 0 = \delta_3(u), \quad \delta_3(z) = u, \quad \delta_3(y) = x,\\
&\delta_4(x) = u, \quad \delta_4(y) = z, \quad \delta_4(z) = \delta_4(u) = 0.
\end{aligned}$$

In fact, $\exp(t\delta_1)$ corresponds to the left action of U_1 on SL(2), and $\exp(t\delta_4)$ to the left action of U_2 on SL(2), where

$$U_1 := \left\{ \begin{pmatrix} 1 & 0 \\ t & 1 \end{pmatrix} \mid t \in k \right\}, \quad U_2 := \left\{ \begin{pmatrix} 1 & t \\ 0 & 1 \end{pmatrix} \mid t \in k \right\}.$$

Meanwhile, $\exp(t\delta_2)$ corresponds to the right action of U_1 on SL(2), and $\exp(t\delta_3)$ to the right action of U_2. Consider the tangential directions of G_a-orbits at a general point P. Since

$$\frac{\partial}{\partial u} = y\left(\frac{1}{z} \cdot \frac{\partial}{\partial x} + \frac{1}{u} \cdot \frac{\partial}{\partial y} + \frac{1}{x} \cdot \frac{\partial}{\partial z}\right),$$

we may assume that $\{\frac{\partial}{\partial x}, \frac{\partial}{\partial y}, \frac{\partial}{\partial z}\}$ is a basis of $T_{\mathrm{SL}(2),P}$. Now we can write

$$\begin{aligned}
\delta_1 &= y\frac{\partial}{\partial z} + x\frac{\partial}{\partial u}, \qquad \delta_2 = y\frac{\partial}{\partial x} + z\frac{\partial}{\partial u},\\
\delta_3 &= u\frac{\partial}{\partial z} + x\frac{\partial}{\partial y}, \qquad \delta_4 = u\frac{\partial}{\partial x} + z\frac{\partial}{\partial y}.
\end{aligned}$$

Let (a,b,c,d) be the coordinates of P with respect to the system (x,y,z,u). Then we have

$$\delta_1 = \left(\frac{ab}{c}, \frac{ab}{d}, 2b\right), \quad \delta_2 = \left(2b, \frac{bc}{d}, \frac{bc}{a}\right), \quad \delta_3 = (0,a,d), \quad \delta_4 = (d,c,0)$$

with respect to the above basis of $T_{SL(2),P}$. It is now clear that $\delta_1, \delta_2, \delta_3$ generate the tangent space $T_{SL(2),P}$. Hence the G_a-actions corresponding to δ_i ($i = 1, 2, 3$) are independent. □

The following result is due to Nagata [174], which we call *Nagata's lemma*.

Lemma 2.4.22. *Let A be a noetherian domain. Then A is factorial if there exists an element $a \in A \setminus \{0\}$ such that*
(1) *a is a prime element of A, i. e., A/aA is an integral domain.*
(2) *The ring $A[a^{-1}]$ is factorial.*

We used the following result (see [154] and Theorem 3.2.5).

Theorem 2.4.23. *Let $A = k[x, y, z]$ be a polynomial ring in three variables and let δ be an lnd of A. If $\delta \neq 0$, $\operatorname{Ker} \delta$ is a polynomial ring in two variables.*

We will introduce an invariant which measures roughly how many independent G_a-actions a given affine variety X has. Let $X = \operatorname{Spec} A$ be an affine variety of dimension n. Considering all locally nilpotent derivations δ of A, take the k-subalgebra $\bigcap_\delta \operatorname{Ker} \delta$, which we call the *Makar-Limanov invariant* and denote by $\mathrm{ML}(X)$. We call the number $\mathrm{tr.deg}_k Q(\mathrm{ML}(X))$ the ML-*dimension* of X and denote it by $\mathrm{MLdim}(X)$. We say that X is an ML_i-*variety* if $\mathrm{MLdim}(X) = i$, where $0 \le i \le n$. The condition $\mathrm{MLdim}(X) = n$ is equivalent to the nonexistence of a nontrivial G_a-action on X. There are many algebraic descriptions and examples of Makar-Limanov invariants in [55] including the following definition [Principle 12, p. 29].

Definition 2.4.24. Let A be an integral domain. Two nonzero lnds δ_1, δ_2 are *rationally equivalent* ($\delta_1 \sim \delta_2$ by notation) if the following two equivalent conditions are satisfied:
(1) $\operatorname{Ker} \delta_1 = \operatorname{Ker} \delta_2$.
(2) $a_1\delta_2 = a_2\delta_1$ for nonzero elements a_1, a_2 of $\operatorname{Ker} \delta_1$.

A geometric interpretation of rational equivalence is given by the following lemma.

Lemma 2.4.25. *Let $X = \operatorname{Spec} A$ be an affine variety equipped with two G_a-actions $\sigma_i : G_a \times X \to X$ ($i = 1, 2$) and let δ_i be an lnd of A corresponding to σ_i. Then $\delta_1 \sim \delta_2$ if and only if the two G_a-orbits of P coincide, i. e., $\sigma_1(G_a, P) = \sigma_2(G_a, P)$, for a general point P of X.*

Proof. Suppose that $\delta_1 \sim \delta_2$. Let $B = \operatorname{Ker} \delta_1 = \operatorname{Ker} \delta_2$. Then there exist elements $x_i \in A$ and $b_i = \delta_i(x_i) \in B$ for $i = 1, 2$ such that $A[b_i^{-1}] = B[b_i^{-1}][x_i]$, where the G_a-actions σ_i are given by $\sigma_i(t, x_i) = x_i + b_i t$ for $t \in G_a$. Let $b = b_1 b_2$. Replacing x_1 and x_2 by $b_2 x_1$ and $b_1 x_2$, respectively, we may assume that $\delta_1(x_1) = \delta_2(x_2) = b$ in $A[b^{-1}] = B[b^{-1}][x_1] = B[b^{-1}][x_2]$. Then $x_2 = cx_1 + d$ with $c \in B[b^{-1}]^*, d \in B[b^{-1}]$. Hence we have the coincidence of two orbits of a general point P.

Suppose that two orbits of P by σ_i coincide for a general point P. This implies that the two $\mathbb{A}^1$-fibrations $\operatorname{Spec} A \to \operatorname{Spec} B_i$ with $B_i = \operatorname{Ker}\delta_i$ coincide on open sets of $\operatorname{Spec} B_1$ and $\operatorname{Spec} B_2$. In particular, it follows that $Q(B_1) = Q(B_2)$. Then $B_1 = B_2$ because $B_i = A \cap Q(B_i)$. Let $B = B_1 = B_2$. With the above notations, the $\mathbb{A}^1$-fibrations $\operatorname{Spec} A[b_i^{-1}] \to \operatorname{Spec} B[b_i^{-1}]$ for $i = 1, 2$ constructed by means of local slices coincide. By the above substitution, we have $A[b^{-1}] = B[b^{-1}][x_1] = B[b^{-1}][x_2]$. Hence $x_2 = cx_1 + d$ with $c \in B[b^{-1}]^*, d \in B[b^{-1}]$. Then $(c\delta_2 - \delta_1)(x_1) = 0$. This implies that $c\delta_2 = \delta_1$, whence $a_1\delta_2 = a_2\delta_1$ for $c = a_1/a_2$ with $a_1, a_2 \in B$. □

Given an affine domain over k, it is not clear yet how $\operatorname{MLdim}(X)$ is related to the maximal number of independent lnds on A. We still have the following result.

Lemma 2.4.26. *Let $X = \operatorname{Spec} A$ be an affine variety of dimension n. The following assertions hold:*

(1) *Suppose that* $\operatorname{MLdim}(X) = n - 1$. *Then there exists a nontrivial lnd δ which is unique up to the rational equivalence relation. Namely, for two nontrivial lnds δ_1 and δ_2, δ_1 is rationally equivalent to δ_2.*
(2) *Suppose that there exist independent lnds $\delta_1, \dots, \delta_i$ for $1 \le i \le n$ on A. Then* $\operatorname{MLdim}(X) \le n - i$.
(3) *If* $\operatorname{MLdim}(X) = n - 2$, *then there exist two independent lnds δ_1, δ_2 such that* $\operatorname{ML}(X) = \operatorname{Ker}\delta_1 \cap \operatorname{Ker}\delta_2$.

Proof. (1) Let $A_i = \operatorname{Ker}\delta_i$ for $i = 1, 2$. We show first that $A_1 = A_2$. By Lemma 2.2.1, $Q(A_i)$ is algebraically closed in $Q(A)$ and $\operatorname{tr.deg}_k Q(A_i) = n - 1$. Since $Q(\operatorname{ML}(X)) \subseteq Q(A_1 \cap A_2)$, the assumption implies that $Q(A_i)$ is algebraic over $Q(A_1 \cap A_2)$. Take any element a_1 of A_1. Then a_1 is algebraic over $Q(A_2)$, whence $a_1 \in Q(A_2) \cap A = A_2$. Hence $A_1 \subseteq A_2$. Similarly, $A_2 \subseteq A_1$. So, $A_1 = A_2$. Now we can choose an element $x \in A$ so that $\delta_1(x) = a_1 \in A_1 \setminus \{0\}$. Hence $A[a_1^{-1}] = A_1[a_1^{-1}][x]$ by Lemma 2.2.1. Since δ_2 is an lnd of A, we have $\delta_2(x) = F(x) \in A_1[a_1^{-1}][x]$. Then $\delta_2^2(x) = F'(x)\delta_2(x) = F'(x)F(x)$. Since δ_2 is locally nilpotent, this implies that $F(x) \in A_1[a_1^{-1}] \cap A = A_1$. Hence we can write $\delta_2(x) = a_2 \in A_1 \setminus \{0\}$. Then we have $a_2\delta_1 = a_1\delta_2$.

(2) Let $B = \bigcap_{j=1}^{i} \operatorname{Ker}\delta_i$ and let $K = Q(B)$. Although B might not be finitely generated over k, the quotient field K is finitely generated over k as a subfield of $Q(A)$. Let $A_K = A \otimes_B K$ and let $X_K = \operatorname{Spec} A_K$. Then A_K is an affine domain over the field K and X_K is the generic fiber of the morphism $q : X \to \operatorname{Spec} B$. In fact, K is algebraically closed in A_K. If $\alpha \in A_K$ is algebraic over K, it follows that $\delta_j(\alpha) = 0$ for $(1 \le j \le i)$ and hence $\alpha \in K$. This implies that X_K is a geometrically integral scheme over K. In order to verify the assertion, it suffices to show that $\dim X_K \ge i$. In fact, let P be a general point of X. Let $\sigma_j : G_a \times X \to X$ be the G_a-action corresponding to δ_j. As in the proof of Theorem 2.4.19, G_j in the action σ_j denotes the additive group G_a. Set

$$F_P = \{g_i(g_{i-1}(\cdots(g_1P))) \mid g_j \in G_j, 1 \le j \le i\}$$

and let $\overline{F}_P$ be the closure of F_P in X. Then $\dim \overline{F}_P = i$ by the independence of $\delta_1, \ldots, \delta_i$. On the other hand, $f(Q) = f(P)$ for every point $Q \in F_P$ and every element $f \in B$ because f is invariant under the σ_j-action of G_a. Hence f is constant on $\overline{F}_P$. This implies that $\overline{F}_P$ is contained in the fiber $q^{-1}(q(P))$. Since q is an equidimensional morphism of algebraic varieties over an affine open set $U = \operatorname{Spec} B[b^{-1}]$ of $\operatorname{Spec} B$ for some $b \in B$, we have $\dim X_K \geq i$.

(3) There exist at least two lnds δ_1, δ_2 of A such that $\delta_1 \not\sim \delta_2$. Otherwise, $\mathrm{MLdim}(X) = n-1$ by assertion (1). Hence δ_1 and δ_2 are independent by Lemma 2.4.25. Suppose that $\operatorname{Ker} \delta_1 \cap \operatorname{Ker} \delta_2 \supsetneq \mathrm{ML}(X)$. Since $\operatorname{tr.deg}(\operatorname{Ker} \delta_1 \cap \operatorname{Ker} \delta_2) \leq n-2$ by assertion (2), an element $\alpha \in (\operatorname{Ker} \delta_1 \cap \operatorname{Ker} \delta_2) \setminus \mathrm{ML}(X)$ is algebraic over $\mathrm{ML}(X)$. Replacing α by $b\alpha$ with $b \in \mathrm{ML}(X)$, we may assume that α satisfies a monic relation

$$\alpha^m + c_1\alpha^{m-1} + \cdots + c_{m-1}\alpha + c_m = 0, \quad c_i \in \mathrm{ML}(X).$$

We can choose α so that m is minimal. Let δ be an arbitrary lnd of A. Since $\delta(c_i) = 0$, we have

$$(m\alpha^{m-1} + (m-2)c_1\alpha^{m-2} + \cdots + c_{m-1})\delta(\alpha) = 0.$$

By the choice of α so that m is minimal, it follows that $\delta(\alpha) = 0$. Hence $\alpha \in \mathrm{ML}(X)$. This is a contradiction. So, $\mathrm{ML}(X) = \operatorname{Ker} \delta_1 \cap \operatorname{Ker} \delta_2$. □

It is not known if the assumption $\mathrm{MLdim}(X) = n - i$ implies the existence of i independent lnds on $X = \operatorname{Spec} A$. The following example due to V. L. Popov is related to Theorem 2.4.19.

Example 2.4.27. Let G be a connected, simply-connected, semisimple algebraic group and let H be a reductive subgroup of G without nontrivial multiplicative characters. Let $X = G/H$. Then $\mathrm{ML}(X) = k$, but X is not isomorphic to $\mathbb{A}^n$, where $n = \dim X$.

Proof. By a theorem of Matsushima, X is a smooth affine variety. By [188, Proposition 1], $\operatorname{Pic} G = 0$, and by [188, Corollary, p. 304], $\operatorname{Pic} X = 0$ since H has no nontrivial characters. On the other hand, by [189, Corollary 1.5], $\mathrm{ML}(X) = k$. Finally, by [130, Corollary, p. 476], X is not isomorphic to $\mathbb{A}^n$. Note that X is simply-connected since so is G by the assumption. □

Let $\delta_1, \ldots, \delta_i$ be lnds of an affine k-domain A. We say that $\delta_1, \ldots, \delta_i$ are *mutually commutative* if $\delta_j\delta_{j'} = \delta_{j'}\delta_j$ for $1 \leq j, j' \leq i$. Let σ_j be the G_a-action on $X = \operatorname{Spec} A$ associated with δ_j. Then $\sigma_1 \times \cdots \times \sigma_i$ defines a U-action on X, where $U \cong G_a^{\times i}$ is a commutative unipotent group, i. e., a *vector group*. The U-action is faithful, i. e., the isotropy group at a general point of X is trivial, if and only if $\delta_1, \ldots, \delta_i$ are independent. The following result is a generalization of [82, Lemma 1.2].

Lemma 2.4.28. *Let $\delta_1, \ldots, \delta_i$ be mutually commutative lnds on an affine domain $A := A_0$ and let $B = \bigcap_{j=1}^{i} \operatorname{Ker} \delta_j$. Set $A_j = \bigcap_{\ell=1}^{j} \operatorname{Ker} \delta_\ell$ for $1 \leq j \leq i$. Then the following assertions hold:*

(1) *Suppose that* $\delta_1, \ldots, \delta_i$ *are independent. Then there exist elements* $b \in B \setminus \{0\}, x_1, \ldots, x_i \in A$ *such that* $A[b^{-1}] = B[b^{-1}][x_1, \ldots, x_i]$, *where* $x_1, \ldots, x_i$ *are algebraically independent over* $B[b^{-1}]$ *and the* x_j*s are taken so that* $x_j \in A_{j-1}$ *for* $1 \leq j \leq i$. *Hence* X *contains an* $\mathbb{A}^i$*-cylinder* $\mathbb{A}^i \times Z$ *as an open set, where* $Z = \operatorname{Spec} B[b^{-1}]$.

(2) *Conversely, suppose that* X *contains an* $\mathbb{A}^i$*-cylinder* $W = \mathbb{A}^i \times Z$ *as an affine open set. If we assume additionally that* X *is factorial, then there exist mutually commutative and independent lnds* $\delta_1, \ldots, \delta_i$ *of* A *such that the cylinder* W *is the one obtained in the above assertion.*

Proof. (1) Since $\delta_1, \ldots, \delta_i$ are independent, we have a descending sequence of k-subdomains

$$A \supsetneq A_1 \supsetneq A_2 \supsetneq \cdots \supsetneq A_{i-1} \supsetneq A_i = B.$$

We prove by induction on i that there exist elements $x_1 \in A$ and $a_1 \in B \setminus \{0\}$ such that $\delta_1(x_1) = a_1$. If $i = 1$ this is well known. Suppose that this is the case for $i - 1$. Take $x_1 \in A$ such that $a_1 = \delta_1(x_1) \in A_{i-1}$. Find an integer s such that $\delta_i^s(a_1) \neq 0$ and $\delta_i^{s+1}(a_1) = 0$. Let $y = \delta_i^s(x_1)$. Since $\delta_1(y) = \delta_i^s(\delta_1(x_1)) = \delta_i^s(a_1) \neq 0$, whence $y \neq 0$. Further, $\delta_j(\delta_i^s(a_1)) = \delta_i^s(\delta_j(a_1)) = 0$ for $1 \leq j < i$. So, replace x_1, a_1 by $\delta_i^s(x_1), \delta_i^s(a_1)$, respectively.

By the local slice construction, $A[a_1^{-1}] = A_1[a_1^{-1}][x_1]$. Since $\delta_j(a_1) = 0$ for $1 < j \leq i$, δ_j extends to the lnds of $A_1[a_1^{-1}]$ which is an affine domain over k. By induction on i, we can find elements $x_2, \ldots, x_i \in A$ and $b \in B$ such that $A_1[(ba_1)^{-1}] = B[(a_1 b)^{-1}][x_2, \ldots, x_n]$. Then $A[(ba_1)^{-1}] = B[(ba_1)^{-1}][x_1, x_2, \ldots, x_n]$.

(2) Since Z is affine, write $Z = \operatorname{Spec} R$. Then $W = \operatorname{Spec} R[x_1, \ldots, x_i]$. Let $B = A \cap R$ as a k-subalgebra of $k(X) = Q(A)$. Since X is factorial and W is an affine open set, the complement $F := X \setminus W$ has pure codimension one by Lemma 2.4.12. Then there exists an element $f \in A$ such that $F = V(f)$. Since f has no zero on W, f is invertible in $R[x_1, \ldots, x_i]$ and hence $f \in R^*$. Let $h \in R$ be an arbitrary element. Since the divisor (h) has poles only on F, it follows that $f^m h \in A$. Hence $f^m h \in R \cap A = B$. This implies that $R = B[f^{-1}]$. Consider x_j which is a coordinate of the $\mathbb{A}^i$-cylinder W. Considering the divisor (x_j) on X, it follows that $f^r x_j$ is an element of A for some $r > 0$. Replacing x_j by $f^r x_j$, we may assume that $x_1, \ldots, x_i \in A$. Now consider a partial derivative $\delta_j' = \partial/\partial x_j$ of $B[f^{-1}][x_1, \ldots, x_i]$. Since A is finitely generated over k, a multiple $\delta_j = f^N \delta_j'$ is an lnd of A for some $N > 0$. It is then clear that $\delta_1, \ldots, \delta_i$ are independent, mutually commutative lnds of A such that $B = \bigcap_{j=1}^i \operatorname{Ker} \delta_j$. □

Lemma 2.4.28 implies the following algebraic characterization of the affine n-space.

Theorem 2.4.29. *Let* $X = \operatorname{Spec} A$ *be a factorial affine variety of dimension* n. *Then* X *is isomorphic to* $\mathbb{A}^n$ *if and only if there exist independent, mutually commuting lnds* $\delta_1, \ldots, \delta_n$ *of* A.

Proof. If $X \cong \mathbb{A}^n$, write $A = k[x_1, \ldots, x_n]$. Let $\delta_i = \partial/\partial x_i$ for $1 \le i \le n$. Then the δ_is are independent, mutually commutative lnds of A. Conversely, suppose that the independent lnds δ_i satisfy the mutual commutativity condition. Then $B = \bigcap_{i=1}^n \operatorname{Ker} \delta_i$ is isomorphic to k because $\mathrm{MLdim}(X) = 0$ by Lemma 2.4.26. By Lemma 2.4.28, (1), X contains an affine open set W isomorphic to $\mathbb{A}^n$. Since $X \setminus W$ has pure codimension one by Lemma 2.4.12, $X \setminus F$ is the closed set defined by $f = 0$ for some $f \in A$. Then $f \in (k[x_1, \ldots, x_n])^* = k^*$. Hence $X \setminus W = \emptyset$. So, $X \cong \mathbb{A}^n$. □

Theorem 2.4.29 can be slightly generalized if some of the assumptions are strengthened.

Proposition 2.4.30. *Let $X = \operatorname{Spec} A$ be an affine variety of dimension n. Then X is isomorphic to $\mathbb{A}^n$ if and only if the following conditions are satisfied:*
(1) *A is factorial and $A^* = k^*$.*
(2) *There exist mutually commutative lnds $\delta_1, \ldots, \delta_{n-1}$ of A which are independent at every closed point P of X.*
(3) *Every closed fiber of the morphism $q : X \to \operatorname{Spec} B$ is factorial, where $B = \bigcap_{i=1}^{n-1} \operatorname{Ker} \delta_i$.*

Proof. With B defined as in the statement, B is a factorial affine k-domain of dimension one with $B^* = k^*$. In fact, $B = A \cap Q(B)$, whence B is an affine domain by Theorem 2.2.4. Hence $B = k[x]$. For every $\lambda \in k$, let $\overline{\delta}_1^{\lambda}, \ldots, \overline{\delta}_{n-1}^{\lambda}$ be the lnds induced by $\delta_1, \ldots, \delta_{n-1}$ on $A/(x - \lambda)$, where $A/(x - \lambda)$ is an affine domain of dimension $n - 1$ because $x - \lambda$ is irreducible in A. By condition (2), $\overline{\delta}_1^{\lambda}, \ldots, \overline{\delta}_{n-1}^{\lambda}$ are independent on $A/(x - \lambda)$, and hence the factorial variety $\operatorname{Spec} A/(x - \lambda)$ is isomorphic to $\mathbb{A}^{n-1}$ by Theorem 2.4.29. Now, by [35, Theorem 3], $A = k[x, x_1, \ldots, x_{n-1}]$, i. e., $X \cong \mathbb{A}^n$. □

In the surface case, structures of ML_i-surfaces will be discussed in detail in Section 2.6.

2.4.4 G_a-actions on projective varieties and orbit stratifications

We begin with a fundamental result of Horrocks [93].

Lemma 2.4.31. *Let G be a unipotent group acting nontrivially on a projective variety V. Then the following assertions hold:*
(1) *The fixed point locus $V^G = \{P \in V \mid {}^gP = P,\ g \in G\}$ is a nonempty closed algebraic set.*
(2) *Let $\text{Ét}(V)$ and $\text{Ét}(V^G)$ be the categories of étale finite coverings of V and V^G, respectively. Then a functor $\text{Ét}(V) \to \text{Ét}(V^G)$ which maps $V' \in \text{Ét}(V)$ to $V' \times_V V^G \in \text{Ét}(V^G)$ is an equivalence of categories.*

(3) *Let $\pi_0(V)$ and $\pi_0(V^G)$ be the sets of connected components of V and V^G, respectively. Then $\pi_0(V) = \pi_0(V^G)$. In particular, if V is connected, so is V^G.*
(4) *Let $\pi_1^{\mathrm{alg}}(V)$ and $\pi_1^{\mathrm{alg}}(V^G)$ be the algebraic fundamental groups of V and V^G, respectively. Then $\pi_1^{\mathrm{alg}}(V) \cong \pi_1^{\mathrm{alg}}(V^G)$. Here $\pi_1^{\mathrm{alg}}(V)$ is the profinite completion of $\pi_1(V)$ if k is of characteristic zero.*

Next we recall necessary results of G-linearizations of line bundles according to [172, 116].

Definition 2.4.32. Let G be a linear algebraic group acting on a projective variety V by $\sigma : G \times V \to V$. Let $\pi : L \to V$ be a line bundle over V which is identified with the associated invertible sheaf $\mathcal{L}$. Then L is *G-linearizable* if there exists a G-action $\sigma_L : G \times L \to L$ such that
(1) The projection $\pi : L \to V$ is G-equivariant, i. e., $\pi \cdot \sigma_L = \sigma \cdot (\mathrm{id}_G \times \pi)$,
(2) The induced action of G on fibers of π are linear, i. e., for $P \in V$ and $g \in G$, $(\sigma_L)_P : L_P \to L_{gP}$ is linear.

The action $\sigma_L : G \times L \to L$ is called a *G-linearization* of L.

The following results are found in [116].

Lemma 2.4.33. *Let $\sigma : G \times V \to V$ and L be the same as above. Assume that V is normal. Then the following assertions hold:*
(1) *L is G-linearizable if and only if two line bundles $p_V^*(L)$ and $\sigma^*(L)$ on $G \times V$ are isomorphic, where $p_V : G \times V \to V$ is the projection.*
(2) *For any line bundle M on $G \times V$, we have $M \cong p_G^*(M|_{G\times\{P_0\}}) \otimes p_V^*(M|_{\{e\}\times V})$, where P_0 is a closed point of V and $p_G : G \times V \to G$ is the projection.*
(3) *$\mathrm{Pic}(G)$ is a finite group.*
(4) *There exists an integer $n > 0$ such that $L^{\otimes n}$ is G-linearizable, where we can take n to be $|\mathrm{Pic}(G)|$. If G is factorial, then L is G-linearizable.*
(5) *If G is a unipotent group then L is G-linearizable.*
(6) *If L is G-linearizable, $H^0(V, L)$ is a G-linear representation by the action $({}^g s)(P) = {}^g(s({}^{g^{-1}}P))$ for $s \in H^0(V, L)$ and $g \in G$.*
(7) *If H is a Cartier divisor on V and G is a unipotent group, then the linear system $|H|$ has a G-fixed point if it is not the empty set. Namely, there exists an effective divisor H_0 such that $H_0 \sim H$ and ${}^g H_0 = H_0$ for $g \in G$.*

Proof. We only prove assertion (7). Let L be the line bundle associated with an invertible sheaf $\mathcal{O}_V(H)$. Since G is unipotent, L is G-linearizable. Hence G acts linearly on $H^0(V, \mathcal{O}(H))$ and accordingly on $\mathbb{P}(H^0(V, \mathcal{O}_V(H)) = |H|$. Since there exists a G-fixed point on the projective space $|H|$, which corresponds to a member H_0 of $|H|$ such that ${}^g H_0 = H_0$ for $g \in G$. □

These results will be used later on.

We consider a characterization of the projective space $\mathbb{P}^n$ in terms of the unipotent group orbit stratification given in Proposition 2.4.15(3). Let V be a smooth projective variety of dimension n and let H be a very ample divisor. Then the following conditions are satisfied:

(1) The linear system $|H|$ has no base points and contains a smooth member.
(2) There exist members $H_1, H_2, \ldots, H_n$ of $|H|$ such that $V_i = H_1 \cap H_2 \cap \cdots \cap H_i$ is a smooth and irreducible subvariety of dimension $n - i$, where $0 \le i \le n$ and $V_0 = V$.

We consider a descending chain of log pairs (V_i, V_{i+1}) for $0 \le i \le n - 1$.

Theorem 2.4.34. *With the above notations, we assume that the pair (V_i, V_{i+1}) has log-Kodaira dimension $-\infty$ for $0 \le i \le n-1$. Then V is isomorphic to $\mathbb{P}^n$ and H is a hyperplane.*

Proof. We consider first the case $n = 2$. Then V_1 is a smooth irreducible curve with $\overline{\kappa}(V_1 - V_2) = -\infty$. Hence $V_1 \cong \mathbb{P}^1$ and V_2 is a point. This implies that the self-intersection number H^2 is equal to one. Let $X = V \setminus V_1$. Since H is ample, X is an affine surface with $\overline{\kappa}(X) = -\infty$. Hence there exists an $\mathbb{A}^1$-fibration $\rho : X \to C$, where C is a smooth curve. Suppose that ρ extends to a $\mathbb{P}^1$-fibration $p : V \to \overline{C}$. Then $V_1 \cong \overline{C}$, and hence V is a rational ruled surface and V_1 is a cross-section of p. If V is not minimal, there exists an irreducible fiber component which is disjoint from V_1. This is absurd because H is ample. So, V is a Hirzebruch surface $\mathbb{F}_d$ of degree d. Let M be a minimal section of $\mathbb{F}_d$. Then $H \sim M + s\ell$, where ℓ is a fiber of p. Since $1 = H^2 = -d + 2s$, we have $d = 2s - 1$. Meanwhile, $H \cdot M = s - d = -s + 1 \ge 0$ and $d \ge 0$, whence $s = 1$ and $d = 1$. Then $H \cdot M = 0$, which contradicts the ampleness of H. This implies that the closures of the fibers of ρ in V form a linear pencil Λ with a base point, say P. The curve V_1 is a member of Λ. Since $H^2 = 1$, the pencil Λ becomes free of base point after a single blowing-up with center P. Then the blown-up surface becomes a Hirzebruch surface $\mathbb{F}_1$ and the exceptional curve is the minimal section. Here we note that if there is a reducible member, say F, of Λ, then $F \cdot V_1 = 1$ and hence there exists an irreducible component of F which is disjoint from V_1. This contradicts the ampleness of H. By contracting the exceptional curve back to the point P, we know that $V \cong \mathbb{P}^2$ and H is a line.

Suppose that $n > 2$. We assume by induction that V_1 is isomorphic to $\mathbb{P}^{n-1}$ and V_2 is a hyperplane. We consider a $\mathbb{Z}$-cohomology exact sequence for a pair (V, V_1):

$$H^{2n-2}(V;\mathbb{Z}) \xrightarrow{i^*_{2n-2}} H^{2n-2}(V_1;\mathbb{Z}) \longrightarrow H^{2n-1}(V, V_1;\mathbb{Z}) \longrightarrow H^{2n-1}(V;\mathbb{Z})$$

where $i : V_1 \to V$ is the canonical inclusion. Since $V_2 = V_1 \cap H_2$ for a general member H_2 of $|H|$ and V_2 is a hyperplane of $\mathbb{P}^{n-1}$, H_2 is also $\mathbb{P}^{n-1}$ and V_2 is a hyperplane of H_2. Then there exists a line L on H_2 such that L intersects V_1 transversally in one point P. By the Poincaré duality, $H^{2n-2}(V_1;\mathbb{Z}) \cong H_0(V_1;\mathbb{Z}) \cong \mathbb{Z}$. We may assume that $H_0(V_1;\mathbb{Z})$ is generated by the point P. Similarly, $H^{2n-2}(V;\mathbb{Z}) \cong H_2(V;\mathbb{Z})$. The class $[L]$ in $H_2(V;\mathbb{Z})$

gives rise to an element α of $H^{2n-2}(V;\mathbb{Z})$ which is mapped by i^*_{2n-2} to the element of $H^{2n-2}(V_1;\mathbb{Z})$ corresponding to $[P]$. By the Poincaré duality, this mapping is simply the intersection $L \cap V_1 = \{P\}$. Hence the mapping i^*_{2n-2} is surjective. On the other hand, $H^{2n-1}(V;\mathbb{Z})$ is isomorphic to $H_1(V;\mathbb{Z})$ by the Poincaré duality, and $H_1(V;\mathbb{Z}) \cong H_1(V_1;\mathbb{Z})$ by the Lefschetz hyperplane section theorem. Then $H_1(V_1;\mathbb{Z}) = (0)$ because $V_1 \cong \mathbb{P}^{n-1}$. Hence $H^{2n-1}(V;\mathbb{Z}) = (0)$. The above exact sequence shows that $H^{2n-1}(V, V_1;\mathbb{Z}) = (0)$. By the Lefschetz duality, it follows that $H_1(V - V_1;\mathbb{Z}) = (0)$. Now we can use the following result of C. P. Ramanujam [191] to conclude that $V \cong \mathbb{P}^n$ and V_1 is a hyperplane. □

The following result is due to C. P. Ramanujam [191].

Lemma 2.4.35. *Let V be a smooth projective variety of dimension $n > 2$ and let H be a divisor of V isomorphic to $\mathbb{P}^{n-1}$ such that $V - H$ is affine and $H_1(V - H;\mathbb{Z})$ is torsion free. Then V is isomorphic to $\mathbb{P}^n$.*

Let V be a smooth projective variety of dimension n defined over k. Let Δ be a regular vector field. We say that Δ has a *locally nilpotent stratification* if V has a decomposition $V = \coprod_i W_i$ into locally closed subsets satisfying the following conditions:

(1) W_0 is an affine open set of V, W_i is an affine open set of the closure $\overline{W}_i$ and $\overline{W}_i \setminus W_i$ is the union of closures $\overline{W}_j$ of several W_j with $j > i$.
(2) For every i, $\Delta|_{W_i}$ is a locally nilpotent derivation on $\Gamma(W_i, \mathcal{O}_{W_i})$ unless W_i is a point.

Note that $V \setminus W_0$ is supported by an ample divisor, and for $i > 0$, $\overline{W}_i \setminus W_i$ is also supported by an ample divisor on $\overline{W}_i$ provided $\overline{W}_i$ is smooth.

Lemma 2.4.36. *Let $U = \operatorname{Spec} R$ be an affine open set. Then $\Delta|_U$, the restriction of Δ onto U, is a k-derivation D of R.*

Proof. Since $\Delta \in \Gamma(V, \mathcal{T}_{V/k})$, it follows that $\Delta|_U \in \Gamma(U, \mathcal{T}_{U/k}) \cong \operatorname{Der}_k(R)$. □

Lemma 2.4.37. *Let $U = \operatorname{Spec} R$ and D be as above. Let $U' = \operatorname{Spec} R'$ be an affine open set such that $U' \subseteq U$ and let D' be the k-derivation corresponding to $\Delta|_{U'}$. Then D is the restriction of D' to R. If D' is locally nilpotent, then so is D on R.*

Proof. Note that R and R' are subalgebras in the function field $k(V)$ and hence that $R \subseteq R'$ in $k(V)$. Define a k-algebra homomorphism $\Phi_D : R \longrightarrow R[[t]]$ by

$$\Phi_D(a) = \sum_{i \geq 0} \frac{1}{i!} D^i(a) t^i.$$

Then we have a commutative diagram

$$\begin{array}{ccc} R & \xrightarrow{\Phi_D} & R[[t]] \\ \big\downarrow{\scriptstyle i} & & \big\downarrow{\scriptstyle i[[t]]} \\ R' & \xrightarrow[\Phi_{D'}]{} & R'[[t]] \end{array}$$

where i and $i[[t]]$ are the canonical inclusions. If D' is locally nilpotent, $\Phi_{D'}$ splits via $R'[t]$. Then Φ_D splits via $R[t]$ because $R[t] = R'[t] \cap R[[t]]$. Hence D is locally nilpotent. □

This result implies that if Δ is locally nilpotent on a nonempty affine open set, there exists a maximal affine open set $U_{\max}$ of V such that Δ induces a locally nilpotent derivation on $U_{\max}$. Note that Aut^0_V is a connected algebraic group and $\mathrm{Lie}(\mathrm{Aut}^0(V)) \cong \Gamma(V, \mathcal{T}_{V/k})$. If there is a G_a-action on V, then the associated group homomorphism $G_a \to \mathrm{Aut}^0(V)$ gives rise to a regular vector field Δ on V.

Theorem 2.4.38. *Let V be a smooth projective variety which has an algebraic G_a-action and let Δ be the regular vector field associated with the G_a-action. The following assertions hold:*
(1) *Let H be an effective ample divisor such that the subset H_{red} of codimension one is G_a-stable and let $W_0 = \mathrm{Spec}\, R_0$ be the complement of H_{red}. Then $D_0 = \Delta|_{W_0}$ is a locally nilpotent derivation.*
(2) *The divisor H is G_a-linearizable, and hence there exists a member of $|H|$ which is G_a-stable. If H_0 is a G_a-stable member of $|H|$, then $V \setminus H_0$ is a G_a-stable affine open set.*

Proof. (1) Since H is ample, the complement $W_0 = V \setminus H_{\mathrm{red}}$ is an affine open set and has the induced G_a-action. Hence the restriction $\Delta|_{W_0}$ is associated to the induced G_a-action on W_0. This implies that D_0 is locally nilpotent.

(2) By Lemma 2.4.33, H is G_a-linearizable and the linear system $|H|$ contains a G_a-stable member H_0. Then it is clear that $V \setminus H_0$ is a G_a-stable affine open set. □

Remark 2.4.39. Let V be a smooth projective variety with a nontrivial G_a-action. Let H be a very ample divisor. By Theorem 2.4.38(2), there is a G_a-stable affine open set W_0 such that $W_0 = V \setminus H_1$ with a G_a-stable member $H_1 \in |H|$. If H_1 is smooth, then we consider a very ample divisor $H|_{H_1}$. If the induced G_a-action on H_1 is nontrivial, we find a G_a-fixed member of $|H|_{H_1}|$. Suppose further that the irregularity $q := h^1(V, \mathcal{O}_V) = 0$. Then the exact sequence

$$0 \longrightarrow \mathcal{O}_V \longrightarrow \mathcal{O}_V(H) \longrightarrow \mathcal{O}_{H_1}(H) \longrightarrow 0$$

induces the surjection $\mathrm{H}^0(V, \mathcal{O}_V(H)) \to \mathrm{H}^0(H_1, \mathcal{O}_{H_1}(H))$. So, the G_a-fixed member of $|H|_{H_1}|$ is written as $H_1 \cap H_2$. Note that the irregularity of H_1 vanishes. In fact, if $\dim V \geq 3$, then the exact sequence

$$\mathrm{H}^1(V, \mathcal{O}_V) \longrightarrow \mathrm{H}^1(H_1, \mathcal{O}_{H_1}) \longrightarrow \mathrm{H}^2(V, \mathcal{O}_V(-H))$$

implies the assertion because $\mathrm{H}^2(V, \mathcal{O}_V(-H)) = 0$ by the Kodaira vanishing theorem. If $\dim V = 2$, then $\mathrm{H}^2(V, \mathcal{O}_V(-H)) \cong \mathrm{H}^0(V, \mathcal{O}(K_V + H)) = 0$ because the existence of a G_a-action implies $\overline{\kappa}(V \setminus H_1) = -\infty$. Suppose that the fixed point locus Γ of the G_a-action on V consists of a single point and inductively that $H_1 \cap H_2 \cap \cdots \cap H_i$ is smooth for $1 \le i \le \dim V - 1$. Then the G_a-action on $H_1 \cap \cdots \cap H_i$ is nontrivial if $i \le \dim V - 1$ since Γ consists of a single point. Thus we reach to the set-up of Theorem 2.4.34. The theorem implies that $V \cong \mathbb{P}^n$. If $V \not\cong \mathbb{P}^n$, it fails to hold that $H_1 \cap \cdots \cap H_i$ is smooth for some $1 \le i \le \dim V - 1$. □

Given a regular vector field Δ on a smooth projective variety V, an irreducible subvariety W of codimension one is called *integral* if for every smooth point P of W and for a system of local parameters $\{u_1, \ldots, u_n\}$ of V at P such that W is defined by $u_1 = 0$, we have $\Delta(u_1) \in u_1\mathcal{O}_{V,P}$. If dimV= 2, we call W an *integral curve* of Δ.

Example 2.4.40. With the notations in Lemma 2.4.1, we consider the case $V = \mathbb{P}^2 = U_0 \coprod H_0$ with $H_0 = \{X_0 = 0\}$, where $U_0 = \operatorname{Spec} k[x, y]$. Write a regular vector field Δ on $\mathbb{P}^2$ as $\Delta = f\frac{\partial}{\partial x} + g\frac{\partial}{\partial y}$, where

$$\begin{aligned} f &= a_0x^2 + a_1xy + c_0x + c_1y + c_2, \\ g &= a_0xy + a_1y^2 + d_0x + d_1y + d_2. \end{aligned}$$

In terms of the coordinates $\{u, v\}$ on $U_1 = \{X_1 \neq 0\}$ and $\{z, w\}$ on $U_2 = \{X_2 \neq 0\}$ (cf. Lemma 2.4.1), we compute $\Delta|_{U_1}$ and $\Delta|_{U_2}$ as follows:

$$\begin{aligned} \Delta|_{U_1} &= -(a_0 + a_1v + c_0u + c_1uv + c_2u^2)\frac{\partial}{\partial u} \\ &\quad + (d_0 + d_1v + d_2u - c_0v - c_1v^2 - c_2uv)\frac{\partial}{\partial v}, \\ \Delta|_{U_2} &= -(a_1 + d_1z + a_0w + d_2z^2 + d_0zw)\frac{\partial}{\partial z} \\ &\quad + (c_1 + c_2z + (c_0 - d_1)w - d_2zw - d_0w^2)\frac{\partial}{\partial w}. \end{aligned}$$

We assume that H_0 is an integral curve of Δ. By the above expression of $\Delta|_{U_1}$ and $\Delta|_{U_2}$, where H_0 is defined by $u = 0$ and $z = 0$, respectively, H_0 is an integral curve if and only if $a_0 = a_1 = 0$. Hence we have

$$\Delta = (c_0x + c_1y + c_2)\frac{\partial}{\partial x} + (d_0x + d_1y + d_2)\frac{\partial}{\partial y}.$$

Namely, we have

$$\Delta\begin{pmatrix} x \\ y \end{pmatrix} = \begin{pmatrix} \Delta(x) \\ \Delta(y) \end{pmatrix} = \begin{pmatrix} c_0 & c_1 \\ d_0 & d_1 \end{pmatrix}\begin{pmatrix} x \\ y \end{pmatrix} + \begin{pmatrix} c_2 \\ d_2 \end{pmatrix}.$$

Furthermore, we have

$$\Delta^n\begin{pmatrix} x \\ y \end{pmatrix} = \begin{pmatrix} c_0 & c_1 \\ d_0 & d_1 \end{pmatrix}^n\begin{pmatrix} x \\ y \end{pmatrix} + \begin{pmatrix} c_0 & c_1 \\ d_0 & d_1 \end{pmatrix}^{n-1}\begin{pmatrix} c_2 \\ d_2 \end{pmatrix}$$

for all $n \geq 1$. This implies that Δ is locally nilpotent if and only if the matrix $\begin{pmatrix} c_0 & c_1 \\ d_0 & d_1 \end{pmatrix}$ is a nilpotent matrix. Hence there exists a matrix $P \in \mathrm{GL}(2, k)$ such that

$$P^{-1}\begin{pmatrix} c_0 & c_1 \\ d_0 & d_1 \end{pmatrix}P = \begin{pmatrix} 0 & \alpha \\ 0 & 0 \end{pmatrix}.$$

Let $\widetilde{P} = \begin{pmatrix} 1 & 0 \\ 0 & P \end{pmatrix}$. By a change of coordinates ${}^t(X_0, X_1, X_2) \mapsto (\widetilde{P}^{-1})^t(X_0, X_1, X_2)$, we may assume that

$$\Delta = (\alpha y + c_2)\frac{\partial}{\partial x} + d_2\frac{\partial}{\partial y},$$

whence $\Delta(x) = \alpha y + c_2$ and $\Delta(y) = d_2$. Now we define a G_a-action on $\mathbb{P}^2$ by

$$t \cdot \begin{pmatrix} X_0 \\ X_1 \\ X_2 \end{pmatrix} = \begin{pmatrix} 1 & 0 & 0 \\ c_2 t + \frac{1}{2!}\alpha d_2 t^2 & 1 & \alpha t \\ d_2 t & 0 & 1 \end{pmatrix}\begin{pmatrix} X_0 \\ X_1 \\ X_2 \end{pmatrix}.$$

It is now clear that Δ is the vector field associated to the G_a-action.

Consider a regular vector field $\Delta = (c_1 y + c_2)\frac{\partial}{\partial x}$ with $c_1 \neq 0$. Then $\Delta|_{U_2 \cap H_0} = c_1\frac{\partial}{\partial w}$. This implies that the G_a-action is effective on $U_2 \cap H_0$ and the fixed point locus Γ consist of a single point $(0, 1, 0)$. In fact, the decomposition $U_0 \coprod (U_2 \cap H_0) \coprod \Gamma$ is a locally nilpotent stratification of Δ.

A similar fact holds in the case $n > 2$. With the notations in Lemma 2.4.1, we have the following.

Remark 2.4.41. Assume that the hyperplane H_0 is integral for Δ as in Example 2.4.40. This implies that $\Delta(\frac{X_0}{X_i})$ is divisible by $\frac{X_0}{X_i}$ for every $i \neq 0$. This condition is equivalent to $a_1 = a_2 = \cdots = a_n = 0$. Hence we have

$$\Delta\begin{pmatrix} x_1 \\ x_2 \\ \vdots \\ x_n \end{pmatrix} = B\begin{pmatrix} x_1 \\ x_2 \\ \vdots \\ x_n \end{pmatrix} + \begin{pmatrix} c_1 \\ c_2 \\ \vdots \\ c_n \end{pmatrix}.$$

Since

$$\Delta^n\begin{pmatrix} x_1 \\ x_2 \\ \vdots \\ x_n \end{pmatrix} = B^n\begin{pmatrix} x_1 \\ x_2 \\ \vdots \\ x_n \end{pmatrix} + B^{n-1}\begin{pmatrix} c_1 \\ c_2 \\ \vdots \\ c_n \end{pmatrix},$$

Δ is locally nilpotent if and only if B is nilpotent. Hence, after a suitable base change of $(x_1, \ldots, x_n)$, we may assume that B is an upper triangular matrix (c_{ij}) with all the diagonal entries zero. Then the vector field Δ is associated with a G_a-action on $\mathbb{P}^n$ which stabilizes the hyperplane H_0. □

The following two definitions are parallel in the spirit.

Definition 2.4.42. Let V be a smooth projective variety of dimension n. A G_a-action on V is called a *stratified action* if there exists a reduced effective divisor $H = H_1 + \cdots + H_r$ (irreducible decomposition) supporting an ample divisor satisfying the following two conditions:
(1) H is G_a-stable, whence there exists the induced effective G_a-action on $X = V \setminus H$. Further, there is an induced G_a-action on each H_i.
(2) For each irreducible component H_i, there exists a reduced effective divisor K_i supporting an ample divisor on H_i such that K_i is G_a-stable and the induced G_a-action on $H_i \setminus K_i$ is effective.

Definition 2.4.43. Let V be a smooth projective variety of dimension n and let Δ be a regular vector field. We call Δ *effectively locally nilpotent with stratification* if there exists a reduced effective divisor $H = H_1 + \cdots + H_r$ supporting an ample divisor satisfying the following two conditions:
(1) Δ induces a nontrivial locally nilpotent derivation on $X = V \setminus H$ and each irreducible component H_i is Δ-integral in the sense that, for each smooth point P of H_i, $\Delta(u_i)$ is divisible by u_i in $\mathcal{O}_{V,P}$, where $u_i = 0$ is a local defining equation of H_i.
(2) For each H_i, there exists a reduced effective divisor K_i on H_i supporting an ample divisor such that $\Delta|_{H_i \setminus K_i}$ induces a nontrivial locally nilpotent derivation.

Remark 2.4.44. In view of Remark 2.4.39, if V is a smooth projective variety with a nontrivial G_a-action and if the G_a-fixed point locus has codimension greater than or equal to 2, then the G_a-action is a stratified G_a-action. □

The following result will explain partly when a given regular vector field comes from a G_a-action on a smooth projective variety.

Theorem 2.4.45. *Let V be a smooth projective variety of dimension $n \geq 2$. Then the following assertions hold:*
(1) *A stratified G_a-action σ on V induces the vector field Δ on V which is effectively locally nilpotent with stratification.*
(2) *Let Δ be a regular vector field on V which is effectively locally nilpotent with stratification. Then there exists a stratified G_a-action on V which induces the vector field Δ.*

Proof. (1) By the hypothesis, the G_a-action σ on X (see the notations in Definition 2.4.42) is effective, whence the associated vector field is nontrivial on X. Similarly, we can extend this vector field, say Δ, to $H_i \setminus K_i$ for each i because it is associated to the induced G_a-action on $H_i \setminus K_i$. Then Δ is an element of $\Gamma(V \setminus (\bigcup_{i=1}^r K_i), \mathcal{T}_{V/k})$, where $\mathrm{codim}_V(\bigcup_{i=1}^r K_i) \geq 2$. Then Δ is defined on V as a regular vector field. It is now clear that Δ is effectively locally nilpotent with stratification because the G_a-action on $H_i \setminus K_i$ is effective.

(2) Let $R = \Gamma(X, \mathcal{O}_V)$. Then Δ is considered as a locally nilpotent derivation of R. Let σ_X be the induced G_a-action, which is given by the coaction

$$\Phi : R \to R[t], \quad \Phi(z) = \sum_{i\geq 0} \frac{1}{i!}\Delta^i(z)t^i.$$

For $\alpha \in k$, define the automorphism φ_α of R by $\varphi_\alpha(z) = \Phi(z)|_{t=\alpha}$. Then $\sigma_\alpha := {}^a\varphi_\alpha$ is the automorphism of X such that $\sigma_\alpha \cdot \sigma_\beta = \sigma_{\alpha+\beta}$ for $\alpha, \beta \in k$. The k-algebra homomorphism Φ extends to a k-homomorphism $\Phi_{k(V)} : k(V) \to k(V)(t)$ such that $\Phi_{k(V)}(\frac{z_2}{z_1}) = \Phi(z_2)/\Phi(z_1)$, where $k(V) = Q(R)$ is the function field of V over k and $z_1, z_2 \in R$. Then σ_α for $\alpha \in k$ is viewed as a birational automorphism of V. Although σ_α is biregular on $X = V \setminus H$, it may not be biregular on the irreducible component $H_1, \dots, H_r$. Suppose that σ_α induces a correspondence between H_i and an algebraic variety of dimension less than or equal to $\dim H_i$. However, the correspondence induces an automorphism on the affine open set $H_i \setminus K_i$ by the hypothesis. Hence σ_α is biregular on all codimension one points of V. This implies that σ_α is a biregular automorphism of V. Since $\sigma_\alpha \cdot \sigma_\beta = \sigma_{\alpha+\beta}$ for $\alpha, \beta \in k$, the collection $\{\sigma_\alpha \mid \alpha \in k\}$ defines a G_a-action on V. It is clear that this G_a-action is stratified and induces the vector field Δ. □

Example 2.4.46. Let $V = \mathbb{F}_n$ with $n \geq 0$. With the notations after Remark 2.4.2, a regular vector field $\Delta = f\frac{\partial}{\partial x} + g\frac{\partial}{\partial y}$ makes the divisor $M + \ell_\infty$ integral if and only if $a_{20} = 0$ for $n > 0$ and $a_{20} = b_{02} = 0$ for $n = 0$. Δ is locally nilpotent on $\mathbb{F}_n \setminus (M + \ell_\infty)$ if and only if $a_{10} = b_{01} = 0$ for $n \geq 0$. If the latter condition is satisfied, Δ is associated with a stratified G_a-action on $\mathbb{F}_n$.

2.5 Fibrations on algebraic surfaces

Let $f : V \to B$ be a fibration from a smooth projective surface to a smooth projective curve. Let F be a fiber of f and let $F_{\text{red}} = \bigcup_i C_i$ be the irreducible decomposition. Let ξ_i be the generic point of C_i. The local ring $\mathcal{O}_{F,\xi}$ is an Artin local ring. The length of $\mathcal{O}_{F,\xi}$ is called the *multiplicity* of the fiber component C_i. Then we simply write $f^{-1}(y) = F = \sum_i m_i C_i$ as a divisor with $m_i = \text{length}\mathcal{O}_{F,\xi}$. Then $(S \cdot F) = \sum_i m_i (S \cdot C_i)$ for a curve S on V. The gcd of the multiplicities m_i is called the *multiplicity* of F. If $f : V \to B$ is a fibration, it is known that $(D \cdot f^{-1}(y))$ is independent of $y \in B$ for every divisor D on V. Since $f^{-1}(y) \approx f^{-1}(y')$ (algebraic equivalence) for $y, y' \in B$, the intersection number is not affected by the algebraic equivalence. The most basic fibration is a $\mathbb{P}^1$-fibration on a smooth projective surface. The following result is well-known (see [159, Lemmas 2.11.2 and 2.12.1]).

Lemma 2.5.1. *Let $f : V \to B$ be a $\mathbb{P}^1$-fibration from a smooth projective surface V to a smooth projective curve B. Then the following assertions hold:*

(1) *f has a cross-section S such that $(S \cdot f^*(y)) = 1$ for every $y \in B$. Hence there are no multiple fibers in the fibration f. Furthermore, every fiber of f is connected.*

(2) *Let $F = \sum_i m_i C_i$ be a singular fiber of f. Then there exists a (-1)-component, say C_1. If $m_1 = 1$ there exists another (-1)-component in the fiber F.*
(3) *Define the dual graph of F which assigns a vertex to each irreducible component C_i and connects two vertices if the corresponding components meet each other. Then the dual graph contains no loops, i. e., circular chains of vertices. Namely, it is a tree.*
(4) *Let $F = \sum_{i=1}^n m_i C_i$ be a singular fiber of f. Then the intersection matrix $((C_i \cdot C_j))_{1 \le i,j \le n}$ is negative semidefinite. Namely, for a $\mathbb{Q}$-divisor $\Delta = \sum_{i=1}^n a_i C_i$ with $a_i \in \mathbb{Q}$, we have $(\Delta^2) \le 0$, and $(\Delta^2) = 0$ if and only if Δ is a $\mathbb{Q}$-multiple of the fiber F.*
(5) *Let $\sigma_1 : V \to V_1$ be the contraction of the (-1)-component C_1 and let $f_1 : V_1 \to B$ be the induced $\mathbb{P}^1$-fibration. Then $(\sigma_1)_* F$ is the fiber $f_1^{-1}(y)$. By repeating this process, we obtain a $\mathbb{P}^1$-fibration without singular fibers.*
(6) *A $\mathbb{P}^1$-bundle $f : V \to B$ is isomorphic over B to $\mathbb{P}(\mathcal{F})$, where $\mathcal{F}$ is a locally free $\mathcal{O}_B$-module of rank 2.*

Lemma 2.5.2. *Let $g : W \to B$ be a $\mathbb{P}^1$-fibration from a normal projective surface W to a smooth projective curve B. Let $\sigma : V \to W$ be a minimal resolution of singularity and let $f = g \cdot \sigma$. Then the following assertions hold:*
(1) *W has only rational singularities, whose resolution graph is a tree of smooth rational curves and is a part of a degenerate fiber of the $\mathbb{P}^1$-fibration f.*
(2) *Every fiber G of g is a union of smooth rational curves, and its intersection dual graph is a tree in the sense that the dual graph of the inverse image F of G by σ in a minimal resolution of singularity is a tree.*
(3) *$H_1(G; \mathbb{Z}) = 0$.*

Proof. (1) Note that the composite $f = g \cdot \sigma : V \to B$ is a $\mathbb{P}^1$-fibration. Since $g^{-1}(U) \cong f^{-1}(U)$ for an open set U of B, each connected component of the exceptional locus of σ is a part of a degenerate fiber of f. Then it is well-known that this part contracts to a rational singular point. In fact, by using the five-term exact sequence associated to a spectral sequence $E_2^{pq} = R^q g_* R^p \sigma_* \mathcal{O}_V \Rightarrow R^n f_* \mathcal{O}_V$, we obtain an exact sequence

$$0 \to g_* R^1 \sigma_* \mathcal{O}_V \to R^1 f_* \mathcal{O}_V \to R^1 g_*(\sigma_* \mathcal{O}_V),$$

where $R^1 f_* \mathcal{O}_V = 0$ by [120, (2.8.6.2), p. 107]. Hence $g_* R^1 \sigma_* \mathcal{O}_V = 0$, whence $R^1 \sigma_* \mathcal{O}_V = 0$ because it is supported by a finite set of W.

(2) Let $F = \sigma^{-1}(G)$. Then G is the surjective image of F. Since F is a union of rational curves, so is the image G. More precisely, $H^1(G_{\text{red}}, \mathcal{O}_{G_{\text{red}}}) = 0$ by [120, (2.8.6.3), p. 107]. If $G_{\text{red}} = \bigcup_{i=1}^r C_i$ is the irreducible decomposition, it follows that $H^1(C_i, \mathcal{O}_{C_i}) = 0$ for every i. This in turn implies that $C_i \cong \mathbb{P}^1$ for every i.

(3) The fiber G is obtained topologically from F by removing the sum of the exceptional loci Γ of singular points $Q_1, \ldots, Q_n$, which are rational singularities lying on G, and adding back the points $T = \{Q_1, \ldots, Q_n\}$. As long exact sequences of integral ho-

mology (cohomology) groups for the pairs (F, Γ) and (G, T), we have

$$\begin{array}{ccccccccc} H_1(\Gamma) & \to & H_1(F) & \to & H^1(F \setminus \Gamma) & \to & H_0(\Gamma) & \to & H_0(F), \\ H_1(T) & \to & H_1(G) & \to & H^1(G \setminus T) & \to & H_0(T) & \to & H_0(G), \end{array}$$

where $H_1(\Gamma) = H_1(T) = 0$, $F \setminus \Gamma$ is homeomorphic to $G \setminus T$ and $H_0(\Gamma) \cong H_0(T) \cong \mathbb{Z}^{\oplus\#(T)}$. Hence $H_1(G; \mathbb{Z}) \cong H_1(F; \mathbb{Z}) = 0$. □

We recall a theorem of M. Artin [9, Theorem 2.3] when V is normal.

Lemma 2.5.3. *Let V be a normal projective surface and let $X = \sum_{i=1}^{s} X_i$ be a reduced effective divisor on V such that V is smooth along X and the intersection matrix $(X_i \cdot X_j)$ is negative definite. Then the following conditions are equivalent:*
(a) *X is contractable and if $\pi : V \to \overline{V}$ is the contraction of X then $\chi(\mathcal{O}_V) = \chi(\mathcal{O}_{\overline{V}})$, where $\chi(\mathcal{O}_V)$ and $\chi(\mathcal{O}_{\overline{V}})$ are the Euler–Poincaré characteristics of V and $\overline{V}$.*
(b) *For every cycle $Z > 0$ with $\mathrm{Supp}(Z) \subseteq X$, we have $p_a(Z) \le 0$, where $p_a(Z) = \frac{1}{2}(Z \cdot Z + K_V) + 1$.*

Moreover, $\overline{V}$ is also projective.

Our purpose is to apply this contractibility criterion to the following situation. Obviously, we can apply this lemma in the following relatively projective case.

Let $f : V \to B$ be a $\mathbb{P}^1$-fibration from a smooth algebraic surface V to a smooth algebraic curve B such that f is surjective. Let F_0 be a reducible fiber of f such that $(F_0)_{\mathrm{red}} = \sum_{i=1}^{n} C_i$ is the irreducible decomposition. Let Γ be a reduced effective divisor such that $\mathrm{Supp}(\Gamma) \subsetneq \mathrm{Supp}(F_0)$ and Γ consists of irreducible components with self-intersection number ≤ -2. We do not need to assume that Γ is connected. Then we have the following result.

Theorem 2.5.4. *There exists a birational morphism $\pi : V \to W$ from V to a normal algebraic surface W with a $\mathbb{P}^1$-fibration $g : W \to B$ such that $\pi(\Gamma)$ consists of finitely many points, $V \setminus \Gamma \cong W \setminus \pi(\Gamma)$ and $f = g \cdot \pi$. The set $\pi(\Gamma)$ consists of rational singular points.*

Proof. By Lemma 2.5.1, the fiber F_0 can be contracted to a smooth fiber by a successive contractions of (-1)-components. Hence F_0 contains an irreducible component, say C_0, such that $m_0 = 1$.

We consider first the case where Γ does not contain the component C_0. Since $(F_0)_{\mathrm{red}} \setminus C_0$ is contractible to smooth points, for any positive cycle Z with $\mathrm{Supp}(Z) \subseteq (F_0)_{\mathrm{red}} \setminus C_0$, we have $p_a(Z) \le 0$ by Lemma 2.5.3. In particular, for any positive cycle Z with $\mathrm{Supp}(Z) \subseteq \mathrm{Supp}(\Gamma)$, we have $p_a(Z) \le 0$. Hence, again by Lemma 2.5.3, Γ is contractible algebraically to normal points. We obtain thus a normal algebraic surface W and a birational morphism $\pi : V \to W$. Since the contraction π does not change fibers of f other than F_0, we have a $\mathbb{P}^1$-fibration $g : W \to B$ such that $f = g \cdot \pi$. The points obtained by the contraction π from Γ are rational singular points by Lemma 2.5.2.

Assume next that every irreducible component of F_0 with multiplicity 1 is contained in Γ. Take one irreducible component m_0C_0 of F_0 such that $C_0 \not\subset \mathrm{Supp}(\Gamma)$. Then $m_0 > 1$. Let $b = f(F_0)$ and replace B by a small affine neighborhood of b so that if t is a local uniformizant of B at b then $t(b') \neq 0$ for every $b' \in B \setminus \{b\}$. Let $\mu : \widetilde{B} \to B$ be a cyclic covering of degree m_0 defined as $\widetilde{B} = \mathrm{Spec}\,\Gamma(B, \mathcal{O}_B)[u]/(u^{m_0} - t)$. Then $\widetilde{B}$ is a smooth affine curve and μ is unramified over $B \setminus \{b\}$ and totally ramified over b. Let $\widetilde{V}$ be the normalization of the fiber product $V \times_B \widetilde{B}$ and let $\widetilde{f} : \widetilde{V} \to \widetilde{B}$ be the composite of the normalization morphism and the projection onto $\widetilde{B}$. Let $\nu : \widetilde{V} \to V$ be also the composite of the normalization morphism and the projection to V which is a finite morphism. Note that $\widetilde{f}$ is a $\mathbb{P}^1$-fibration. Then we have a splitting $\nu^{-1}(C_0) = \widetilde{C}_0^{(1)} + \cdots + \widetilde{C}_0^{(m_0)}$ and each $\widetilde{C}_0^{(i)}$ has multiplicity 1 in the fiber of $\widetilde{F}_0 := \widetilde{f}^{-1}(\widetilde{b})$, where $\mu^{-1}(b) = \{\widetilde{b}\}$. Let $\widetilde{\Gamma} = \nu^{-1}(\Gamma)$. Then $\widetilde{C}^{(i)} \not\subset \widetilde{\Gamma}$. Hence by the first step, we have a contraction $\widetilde{\pi} : \widetilde{V} \to \widetilde{W}$ of $\widetilde{\Gamma}$. In fact, $\widetilde{V}$ might have rational singular points on the fiber $\widetilde{F}_0$. If so, $\widetilde{\Gamma}$ with singular points herein minimally resolved contracts algebraically by Lemma 2.5.3. Now the finite group $G := \mathbb{Z}/m_0\mathbb{Z}$ acts on $\widetilde{W}$ whose action preserves the induced $\mathbb{P}^1$-fibration. Let $W = \widetilde{W}/G$ be the geometric quotient. The morphism $\widetilde{\pi}$ induces a morphism $\pi : V \to W$, and W has a $\mathbb{P}^1$-fibration $g : W \to B$. Since W is normal, Lemma 2.5.2 says that the set of points on W which is the image of $\widetilde{\pi}(\widetilde{\Gamma})$ by the quotient morphism $\widetilde{W} \to W$ consists of rational singular points. □

Remark 2.5.5. In the case of $\mathbb{P}^1$-fibrations on normal surfaces, we might have singular fibers which never occur in the smooth case. Let $g : W \to B$ be a $\mathbb{P}^1$-fibration as in Lemma 2.5.2. If $\mathrm{Sing}(W) \neq \emptyset$, we may have either a multiple fiber or a fiber with more than two irreducible fiber components meeting in one point. We exhibit these phenomena by the following examples. We denote by V the surface obtained by the blowing-ups.

(1) Let L be a smooth fiber of a $\mathbb{P}^1$-fibration on a smooth surface which is isomorphic to $\mathbb{P}^1$. Blow up a point P_0 on L to obtain the exceptional curve E_1. Then blow up the intersection point P_1 of E_1 and the proper transform of L. Let E_2 be the exceptional curve. Then the inverse image of L is $L' + 2E_2 + E_1'$ with $(L')^2 = (E_1')^2 = -2$, where L' and E_1' are the proper transforms of L and E_1. Now contract L' and E_1' to rational double points of type A_1. Let $\overline{E}_2$ be the image of E_2. Then $2\overline{E}_2$ is a fiber of the induced $\mathbb{P}^1$-fibration.

(2) In the above example, let P_2 be a point on E_2 which is different from the intersection points $L' \cap E_2$ and $E_1' \cap E_2$. Blow up P_2 and let E_3 be the exceptional curve. Then the obtained fiber is $L' + E_1' + 2E_2 + 2E_3$, where we denote the proper transforms of L', E_1', E_2 by the same letters. We have $(L')^2 = (E_1')^2 = (E_2)^2 = -2$. We contract the (-2)-curve E_2 to a rational double point of type A_1. Then the images $\overline{L}', \overline{E}_1', \overline{E}_3$ of L', E_1', E_3 meet in the contracted singular point. According to [202], the intersection numbers are given as follows: $(\overline{E}_1')^2 = (\overline{L}')^2 = -\frac{3}{2}, (\overline{E}_3)^2 = -\frac{1}{2}$, and $(\overline{E}_1' \cdot \overline{L}') = (\overline{E}_3 \cdot \overline{L}') = (\overline{E}_1' \cdot \overline{E}_3) = \frac{1}{2}$.

(3) In assertions (1) and (2), the contraction is algebraic by Theorem 2.5.4.

As an application of the above results, we consider what singular points are admitted on a normal affine surface X with either an $\mathbb{A}^1$-fibration or an $\mathbb{A}^1_*$-fibration $\pi : X \to C$. The fibration $\pi : X \to C$ can be extended to a $\mathbb{P}^1$-fibration $f : V \to B$ such that the following conditions are satisfied:

(1) V is a normal projective surface containing X as an open set and V is smooth along the closed connected set $D := V \setminus X$ of pure codimension one; D is identified with a reduced effective divisor with simple normal crossings such that $\mathrm{Supp}(D) = V \setminus X$; C is an open set of a smooth projective curve B. Further, f is a $\mathbb{P}^1$-fibration such that $\pi = f|_X$.
(2) If π is an $\mathbb{A}^1$-fibration, then D contains a section S as an irreducible component and other components of D are fiber components of f.
(3) If π is an $\mathbb{A}^1_*$-fibration, D contains either one irreducible component T which is a 2-section of f (the twisted case) or two irreducible components $S_1 + S_2$ which are sections of f (the untwisted case), and other components of D are fiber components of f. See Remark 2.1.13.

The following result was originally given in [153].

Theorem 2.5.6. *Let $\pi : X \to C$ be an $\mathbb{A}^1$-fibration from a normal affine surface to a smooth algebraic curve. Then the following assertions hold:*

(1) X has only cyclic quotient singularities.
(2) A reducible fiber of π is a disjoint union of irreducible components which are isomorphic to $\mathbb{A}^1$.
(3) If P is a singular point of X lying on an irreducible component Z of the fiber $\pi^{-1}(\pi(P))$ and E is the exceptional locus of minimal resolution of singularity at P, then the proper transform Z' of Z on the minimal resolution of P meets E transversally at one of the end components of the linear chain E.

Proof. (1) Let $f : V \to B$ be a $\mathbb{P}^1$-fibration as fixed before the statement of the theorem. Let $\sigma : \widetilde{V} \to V$ be the minimal resolution of singularities on V (hence achieved only on X by the choice of V). Let $\widetilde{X} = \sigma^{-1}(X)$. Then $\widetilde{V} = \widetilde{X} \cup (V \setminus X)$. Let $\widetilde{f} : \widetilde{V} \to B$ be the induced $\mathbb{P}^1$-fibration. Then $\widetilde{\pi} = \widetilde{f}|_{\widetilde{X}} : \widetilde{X} \to C$ is the induced $\mathbb{A}^1$-fibration. Let $\widetilde{F} = \widetilde{f}^{-1}(\pi(P))$ and $\widetilde{G} = \widetilde{\pi}^{-1}(\pi(P))$ for $P \in X$. The boundary $\widetilde{V} \setminus \widetilde{X}$, which is identified with $V \setminus X$, contains a section S of the fibration $\widetilde{f}$ and the fiber $\widetilde{F}$ meets S with $(S \cdot \widetilde{F}) = 1$. We make use of the following observations.

(i) By [61, Theorem 2], there exists an effective ample divisor of V whose support is $D = V \setminus X$. In particular, D is connected. Since $D = S + \sum_{Q \in C}(f^{-1}(Q) \setminus (f^{-1}(Q) \cap X))$, the boundary part of a fiber F of f, i. e., $F \setminus (F \cap X)$ is connected if it is not the empty set.
(ii) The fiber $\widetilde{F}$ is connected, and each irreducible component of $\widetilde{F}$ is linked to the unique irreducible component of $\widetilde{F}$ which meets the section S. By Lemma 2.5.1(3), every irreducible component of $\widetilde{F}$ is isomorphic to $\mathbb{P}^1$ and $\widetilde{F}$ as a divisor has only normal crossings and contains no loops.

(iii) Since X is affine, every irreducible component of $F \cap X$ if taken the closure in V is linked to an irreducible component of $F \setminus (F \cap X)$.

Let P be a singular point of X lying on an irreducible component Z of the fiber F. Let $E = \sum_i E_i$ be the irreducible decomposition of the exceptional locus of the minimal resolution of singularity at P. Since the resolution is minimal, $(E_i^2) \le -2$ for every i. If E has a branching component, i. e., an irreducible component of E meeting more than two other components of E, by Lemma 2.5.1, the fiber $\widetilde{F}$ is contracted by successive contractions of (−1)-components to the irreducible component which meets S and hence has multiplicity one. In the course of these contractions, the component $\overline{Z}$ (the closure of Z' in $\widetilde{V}$) must be contracted first among the irreducible components of $\overline{Z} + E$. When we contract the branching component, we have three or more components meeting in one point. This cannot happen by Lemma 2.5.1. So, E has no branching components, hence E is a linear chain. If $\overline{Z}$ meets more than one irreducible components of E or meets an irreducible component which is not the end components of E, then either the fiber $\overline{F}$ contains a loop or has three components meeting in one point in the process of successive contractions. This implies that Z' meets an end component of E, and the singularity at P is a cyclic quotient singularity. Further, Z has at most one singular point of X lying on it.

By a similar argument, we can show that $F \cap X$ is a disjoint union of irreducible components Z's and each irreducible component Z has only one place at infinity. This implies that Z is isomorphic to $\mathbb{A}^1$ if it has no singular point of X lying on it. Even if there is a singular point P on Z, the proper transform Z' on $\widetilde{V}$ meets an end component of E transversally, hence $Z \cong \mathbb{A}^1$. □

Next we consider possible singularities on a normal affine surface X with an $\mathbb{A}^1_*$-fibration $\pi : X \to C$. The fibration π is extended to a $\mathbb{P}^1$-fibration $f : V \to B$, where V is a normal projective surface which is smooth along $D := V \setminus X$. Let $\sigma : \widetilde{V} \to V$ be the minimal resolution of singularities. Then $\widetilde{f} = f \cdot \sigma : \widetilde{V} \to B$ is a $\mathbb{P}^1$-fibration on a smooth surface $\widetilde{V}$. We assume that D contains no (−1) fiber components of $\widetilde{f}$ possibly except for a (−1) component meeting either two cross-sections S_1, S_2 (the untwisted case) or a 2-section T contained in D (the twisted case).

Theorem 2.5.7. *With the above settings, let F be a fiber of π. Then the following assertions hold:*

(1) *$F = \Gamma + \Delta$, where $\Gamma_{\mathrm{red}} = \emptyset, \mathbb{A}^1_*$ or $C_1 + C_2$ with $C_1 \cong C_2 \cong \mathbb{A}^1$ meeting transversally in one point, and Δ_{red} is a disjoint union of the affine lines.*

(2) *If $\Gamma_{\mathrm{red}} = \mathbb{A}^1_*$, there are no singular points lying on Γ.*

(3) *If $\Gamma_{\mathrm{red}} = C_1 + C_2$, then $P := C_1 \cap C_2$ can be a cyclic quotient singular point.*

(4) *Each component of Δ_{red} has at most one cyclic quotient singular point except in the cases:*

(4-1) *In the twisted case, suppose a smooth fiber L touches a 2-section T in a point Q. Blowing up Q and its infinitely near point of the first order lying on T, we have* *a degenerate fiber $E_1 + E_2 + 2E$ such that $(E_1^2) = (E_2^2) = -2, (E^2) = -1$, and E meets the proper transform T' of T. Starting with a point $P_0 = E \cap T'$, apply a succession of blowing-ups to produce a chain of curves $\Sigma_0 + E_n + \Sigma_1$ connecting E and T', where $E \cap \Sigma_0 \neq \emptyset, \Sigma_1 \cap T' \neq \emptyset$ and $(E_n^2) = -1$, and contract the curves $E_1 + E_2 + E + \Sigma_0$ to produce a quotient singular point $\overline{P}$ of Dynkin type D_r (not necessarily a double point). Then the image of E_n gives a curve in X isomorphic to $\mathbb{A}^1$ with a singular point $\overline{P}$.*

(4-2) *In the situation of* (4-1), *contract E_1 and E_2 to rational double points of type A_1 without producing a string of exceptional curves $\Sigma_0 + E_n + \Sigma_1$. Then E gives a fiber isomorphic to $\mathbb{A}^1$ with these two singular points.*

(5) *Let P be a cyclic quotient singular point of type d/e lying on a component Z of Δ.*[12] *Let m be the multiplicity of Z in F. Then d divides m.*

Proof. Let $\tilde{F} = \tilde{f}^{-1}(\pi(F))$. All (-1) components of $\tilde{F}$ then come from the components of F except for a (-1) component meeting either two cross-sections S_1, S_2 or a 2-section T in D. Furthermore, since X is affine, D is connected and every connected component of $\tilde{F} - F$ which is disjoint from D is contracted to a singular point. Hence every irreducible component of F gives rise to a (-1) component of $\tilde{F}$ except for C_1 or C_2 of $\Gamma_{\text{red}} = C_1 + C_2$. Note that an irreducible component meeting S_i $(i = 1, 2)$ or T gives a component in the part Γ except for the case where a component touches T at a single point. On the other hand, it is well-known that any (-1) component of a degenerate fiber of a $\mathbb{P}^1$-fibration does not meet three other fiber components and that if a (-1) component has multiplicity 1 there is another (-1) component. All the stated assertions can be verified without difficulty by virtue of these observations. □

A fibration $\pi : X \to C$ from a normal affine surface X to a smooth algebraic curve C is an $\mathbb{A}^1_{(n*)}$*-fibration* for an integer $n \geq 0$ if there exists a $\mathbb{P}^1$-fibration $f : V \to B$ extending the fibration π as explained in condition (1) before Theorem 2.5.6 such that the following conditions are satisfied:

(n-1) A general fiber F of π has $n + 1$ distinct places at infinity, i. e., $\overline{F} \setminus F$ consists of $(n + 1)$ distinct smooth points, where $\overline{F}$ is the closure of F in V, hence $\overline{F} \cong \mathbb{P}^1$. This condition is equivalent to saying that the generic fiber V_η of f is a rational curve over the function field $k(C) = k(B)$ by Lemma 2.1.12(3).

(n-2) We say that π is *untwisted* (or *twisted*) if the generic fiber X_η of π has $n+1$ distinct points at infinity which are rational over $k(C)$ (or otherwise).

We call the fibration $f : V \to B$ a *projective completion* of π.

Example 2.5.8.

(1) A polynomial $f(x, y) \in k[x, y]$ is called *generically rational* with $(n + 1)$-punctures if the fibration $\pi : \mathbb{A}^2 \to \mathbb{A}^1$ given by $P \in \mathbb{A}^2 \mapsto f(P) \in \mathbb{A}^1$ is an $\mathbb{A}^1_{(n*)}$-fibration for

12 It is the quotient singularity obtained by a $\mathbb{Z}/d\mathbb{Z}$-action on the (x, y)-plane by $(x, y) \mapsto (\zeta x, \zeta^e y)$, where $\mathbb{Z}/d\mathbb{Z}$ is identified with the multiplicative group of the dth roots ζ^i of unity.

some integer $n \geq 0$. A generically rational polynomial $f(x,y)$ is *of simple type* if π is untwisted. A classification of generically rational polynomials is given in [165, 181, 23] in the case of simple type. For an $\mathbb{A}^1_{n*}$-fibration $\pi : X \to C$ for a factorial affine surface, the generic fiber X_η is factorial as it is obtained by a localization of X. Hence $\mathrm{Pic}(X_\eta) = 0$. If $n = 1$ and X is factorial, the fibration π is untwisted, for otherwise $\mathrm{Pic}(X_\eta) \cong \mathbb{Z}/2\mathbb{Z}$. Generically rational polynomials $f(x,y)$ with two punctures are necessarily of simple type. Such polynomials are classified by Saito [201]. Needless to say, a generically rational polynomial $f(x,y)$ with one puncture is a coordinate of $k[x,y]$ by Abhyankar–Moh–Suzuki theorem.

(2) An example of a twisted $\mathbb{A}^1_*$-fibration is given as follows. Let C and L_0 respectively be a conic and a line in $\mathbb{P}^2$ such that $C \cap L_0 = \{P_0\}$. Let P_1 be a point on L_0 different from P_0. Let V be the blowing-up of the point P_1. Then the linear pencil of lines on $\mathbb{P}^2$ through P_1 gives rise to a $\mathbb{P}^1$-fibration $f : V \to \mathbb{P}^1$ for which the proper transform T of C is a 2-section. Let $X = V \setminus (T \cup L'_0)$, where L'_0 is the proper transform of L_0 and let $\pi = f|_X : X \to \mathbb{P}^1$. Then π is a twisted $\mathbb{A}^1_*$-fibration. By the construction explained in assertion (4) of Theorem 2.5.7, one can produce an $\mathbb{A}^1_*$-fibration $\pi' : X' \to \mathbb{P}^1$ on a normal affine surface with a noncyclic quotient singular point or two singular points of type A_1 on the fiber component isomorphic to $\mathbb{A}^1$.

(3) We can produce similar examples of twisted $\mathbb{A}^1_{(n*)}$-fibrations with $n = 2$ or 3 on a smooth affine surface or on a normal affine surface with rational singular points which are not even quotient singular points. Namely, take a smooth cubic C and an inflectional tangent line L_0 such that $C \cap L_0 = \{P_0\}$. Take a point $P_1 \neq P_0$ on L_0. With the same notations as above, $\pi : X \to \mathbb{P}^1$ is a twisted $\mathbb{A}^1_{(2*)}$-fibration with a 3-section T. A construction similar to (4) of Theorem 2.5.7 is possible. If we take a smooth quartic curve C and a bitangent L_0 such that $C \cap L_0 = \{P_0, Q_0\}$, $\pi : X \to \mathbb{P}^1$ is a twisted $\mathbb{A}^1_{(3*)}$-fibration with a 4-section T.

(4) Compared to the case of simple type, a thorough classification of generically rational polynomials in the twisted case is complicated indeed. Sasao [203] gave a classification of such polynomials in the case of quasisimple type, i. e., all the boundary components of X which are transversal to the $\mathbb{P}^1$-fibration $f : V \to \mathbb{P}^1$ are cross-sections of f except for one a-section with $a \geq 2$.

(5) Even in the untwisted case, if $n \geq 2$, a normal affine surface with an $\mathbb{A}^1_{(n*)}$-fibration has a noncyclic singularity. Here is an example. Let $V_0 = \mathbb{P}^1 \times \mathbb{P}^1$. Denote by $p_i : V_0 \to \mathbb{P}^1$ for $i = 1, 2$ the projections to the first and second factors. Let L_0, L_∞ be the fibers of p_1 and let M_i $(0 \leq i \leq n)$ the fibers of p_2. Let $P_i = L_0 \cap M_i$. Let $\tau : \widetilde{V} \to V_0$ be the blowing-ups of the point P_i and infinitely near points $P_i^{(j)}$ $(1 \leq j \leq m_i)$ of P_i lying on the proper transform of M_i. Let E_i be the last (-1)-curve of the blowing-up with center $P_i^{(m_i)}$ and let L'_0 be the proper transform of L_0. Then $({L'}_0^2) = -(n+1)$ and the divisor $E := {L'}_0 + \sum_{i=0}^{n} \sum_{j=1}^{m_i} E_i^{(j)}$ has the star-shaped dual graph which has $(n+1)$ linear chains of (-2)-curves $E_i^{(j)}$ of length m_i sprouting out of the central component L'_0. Let M'_i $(0 \leq i \leq n)$ be the proper transform of M_i. Then

the divisor E is contracted to a rational singular point P on a normal projective surface V, where $\sigma : \widetilde{V} \to V$ is the contraction of E. In fact, the divisor $D := L_\infty + \sum_{i=0}^{n} M_i'$ supports an ample divisor on V. Let $X = V \setminus D$. Then the projection $p_1 : V_0 \to \mathbb{P}^1$ induces an $\mathbb{A}^1_{(n*)}$-fibration on X, where the point P lies on the fiber $F_0 = \sum_{i=0}^{n} A_i$ and A_i is the image of E_i with the point $E_i \cap M_i'$ deleted. Hence $A_i \cong \mathbb{A}^1$ and passes through the point P. Note that if $n = 1$ the point has a cyclic quotient singularity.

2.6 ML_0-surfaces

This section deals with ML_0-properties concerning the existence of independent G_a-actions on a smooth affine surface X. The ML_i-property depends, in general, on the shape (more precisely, the weighted dual graph) of the boundary divisor D of a smooth normal completion $X \hookrightarrow V$, where $D = V - X$. The ML_0-property follows from the condition that D is a linear chain. Since D supports an effective ample divisor on V, D has an irreducible component, say D_i, with self-intersection number greater than -1. Then, by applying the blowing-ups and blowing-downs with centers on D, we can shift the component D_i to either end of the linear chain D. Hence we can produce two independent G_a-actions. This kind of phenomena was first discovered by M. H. Gizatullin in his research on quasihomogeneous surfaces (see [59, 60]).

In Section 2.7 we see that the ML_1-property depends on the condition that D is a tree of smooth rational curves such that D is not a linear chain and one of its maximal twigs has an irreducible component of self-intersection number greater than -1. Then this component can be shifted to the end component of the maximal twig and produce a G_a-action.

The existence of $\mathbb{A}^1$-fibrations of complete type seems to have nothing to do with the ML_i-property. Only the existence of $\mathbb{A}^1$-fibrations of affine type matters.

2.6.1 Gizatullin theorem

Let X be a smooth affine surface defined over k. In Section 2.5 we defined the Makar-Limanov invariant $\mathrm{ML}(X)$ and ML_i-varieties as those with $\mathrm{tr.deg}_k Q(\mathrm{ML}(X)) = i$. In the present section, we are mostly interested in ML_0-surfaces and the geometry of curves on such surfaces. The presentation is based on our joint paper with P. Russell [83], which is a good example of investigation on structures of affine surfaces via the existence of $\mathbb{A}^1$-fibrations.

The simplest example of an ML_0-surface is the affine plane $\mathbb{A}^2$, and we have the well-known Abhyankar–Moh–Suzuki theorem (AMS theorem, for future use) which states that any curve in $\mathbb{A}^2$ isomorphic to the affine line $\mathbb{A}^1$ is a fiber of an

$\mathbb{A}^1$-fibration. On the other hand, the affine plane $\mathbb{A}^2$ is characterized as an ML_0-surface with $\mathrm{Pic}(X) = 0$.

We set

$$\rho(X) = \operatorname{rank} \mathrm{Pic}(X)_{\mathbb{Q}}, \quad \gamma(X) = \operatorname{rank} \Gamma(X, \mathcal{O}_X)^*/k^*,$$

where $\operatorname{rank} \Gamma(X, \mathcal{O}_X)^*/k^*$ is a free abelian group of finite rank. In fact, if (V, D) is a pair of a smooth projective surface V containing X as an open set and $D := V - X$ is a divisor with simple normal crossings, there exists a canonical exact sequence

$$0 \to \Gamma(X, \mathcal{O})^*/k^* \to \bigoplus_{i=1}^{r} \mathbb{Z}D_i \to \mathrm{Pic}(V),$$

where $D = \sum_{i=1}^{r} D_i$ is the irreducible decomposition of D and $\bigoplus_{i=1}^{r} \mathbb{Z}D_i$ is a free abelian group generated by the components of D. Note that $\gamma(X) = 0$ if and only if every unit of $\Gamma(X, \mathcal{O}_X)$ (i. e., invertible function on X) is a constant (i. e., an element of k). We are interested in the following.

Question 2.6.1. Let X be an ML_0-surface and let C be a curve on X isomorphic to the affine line $\mathbb{A}^1$. Does there exist an $\mathbb{A}^1$-fibration $f : X \to B$ such that C is a fiber-component of f, where $B \cong \mathbb{A}^1$?

An affirmative answer to this question would have been a good generalization Abhyankar–Moh–Suzuki theorem. But we will see that the answer is negative if $\rho(X) \geq 1$, and affirmative if $\rho(X) = 0$. Moreover, we note that there are plenty of ML_0-surfaces with $\rho(X) = 0$ but $\mathrm{Pic}(X)_{\mathrm{tor}} \neq 0$ (see Lemma 2.6.9).

The main results in this section are Theorems 2.6.2 and 2.6.12. We have also given several examples where the analogue of the AMS theorem is false for ML_0-surfaces with $\rho > 0$ (see Propositions 2.6.18 and 2.6.19) and for ML_1-surfaces (see Propositions 2.7.9 and 2.7.11). In addition, an example of a simply-connected ML_0-surface with $\rho > 0$ which does not contain $\mathbb{A}^2$ as a Zariski-open subset is given.

An (n)-curve on a smooth surface X is a smooth projective rational curve C with $(C^2) = n$. A linear chain $C_1 + \cdots + C_n$ of smooth rational curves is called *admissible* if $(C_i^2) \leq -2$ for $1 \leq i \leq n$. For related definitions and relevant results we refer to [159].

Let X be a smooth irreducible quasiprojective surface. A projective completion $X \subset V$ is called a *normal completion* of X if V is smooth and $D := V - X$ is a divisor with simple normal crossings. If further any (-1)-curve in D intersects at least three other irreducible components of D then we say that this completion is a *minimal normal* completion of X. If X is an ML_0-surface then X is rational and the boundary divisor $V - X$ in any normal completion of X is a tree of $\mathbb{P}^1$s. In fact, two independent G_a-actions give a dominant morphism $G_a \times G_a \to X$, whence X is rational. Furthermore, one G_a-action on X gives an $\mathbb{A}^1$-fibration on X. After performing finitely many blowing-ups on V, the $\mathbb{A}^1$-fibration extends to a $\mathbb{P}^1$-fibration on the blown-up projective surface $\widetilde{V}$. Then

$\widetilde{V} - X$ contains no loops. Hence $V - X$ contains no loops. In particular, the boundary divisor $V - X$ is simply connected.

We have the following theorem due to M. H. Gizatullin which is called *Gizatullin theorem* and characterizes ML_0-surface in terms of the boundary graph [59, 60, 32, 74].

Theorem 2.6.2. *Let X be a smooth affine surface and let V be a minimal normal completion of X. Then X is an ML_0-surface if and only if $\gamma(X) = 0$ and the dual graph of the boundary divisor $D := V - X$ is a linear chain of smooth rational curves.*

Proof. We prove the "if" part. Suppose that $q := h^1(\mathcal{O}_V)$ is positive. Let $g : V \to \mathrm{Alb}(V)$ be the Albanese morphism which is a proper morphism. Then $g(D)$ is a point because D consists of rational components. Since the image $g(V)$ generates $\mathrm{Alb}(V)$, we have $\dim g(V) \geq 1$. If $\dim g(V) = 1$, then the inverse image $g^{-1}(Q)$ of a general point of $g(V)$ is a complete curve in an affine variety X. This is absurd. Hence $\dim g(V) = 2$. Then there exists a finite set S of closed points in $g(V)$ such that $g|_{g^{-1}(W)} : g^{-1}(W) \to W$ is a finite morphism, where $W = g(V) \setminus S$. Let A be an effective ample divisor on $g(V)$ such that $A \cap S = \emptyset$. Then $(g^*(A))^2 > 0$ and D is disjoint from $g^*(A)$. This is a contradiction because D supports an ample divisor on V. Hence $q = 0$.

Write the boundary divisor D as $D = \sum_{i=1}^r D_i$, where the irreducible components D_i are arranged in this order. Set $a_i = -(D_i^2)$. We may assume that D is minimal, i. e., no a_i is equal to 1. If $a_i \geq 2$ for every i, then the intersection matrix of D is negative definite, hence D cannot support an ample divisor. So, there is a component D_i with $(D_i^2) = -a_i \geq 0$. Take such a D_{i_0} located in the leftmost position on the linear chain D. By applying blowing-ups and blowing-downs with centers in the boundary D, we can shift the component D_{i_0} to the leftmost end component of D and assume that $(D_1^2) = -a_1 = 0$ (see [152, Corollary 2.4.3]) and $r \geq 2$. Since $(D_1 \cdot K_V) = -2$, we have $p_n(V) := h^0(\mathcal{O}_V(nK_V)) = 0$ for every $n > 0$. In fact, if $p_n > 0$, then $(nK_V \cdot D_1) \geq 0$ because $|nK_V| \neq \emptyset$ and $(D_1^2) = 0$. Meanwhile, $(nK_V \cdot D_1) = -2n < 0$. This is a contradiction. Hence V is a ruled surface by Enriques criterion. Since $q = 0$, V is a rational surface. By the Riemann–Roch theorem, we have

$$\begin{aligned}\chi(\mathcal{O}(D_1)) &= h^0(\mathcal{O}(D_1)) - h^1(\mathcal{O}(D_1)) = \frac{1}{2}(D_1 \cdot D_1 - K_V) + \chi(\mathcal{O}_V)\\ &= 1 + \chi(\mathcal{O}_V) = 2.\end{aligned}$$

Hence $h^0(\mathcal{O}(D_1)) \geq 2$. This implies that $\dim |D_1| \geq 1$. In fact, $\dim |D_1| = 1$ since $(D_1^2) = 0$. Hence $\Phi_{|D_1|}$ defines a $\mathbb{P}^1$-fibration $p : V \to \mathbb{P}^1$ such that D_1 is a fiber. Note that D has a adjacent component D_2 which is a cross-section of p. For otherwise, D cannot support an ample divisor. Hence p restricted onto X induces an $\mathbb{A}^1$-fibration $p_X : X \to \mathbb{A}^1$, where p_X is surjective because $\gamma(X) = 0$. By Lemma 2.2.7, there exists a G_a-action σ on X such that p_X is the quotient morphism by the action σ. Suppose $r \geq 3$. Then $D_3 + \cdots + D_r$ with several curves lying in X forms a single fiber of the $\mathbb{P}^1$-fibration p, where several curves lying in X are all isomorphic to $\mathbb{A}^1$ and their closures in V are (-1)-curves meeting one of $D_3, \ldots, D_r$. Since we may assume that $D_3 + \cdots + D_r$ contains no (−1) components, $(D_i^2) \le -2$ for $3 \le i \le r$. This observation follows from the fact that D_3 has multiplicity one in the corresponding fiber of p and hence all other components are contracted to the points on D_3.

Applying also blowing-ups and blowing-downs with centers in the boundary D, we can shift the component D_{i_0} to the rightmost component of the linear chain D. So, we may assume that D_r satisfies $(D_r^2) = 0$. Hence $|D_r|$ induces a $\mathbb{P}^1$-fibration $p' : V \to \mathbb{P}^1$ and an $\mathbb{A}^1$-fibration $p'_X : X \to \mathbb{A}^1$. Further, p'_X gives rise to a G_a-action σ'. We have to show that the G_a-actions σ and σ' are independent. This is done by comparing the sections of p_X and p'_X lying in the boundary D, i. e., D_2 for p_X and D_{r-1} for p'_X. Noncoincidence of these sections at boundary can be seen by chasing the process of shifting the 0-component D_0 to the 0-component D_r by blowing-ups and blowing-downs. To exhibit the idea by an example, we consider the case where $D = D_1 + D_2$ with $(D_1^2) = 0$ and $(D_2^2) = -2$. The case $r \ge 3$ can be treated in a similar way. Blow up first a point P_1 which is not the point $D_1 \cap D_2$. Let E be the exceptional curve and D'_1 the proper transform of D_1. Blow down D'_1. Then the image E' of E becomes a 0-component and the image D'_2 of D_2 is a (−1)-component. After blowing down D'_1, we repeat the above process once again to obtain the boundary chain $D'' = D''_1 + D''_2$, where $({D''_1}^2) = ({D''_2}^2) = 0$. Then $|D''_2|$ gives a $\mathbb{P}^1$-fibration and D''_1 is its section. It is clear that D''_1 is different from D_2. □

The *"only if"* part of Theorem 2.6.2 follows from the following result. In fact, $\mathrm{MLdim}(X) = 0$ implies the existence of two independent G_a-actions. Since there exists then a dominant morphism $\mathbb{A}^2 \to X$, it follows that $\gamma(X) = 0$.

Lemma 2.6.3. *Let X be an ML_0-surface. Then X has a minimal normal completion V such that $D := V - X$ is a linear chain of rational curves. In particular, $\pi_1^\infty(X)$ is a finite cyclic group.*

Proof. For the proof we will use some arguments from [143, Lemma 2.6, Theorem 2.7]. If X is isomorphic to the affine plane $\mathbb{A}^2$, it is well-known by [192] that the boundary divisor of any minimal normal completion of $\mathbb{A}^2$ is a linear chain. Hence we may and shall assume that X is not isomorphic to $\mathbb{A}^2$. Let σ, σ' be two G_a-actions on X. By making use of one G_a-action σ on X, we consider an associated $\mathbb{A}^1$-fibration $\rho : X \to B$. We claim that $B \cong \mathbb{A}^1$. First of all, by Lemma 2.2.7, B is an affine curve. Since a general fiber of the $\mathbb{A}^1$-fibration corresponding to σ' dominates B, we conclude that $B \cong \mathbb{A}^1$.

For a suitable smooth completion V of X, we can extend ρ to a $\mathbb{P}^1$-fibration $p : V \to \overline{B} \cong \mathbb{P}^1$ such that $D := V - X$ consists of a smooth fiber F_∞, a cross-section S, and a union G_1 of irreducible components contained in a degenerate fiber of p. By using Lemma 2.5.1 repeatedly, we can assume that $(C^2) < -1$ for any component C of G_1.

Let Λ' be the pencil of rational curves corresponding to σ' and let T' be the closure of a general orbit of σ'. If $T' \cap F_\infty = \emptyset$ then the $\mathbb{A}^1$-fibrations corresponding to σ, σ' are the same. Hence T' meets F_∞. Suppose that Λ' has no base point on F_∞. Then we get another $\mathbb{P}^1$-fibration p' on V such that F_∞ is a cross-section for p'. We then claim that $X \cong \mathbb{A}^2$. Since a general fiber of p' is disjoint from S, by Hodge index theorem, $(S^2) \le 0$.

If $(S^2) = 0$ then S is a member of the pencil Λ'. In this case $D = F_\infty \cup S$ and we see that $X \cong \mathbb{A}^2$. Suppose $(S^2) < 0$. Then $S \cup G_1$ is contained in a fiber of p'. In fact, since the base of the $\mathbb{A}^1$-fibration on X corresponding to p' is also isomorphic to $\mathbb{A}^1$ the union $S \cup G_1$ is a full fiber of p'. It follows that $(S^2) = -1$ and, starting with the contraction of S, we can contract $S \cup G_1$ to a smooth rational curve with self-intersection 0. Then again we see that $X \cong \mathbb{A}^2$.

Now we know that Λ' has a base point on F_∞. By performing elementary transformations[13] at $F_\infty \cap S$, we can further assume that this base point is not the point $F_\infty \cap S$. By Hodge index theorem, $(S^2) < 0$. By blowing up the base point of Λ' and its infinitely near points, we get a surface V' which admits a $\mathbb{P}^1$-fibration p' such that the proper transform of F_∞, S, G_1 and some of exceptional curves obtained by blowing-ups form a single fiber, say G' of p'. By Lemma 2.5.1, we can contract G' to a regular fiber. Since no irreducible component of G_1 is a (-1)-curve, the first (-1)-curve to be contracted is the proper transform of F_∞ or S. By using Lemma 2.5.1, we deduce that $D = F_\infty \cup S \cup G_1$ is linear. In particular, $\pi_1^\infty(X)$ is a finite cyclic group. □

Remark 2.6.4. Without the condition $\gamma(X) = 0$, Theorem 2.6.2 does not hold. Here is an example. Let $V = \mathbb{F}_n$ be the Hirzebruch surface with the minimal section M and two distinct fibers F_0, F_∞. Let $D = F_0 + M + F_\infty$ and $X = V \setminus D$. Then $\gamma(X) = 1$ because $F_0 \sim F_\infty$. Since the end components of D have self-intersection numbers 0, they define G_a-actions on X, which are identical because F_0, F_∞ define the same linear pencil on $\mathbb{F}_n$.

2.6.2 Topology and structure of ML_0-surfaces

In what follows throughout the present subsection, we assume that the ground field is the complex number field $\mathbb{C}$ to use the topological arguments. But most results can be stated over an algebraically closed field k of characteristic zero by the Lefschetz principle and by replacing the fundamental groups by the étale fundamental groups. This replacement of the fundamental groups is possible because the fundamental groups are finite groups in most cases.

We begin with some preparations of results which we need in the subsequent arguments.

Lemma 2.6.5. *Let C be an irreducible curve on a smooth affine surface X and let $X' = X - C$. Then we have*

$$\gamma(X') - \rho(X') = \gamma(X) - \rho(X) + 1.$$

13 Let $f : V \to \overline{B}$ be a $\mathbb{P}^1$-fibration and let ℓ is a smooth fiber. Choose a point P on ℓ and blow up the point P with the exceptional curve E. Then the proper transform ℓ' of ℓ is a (-1)-curve and hence contractible. As a result, we have a $\mathbb{P}^1$-fibration $f' : V' \to \overline{B}$ for which the image of E is a smooth fiber. This birational transformation V to V' is called an *elementary transformation* with center P.

Proof. In the case where mC is a principal divisor for some $m > 0$, write $mC = (u)$. Then $\gamma(X') = \gamma(X) + 1$ and $\rho(X') = \rho(X)$. In the case where the divisor class $[C]$ is not zero in $\mathrm{Pic}(X)_{\mathbb{Q}}$, we have $\rho(X') = \rho(X) - 1$ and $\gamma(X') = \gamma(X)$. In both cases the result follows. □

Lemma 2.6.6. *Let X be an ML_0-surface. Then the following assertions hold:*
(1) $\gamma(X) = 0$.
(2) *The torsion part $\mathrm{Pic}(X)_{\mathrm{tor}}$ is isomorphic to $\pi_1(X)$ and it is a finite cyclic group, while $H_2(X)$ is the free part of $\mathrm{Pic}(X)$.*
(3) *Any $\mathbb{A}^1$-fibration $f : X \to B$ has at most two (resp., one) multiple fibers if $B \cong \mathbb{P}^1$ (resp., $B \cong \mathbb{A}^1$).*

Proof. (1) The assertion is redundant with Theorem 2.6.2. It is included here again for the convenience of reference. Since there are two independent G_a-actions on X, there is a dominant morphism $\varphi : \mathbb{A}^2 \to X$. Hence, for any nonconstant unit u on X, $\varphi^*(u)$ is a nonconstant unit on $\mathbb{A}^2$. But this is a contradiction. Hence $\gamma(X) = 0$.

(2) Let $X \hookrightarrow V$ be a normal completion. Let $D := V - X$. Then V is a rational surface and the dual graph of the divisor D is a tree because its minimization by contraction of all (-1) components in D is a linear chain by Lemma 2.6.3. Consider an exact sequence of sheaves on V

$$0 \longrightarrow \mathbb{Z} \longrightarrow \mathcal{O}_V \overset{\exp}{\longrightarrow} \mathcal{O}_V^* \longrightarrow 1.$$

Since $H^1(\mathcal{O}_V) = \mathrm{H}^2(\mathcal{O}_V) = 0$ as V is rational, the associated cohomology exact sequence shows that $\mathrm{Pic}(V) = H^1(\mathcal{O}_V^*) \cong H^2(V;\mathbb{Z})$. On the other hand, we have an exact sequence of cohomology groups with $\mathbb{Z}$-coefficients associated to the pair (V, D),

$$H^1(D) \to H^2(V,D) \to H^2(V) \to H^2(D) \to H^3(V,D) \to H^3(V), \tag{2.12}$$

where $H^1(D) = (0)$ since D is simply-connected, $H^2(V) \cong \mathrm{Pic}(V)$, $H^3(V,D) \cong H_1(X)$ by the Lefschetz duality and $H^3(V) \cong H_1(V) = (0)$ by the Poincaré duality. Meanwhile, since $H_2(D)$ is a free abelian group generated by the components of D and $H_2(V) \cong H^2(V) \cong \mathrm{Pic}(V)$, we have an exact sequence

$$0 \to H_2(D) \to H_2(V) \to \mathrm{Pic}(X) \to 0.$$

Taking the dual of the last exact sequence, we have an exact sequence

$$\begin{aligned} 0 &\to \mathrm{Hom}(\mathrm{Pic}(X), \mathbb{Z}) \to \mathrm{Hom}(H_2(V), \mathbb{Z}) \to \mathrm{Hom}(H_2(D), \mathbb{Z}) \\ &\to \mathrm{Ext}^1(\mathrm{Pic}(X), \mathbb{Z}) \to \mathrm{Ext}^1(H_2(V), \mathbb{Z}), \end{aligned}$$

where $\mathrm{Hom}(\mathrm{Pic}(X), \mathbb{Z})$ is the dual of the free part of $\mathrm{Pic}(X)$, $\mathrm{Hom}(H_2(V), \mathbb{Z}) \cong H^2(V)$ by the universal coefficient theorem, and we use the fact that

$$H_1(V) = (0), \quad \mathrm{Hom}(H_2(D), \mathbb{Z}) \cong H^2(D), \quad \mathrm{Ext}^1(\mathrm{Pic}\,(X), \mathbb{Z}) \cong \mathrm{Pic}\,(X)_{\mathrm{tor}},$$

and $\mathrm{Ext}^1(H_2(V), \mathbb{Z}) = (0)$ because $H_2(V)$ is free. Hence we obtain an exact sequence

$$0 \to \mathrm{Hom}(\mathrm{Pic}(X), \mathbb{Z}) \to H^2(V) \to H^2(D) \to \mathrm{Pic}\,(X)_{\mathrm{tor}} \to 0. \tag{2.13}$$

Hence it follows from (2.12) that $H_1(X) \cong \mathrm{Pic}(X)_{\mathrm{tor}}$ and $H_2(X)$ is the free part of $\mathrm{Pic}(X)$ because $H^2(V, D) \cong H_2(X)$.

(3) Let $f : X \to B$ be an $\mathbb{A}^1$-fibration. Let $X \hookrightarrow V$ be a smooth normal completion. We may assume that the $\mathbb{A}^1$-fibration f extends to a $\mathbb{P}^1$-fibration $\overline{f} : V \to \overline{B}$, where $\overline{B} \cong \mathbb{P}^1$. Let M be the cross-section of $\overline{f}$ lying in the boundary divisor $D := V - X$. Suppose that there exist two or more multiple fibers, say $m_1F_1, \dots, m_rF_r$, in the fibration f. We include the fiber of $\overline{f}$ over the point $\overline{B} - B$ as m_rF_r (even though it is reduced) if $B \cong \mathbb{A}^1$. The corresponding fibers $\Phi_1, \dots, \Phi_r$ of $\overline{f}$ have respectively the components $B_1, \dots, B_r$ in D meeting the section M. Then M is a branching component of D, i. e., a component meeting three or more other components of D. If we make the boundary D minimal by contracting successively all (-1)-curves in D which lie in fibers of $\overline{f}$ so that the resulting boundary divisor has still normal crossings, then the resulting divisor must be a linear chain by Theorem 2.6.2. But this is impossible as long as $r \geq 3$. So, f has at most two (resp., one) multiple fibers if $B \cong \mathbb{P}^1$ (resp., $B \cong \mathbb{A}^1$). □

With the above notations, let Φ be a fiber of $\overline{f}$ which cuts out a singular fiber of f. Then Φ consists of a boundary part $C_1 + C_2 + \cdots + C_n$ with C_1 meeting M and several components $A_1, \dots, A_m$ such that $A_i \cap X$ is isomorphic to $\mathbb{A}^1$ for $1 \leq i \leq m$ (see Lemma 2.6.7 below). We sometimes call any of $A_1 \cap X, \dots, A_m \cap X$ a *feather* of the $\mathbb{A}^1$-fibration f. If there is no fear of confusion, we also call any of $A_1, \dots, A_m$ a feather of X.

For later use, we shall look into a more systematic construction of ML_0-surfaces. The following result is a part of Theorem 2.5.6.

Lemma 2.6.7. *Let $U \subset X$ be smooth affine surfaces such that U is an open set of X and U has an $\mathbb{A}^1$-fibration. Then $X - U$ is a disjoint union of irreducible curves which are isomorphic to $\mathbb{A}^1$.*

Recall that an *admissible chain* is a linear chain of smooth rational curves such that each irreducible component has self-intersection number ≤ -2.

Lemma 2.6.8. *Let X be an ML_0-surface and $f : X \to \mathbb{A}^1$ an $\mathbb{A}^1$-fibration. Then there exists a smooth normal completion $\overline{X} := \overline{X}_{f,n}$ of X such that the boundary divisor $D := \overline{X} - X$ consists of a linear chain*

$$D = \ell + M + A,$$

where ℓ is a (0)-curve, M is an (n)-curve and A is an admissible linear chain. The fibration f extends to a $\mathbb{P}^1$-fibration $\overline{f} : \overline{X}_{f,n} \to \mathbb{P}^1$ such that $\ell = \overline{f}^{-1}(\infty)(:= F_\infty)$, M is a cross-section of $\overline{f}$ and $F_0 := \overline{f}^{-1}(0) = A + A_1 + \cdots + A_r$, where each A_i is isomorphic to $\mathbb{P}^1$ and meets $\overline{X} - X$ normally in a component A_i' of A, hence A_i is a feather of X. Furthermore, the following assertions hold:

(1) *Given an* ML_0*-surface* $X = X_{f,n}$, *an* ML_0*-surface* $X_{f,n'}$ *with an arbitrary* n' *is obtained by elementary transformations on* $\overline{f}^{-1}(\infty)$ *of* $\overline{X}_{f,n}$, *where* ℓ *(resp.,* M*) is replaced by a* (0)*-curve (resp.,* (n')*-curve)* ℓ' *(resp.,* M'*) and* A, X *are not touched in the process.*
(2) *We have* $r \geq 1$ *and* $\rho(X) = r - 1$. *Let* m_i *be the multiplicity of* A_i *in* $\overline{f}^{-1}(0)$. *Then* $\mathrm{Pic}(X)$ *is generated by the classes of* $A_1, \ldots, A_r$ *which are subject to the relation* $\sum_{i=1}^{r} m_i A_i = 0$. *Hence* $\mathrm{Pic}(X)_{\mathrm{tor}}$ *is a finite cyclic group of order* $d = \gcd(m_1, \ldots, m_r)$ *(see Lemma 2.6.5).*
(3) *At least one* A_i *is a* (-1)*-curve.*
(4) *Every fiber of* f *except for* $f^{-1}(0)$ *is irreducible and reduced, hence isomorphic to* $\mathbb{A}^1$.

Proof. The existence of a normal completion follows from the proof of Lemma 2.6.3. Assertions (1), (2), (3), and (4) are straightforwardly verified. □

We call a chain $D = \ell + M + A$ as above a *standard chain* and the completion $\overline{X}_{f,n}$ of X attaching D as the boundary divisor a *standard completion* of X. Note that for any standard chain on a smooth projective rational surface $\overline{X}$ the curve ℓ induces an $\mathbb{A}^1$-fibration on $X := \overline{X} - D$. X is an ML_0-surface precisely when X is affine by Theorem 2.6.2.

We shall explain a method of attaching feathers to construct any ML_0-surface. Let $\mathbb{F}_s$ be the Hirzebruch surface with a $\mathbb{P}^1$-fibration $p : \mathbb{F}_s \to \mathbb{P}^1$. Let $\ell = \ell_\infty$ and ℓ_0 be two fibers of p and let M be a cross-section with $(M^2) = n$. We apply the following two operations.

(1) Blow up above a point on ℓ_0 not on M so that the proper transform ℓ_0' of ℓ_0 together with the exceptional curves forms a linear chain $B^{(1)}$. Suppose $B^{(1)}$ has length ≥ 2. Let $B_0^{(1)}$ be the tip of $B^{(1)}$ not meeting M and $B_1^{(1)}$ the component meeting $B_0^{(1)}$. Blow up above a point on $B_1^{(1)}$ not on any other component of $B^{(1)}$ to produce a chain $B^{(2)}$ attached to $B_1^{(1)}$. Continuing this way we produce a linear chain
$$A' = (B^{(1)} \setminus B_0^{(1)}) + (B^{(2)} \setminus B_0^{(2)}) + \cdots + (B^{(s-1)} \setminus B_0^{(s-1)}) + B^{(s)}.$$
We allow $B^{(s)}$ to have length 1 or to be empty.
(2) Choose irreducible components $A_1', \ldots, A_t'$ of A', including all (-1)-components, and blow up n_i ordinary points of A_i' not on any other component of A' to produce exceptional curves A_{ij}. Let A be the same chain as A' with $(A_i'^2)$ replaced by $(A_i'^2) - n_i$. Relabel $A_{11}, \ldots, A_{tn_t}$ as $A_1, \ldots, A_{r'}$ and $B_0^{(1)}, \ldots, B_0^{(s-1)}$ as $A_{r'+1}, \ldots, A_r$.

We thus produce a projective surface $\overline{X}$ with a $\mathbb{P}^1$-fibration $\overline{f} : \overline{X} \to \mathbb{P}^1$ induced by p such that $\overline{f}^{-1}(\infty) = \ell$, M is a cross-section of $\overline{f}$ and $\overline{f}^{-1}(0) = A + A_1 + \cdots + A_r$. If we assume $r \geq 2$ in case $B^{(1)}$ has length 1 (i. e., $A = \ell_0$), A is an admissible chain.

Lemma 2.6.9. *Let* $\overline{X}$ *be as constructed as above and let* $D = \ell + M + A$. *Then* $X := \overline{X} - D$ *is an* ML_0*-surface and* $\overline{f}$ *induces an* $\mathbb{A}^1$*-fibration* $f : X \to \mathbb{A}^1$, *where* D *is a standard linear chain and* $X = X_{f,n}$. *Furthermore, every* ML_0*-surface arises in this way.*

Proof. Steps (1) and (2) above are the only way to create standard chains D on a smooth projective rational surface $\overline{X}$ such that $\overline{X} - D$ is affine. □

The following result shows that if an ML_0-surface has positive Picard rank then it contains an affine open set which is an ML_0-surface with smaller Picard rank.

Lemma 2.6.10. *Let X be an ML_0-surface with $\rho(X) > 0$. Then there exists an irreducible curve A such that:*
(1) *A is isomorphic to $\mathbb{A}^1$,*
(2) *$X' := X - A$ is an ML_0-surface with $\rho(X') = \rho(X) - 1$.*

In particular, if $\rho(X) = 1$ then X' is a $\mathbb{Q}$-homology plane.

Proof. We choose a standard completion of X as defined in Lemma 2.6.8 and constructed concretely in Lemma 2.6.9. Since $\rho \geq 1$, we have $r \geq 2$. Let A_1 be a (-1)-curve. Let $D' = D + A_1$ and make D' minimal by the contraction of A_1 and successively contractible curves. If a component of A becomes a (-1)-curve in this process, then no feather A_i, $i \geq 2$, is attached to it unless A is a (-2)-curve and $r = 2$. In that case $X - A_1 \cong \mathbb{A}^2$. In the general case we end up with a standard minimal boundary $D'' = \ell'' - M'' - A''$ of $X - A_1$ with $A_2, \ldots, A_r$ attached to A''. We may assume that D'' is minimal by Lemma 2.6.8(1). We have $\rho(X - A_1) = r - 1$ and hence $\gamma(X - A_1) = \gamma(X) = 0$ by Lemma 2.6.5. By Theorem 2.6.2, $X - A_1$ is an ML_0-surface. □

Recall that a smooth affine surface X is a $\mathbb{Q}$*-homology plane* if $H_i(X; \mathbb{Q}) = 0$ for all $i > 0$. In order to prove the last assertion of Lemma 2.6.10, we use the following result.

Lemma 2.6.11. *Let X be a smooth affine surface and let (V, D) be a smooth normal completion of X, where $D = V - X$. Then the following conditions are equivalent:*
(1) *X is a $\mathbb{Q}$-homology plane.*
(2) *V is a rational surface, D is a tree of smooth rational curves and $\rho(X) := \dim_{\mathbb{Q}} \mathrm{Pic}(X) \otimes_{\mathbb{Z}} \mathbb{Q} = 0$.*

Proof. Assume that X is a $\mathbb{Q}$-homology plane. Then X is rational by [78]. Further, D is a tree of smooth rational curves and $\rho(X) = 0$ by [166, Lemma 1.1]. Assume condition (2). In the exact sequence (2.12) considered with $\mathbb{Q}$-coefficients, we have $H^1(D; \mathbb{Q}) = 0$ and $H^3(V; \mathbb{Q}) \cong H_1(V; \mathbb{Q}) = 0$ because D and V are simply connected. The exact sequence (2.13) considered over $\mathbb{Q}$ gives rise to an exact sequence

$$0 \longrightarrow \mathrm{Hom}_{\mathbb{Q}}(\mathrm{Pic}(X) \otimes \mathbb{Q}, \mathbb{Q}) \longrightarrow H^2(V; \mathbb{Q}) \longrightarrow H^2(D; \mathbb{Q}) \longrightarrow 0.$$

Since $\rho(X) = 0$, it follows that $H^2(V; \mathbb{Q}) \cong \mathrm{H}^2(D; \mathbb{Q})$. Then the exact sequence (2.12) over $\mathbb{Q}$ yields that $H_1(X; \mathbb{Q}) = H_2(X; \mathbb{Q}) = 0$. On the other hand, $H_i(X; \mathbb{Q}) = 0$ for $i > 2 = \dim X$ by Theorem 1.1.4 . Hence X is a $\mathbb{Q}$-homology plane. □

We note here that all ML_0-surfaces with $\rho(X) = 0$ are $\mathbb{Q}$-homology planes, and structures of such surfaces are described in [143, 33].

2.6.3 An analogue of Abhyankar–Moh–Suzuki theorem

We shall prove the following result which is a generalization of Abhyankar–Moh–Suzuki theorem (AMS theorem for short). The proof below is very similar to a proof of AMS theorem given in Subsection 1.4.2. The arguments in the proof below are used again in the proof of Theorem 2.7.12.

Theorem 2.6.12. *Let X be an ML_0-surface with $\rho(X) = 0$. Let C be a curve isomorphic to the affine line on X. Then there exists an $\mathbb{A}^1$-fibration $f : X \to B$ such that $B \cong \mathbb{A}^1$ and C is a fiber component of f.*

The proof uses Theorem 2.6.2 in an essential way and follows from the subsequent four lemmas. We assume that X satisfies the assumption in Theorem 2.6.12.

Lemma 2.6.13. *We have $e(X - C) = 0$ and $\overline{\kappa}(X - C) \le 1$.*

Proof. By Lemma 2.6.11, X is a $\mathbb{Q}$-homology plane. Hence the Euler number $e(X)$ is 1. Since C is isomorphic to the affine line, we have $e(C) = 1$. Hence we have $e(X - C) = 0$. By a logarithmic analogue of the Miyaoka–Yau inequality due to R. Kobayashi [117], we have $\overline{\kappa}(X - C) \le 1$ for otherwise $e(X - C) > 0$. For details, see [159, Theorem 6.7.1]. □

We consider below each of the cases $\kappa(X - C) = -\infty, 0, 1$ separately.

Lemma 2.6.14. *If $\overline{\kappa}(X - C) = -\infty$, then there exists an $\mathbb{A}^1$-fibration $f : X \to B$ such that C is a fiber component of f.*

Proof. Since $X - C$ is an affine surface with $\overline{\kappa}(X - C) = -\infty$, there exists an $\mathbb{A}^1$-fibration $f' : X - C \to B'$ which extends to an $\mathbb{A}^1$-fibration $f : X \to B$ such that C is a fiber component of f. In fact, the closures of general fibers of f' in V generate a linear pencil Λ. Suppose that Λ has a base point. Then the base point is a one-place point for a general member of Λ. Let $\sigma : \widetilde{V} \to V$ be the shortest sequence of blowing-ups which eliminate the base points of Λ. Let E be the last (-1)-curve appearing in the process σ. Then E is a cross-section of a $\mathbb{P}^1$-fibration $\widetilde{f} : \widetilde{V} \to \overline{B'}$ on $\widetilde{V}$ and the proper transform of the closure $\overline{C}$ of C is contained in a fiber of $\widetilde{f}$. Since X is an affine open set of $\widetilde{V}$, the restriction $f := \widetilde{f}|_X : X \to B := \widetilde{f}(X)$ is an $\mathbb{A}^1$-fibration such that f is an extension of f' and C is a fiber component of f. Suppose that Λ has no base points. Then Λ defines a $\mathbb{P}^1$-fibration $\overline{f'} : V \to \overline{B'}$, and a general fiber of $\overline{f'}$ does not meet C because X contains no complete curves. Hence the closure $\overline{C}$ of C in V is a fiber component of $\overline{f'}$. Then the restriction $f := \overline{f'}|_X$ is an $\mathbb{A}^1$-fibration for which C is a fiber component. □

Lemma 2.6.15. *The case $\overline{\kappa}(X - C) = 0$ does not occur.*

Proof. Suppose on the contrary that the case $\overline{\kappa}(X-C) = 0$ occurs. Since $\mathrm{Pic}(X)$ is a finite cyclic group by Lemma 2.6.6, there exists an element $u \in \Gamma(X, \mathcal{O}_X)$ such that $\mathrm{Supp}(u) = C$. Let $X' = X - C$. Consider the morphism $X' \to \mathbb{A}^1_*$ which is defined by the function

u and hence denoted by u, and consider the Stein factorization $u : X' \xrightarrow{v} B' \to \mathbb{A}^1_*$ of the morphism u, where B' is necessarily isomorphic to $\mathbb{A}^1_*$. In fact, since $\gamma(X) = 0$ we have $\Gamma(X, \mathcal{O}_X)^* = k^*$. Since C is irreducible, it follows that $\Gamma(X', \mathcal{O}_{X'})^* \cong k^* \times \{v^i \mid i \in \mathbb{Z}\}$, where we can take $v \in \Gamma(X', \mathcal{O}_{X'})$ so that u is a positive power of v. On the other hand, X' is rational, hence B' is a smooth affine rational curve such that $\Gamma(B', \mathcal{O}_{B'})^*$ is a subgroup of $\Gamma(X', \mathcal{O}_{X'})^*$ and B' dominates $\mathbb{A}^1_*$. This implies that $B' \cong \mathbb{A}^1_*$. By replacing the morphism u by v, we may assume that the general fibers of the morphism u are irreducible. By Kawamata's addition formula [159, Lemma 1.14.1], we have

$$0 = \overline{\kappa}(X - C) \geq \overline{\kappa}(\mathbb{A}^1_*) + \overline{\kappa}(F) = \overline{\kappa}(F),$$

where F is a general fiber of u. Then $\overline{\kappa}(F)$ is either $-\infty$ or 0. If $\overline{\kappa}(F) = -\infty$, then u is an $\mathbb{A}^1$-fibration and $\overline{\kappa}(X - C) = -\infty$, which is not the case.

So $\overline{\kappa}(F) = 0$, i. e., $F \cong \mathbb{A}^1_*$. By the construction of the morphism u, it extends to a morphism $u : X \to B$ such that $u^*(0)_{\text{red}} = C$, where $B \cong \mathbb{A}^1$. In fact, we replaced u by an element $v \in \Gamma(X', \mathcal{O}_{X'})^*$. But if we take the Stein factorization $u : X \to B \to \mathbb{A}^1$, the morphism $X \to B$ decomposing u is defined by v, and hence v extends to an element of $\Gamma(X, \mathcal{O}_X)$. By the theorem of Zaidenberg–Suzuki (see Theorem 1.1.17) applied to $u : X - C \to \mathbb{A}^1_*$, we have

$$0 = e(X - C) = e(\mathbb{A}^1_*)e(F) + \sum_i (e(F_i) - e(F)),$$

where the F_i exhaust all singular fibers of u. Hence $e(F_i) = e(F)$ for all i. This implies that all the fibers of $u : X - C \to \mathbb{A}^1_*$ are isomorphic to $\mathbb{A}^1_*$ if taken with reduced structures. Now we consider a smooth normal completion $X \hookrightarrow V$ such that u extends to a $\mathbb{P}^1$-fibration $p : V \to \overline{B}$, where $\overline{B} \cong \mathbb{P}^1$. We have two cases to consider:

Case u is untwisted. This means that there exist two cross-sections H_1, H_2 of p lying in the boundary $D := V - X$. We may assume that the fiber $\ell_\infty = p^{-1}(P_\infty)$ is a smooth fiber and meets H_1 and H_2 in distinct points, where $\{P_\infty\} = \overline{B} - B$. In fact, the intersection point $H_1 \cap H_2$ is a one-place point for H_1 and H_2 if it exists. Since ℓ_∞ is a component of D, we can separate H_1 and H_2 by the repeated use of elementary transformation with center $H_1 \cap H_2$. The fiber $p^{-1}(P_0)$ for $P_0 = \{u = 0\}$ contains the closure $\overline{C}$ of C, and $\mathrm{Supp}(p^{-1}(P_0)) - \overline{C}$ is connected and meets both H_1 and H_2 because $C \cong \mathbb{A}^1$. Then the boundary divisor D contains a loop, which is a contradiction because D is a tree. See Lemma 2.6.8 or use the fact that the boundary divisor of X is a tree if X is a $\mathbb{Q}$-homology plane [164].

Case u is twisted. There exists a 2-section H of p contained in D. Since $p|_H : H \to \overline{B}$ ramifies in two points and $p^{-1}(P)_{\text{red}} \cap X \cong \mathbb{A}^1_*$ if $P \neq P_0, P_\infty$, it follows that $p|_H$ ramifies over the points P_0 and P_∞ of $\overline{B}$. So, we may assume that $p^{-1}(P_\infty) + H$ looks like

$$\begin{array}{ccccc} & & (-2) & \ell'_\infty & \\ & & | & & \\ (-2) & - & (-1) & - & H, \\ E_1 & & E_2 & & \end{array}$$

where $(-i)$ represents a $(-i)$-curve, i. e., a curve isomorphic to $\mathbb{P}^1$ and having self-intersection number $-i$ for $i = 1, 2$.

The divisor $H + p^*(P_0)_{\text{red}} - \overline{C}$ looks like $H + \Gamma$, where there are three possibilities:
(i) Γ contains at least one branch component for D.
(ii) $\Gamma = F_1 + F_2$ and $H + \Gamma + \overline{C}$ looks like

$$\begin{array}{ccccc} & & (-2) & \overline{C} & \\ & & | & & \\ H & - & (-1) & - & (-2) \\ & & F_2 & & F_1 \end{array}$$

(iii) $\Gamma = \emptyset$ and $\overline{C}$ touches H in a smooth point with order of contact 2.

In the case (i), the boundary D has a configuration

$$\begin{array}{cccc} (-2)- & (-1) & -H-\Gamma & \\ & | & & \\ & (-2) & \ell'_\infty & \end{array}$$

and D cannot be minimized to a linear chain because Γ contains a branching component. Hence X is not an ML_0-surface.

In the case (ii), contract E_2, E_1, F_2, and F_1 to obtain a relatively minimal $\mathbb{P}^1$-fibration $\overline{p} : \overline{V} \to \overline{B}$. Let M_0 and ℓ be respectively the minimal section and a general fiber of $\overline{p}$. Then the image $\overline{H}$ of H is linearly equivalent to $2M_0 + a\ell$ with $a \geq 2n$, where $n = -(M_0{}^2)$. Then $(\overline{H}^2) = 4(a-n)$ and $(H^2) = 4(a-n-1)$. On the other hand, X is ML_0-surface if and only if $H + F_2 + F_1$ is contractible to a point, i. e., $(H^2) = -3$. This implies that D does not minimize to a linear chain.

In the case (iii), D minimizes to a linear chain if and only if H is a (-1)-curve. With notations as in the case (ii), contract E_1 and E_2 to obtain a relatively minimal $\mathbb{P}^1$-fibration $\overline{p} : \overline{V} \to \overline{B}$. As above, $\overline{H} \sim 2M_0 + a\ell$ and hence $(\overline{H}^2) = 4(a-n)$. Meanwhile, $(\overline{H}^2) = 1$ because $(H^2) = -1$, which is a contradiction. □

Lemma 2.6.16. *The case $\overline{\kappa}(X - C) = 1$ does not occur.*

Proof. Suppose now that $\overline{\kappa}(X - C) = 1$. Then there exists an $\mathbb{A}^1_*$-fibration $f' : X - C \to B'$ by a theorem of Kawamata (see Theorem 1.2.5). We claim that $B' \cong \mathbb{A}^1$ or $\mathbb{A}^1_*$. First of all, since $X - C$ is rational, B' is a smooth rational curve. If $B' \cong \mathbb{A}^1_{n*}$

($n \geq 2$), which is $\mathbb{A}^1$ with n points punctured, then $\gamma(X - C) \geq n$, while $\gamma(X - C) = 1$. Hence $B' \cong \mathbb{A}^1_*, \mathbb{A}^1$ or $\mathbb{P}^1$. If $B' \cong \mathbb{P}^1$ then $\rho(X - C) > 0$, while $\rho(X - C) = 0$ by Lemma 2.6.5. So, $B' \cong \mathbb{A}^1$ or $\mathbb{A}^1_*$. We consider these cases separately. By Theorem 1.1.17, it follows that all fibers of f' are isomorphic to $\mathbb{A}^1_*$ if taken with the reduced structures.

Case $B' \cong \mathbb{A}^1$. We consider the linear pencil Λ spanned by the closures of general fibers of f'. Suppose that the general fibers of f' are closed in X. Then f' extends to an $\mathbb{A}^1_*$-fibration $f : X \to B$. We have the following four cases to consider:
(i) $B \cong \mathbb{P}^1$ and f is untwisted.
(ii) $B \cong \mathbb{P}^1$ and f is twisted.
(iii) $B \cong \mathbb{A}^1$ and f is untwisted.
(iv) $B \cong \mathbb{A}^1$ and f is twisted.

Let $X \hookrightarrow V$ be a smooth normal completion and let $D := V - X$. We may assume that there is a $\mathbb{P}^1$-fibration $p : V \to \overline{B}$ such that $p\,|_X = f$. In the untwisted (resp., twisted) case, there are two cross-sections H_1, H_2 (resp., one 2-section H) in D.

In the case (i), $\overline{B} = B$ and H_1 and H_2 do not meet on the fiber $p^{-1}(P)$ ($P \neq P_0$), where $\{P_0\} = B - B'$, because $f^{-1}(P)_{\text{red}} \cong \mathbb{A}^1_*$ for every $P \neq P_0$ as remarked above. We may assume that the boundary $D := V - X$ contains no (-1)-curves as fiber components. Let $\overline{C}$ be the closure of the given affine line C. Then $\overline{C}$ is a unique (-1)-curve in the fiber $p^{-1}(P_0)$ and a tip (= an end component) in the fiber. Since D minimizes to a linear chain, it follows that H_1 and H_2 are not branching components in D. This implies that $f' : X - C \to B'$ has at most one multiple fiber. In fact, if $f^{-1}(P) = mF$ with $m > 1$ and $F \cong \mathbb{A}^1_*$, then the closure $\overline{F}$ is a (-1)-component and $\overline{F}$ is connected to H_1 and H_2 by nonempty linear chains in $p^{-1}(P)$. If there are two multiple fibers in f', H_1 and H_2 are branching components. Let $V \to V'$ be a sequence of contractions which brings $p^{-1}(P_0)$ to a smooth fiber ℓ_0 and let D' be the reduced image of $D + \overline{C}$ on V'. Then $\kappa(V, D + \overline{C} + K_V)$ is equal to $\kappa(V', D' + K_{V'})$ and we have by [152, p. 81]

$$D' + K_{V'} \sim_{\mathbb{Q}} \left\{-2 + 1 + \left(1 - \frac{1}{m}\right)\right\}\ell + \varepsilon,$$

where ℓ represents a general fiber of p, m is the multiplicity of a possible multiple fiber of f' and ε is an effective $\mathbb{Q}$-divisor with negative-definite intersection matrix which does not affect the calculation of $\kappa(V', D' + K_{V'})$. In fact, $\varepsilon = \frac{1}{m}p^*(P_1) - F_1$, where $m(F_1 \cap X)$ is a multiple fiber of f' and $P_1 = p(F_1)$. If f' has no multiple fiber, we consider m to be 1. Then $\overline{\kappa}(X - C) = -\infty$. This is a contradiction. A simpler argument is to note that we have $\kappa(V, D + \overline{C} + K_V) = \kappa(V, D + K_V) = \overline{\kappa}(X) = -\infty$ since $\overline{C}$ is a (-1)-curve sprouting from D. This contradicts the hypothesis that $\overline{\kappa}(X - C) = 1$.

In the case (ii), $p|_H\colon H \to B$ branches over two points P_0, P_1 of B. We may assume that $f^{-1}(P_0)_{\mathrm{red}} = C \cong \mathbb{A}^1$. Since D is connected, it follows that $f'^{-1}(P_1)_{\mathrm{red}} \cong \mathbb{A}^1$, which is a contradiction because $f^{-1}(P_1)_{\mathrm{red}}$ must be isomorphic to $\mathbb{A}^1_*$.

In case (iii), since $B = B' \cong \mathbb{A}^1$, D contains a complete fiber $p^{-1}(P_\infty)$ of p, which we may assume to be a smooth fiber. The fiber $p^{-1}(P_0)$ containing $\overline{C}$ cuts out a fiber of the form either $m_1\mathbb{A}^1_* + n_1C$ or $m_1C' + n_1C$ on X, where $C' \cong C \cong \mathbb{A}^1$ and C', C meet in one point transversally (see Theorem 2.5.7). We may further assume that there are no (-1) fiber components of p in D.

Consider first the case $p^{-1}(P_0) \cap X = m_1\mathbb{A}^1_* + n_1C$. If $m_1 > 1$ there are no other multiple fibers of f', for otherwise both H_1 and H_2 are branching components of D and D cannot be minimized to a linear chain. If $m_1 = 1$, there is at most one multiple fiber $m_2\mathbb{A}^1_*$ in f' by the same reason as above. Since $\overline{C}$ is a (-1)-curve, contract $\overline{C}$ and consecutively contractible fiber components in $D \cap p^{-1}(P_0)$ and obtain the images V', D' of V, D. Then $\kappa(V, D + \overline{C} + K_V) = \kappa(V', D' + K_{V'})$ and

$$D' + K_{V'} \sim_{\mathbb{Q}} \left\{-2 + 1 + \left(1 - \frac{1}{m_i}\right)\right\}\ell + \varepsilon = \left(-\frac{1}{m_i}\right)\ell + \varepsilon,$$

where $i = 1$ or 2 according to $m_1 > 1$ or $m_1 = 1$ and ε is, as in the case (i) above, an effective $\mathbb{Q}$-divisor which does not affect the calculation of $\kappa(V', D' + K_{V'})$. Then $\overline{\kappa}(X - C) = -\infty$. This is a contradiction.

Consider next the case $p^{-1}(P_0) \cap X = m_1C' + n_1C$. Suppose $\min(m_1, n_1) > 1$. If there is at least one multiple fiber $m_2\mathbb{A}^1_*$ in f', then H_1 and H_2 are branching components in D, and D cannot be minimized to a linear chain. So this case does not occur. If there is no multiple fiber in f' other than $m_1\mathbb{A}^1_*$ which is cut out by m_1C' on $X - C$, a computation similar to the above shows that $\overline{\kappa}(X - C) = -\infty$, which is a contradiction. Hence $\min(m_1, n_1) = 1$. Suppose $m_1 = n_1 = 1$. Let $\overline{C'}$ be the closure of C' on V. Then each of $\overline{C'}$ and $\overline{C}$ meets one of H_1 and H_2. Assume that $\overline{C'}$ meets H_1 and $\overline{C}$ does H_2. In this case, f' has at most one multiple fiber $m_2\mathbb{A}^1_*$ and $\overline{\kappa}(X - C) = -\infty$, a contradiction. Hence either $m_1 = 1, n_1 > 1$ or $m_1 > 1, n_1 = 1$. Suppose that $m_1 = 1$ and $n_1 > 1$. Then the fiber $p^{-1}(P_0)$ is a linear chain and there is a nonempty linear chain connecting $\overline{C}$ to H_2. Hence there is no other multiple fiber in f', for otherwise H_2 is a branching component and D is not minimized to a linear chain because $f^{-1}(P_\infty)$ is a smooth fiber with self-intersection number 0. Then $\overline{\kappa}(X - C) = -\infty$ by a computation similar to the above. Hence $m_1 > 1$ and $n_1 = 1$. Then there is a nonempty linear chain in $p^{-1}(P_0)$ connecting $\overline{C'}$ to H_1 and hence there is no other multiple fiber in f'. By the computation of $\overline{\kappa}(X - C)$, this leads to a contradiction as well.

In the case (iv), one fiber $f^{-1}(P_0)$ of f must be of the form either $m_1\mathbb{A}^1_* + n_1C$ or $m_1C' + n_1C$ as in the case (iii). Furthermore, we may assume that the fiber at infinity $p^{-1}(P_\infty)$ together with H has the following configuration:

$$\begin{array}{ccccc} (-2) & - & (-1) & - & H \\ & & | & & \\ & & (-2) & \ell'_\infty & \end{array}$$

Note that $p\,|_H\colon H \to \mathbb{P}^1$ ramifies over P_∞ and a point P_1 of B. Since D is connected, the fiber $p^*(P_1)$ cannot produce a fiber of f which has the form $m_1\mathbb{A}^1_* + n_1C, m_1C' + n_1C$ (case $P_1 = P_0$) or $m\mathbb{A}^1_*$. This is a contradiction.

Now suppose that the pencil Λ has a base point P on X. Then P is a one-place point for the general members of Λ, $P \in C$ and C is contained in a member of Λ. The pencil must be parametrized by $\mathbb{P}^1$ because it has a base point in the affine part X. So, mC is a member of Λ for some $m \geq 1$. We consider elimination of base points of Λ as well as a smooth normal completion $X \hookrightarrow V$. If Λ extended to a pencil on V has base points on $V - X$, we replace V by a suitable smooth surface obtained from V by elimination of the base points outside X. Then we obtain a $\mathbb{P}^1$-fibration $p : V \to \overline{B}$ satisfying the following conditions:

(1) $\overline{B} \cong \mathbb{P}^1$.
(2) The exceptional locus Γ arising from the elimination of the base point P consists of a cross-section of p, which is the last (-1)-curve in the elimination process, and several trees sprouting out of the cross-section.
(3) The boundary divisor $D := V - X$ has one cross-section H and several trees sprouting out of H.
(4) The closure $\overline{C}$ of C together with the trees from Γ and D forms a fiber of p.

Furthermore, we may assume that any fiber of p contains no (-1)-curves in $\Gamma + D$. Then we have:

(5) Every degenerate fiber consists of a linear chain of components.
(6) There is at most one degenerate fiber besides the one containing $\overline{C}$. If it exists, let $m'C'$ be the multiple fiber of f' cut out by this degenerate fiber, where $C' \cong \mathbb{A}^1_*$.

We shall compute $\overline{\kappa}(X - C)$. We have

$$\begin{aligned}&\overline{C} + D + \Gamma + K_V\\ &\quad\sim_{\mathbb{Q}} \varepsilon + \begin{cases} -\ell & \text{(no multiple fibers case)},\\ (-2 + 1 + (1 - \frac{1}{m'}))\ell & \text{(one multiple fiber case)},\end{cases}\end{aligned}$$

where ℓ is a general fiber of p and ε is an effective divisor which does not affect the calculation of $\kappa(V, \overline{C}+D+\Gamma+K_V)$. In both cases, $\overline{\kappa}(X-C) = -\infty$ which is a contradiction to the assumption $\overline{\kappa}(X - C) = 1$.

Suppose that Λ has no base points on C. Then C is a cross-section of Λ and there are no multiple fibers in Λ. Hence Λ defines an $\mathbb{A}^1$-bundle structure on X parametrized by $\mathbb{A}^1$. Hence X is isomorphic to $\mathbb{A}^2$. In this case, $\overline{\kappa}(X - C) = -\infty$, which is a contradiction.

Case $B' \cong \mathbb{A}^1_*$. Consider again the pencil Λ generated by the closures on X of general fibers of f'. Then Λ has no base points. In fact, if Λ has a base point, Λ should be parametrized by $\mathbb{P}^1$. Meanwhile, we have only as many members as parametrized by $\mathbb{A}^1_*$ and one more member corresponding to C. This is a contradiction. If the general members of Λ meet the curve C, then C is a cross-section. But we have one point to which corresponds no member of Λ. So, general members of Λ do not meet the curve C. In this case we argue as in the case $\overline{\kappa}(X - C) = 0$. □

Thus the proof of Theorem 2.6.12 is completed. We note that Theorem 2.6.12 is a generalization of Abhyankar–Moh–Suzuki theorem. In fact, if $X \cong \mathbb{A}^2$, it is clearly an ML_0-surface with $\rho(X) = 0$.

Before finishing this subsection we shall construct a smooth affine surface X satisfying the following conditions:

(1) X is an ML_0-surface with $\rho(X) = r > 0$.
(2) There exists an affine line C such that $\overline{\kappa}(X - C) \geq 0$. We call such an affine line C an *anomalous* affine line.

In fact, the existence of such an example shows that the analogue of Abhyankar–Moh–Suzuki theorem fails to hold.

We begin with constructing a basic example in the case $\rho(X) = 1$.

Example 2.6.17. Let $\mathbb{F}_0$ be the Hirzebruch surface $\mathbb{P}^1 \times \mathbb{P}^1$. We denote any fiber of the vertical (resp., the horizontal) $\mathbb{P}^1$-fibration by ℓ (resp., M) and call it a *fiber* (resp., *a section*). Take two horizontal sections M_0, M_1 and three fibers $\ell_0, \ell_1, \ell_\infty$. Let $P_0 := M_0 \cap \ell_0$ and $P_1 := M_1 \cap \ell_1$. Let A be a smooth irreducible curve such that $A \sim M + 2\ell$ and touches M_0 (resp., M_1) at P_0 (resp., P_1) with order of contact 2. Clearly, A is a cross-section of one of the $\mathbb{P}^1$-fibrations on $\mathbb{F}_0$ and a 2-section for the other $\mathbb{P}^1$-fibration and $A^2 = 4$. Now A meets ℓ_∞ at a point other than $M_0 \cap \ell_\infty$ and $M_1 \cap \ell_\infty$. Blow up the point P_0 (resp., P_1) and its infinitely near point of the first order lying on M_0 (resp., M_1) to produce irreducible exceptional curves E_1, E_2 (resp., F_1, F_2), where $({E_1}^2) = ({F_1}^2) = -2$ and $({E_2}^2) = ({F_2}^2) = -1$. Then the proper transform A' of A meets E_2 and F_2. We blow up these two intersection points to obtain the exceptional curves E_3 and F_3. We denote the proper transforms of E_i, F_i $(i = 1, 2)$ by the same letters. Now we have $({E_i}^2) = ({F_i}^2) = -2$ for $i = 1, 2$ and $({E_3}^2) = ({F_3}^2) = -1$. Let A' be anew the proper transform of A'. Then $({A'}^2) = -2$. Let $\sigma : V \to \mathbb{F}_0$ be the composite of these blowing-ups. Let $M_0', M_1', \ell_0', \ell_1', A'$ signify the proper transforms of $M_0, M_1, \ell_0, \ell_1, A$ on V. We set $X := V - D$, where $D := M_0' + M_1' + \sigma^*(\ell_\infty) + E_1 + E_2 + F_1 + F_2$ and $C := A' \cap X$.

Then we prove the following result for this example.

Proposition 2.6.18. *The following assertions hold:*

(1) *X is an ML_0-surface with $\rho(X) = 1$.*
(2) *$\overline{\kappa}(X - C) = 0$ and hence C is an anomalous affine line.*
(3) *$X - C$ is a $\mathbb{Q}$-homology plane.*

Proof. The boundary $D := V - X$ consists of a linear chain

$$\begin{array}{ccccccccccccc} E_1 & + & E_2 & + & M_0' & + & \sigma^*(\ell_\infty) & + & M_1' & + & F_2 & + & F_1 \\ (-2) & & (-2) & & (-2) & & (0) & & (-2) & & (-2) & & (-2) \end{array}$$

Since D is a linear chain with a component $\sigma^*(\ell_\infty)$ with self-intersection number zero we see that X is an ML_0-surface (see the proof of Theorem 2.6.2) with $\rho(X) = 1$, and C is an affine line on X. In fact, $\mathrm{Pic}(X)$ is generated by $E_3^* = E_3 - D$ and $F_3^* = F_3 - D$ with $2E_3^* = 2F_3^* = C$. Hence $\rho(X) = 1$ and $\rho(X - C) = 0$. This gives (1) and (3). By the adjunction formula we check easily that

$$4K_V \sim -3M_0' - 3M_1' - E_1 - 2E_2 - F_1 - 2F_2 - 4l_\infty' - 2A'.$$

Hence

$$4(K_V + D + A') \sim M_0' + M_1' + 3E_1 + 2E_2 + 3F_1 + 2F_2 + 2A'.$$

Since the reduced divisors $M_0' + E_1 + E_2, M_1' + F_1 + F_2$ and A' are disjoint from each other and since each of them has negative definite intersection matrix, it follows that $\dim |4(A' + D + K_V)| = 0$. Hence $\overline{\kappa}(X - C) = 0$ and C is an anomalous affine line as asserted in (2). □

We recall the definition of *a half-point attachment* to produce further examples of ML_0-surfaces admitting anomalous affine lines. Let $X \hookrightarrow V$ be a normal completion of a smooth affine surface and let $D := V - X$. Let P be a point on D which is not an intersection point of two irreducible components of D. Let $\sigma : V' \to V$ be the blowing-up of P, $E = \sigma^{-1}(P)$, $D' := \sigma'(D)$ the proper transform of D, and let $X' := V' - D'$. We say that the affine line $E \cap X'$ is a *half-point* and that X' is obtained by a half-point attachment with center P. In fact, X is a Zariski open set of X'. We are interested only in the case where X' is again affine. An operation of attaching a feather, which we defined before Lemma 2.6.7, is a kind of half-point attachment.

Proposition 2.6.19. *Let $X \hookrightarrow V$ be the same as in Example 2.6.17. Choose r points $P_1, \dots, P_r$ on the boundary component E_1 and let X' be a smooth surface obtained by half-point attachments with centers $P_1, \dots, P_r$. Then the following assertions hold:*
(1) *X' is an ML_0-surface with $\rho(X) = r + 1$ for any $r \geq 0$.*
(2) *The curve C in X is also an anomalous affine curve in X'.*

Proof. (1) Let $\tau : V' \to V$ be the blowing-ups of $P_1, \dots, P_r$. We denote the proper transforms on V' of the components of D by the same letters. Then $({E_1}^2) = -(r + 2)$. Let

$$\begin{aligned} L = {} & 4E_1 + (4r + 9)E_2 + (8r + 15)M_0' + (12r + 22)\sigma^*(\ell_\infty) \\ & + (9r + 15)M_1' + (6r + 9)F_2 + (3r + 4)F_1. \end{aligned}$$

By the Nakai–Moishezon test, L is an effective ample divisor on V' with Supp $L = \tau'(D)$, the proper transform of D. Hence X' is affine. Since $\tau'(D)$ is a linear chain, X' is an ML_0-surface with $\gamma(X') = 0$. Hence by Lemma 2.6.5, we have $\rho(X') = \rho(X) = r + 1$.

(2) It is known that a half-point attachment does not change the logarithmic Kodaira dimension (cf. [95]). So, $\bar{\kappa}(X' - C) = \bar{\kappa}(X - C) = 0$ and C is anomalous in X', too. □

The above propositions give counterexamples to Question 2.6.1 at the beginning of Section 2.6. The divisor class of C has infinite order in Pic(X) by (3) of Proposition 2.6.18. Similarly, C has infinite order in Pic(X').

So we shall reformulate Question 2.6.1 as follows.

Question 2.6.20. Let X be an ML_0-surface and let C be a curve on X isomorphic to the affine line $\mathbb{A}^1$. Suppose that C has torsion divisor class in Pic(X). Does there exist an $\mathbb{A}^1$-fibration $f : X \to B$ such that C is a fiber component of f, where $B \cong \mathbb{A}^1$?

2.7 ML_1-surfaces

Let X be a smooth affine surface. We say that X is *unruled* (resp., *simply ruled*, *multiruled*) if X has no $\mathbb{A}^1$-fibrations (resp., only one $\mathbb{A}^1$-fibration, two independent $\mathbb{A}^1$-fibrations). If there is no fear of confusion, we say that X is 0-*ruled*, 1-*ruled*, 2-*ruled* if X is unruled, simply-ruled, multiruled, respectively. Note that an $\mathbb{A}^1$-fibration on X does not necessarily have an affine curve as the base curve. If X is an ML_0-surface then it is 2-ruled. We shall see later that the converse is not necessarily true.

2.7.1 Boundary divisors of ML_1-surfaces

We shall begin with giving a criterion for an affine surface to be an ML_1-surface. Let V be a smooth projective surface and let D be an effective reduced divisor with simple normal crossings. We assume that D consists of smooth rational curves. Let $\Gamma(D)$ be the associated weighted graph. In the following, *we only consider D which is a tree.* This condition is satisfied if $X = V - D$ is an ML_1-surface. Blowing up an intersection point of two components will add one more component of weight -1 and decrease by -1 the weights of the components concerned; we call this blowing-up *subdivisional*. Blowing up a point on a single component also adds one component of weight (-1) and decrease by (-1) the weight of the concerned component; we call this blowing-up *sprouting*. See [56, 152] for further details of the terminology. Modeled on $\Gamma(D)$, we consider a connected weighted graph Γ in general. We say that Γ is *minimal* if any (-1)-component meets at least three other components, i. e., it is a branching component. Let Γ, Γ' be two weighted graphs. If there exist a weighted graph Γ'' and morphisms $\varphi : \Gamma'' \to \Gamma, \varphi' : \Gamma'' \to \Gamma'$, which are, by definition, composites of subdivisional or sprouting blowing-ups, we say that Γ and Γ' are *pre-equivalent*. Let $\mathfrak{G}$ be the set of all connected weighted graphs, which are trees, together with the equivalence relation generated by the pre-equivalence relations. Given two graphs Γ, Γ' which are equivalent, we say that Γ' is a *modification* of Γ, and vice versa.

Lemma 2.7.1. *Let $\Gamma, \Gamma', \Gamma''$ be the connected weighted graphs and let $\varphi : \Gamma'' \to \Gamma, \varphi' : \Gamma'' \to \Gamma'$ be morphisms. Assume that any (-1)-component of Γ'' which is φ-exceptional is not φ'-exceptional, and vice versa. Then the following assertions hold:*
(1) *A branch component of Γ does not become φ'-exceptional on Γ''.*
(2) *If φ contains a sprouting blowing-up on a nontip component D_i, then D_i becomes a branching component and hence does not become φ'-exceptional.*
(3) *One can create φ'-exceptional (-1)-components on Γ'' by applying successively one of the following operations:*
  (i) *subdivisional blowing-up on a nonbranching component with nonnegative weight;*
  (ii) *sprouting blowing-up on a tip component with nonnegative weight.*

Proof. (1) On a component D_i of Γ or in the midst of the step φ, a sprouting (resp., subdivisional) blowing-up on D_i increases one component (resp., does not change the number of components) meeting the proper transform of D_i. This implies the assertion. Claims (2) and (3) are left to the readers to fill the details of their proof. □

Let Γ be a weighted graph and let L be a connected part of Γ which is a linear chain. We say that L is *admissible* if all components of L have weight ≤ -2.

Corollary 2.7.2. *Let Γ and Γ' be minimal weighted graphs. The following assertions hold:*
(1) *The modifications between Γ and Γ' occur on*
  (a) *nonadmissible linear chains;*
  (b) *nonadmissible maximal twigs T_i or*
  (c) *nonadmissible linear chains L_j connecting two branching components.*
  Under the modifications, admissible maximal twigs and admissible maximal linear chains between branching components are unaffected.
(2) *If all the maximal twigs of Γ are admissible, then the same is true for Γ'.*
(3) *If Γ is a nonadmissible linear chain and if Γ' is equivalent to Γ, then Γ' is also a nonadmissible linear chain.*

Proof. Assertions (1) and (2) are more or less obvious. As for assertion (3), if one tries to modify the graph, a nonadmissible component is a tip component or has two components adjacent to it. If it is a tip component, it is clear that Γ' is a linear chain. If it has two adjacent components and the blowing-up is sprouting, then it becomes a branch component and hence not contractible. This means that we have to contract back exactly the same exceptional components we obtained by the blowing-ups. If the blowing-up is subdivisional, we obtain a linear chain. □

We need one more lemma.

Lemma 2.7.3. *Suppose that the weighted graph Γ is associated to an effective reduced divisor with simple normal crossings consisting of rational curves on a smooth projective rational surface V. If D_i and D_j are two components of D such that $(D_i^2) \geq 0$ and $(D_j^2) \geq 0$, then either D_i and D_j are adjacent or $(D_i^2) = (D_j^2) = 0$ and $D_i \sim D_j$.*

Proof. Suppose that D_i and D_j are not adjacent. By the Hodge index theorem, $(D_i^2) = (D_j^2) = 0$. Since $|D_i|$ is a linear pencil and $D_i \cap D_j = \emptyset$, D_j is a member of $|D_i|$. Hence $D_i \sim D_j$. □

We can now state a result which leads to a criterion for ML_1-surfaces.

Theorem 2.7.4. *Let X be a smooth affine rational surface and let $X \hookrightarrow V$ be a minimal normal completion. Let $D := V - X$ and $\Gamma = \Gamma(D)$. Then the following conditions are equivalent:*

(1) *X has an $\mathbb{A}^1$-fibration $f : X \to B$, where B is an open set of $\mathbb{A}^1$.*
(2) *Γ as well as any other minimal modification of Γ has a nonadmissible twig.*
(3) *There exists a modification Γ' of Γ which has a tip with weight 0.*

Proof. (1) ⇒ (2). We may assume that the $\mathbb{A}^1$-fibration extends to a $\mathbb{P}^1$-fibration $p : V \to \overline{B}$, where $\overline{B} \cong \mathbb{P}^1$. We may also assume that the fibers of p over the points $\overline{B} - B$ are smooth fibers. Then, in $\Gamma(D)$, these fibers represent nonadmissible twigs.

(2) ⇒ (3). If a nonadmissible twig has a nonnegative component which is not a tip component of the twig, then we can modify the graph by subdivisional blowing-ups and blowing-downs so that the tip component has weight 0. Let D_0 be a nonadmissible tip component and let D_1 be the component adjacent to D_0. Then applying subdivisional blowing-ups with centers at $D_0 \cap D_1$ and its infinitely near points lying on D_0, we can make the proper transform D_0' has weight 0. Then the obtained weighted graph Γ' has D_0' as a tip (see the proof of Theorem 2.6.2).

(3) ⇒ (1). We may assume that D has a tip component D_0 with weight 0. Then $|D_0|$ is a linear pencil of rational curves and the adjacent component, say D_1, is a cross-section. Hence the $\mathbb{P}^1$-fibration $p = \Psi_{|D_0|} : V \to \overline{B}$ restricts to an $\mathbb{A}^1$-fibration $f : X \to B$ with $B \subset \mathbb{A}^1$. □

Corollary 2.7.5. *Let the notations and assumptions be the same as in Theorem 2.7.4. Then the following conditions are equivalent:*

(1) *X is an ML_1-surface which is not isomorphic to $\mathbb{A}^1 \times \mathbb{A}^1_*$.*
(2) *Let $X \hookrightarrow V$ be a smooth minimal normal completion and let $D := X - V$. Then $\Gamma(D)$ has a nonadmissible twig and $\Gamma(D)$ is not a linear chain.*

Proof. (1) ⇒ (2). Since X is an ML_1-surface, there is an $\mathbb{A}^1$-fibration $f : X \to B$ with an open set $B \subset \mathbb{A}^1$. Then, by Theorem 2.7.4, $\Gamma(D)$ has a nonadmissible twig. Suppose that $\Gamma(D)$ is a linear chain. Then we take a different completion of the same kind if necessary and assume that a tip component, say D_1, has weight 0. Consider the $\mathbb{A}^1$-fibration f defined by $|D_1|$. Name the components of $\Gamma(D)$ as $D_1+D_2+\cdots+D_n$. We may assume that $(D_2{}^2)=0$ by elementary transformations with centers on D_1 and $D_3+\cdots+D_n$ contains no (-1)-components if $n\geq 3$. If $n=2$, $X\cong\mathbb{A}^2$ and X is an ML_0-surface. Suppose $n\geq 3$. If $D_3+\cdots+D_n$ is not negative-definite, then $n=3$ and $(D_3{}^2)=0$ because $D_3+\cdots+D_n$ is contained in one and the same fiber of f. Hence $X\cong\mathbb{A}^1\times\mathbb{A}^1_*$. This case is excluded. Hence $D_3+\cdots+D_n$ is negative-definite. Then the base of f is isomorphic to $\mathbb{A}^1$. Hence $\gamma(X)=0$. Then X is an ML_0-surface by Theorem 2.6.2. So, $\Gamma(D)$ is not a linear chain.

(2) $\Rightarrow$ (1). By changing the completion V, we may assume that a tip component, say D_1, has weight 0. Let $f:X\to B$ be the $\mathbb{A}^1$-fibration defined by $|D_1|$. Then B is an open set of $\mathbb{A}^1$. If $B\cong\mathbb{A}^1$ then $\gamma(X)=0$ and X is not an ML_0-surface because $\Gamma(D)$ is not a linear chain. If $B\subsetneq\mathbb{A}^1$ then $\Gamma(D)-(D_1+D_2)$ contains a nonadmissible connected component, say Γ_2, where D_2 is the component adjacent to D_1. Since Γ_2 is contained in a member of $|D_1|$, the minimality of $\Gamma(D)$ shows that Γ_2 consists of a single component of weight 0. Hence $\gamma(X)\neq 0$ and X is not an ML_0-surface by Lemma 2.6.7. Since $\Gamma(D)$ is not a linear chain, $X\not\cong\mathbb{A}^1\times\mathbb{A}^1_*$ as well. □

2.7.2 ML_i-property and j-ruledness

Our interest lies in clarifying the interdependence between the ML_i-property and j-ruledness on a smooth rational affine surface X. We shall first consider a smooth affine rational surface X which is an ML_2-surface, i. e., X has no G_a actions. Note that almost all smooth affine rational surfaces are ML_2 and 0-ruled. For example, if $\overline{\kappa}(X)\geq 0$, X satisfies these properties.

As for an example of X which is ML_2 and 1-ruled, we have the following result [74, Theorem 4.1].

Theorem 2.7.6. *Let $f:X\to B$ be an $\mathbb{A}^1$-fibration on a smooth affine surface X with base B a smooth curve such that every fiber of f is irreducible. Assume further that B is isomorphic to $\mathbb{A}^1$ or $\mathbb{P}^1$ and that f has at least two (resp., three) multiple fibers if $B\cong\mathbb{A}^1$ (resp., $B\cong\mathbb{P}^1$). Then X has no other $\mathbb{A}^1$-fibrations whose general fibers are transverse to f.*

In this theorem, if we take B to be $\mathbb{P}^1$, X is ML_2 and 1-ruled since the $\mathbb{A}^1$-fibration f is of complete type. As an analogy of Platonic $\mathbb{A}^1_*$-fiber space, we define a *Platonic $\mathbb{A}^1$-fiber space* as a smooth affine surface X with an $\mathbb{A}^1$-fibration $f:X\to B$ such that $B\cong\mathbb{P}^1$, all fibers of f are irreducible and there are three multiple fibers m_1F_1, m_2F_2, m_3F_3, where $\{m_1,m_2,m_3\}$ is a Platonic triplet $\{2,2,n\}$, $\{2,3,3\}$, $\{2,3,4\}$, $\{2,3,5\}$ up to permutations.

By modifying the surface X in Example 2.6.17, we can construct a smooth affine surface X_1 which is ML_2 and at least 1-ruled.

Example 2.7.7. With the notations before Proposition 2.6.18, we blow up the points $E_3 \cap A'$ and $F_3 \cap A'$. Let V_1 be the obtained projective surface. To avoid complicated notations, we denote the proper transform of $\sigma^*(\ell_\infty)$ by ℓ_∞ and those of M_0', M_1', E_i, F_i, A' by the same letters, where $1 \le i \le 3$. Now $(E_3{}^2) = (F_3{}^2) = -2$. Let E_4 and F_4 be the new (-1)-curves. Let $D_1 = D + E_3 + F_3$ and $X_1 := V_1 - D_1$. Now E_2 and F_2 are branching in D_1 and there are no nonadmissible maximal twigs in D_1. Hence X is ML_2. On the other hand, we have

$$M_0' + E_1 + 2E_2 + 2E_3 + 2E_4 \sim M_1' + F_1 + 2F_2 + 2F_3 + 2F_4.$$

Hence $\Lambda := |M_0' + E_1 + 2E_2 + 2E_3 + 2E_4|$ is a linear pencil with a section ℓ_∞, and the $\mathbb{P}^1$-fibration $\Phi_\Lambda : V_1 \to \mathbb{P}^1$ restricts on X_1 an $\mathbb{A}^1$-fibration with base isomorphic to $\mathbb{P}^1$ because $E_4 \cap X_1 \cong \mathbb{A}^1$ and $F_4 \cap X_1 \cong \mathbb{A}^1$. So, X_1 is at least 1-ruled.

We consider Question 2.6.1 about the movability of an affine line on an ML_1-surface. We first construct an example.

Example 2.7.8. Let V_0 be a Hirzebruch surface of degree $n = 0$ or 1 with the $\mathbb{P}^1$-fibration $p_0 : V_0 \to \mathbb{P}^1$. Let M_0 and ℓ be respectively the minimal section and a general fiber. Let H_0 be a smooth curve such that $H_0 \sim 2M_0 + \ell$ (resp., $H_0 \sim 2(M_0 + \ell)$) if $n = 0$ (resp., $n = 1$). Let P_0, P_∞ be the points of the base curve of p_0 over which $p_0|_{H_0} : H_0 \to \mathbb{P}^1$ ramifies and let $\ell_0 = p_0^{-1}(P_0)$ and $\ell_\infty = p_0^{-1}(P_\infty)$. Let $\sigma : V \to V_0$ be the blowing-ups of the point $\ell_\infty \cap H_0$ and its infinitely near point on H_0 which produce a (-2)-curve E_1 and a (-1)-curve E_2. Let $H = \sigma'(H_0), L = \sigma'(\ell_\infty)$ and $\overline{C} = \sigma'(\ell_0)$. Let $X = V - (H + E_1 + E_2 + L)$ and let $C = \overline{C} \cap X$. Then it is clear that $(H^2) = 2$. Let $\tau : V' \to V$ be the blowing-ups of the point $H \cap E_2$ and its infinitely near point on H which produce a (-2)-curve E_3 and a (-1)-curve E_4. Denote $\tau'(E_1), \tau'(E_2)$ again by E_1, E_2 and let $H' = \tau'(H)$. Then $(H'^2) = 0$ and $|H'|$ defines a $\mathbb{P}^1$-fibration $\overline{f} : V' \to \mathbb{P}^1$ such that $f = \overline{f}|_X : X \to \mathbb{A}^1$ is an $\mathbb{A}^1$-fibration. In the fibration $\overline{f}$, E_4 is a cross-section and $E_3 + L + 2(E_2 + E_1 + A)$ (resp., $E_3 + E_1 + 2(E_2 + L + A)$) is a fiber of $\overline{f}$ if $n = 0$ (resp., $n = 1$), where A is a certain (-1)-curve meeting X. In fact, if $n = 0$, let M be a section of p_0 such that $M \sim M_0$ and M passes through the point $H_0 \cap \ell_\infty$. Then M meets ℓ_0 in a point other than $H_0 \cap \ell_0$. If $n = 1$, let $M = M_0$. Then the proper transform of M on V is the (-1)-curve A in the above claim.

With this example we can verify the following result.

Proposition 2.7.9. *The following assertions hold:*
1. *X is an ML_1-surface with $\rho(X) = 0$ and one multiple fiber of multiplicity 2.*
2. *C is an affine line lying transversally to f and $\overline{\kappa}(X - C) = 0$. Hence C cannot move in an $\mathbb{A}^1$-fibration on X.*

Proof. (1) The boundary divisor $D = V - X$ is $H + E_1 + E_2 + L$, where E_2 is a branching component and H is a nonadmissible twig. Hence X is an ML_1-surface by Corollary 2.7.5.

A unique $\mathbb{A}^1$-fibration on X is given by f, and $2A$ is the unique multiple fiber of multiplicity 2.

(2) In fact, C is a 2-section of f. It is easy to see that $\overline{\kappa}(X - C) = 0$ since $X - C$ has an $\mathbb{A}^1_*$-fibration over $\mathbb{A}^1_*$ with all fibers reduced and irreducible. □

Example 2.7.10. Let V_0 be a Hirzebruch surface of degree $n \geq 0$ with the $\mathbb{P}^1$-fibration $p_0 : V_0 \to \mathbb{P}^1$ with general fiber ℓ. Let M_0 and M_1 be disjoint sections (so $(M_0{}^2) = -(M_1{}^2)$ and $|(M_i{}^2)| = n$). Choose three fibers $\ell_0, \ell_1, \ell_\infty$. Let $\sigma : V \to V_0$ be a sequence of blowing-ups which produce the following degenerate fibers Γ_i from ℓ_i for $i = 0, 1$:

$$\begin{array}{lccccccccccccc} \Gamma_0 : & M_0' & + & \overline{C} & + & E_0 & + & E_1 & + & \cdots & + & E_{m_1-1} & + & M_1' \\ & & & (-m_1) & & (-1) & & (-2) & & & & (-2) & & \end{array}$$

$$\begin{array}{lccccc} \Gamma_1 : & M_0' + (-a_1) + \cdots + (-a_s) & + & F_0 & + & (-b_t) + \cdots + (-b_1) + M_1' \\ & & & (-1) & & \end{array}$$

where $a_i \geq 2\ (1 \leq i \leq s), b_j \geq 2\ (1 \leq j \leq t), \overline{C} = \sigma'(\ell_0)$ and $M_k' = \sigma'(M_k)$ for $k = 0, 1$. Let m_2 be the multiplicity of the component F_0 in the fiber $\sigma^*(\ell_1)$, let $D = M_0' + M_1' + \ell_\infty + (\sigma^*(\ell_0)_{\mathrm{red}} - (\overline{C} + E_0)) + (\sigma^*(\ell_1)_{\mathrm{red}} - F_0)$ and let $X = V - D$. Let $C = \overline{C} \cap X$. Suppose that $m_1 \geq 2$ and $m_2 \geq 2$.

Proposition 2.7.11. *With Example 2.7.10, the following assertions hold:*

(1) *X is an ML_1-surface.*

(2) *C is an affine line, and it lies transversally to a unique $\mathbb{A}^1$-fibration $f : X \to \mathbb{A}^1$.*

(3) *$\overline{\kappa}(X - C) = 0$ if and only if $m_1 = m_2 = 2$ and $\overline{\kappa}(X - C) = 1$ otherwise. Hence C is an anomalous affine line on X.*

(4) *If $m_1 = m_2 = 2$, X is isomorphic to the surface constructed in Example 2.7.8 for $n = 0$.*

Proof. (1) In the divisor D, the component M_1' is a branching component and has a nonadmissible twig

$$(-a_s) + \cdots + (-a_1) + M_0' + \ell_\infty(+M_1')$$

where $(\ell_\infty{}^2) = 0$. Hence X is an ML_1-surface by Corollary 2.7.5.

(2) When we make the end component of the above twig a (0)-curve A by blowing-ups and blowing-downs (see [152, Corollary 2.4.3]), the image $\widetilde{C}$ of $\overline{C}$ meets the end component A. Hence the pencil $|A|$ defines a unique $\mathbb{A}^1$-fibration $f : X \to \mathbb{A}^1$ and $\widetilde{C} \cap X$ is an affine line which lies transversally to the $\mathbb{A}^1$-fibration f, where $\widetilde{C} \cap X = C$ by the construction.

(3) Consider the $\mathbb{P}^1$-fibration on V defined by the pencil $|\ell_\infty|$. It induces an $\mathbb{A}^1_*$-fibration $f' : X - C \to \mathbb{A}^1$ which has two multiple fibers $m_1\mathbb{A}^1_*, m_2\mathbb{A}^1_*$. By the formula used in the proof of Theorem 2.6.12, $\overline{\kappa}(X - C) = 1$ if and only if

$$-2 + 1 + \left(1 - \frac{1}{m_1}\right) + \left(1 - \frac{1}{m_2}\right) > 0.$$

Namely, $\overline{\kappa}(X - C) = 1$ if and only if $(m_1 - 1)(m_2 - 1) > 1$. Similarly, $\overline{\kappa}(X - C) = 0$ if and only if $m_1 = m_2 = 2$.

(4) In fact, we can blow down M_0', M_1' and two more components in the boundary D so that the image of ℓ_∞ is the curve H and the image of $\overline{C}$ is the curve ℓ_0 in Example 2.7.8. □

2.7.3 Exhaustion of affine lines in ML_1-surfaces

Notwithstanding the above examples, we can prove the following result.

Theorem 2.7.12. *Let X be a $\mathbb{Q}$-homology plane. Suppose that X is an ML_1-surface and not isomorphic to one of the surfaces constructed in Examples 2.7.8 and 2.7.10. Then any affine line on X is a fiber of the unique $\mathbb{A}^1$-fibration $f : X \to \mathbb{A}^1$. In other words, there are no affine lines which lie transversally to the unique $\mathbb{A}^1$-fibration $f : X \to \mathbb{A}^1$.*

Proof. Our proof follows essentially the same arguments as in the proof of Theorem 2.6.12. The reason why the arguments there can be applied to the present case is that if the boundary divisor D becomes a linear chain when minimized, we obtain contradictions by the computations of $\overline{\kappa}(X - C)$, and that if D is not a linear chain when minimized then all maximal twigs of D are admissible, whence X is not even an ML_1-surface by Corollary 2.7.5. We shall use the same notations as in the proof of Theorem 2.6.12 and indicate the points which need special attentions.

Step I. We note that $\mathrm{Pic}(X)$ is a finite group by [164, Lemma 1.1]. Let C be an affine line on X. Then there exists an element $u \in \Gamma(X, \mathcal{O}_X)$ such that $mC = (u)$ for some $m > 0$, which we take to be minimal. Then u defines a morphism $u : X \to \mathbb{A}^1$ such that $u^*(0) = mC$ and the general fibers are irreducible. Since $e(X - C) = 0$, we have $\overline{\kappa}(X - C) \le 1$ by Theorem 1.2.6. If $\overline{\kappa}(X - C) = -\infty$, then $X - C$ has an $\mathbb{A}^1$-fibration which extends to such an $\mathbb{A}^1$-fibration on X with C as a fiber component. The base of the fibration is $\mathbb{A}^1$ since X is a $\mathbb{Q}$-homology plane. By the uniqueness of $\mathbb{A}^1$-fibration on X, it coincides with the given f. Hence C is a fiber of f. We consider below the cases $\overline{\kappa}(X - C) = 0, 1$ separately.

Step II. Suppose that $\overline{\kappa}(X - C) = 0$. Consider $u : X - C \to B'$, where u is taken as in the step I and $B' \cong \mathbb{A}^1_*$. With the notations in Step II in the proof of Theorem 2.6.12, if $\overline{\kappa}(F) = -\infty$, then u is an $\mathbb{A}^1$-fibration over B' which, extended to X, coincides with f and hence C is a fiber of f. Suppose that $\overline{\kappa}(F) = 0$. Then u is an $\mathbb{A}^1_*$-fibration. The case u is untwisted is treated in the same way as in the proof of Theorem 2.6.12. In the case where u is twisted and Γ contains at least one branching component (case (i)), we see readily that D has only admissible twigs. So, X is not ML_1 by Corollary 2.7.5. In the case where u is twisted and the case (ii) or (iii) occurs, contract the curves E_2, E_1 (and F_2, F_1 in the case (ii)) to obtain a relatively minimal $\mathbb{P}^1$-fibration $\overline{p} : \overline{V} \to \overline{B}$. Then $\kappa(V, D + K_V) = -\infty$ and $\kappa(V, D + \overline{C} + K_V) = 0$ if and only if $a = n + 1$, where $a \ge 2n$

because $\overline{H} \cdot M_0 \geq 0$. Hence $n = 0, 1$. Then it is easy to show that X is isomorphic to the surface constructed in Example 2.7.8 and C is as given in the same example.

STEP III. Suppose that $\overline{\kappa}(X - C) = 1$. The arguments for the proof of Theorem 2.6.12 can be applied without change when D is minimized to a linear chain and the criterion in Corollary 2.7.5 can be applied to the non-ML_1-property of X when D is not minimized to a linear chain. The only exception occurs in the case (iii) where $B \cong \mathbb{A}^1$, f is untwisted and $p^{-1}(P_0) \cap X = m_1 C' + n_1 C$. Suppose that one of the cases (1) $\min(m_1, n_1) > 1$ and (2) $m_1 = 1, n_1 \geq 1$ occurs. In order that $\overline{\kappa}(X - C) = 1$, we need multiple fibers $m_2 \mathbb{A}^1_*, \ldots, m_r \mathbb{A}^1_*$ such that the following inequality holds:

$$-2 + 1 + \left(1 - \frac{1}{m_1}\right) + \sum_{i=2}^{r} \left(1 - \frac{1}{m_i}\right) > 0.$$

Hence $r \geq 2$ in case (1) and $r \geq 3$ in case (2). Then H_1 and H_2 are branching components (with the fiber at infinity ℓ_∞ taken into account) and all the twigs of D are admissible. So X is not an ML_1-surface by Corollary 2.7.5. Suppose that $m_1 > 1, n_1 = 1$. Since $p^{-1}(P_0)$ is a linear chain, its graph looks like Γ_0 in Example 2.7.10 with E_0, M_0', M_1' corresponding respectively to $\overline{C'}, H_2, H_1$.

Since $\overline{\kappa}(X - C) = 1$, we have $r \geq 2$ by the above inequality. If $r \geq 3$ then H_1 and H_2 are branching components of D and there are only admissible twigs in D. Hence X is not an ML_1-surface. So $r = 2$. Let $m_2 \mathbb{A}^1_*$ be the second multiple fiber of f' over a point $P_1 \in B'$. Then $\overline{\kappa}(X - C) = 1$ if and only if $(m_1 - 1)(m_2 - 1) > 1$. Furthermore, the fiber $p^{-1}(P_1)$ looks like the fiber Γ_1 in Example 2.7.10 with M_0', M_1' corresponding to H_2, H_1, respectively, and F_0 is the component with multiplicity m_2 and meets X. Now we are in the same situation as in Example 2.7.10. Hence X is isomorphic to the surface constructed in that example. □

2.8 Topology and geometry of ML_0-surfaces

We discussed the liftability of a G_a-action under a finite morphism $\varphi : X \to Y$ when φ is étale or the G_a-action preserves the branch locus of φ (see Subsection 2.2.4 and Theorem 2.2.21). We continue these arguments if X or Y is an ML_0-surface.

2.8.1 Ascent and descent of the ML_0-property

We begin with recalling the following result [143].

Lemma 2.8.1. *Let $\varphi : X \to Y$ be an étale finite morphism of smooth affine surfaces. Then any G_a-action on Y lifts to a G_a-action on X. In particular, if Y is ML_0 then so is X.*

Proof. By Theorem 2.2.18, any G_a-action on Y lifts to a G_a-action on X which commutes with the morphism φ. By Lemma 2.4.26, there exist two independent lnds δ_1, δ_2 on $B :=$

$\Gamma(Y, \mathcal{O}_Y)$ such that $\operatorname{Ker}\delta_1 \cap \operatorname{Ker}\delta_2 = k$. Let $\widetilde{\delta}_1, \widetilde{\delta}_2$ be respectively the lifts of δ_1, δ_2 to $A := \Gamma(X, \mathcal{O}_X)$. Since the G_a-actions associated with $\widetilde{\delta}_1, \widetilde{\delta}_2$ are independent, $\operatorname{Ker}\widetilde{\delta}_1 \cap \operatorname{Ker}\widetilde{\delta}_2$ is algebraic over k. Hence it is equal to k. This implies that X is ML_0, too. □

We ask if the converse holds and obtain the following descent result for the ML_0-property. Note that there is no restriction on the Picard number ρ of X or Y (cf. Theorem 2.8.3 below).

Theorem 2.8.2. *Let $\varphi : X \to Y$ be a finite morphism of smooth affine surfaces with X an ML_0-surface. Assume that either φ is étale or φ is a cyclic Galois (possibly ramified) covering. Then Y is an ML_0-surface.*

Proof. If φ is étale then $\pi_1(X)$ is a subgroup of finite index in $\pi_1(Y)$. Hence by taking a normal subgroup of finite index in $\pi_1(Y)$ contained in $\pi_1(X)$, we can find a smooth affine surface Z and an étale finite morphism $\psi : Z \to X$ such that $\varphi \circ \psi : Z \to Y$ is an étale Galois covering. Since X is ML_0, it follows by Lemma 2.8.1 that Z is also an ML_0-surface. After replacing X by Z if necessary, we can assume that φ is a (possibly ramified) Galois covering with Galois group G.

By the equivariant completion theorem of Sumihiro [216] and G-equivariant resolution of singularities, we can find a smooth normal G-completion $X \hookrightarrow V$, where G acts on the boundary divisor $D := V - X$. If the completion is minimal, then D is a linear chain because X is ML_0 (cf. Theorem 2.6.2). We shall show that V can be chosen so that D is linear.

Assume that D is not minimal. Then D has an irreducible component D_1 such that D_1 is a (-1)-curve and meets at most two other components. Then all the conjugates of D_1 in D have the same property. Let $D_1, D_2, \ldots, D_r$ be all the conjugates of D_1. If $D_i \cap D_j = \emptyset$ for every pair (i, j) with $1 \le i < j \le r$, then we can contract all of them simultaneously and obtain a new normal G-completion. Assume that $D_i \cap D_j \ne \emptyset$ for some pair (i, j), say $(i, j) = (1, 2)$. Let Γ_1 be the connected component of $D - D_2$ containing D_1 and let Γ_2 be the connected component of $D - D_1$ containing D_2. Then Γ_1 (resp., Γ_2) does not contain D_2 (resp., D_1), and Γ_1 and Γ_2 are also conjugate. By the assumption, D_1 meets only one other irreducible component of Γ_1 and similarly for D_2. If Γ_1 contains a branch component of D then so does Γ_2. If we contract D_1 and any subsequent (-1)-curves which meet at most two other irreducible components, then we reach a minimal divisor with simple normal crossings which is a tree but still has two branching components. By Theorem 2.6.2, this contradicts the assumption that X is ML_0. Hence we can assume that D is a G-stable linear chain as $\Gamma_1 + \Gamma_2$ exhausts D since X is affine.

Consider the quotient surface V/G which contains Y as an open set. Then V/G is normal and D/G is a simply-connected divisor. Furthermore, V/G has at most quotient singular points on D/G. See Problem 14 at the end of this chapter.

We shall show that Y has a smooth normal completion W such that $W - Y$ is a linear chain of smooth rational curves. First consider the case when D is irreducible. Then the claim follows from Problem 13 at the end of the chapter.

Now consider the case where D is reducible. Let H be a subgroup of G which keeps all the irreducible components of D stable. Then H has index at most 2 in G because $g \in G$ belongs to H if g stabilizes an end component of D. Any intersection point of two irreducible components of D is fixed by H. From this we deduce that H is a finite cyclic group (see Problem 13) and V/H has only cyclic quotient singularities. More precisely, write $D = D_1 + D_2 + \cdots + D_r$, where D_1 and D_r are the tips. Let K_i be the subgroup of H which consists of elements g fixing D_i pointwise. If $H/K_1 \neq 1$, the quotient morphism $D_1 \to D_1/H$ is a cyclic covering with group H/K_1. Hence there is exactly one more point on D_1 which is fixed by H/K_1. This point can be a cyclic quotient singularity on the surface V/H. If $H/K_1 \neq 1$ and $H/K_2 \neq 1$ then the intersection point $D_1 \cap D_2$ gives a cyclic quotient singular point on V/H. Hence cyclic singular points on V/H come from the intersection points of the components of D (not necessarily all) and possibly two points on D_1 and D_r which are not the intersection points. By taking a minimal G/H-equivariant resolution of singularities on D/H, we obtain a normal completion $X/H \hookrightarrow U$ such that U is smooth along $U - X/H$ and $U - X/H$ is a linear chain of smooth rational curves. We note here that X/H may have singular points if φ is not étale.

Now we can assume that $G \cong \mathbb{Z}/2\mathbb{Z}$ and the generator of G permutes the end components of D which is a linear chain with two or more components. Then a local analysis at a possible fixed point on D shows that G-action is given by $(x, y) \mapsto (y, x)$ with respect to a suitable system of local coordinates at the fixed point and hence that $U/(\mathbb{Z}/2\mathbb{Z})$ is, in fact, smooth. In fact, $x + y, xy$ generates a system of regular coordinates of the point below. Let $W := U/(\mathbb{Z}/2\mathbb{Z})$. Then $W - Y$ is a linear chain of smooth rational curves. Hence Y is an ML_0-surface by Theorem 2.6.2 because $\gamma(X) = 0$ implies $\gamma(Y) = 0$. □

In the case of ML_0-surfaces with $\rho = 0$ we have the following general result. This is also a generalization of the result in [67, 151] which states that if $X \cong \mathbb{A}^2$ then $Y \cong \mathbb{A}^2$ (see Theorem 1.3.23).

Theorem 2.8.3. *Let $f : X \to Y$ be a finite morphism of smooth affine surfaces. Suppose that X is an ML_0-surface with $\rho(X) = 0$. Then Y is also an ML_0-surface with $\rho(Y) = 0$.*

Our proof consists of subsequent three lemmas by which we show that:
(1) Y is a $\mathbb{Q}$-homology plane, i. e., $H_i(Y; \mathbb{Q}) = 0$ for $i = 1, 2$.
(2) The fundamental group $\pi_1^\infty(Y)$ is a finite cyclic group.
(3) If Y satisfies the above conditions (1) and (2), then Y is an ML_0-surface.

The first assertion is proved in the following lemma.

Lemma 2.8.4. *With the notations and assumptions of Theorem 2.8.3, Y is a $\mathbb{Q}$-homology plane.*

Proof. The natural homomorphism $\pi_1(f) : \pi_1(X) \to \pi_1(Y)$ has the image of finite index in $\pi_1(Y)$. For a transparent argument of this fact, we refer to the proof of the assertion B in [182, Lemma 1.5]. In fact, there exists a dense Zariski open set U of Y, which is by definition the complement of proper closed set such that $f^{-1}(U) \to U$ is an analytic fiber bundle. We have then a commutative diagram of homotopy exact sequences

$$\begin{array}{ccccccc} \pi_1(F) & \longrightarrow & \pi_1(f^{-1}(U)) & \longrightarrow & \pi_1(U) & \longrightarrow & \pi_0(F) \\ & & \downarrow & & \downarrow & & \\ & & \pi_1(X) & \xrightarrow{\pi_1(f)} & \pi_1(Y) & & \end{array}$$

where F is a fiber of the fiber bundle, $\pi_0(F)$ is finite, and two vertical arrows are surjections. This implies the assertion. Since $\pi_1(X)$ is a finite cyclic group by Lemma 2.6.6, $\pi_1(Y)$ is a finite group. Hence $H_1(Y;\mathbb{Q}) \cong \pi_1(Y)/[\pi_1(Y), \pi_1(Y)]$ is a finite abelian group, and $H_1(Y;\mathbb{Q}) \cong H_1(Y;\mathbb{Q}) \otimes_{\mathbb{Z}} \mathbb{Q} = 0$.

Since X is a rational surface with $\overline{\kappa}(X) = -\infty$, so is Y. Namely, Y is a rational surface with an $\mathbb{A}^1$-fibration $q : Y \to C$. Then there exists a smooth normal completion $Y \hookrightarrow W$ such that $\Delta := W - Y$ is a divisor with simple normal crossings. We may assume that the $\mathbb{A}^1$-fibration q extends to a $\mathbb{P}^1$-fibration $\overline{q} : W \to \overline{C}$ with $\overline{C} \cong \mathbb{P}^1$. This implies that Δ has a component which is a cross-section of $\overline{q}$ and the other components of Δ consist of fiber components of $\overline{q}$. Hence Δ is a tree of rational curves, which implies that Δ is simply-connected. We now apply the same arguments as in the proof of Lemma 2.6.6 to obtain $H^2(W, \Delta; \mathbb{Q}) = 0$. In fact, since f is a finite morphism, the projection formula and the assumption that $\rho(X) = 0$ imply that $\rho(Y) = 0$. Then $H_2(\Delta;\mathbb{Q}) \cong H_2(W;\mathbb{Q})$. Hence $H^2(W;\mathbb{Q}) \cong H^2(\Delta;\mathbb{Q})$ by the Poincaré duality. By the long exact sequence of $\mathbb{Q}$-cohomology groups for a pair (W, Δ), we have $H^2(W, \Delta; \mathbb{Q}) = 0$. By the Lefschetz duality, we have $H_2(Y;\mathbb{Q}) = 0$. This proves that Y is a $\mathbb{Q}$-homology plane. □

For any smooth affine surface S let $\pi_1^\infty(S)$ denote the fundamental group at infinity for S as defined in [192] (see the definition before Theorem 1.2.11). Since X is a $\mathbb{Q}$-homology plane, it follows from Lemma 2.6.3 that $\pi_1^\infty(X)$ is a finite cyclic group. Hence the above assertion (2) is a consequence of the following lemma.

Lemma 2.8.5. *The natural homomorphism $\pi_1^\infty(X) \to \pi_1^\infty(Y)$ is a surjection.*

Proof. Since f is proper, there is a natural induced homomorphism $\pi_1^\infty(X) \to \pi_1^\infty(Y)$. The finiteness of f also implies that the image H of this homomorphism has finite index in $\pi_1^\infty(Y)$ (cf. [182, Lemma 1.5]). This implies that $\pi_1^\infty(Y)$ is a finite group. We need to show that it is a cyclic group.

Suppose that H is a proper subgroup of $\pi_1^\infty(Y)$. Let $K \subset Y$ be a suitable big compact set such that the closure of $U := Y \setminus K$ in Y is an orientable real 4-manifold with boundary a compact 3-manifold whose fundamental group is $\pi_1^\infty(Y)$. Let N be the inverse image of U in X. We can assume that $\pi_1(N) = \pi_1^\infty(X)$. There is a covering $\widetilde{U} \to U$

corresponding to the subgroup H of $\pi_1(U)$. By covering space theory, we have a complex analytic map $N \to \widetilde{U}$ which factors the map $N \to U$.

Let $Y \subset W$ be a smooth projective embedding such that $D := W - Y$ is a simple normal crossing divisor. The complex space $W \setminus K$ is 0-concave in the sense of [6]. We can find an open embedding $\widetilde{U} \subset T$ such that T is a normal complex analytic space with a proper analytic map with finite fibers $\varphi : T \to W \setminus K$ extending the map $\widetilde{U} \to U$. Since Y is affine, D supports an effective divisor Δ which is very ample for W. Let $B := \varphi^{-1}(D)$ and $\widetilde{B} = \varphi^*\Delta$. Our aim is to show that T is an open subset of a normal projective surface Z such that $Z \setminus \widetilde{B}$ is a smooth affine surface.

Let V be the normalization of W in the function field of X and let $f' : V \to W$ be the natural proper morphism with finite fibers. Let $\overline{N} = N \cup A$, where A is the inverse image of D in V. Now $\overline{N}$ is also 0-concave. There exists a natural complex analytic map $\psi : \overline{N} \to T$ which factors $\overline{N} \to W \setminus K$. The divisor $f'^{-1}(D) = \psi^{-1}(B)$ supports the ample divisor $\psi^*(\widetilde{B})$ on V.

Let p, q be distinct points in T and let p' be a point in $\overline{N}$ lying over p. There exists $m \gg 0$ and a section $L \in H^0(V, \psi^*(m\widetilde{B}))$ such that $L(p') = 0$ and L is not zero at any point lying over q. If $d = \deg \psi$, then $\psi_*(L|_{\overline{N}}) \sim dm\widetilde{B}$ and this divisor vanishes at p but not at q. This shows that sections of multiples of $\widetilde{B}$ separate points in T.

Since T is *essentially compact*, we can find a finite number of open sets T_i in T such that for each i there exists m_i and a section $\sigma_i \in H^0(T, m_i\widetilde{B})$ which does not vanish at any point in T_i. Taking M to be product of all the m_i we can find sections $h_0, \dots, h_R$ of $|M\widetilde{B}|$ such that the map $h : T \to \mathbb{P}^R$ given by sending any point $p \in T$ to $(h_0(p), \dots, h_R(p))$ is holomorphic. Since $\varphi : T \to W \setminus K$ is finite and W is projective, we know that the field of meromorphic functions on T has transcendence degree 2 over $\mathbb{C}$. The meromorphic functions h_i/h_0 generate a field of transcendence degree 2 over $\mathbb{C}$. From this and the fact that $\widetilde{B} = \varphi^*\Delta$ we can further assume that M is so large that the image of h is contained in a projective algebraic surface $S \subset \mathbb{P}^R$. Let q be a general point in $h(T)$. By the argument above, we can find a multiple M' of M and sections $h'_0, \dots, h_{R'}$ of $|M'\widetilde{B}|$ such that the analogous map $h' : T \to \mathbb{P}^{R'}$ is injective on the inverse image of q in T. It is clear that the map h' is bimeromorphic. Since φ is a finite map, by the projection formula no closed curve in T can contract to a point under h'. From these observations we conclude that $h' : T \to h'(T)$ is both finite and bimeromorphic. Further, $h'(T)$ is contained in a projective surface $S_n \subset \mathbb{P}^n$. Then the normalization Z of S_n is biholomorphic to T. Hence T can be embedded as an open subset in a normal projective surface.

By Hartog's theorem, the analytic map ψ extends to a proper analytic map with finite fibers (which is then also algebraic) $Z \to W$. Let S be anew the inverse image of Y in Z. Since $\widetilde{U} \to U$ is an étale map, the morphism $S \to Y$ is ramified only at finitely many points in S. By the purity of branch locus, it follows that the morphism $S \to Y$ is actually étale. Hence S is a smooth affine surface. We have a finite morphism $X \to S$ which induces a surjection $\pi_1^\infty(X) \to \pi_1^\infty(S)$. It follows that $\pi_1^\infty(S)$ is finite cyclic.

Since X is a $\mathbb{Q}$-homology plane, so is S. Hence $e(S) = 1$. But the degree of the étale finite morphism $S \to Y$ is greater than one. Hence $e(S) > 1$. This contradiction shows that $H = \pi_1^\infty(Y)$. Hence $\pi_1^\infty(Y)$ is finite cyclic. □

The following lemma will verify the third assertion.

Lemma 2.8.6. *Let X be a $\mathbb{Q}$-homology plane. Suppose that $\pi_1^\infty(X)$ is a finite cyclic group. Then X has a minimal normal compactification V such that $D := V - X$ is a linear chain of smooth rational curves. Further, X is an* ML_0*-surface.*

In order to prove this lemma we need the following auxiliary result from [211] (see also Subsection 1.3.12).

Lemma 2.8.7. *Let X be a smooth, affine surface with $\pi_1^\infty(X)$ a finite cyclic group. Then X has a minimal normal completion V such that $D := V - X$ is a tree of rational curves as in the case (i) in subection* 1.3.12 *with a twig* 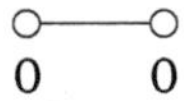

consisting of components D_1 (a tip) and D_2 such that, after removing this twig from D, we get a connected linear chain whose intersection form is negative-definite. Further, V is rational.

Proof of Lemma 2.8.6. With the notation of Lemma 2.8.6, D supports a divisor with strictly positive self-intersection since X is affine. Since V is rational the linear system $|D_1|$ gives a $\mathbb{P}^1$-fibration $p : V \to \mathbb{P}^1$ on V such that D_2 is a cross-section of p and all the other components of D are contained in a singular fiber G of p. Since the Picard number $\rho(X) = 0$, p has no singular fiber other than G. The part of D contained in G is the linear chain with negative-definite intersection form.

If X is isomorphic to $\mathbb{A}^2$, all the assertions hold. Hence we may and shall assume that X is not isomorphic to $\mathbb{A}^2$. In particular, $\mathrm{Pic}(X) \neq 0$. The restriction of p to X is an $\mathbb{A}^1$-fibration $\rho : X \to B$, where $B \cong \mathbb{A}^1$. Since X is a $\mathbb{Q}$-homology plane we see easily that every fiber of ρ is irreducible. In particular, $G = G_1 \cup A_1$, where $G_1 = D \cap G$ is the linear chain and $A := A_1 \cap X \cong \mathbb{A}^1$. Since $X \hookrightarrow V$ is a minimal normal completion, no component of G_1 is a (-1)-curve and hence A_1 is a unique (-1)-curve in G. Further, the multiplicity m of A_1 in G is greater than one. Let $P_0 := p(G)$. Let D_3 be the component of G_1 which meets D_2. We claim that D_3 meets at most one other component of G_1. Suppose that D_3 meets two components of D, say D_4, D_5. Then $G_1 - D_3$ has exactly two connected components, say Δ_1, Δ_2. Starting with A_1 we can successively contract (-1)-curves in G. Suppose that at some stage the image of D_3, say D_3', becomes a (-1)-curve and that D_3' still meets two other components of the image G' of G. Then the multiplicity of D_3' in G' is at least 2. This is a contradiction since D_3 meets the cross-section D_2. Hence if D_3' is a (-1)-curve then it meets only one other component of G'. Further, all the other components of G' have self-intersection less than -1. Since G' is still a linear chain, we can see that G' cannot be contracted to a (0)-curve. Now we know that D is a linear chain of rational curves. This proves the first part of Lemma 2.8.6. Since there is no linear equivalence relation among the components of D, it follows from Theorem 2.6.2 that X is an ML_0-surface. □

This completes the proof of Theorem 2.8.3. The argument in Lemma 2.8.5 proves the following general result.

Proposition 2.8.8. *Let $f : X_1 \to X_2$ be a proper morphism between normal affine surfaces. Then there is a factorization of f in the form $X_1 \to S \to X_2$, where S is a normal affine surface and $X_1 \to S$ is a proper morphism such that $\pi_1^\infty(S)$ is isomorphic to the image of the homomorphism $\pi_1^\infty(X_1) \to \pi_1^\infty(X_2)$.*

We deduce the following result using the proof of Theorem 2.8.3. This result was proved in [151] using the proof of the cancelation theorem for $\mathbb{A}^2$ and in [77] using the Mumford–Ramanujam method and Milnor's classification of finite groups acting freely on a homotopy 3-sphere.

Corollary 2.8.9. *Let Y be a smooth affine surface with a finite morphism $f : \mathbb{A}^2 \to Y$. Then Y is isomorphic to $\mathbb{A}^2$.*

Proof. Suppose first that $\pi_1(Y) = 1$. By Lemma 2.6.5, $\pi_1(Y) \cong \operatorname{Pic}(Y)_{\text{tor}}$ and $H_2(Y) \cong \operatorname{Pic}(Y)/\operatorname{Pic}(Y)_{\text{tor}}$. Since Y is a $\mathbb{Q}$-homology plane, it follows that $\operatorname{Pic}(Y) = 0$ and $H_1(Y) = H_2(Y) = 0$. Then Y is topologically contractible by a theorem of J. H. C. Whitehead. On the other hand, Lemma 2.8.5 shows that $\pi_1^\infty(Y)$ is trivial. Then, by Ramanujam theorem (see Theorem 1.3.19), Y is isomorphic to $\mathbb{A}^2$.

In general, the argument in the proof of Lemma 2.8.4 shows that $\pi_1(Y)$ is a finite group. Let $\widetilde{Y} \xrightarrow{p} Y$ be the universal covering. Then we have a factorization $\mathbb{A}^2 \xrightarrow{q} \widetilde{Y} \xrightarrow{p} Y$. By the previous case, $\widetilde{Y} \cong \mathbb{A}^2$. Hence we may assume that the given finite morphism $\mathbb{A}^2 \to Y$ is a finite Galois covering. Then Y is topologically contractible by [131]. Since $\pi_1^\infty(Y) = 1$ by Lemma 2.8.5, we conclude again that $Y \cong \mathbb{A}^2$. □

Remark 2.8.10. More generally, if $\mathbb{A}^2 \to Y$ is a finite morphism onto a normal affine surface Y, then we can show by similar arguments that Y is isomorphic to a quotient $\mathbb{A}^2/G$ for a finite group of automorphisms of $\mathbb{A}^2$ (see Theorem 1.3.23).

Next we shall consider the ascent and descent of the ML_i-property for $i = 1, 2$. We have the following result.

Theorem 2.8.11. *Let $\varphi : X \to Y$ be an étale finite morphism. Then Y is ML_i ($i = 1, 2$) if and only if so is X.*

Proof. Consider first the ML_1-property. Suppose that Y is ML_1. Then X is ML_1 or ML_0 by Lemma 2.8.1. If X is ML_0 then Y is ML_0 by Theorem 2.8.2. This is a contradiction. Hence X is ML_1. Conversely, suppose that X is ML_1. As in the proof of Theorem 2.8.2, there exists a Galois étale finite covering $\psi : Z \to X$ such that $\varphi \circ \psi : Z \to Y$ is a Galois étale covering. Since Z is ML_1 by what we have just proved, we may assume that $\varphi : X \to Y$ is a Galois étale finite covering with group G. Let $f : X \to B$ be an $\mathbb{A}^1$-fibration with $B \cong \mathbb{A}^1$. Since this $\mathbb{A}^1$-fibration is unique on X, the G-action preserves f. Namely, the G-translates of a fiber of f are again fibers of f. Hence f induces an $\mathbb{A}^1$-fibration g :

$Y \to \mathbb{A}^1$. So, Y is ML_1 or ML_0. If Y is ML_0 then X is ML_0 by Theorem 2.8.2, a contradiction. Hence Y is ML_1.

The case for the ML_2-property follows readily if one uses the ascent and descent of ML_i-property for $i = 0, 1$. □

Remark 2.8.12. The ML_1-property does not descend under a ramified Galois covering morphism. In fact, let X be the hypersurface in $\mathbb{A}^3$ defined by $x^r y = z^n - 1$, where $r, n \geq 2$. Then X is an ML_1-surface. Meanwhile, the projection $(x, y, z) \mapsto (x, y)$ defines a ramified Galois covering morphism $X \to \mathbb{A}^2 = \operatorname{Spec} k[x, y]$ with group $\mathbb{Z}/n\mathbb{Z}$. Hence the ML_1-property is not preserved because $\mathbb{A}^2$ is an ML_0-surface.

2.8.2 Universal coverings of ML_0-surfaces with $\rho = 0$

Let X be an ML_0-surface with $\rho(X) = 0$. By Lemma 2.6.3, the boundary divisor of X is a linear chain. By Lemma 2.6.8, it is of the form $\ell + M + A$, where A is an admissible linear chain. Since $\rho(X) = 0$, there is one component A_1 meeting an irreducible component of A, which gives a multiple fiber on the induced $\mathbb{A}^1$-fibration on X. By Lemma 2.6.6, $\pi_1(X)$ is a finite cyclic group $\mathbb{Z}/m\mathbb{Z}$, where m is the multiplicity of A_1. We then consider what the universal covering looks like.

Theorem 2.8.13. *Let X be an ML_0-surface with the Picard number zero and let $f : X \to B$ be an $\mathbb{A}^1$-fibration with a unique multiple fiber mA_1 of multiplicity $m > 1$, where $B \cong \mathbb{A}^1$. Let $B' \to B$ be a cyclic Galois covering of order m ramifying totally over the point $P_0 = \rho(A)$ and let $\widetilde{X}$ be the normalization of the fiber product $X \times_B B'$. Then $\widetilde{X}$ is isomorphic to a hypersurface $xy = p(z)$, where $p(z)$ is a polynomial of degree m in z with distinct linear factors. Hence $\widetilde{X}$ is smooth and is the universal covering space of X. The given ML_0-surface X is regained as the quotient of $\widetilde{X}$ with respect to a $\mathbb{Z}/m\mathbb{Z}$-action.*

Proof. We use the projective embedding $X \hookrightarrow \overline{X}$ considered before Lemma 2.6.8. We denote $\overline{X}$ by V. In particular, the fiber ℓ_0 of $p : V \to \overline{B}$ over the point P_0 is supported by $A + \overline{A}_1$, where $p := \overline{f}$, the dual graph of A is a linear chain and $\overline{A}_1$ is the closure of A_1 in V. Let A_2 be the irreducible component of A such that $A_2 \cdot \overline{A}_1 = 1$. Let $\sigma : \overline{B}' \to \overline{B}$ be a cyclic Galois covering of order m ramifying totally over the points P_0 and $P_\infty = \overline{f}(\ell_\infty)$. Let W' be the normalization of V in the function field of $\widetilde{X}$ and let $\tau' : W' \to V$ be the normalization morphism. Then the branch locus of τ' contains ℓ_∞ and is contained in the sum $\ell_\infty + A$. Hence W' has a $\mathbb{P}^1$-fibration $q' : W' \to \overline{B}'$. The singularity of W' are at most cyclic quotient singularities which arise from the intersection points of the branch locus and lie on the fiber $q'^{-1}(P_0')$, where P_0' is the point of $\overline{B}'$ lying over P_0. Let $\nu : W \to W'$ be the minimal resolution of the singular points of W' and let $\tau = \tau' \cdot \nu : W \to V$. Then there is an induced $\mathbb{P}^1$-fibration $q : W \to \overline{B}'$, which satisfies $\sigma \cdot q = p \cdot \tau$. Note that the component A_1 splits into a disjoint union of m affine lines $L_1, \ldots, L_m$. This implies that the component A_2 is not contained in the branch locus of τ' and hence τ. Let H_2 be the irreducible component of $q^{-1}(P'_0)$ lying over A_2. Then $\tau|_{H_2}: H_2 \to A_2$ is a cyclic covering of order m, and there are m irreducible components $\overline{L}_1, \ldots, \overline{L}_m$ of $q^{-1}(P'_0)$ such that $H_2 \cdot \overline{L}_i = 1$ and $\overline{L}_i \cap \widetilde{X} = L_i$ for $1 \le i \le m$. Since $\overline{L}_1, \ldots, \overline{L}_m$ are reduced in $q^{-1}(P'_0)$, the multiplicity of H_2 in $q^{-1}(P'_0)$ is accordingly equal to 1. So, we can contract all the components of $q^{-1}(P'_0)$ except for H_2 and $\overline{L}_1, \ldots, \overline{L}_m$. Let $\widetilde{W}$ be the surface thus obtained from W. Then $\widetilde{W}$ has a $\mathbb{P}^1$-fibration $\widetilde{q} : \widetilde{W} \to \overline{B}'$ and $\widetilde{X}$ is embedded into $\widetilde{W}$ as an open set, and the boundary divisor $\widetilde{D} := \widetilde{W} - \widetilde{X}$ consists of the cross-section $\widetilde{S}$ of $\widetilde{q}$, the fiber $\widetilde{\ell}_\infty$ lying above the point at infinity P'_∞, and the component $\widetilde{H}_2$ of the fiber $\widetilde{\ell}_0 = \widetilde{H}_2 + \sum_{i=1}^m \widetilde{L}_i$, where P'_∞ is a unique point of $\overline{B}'$ lying above P_∞, $\widetilde{X}$ is the inverse image of X and $\widetilde{H}_2, \widetilde{L}_1, \ldots, \widetilde{L}_m$ are respectively the proper transforms of $H_2, \overline{L}_1, \ldots, \overline{L}_m$. Then it is straightforward to see that the canonical divisor $K_{\widetilde{X}}$, that is to say, the restriction of $K_{\widetilde{W}}$ onto $\widetilde{X}$ is trivial. On the other hand, since all the G_a-actions on X lift up to $\widetilde{X}$ by Theorem 2.2.18, Y is a smooth affine surface with trivial Makar-Limanov invariant. Hence, by [12], Y is isomorphic to a hypersurface $xy = p(z)$ with $\deg p(z) = m$. □

In fact, it is shown in [143, 33] that $p(z)$ can be taken to be $z^m - 1$.

2.8.3 Derksen invariants and ML_0-surfaces

We define the *Derksen invariant* $\mathrm{Dk}(X)$ of a normal affine variety $X = \operatorname{Spec} R$ as the subalgebra of R generated over k by all $\operatorname{Ker} \delta$, where δ runs over all the locally nilpotent derivations of R. It was originally defined in [34]. If X is a rational normal affine surface with $\gamma(X) = 0$, which we always assume tacitly in this subsection, then $\operatorname{Ker} \delta$ is a polynomial ring $k[u]$, where $\operatorname{Spec} k[u]$ is the parameter space of the $\mathbb{A}^1$-fibration on X associated to δ. Note that an $\mathbb{A}^1$-fibration on X parametrized by $\mathbb{A}^1$ is always associated with a G_a-action on X, hence a locally nilpotent derivation on R. Therefore $\mathrm{Dk}(X)$ is generated by all the elements $u \in R$ such that $\operatorname{Spec} k[u]$ parametrizes an $\mathbb{A}^1$-fibration on X. We sometimes write $\mathrm{Dk}(R)$ instead of $\mathrm{Dk}(X)$ when R is the coordinate ring of X. We do not know if $\mathrm{Dk}(X)$ is finitely generated over k. On the other hand, we also do not know of an ML_0-surface X with $\mathrm{Dk}(X) \neq R$.

Our aim to introduce here the Derksen invariant is to prove the following somewhat surprising result.

Theorem 2.8.14. *Let X be an ML_0-surface with $\rho(X) > 0$. Then there exists a surjective $\mathbb{A}^1$-fibration $X \to \mathbb{P}^1$.*

We prove first a preparatory result.

Lemma 2.8.15. *Let X be an ML_0-surface with $\rho(X) > 0$. Let $C_1, \ldots, C_r$ be affine lines such that $U := X - (C_1 \cup \cdots \cup C_r)$ is also an ML_0-surface. Assume that $\Gamma(U, \mathcal{O}_U)$ is integral over $\mathrm{Dk}(U)$. Then there exists an $\mathbb{A}^1$-fibration $\widetilde{f} : X \to \mathbb{P}^1$.*

Proof. We shall prove the case $r = 1$ and write $C_1 = C$. The general case is treated in a similar way. Let $f : U \to B$ be an $\mathbb{A}^1$-fibration, where B is isomorphic to $\mathbb{A}^1$ or $\mathbb{P}^1$ by the assumption that U is an ML_0-surface. Then it extends to an $\mathbb{A}^1$-fibration $\widetilde{f} : X \to \widetilde{B}$. If either $B \cong \mathbb{P}^1$ or $B \cong \mathbb{A}^1$ and $\widetilde{B} \cong \mathbb{P}^1$, then we are done. Suppose that all $\mathbb{A}^1$-fibrations on U are parametrized by $\mathbb{A}^1$ and such $\mathbb{A}^1$-fibrations extend to $\mathbb{A}^1$-fibrations on X parametrized by $\mathbb{A}^1$. Let $f : U \to B$ be an $\mathbb{A}^1$-fibration with $B \cong \mathbb{A}^1$ and let $\widetilde{f} : X \to B$ be its extension on X. Then C is contained in some fiber of $\widetilde{f}$. This implies that if $B = \operatorname{Spec} k[u]$ the function u is constant along C. In particular, any function of $\mathrm{Dk}(U)$ is constant on C. On the other hand, $\Gamma(X, \mathcal{O}_X) \subset \Gamma(U, \mathcal{O}_U)$, and $\Gamma(U, \mathcal{O}_U)$ is integral over $\mathrm{Dk}(U)$ by the assumption. Hence $\Gamma(X, \mathcal{O}_X)$ is integral over $\mathrm{Dk}(U)$. Let ξ be any element of $\Gamma(X, \mathcal{O}_X)$. Then we have a monic relation

$$\xi^n + a_1\xi^{n-1} + \cdots + a_{n-1}\xi + a_n = 0,$$

where $a_1, \ldots, a_n \in \mathrm{Dk}(U)$, which are constants on C. Hence ξ is also a constant on C. This is a contradiction because any two points of C are separated by a function of $\Gamma(X, \mathcal{O}_X)$. □

We know by Lemma 2.6.9 that if X is an ML_0-surface with $\rho(X) > 0$ then there exists an affine line C such that $X - C$ is still an ML_0-surface. So, in order to prove Theorem 2.8.14, we have only to show the following result.

Lemma 2.8.16. *Let X be an ML_0-surface of $\rho(X) = 0$. Then $\Gamma(X, \mathcal{O}_X)$ is integral over* $\mathrm{Dk}(X)$.

Proof. Let $\widetilde{X}$ be the universal covering of X. It is known by [143, 33] that $\widetilde{X}$ is a hypersurface in $\mathbb{A}^3$ defined by an equation $xy = z^n - 1$ and that X is the quotient of $\widetilde{X}$ by the Galois group $G \cong \mathbb{Z}/n\mathbb{Z}$ which acts as $(x, y, z) \mapsto (\zeta x, \zeta^{-1}y, \zeta^i z)$ for some $0 < i < n, \gcd(n, i) = 1$, where ζ is a primitive nth root of unity and G is identified with the multiplicative group $\{\zeta^j \mid 1 \le j \le n\}$. Furthermore, $\widetilde{X}$ has two independent $\mathbb{A}^1$-fibrations $f_x : \widetilde{X} \to \mathbb{A}^1$ and $f_y : \widetilde{X} \to \mathbb{A}^1$ which are defined by locally nilpotent derivations δ_x and δ_y, respectively, where

$$\begin{array}{lll} \delta_x(x) = 0, & \delta_x(y) = nz^{n-1}, & \delta_x(z) = x, \\ \delta_y(x) = nz^{n-1}, & \delta_y(y) = 0, & \delta_y(z) = y. \end{array}$$

So $k[x, y] \subseteq \mathrm{Dk}(\widetilde{X})$. Hence $\Gamma(\widetilde{X}, \mathcal{O}_{\widetilde{X}})$ is integral over $\mathrm{Dk}(\widetilde{X})$.

On the other hand, the G-action maps a fiber $f_x^{-1}(a)$ to $f_x^{-1}(\zeta a)$ if $a \ne 0$ and permutes the n lines $A_1, \ldots, A_n$ of $f_x^{-1}(0)$, where A_j is defined by $x = z - \zeta^j = 0$ for $1 \le j \le n$. Hence the $\mathbb{A}^1$-fibration f_x descends down to an $\mathbb{A}^1$-fibration on X. So $x^n \in \mathrm{Dk}(X)$. Similarly, $y^n \in \mathrm{Dk}(X)$. Thus, $k[x^n, y^n] \subseteq \mathrm{Dk}(X)$. Note that we have the natural inclusion $\mathrm{Dk}(X) \subseteq \mathrm{Dk}(\widetilde{X})$ because any G_a-action lifts to a G_a-action on $\widetilde{X}$.

Now we have the following inclusion relations:

$$\begin{array}{ccccccc} k[x^n, y^n] & \subseteq & k[x, y] & \subseteq & \mathrm{Dk}(\widetilde{X}) & \subseteq & \Gamma(\widetilde{X}, \mathcal{O}_{\widetilde{X}}), \\ k[x^n, y^n] & \subseteq & \mathrm{Dk}(X) & \subseteq & \Gamma(X, \mathcal{O}_X) & \subseteq & \Gamma(\widetilde{X}, \mathcal{O}_{\widetilde{X}}). \end{array}$$

Since $\Gamma(\widetilde{X}, \mathcal{O}_{\widetilde{X}})$ is integral over $k[x,y]$, it is such also over $k[x^n, y^n]$. Hence $\Gamma(X, \mathcal{O}_X)$ is integral over $\mathrm{Dk}(X)$. □

2.8.4 ML_0-surfaces not containing $\mathbb{A}^2$

Let X be an ML_0-surface. In Theorems 2.6.12 and 2.8.14, we saw that there is a difference between the cases $\rho(X) = 0$ and $\rho(X) > 0$. We shall in the present subsection point out some more differences in these cases.

We recall that a smooth algebraic surface X is isomorphic to the affine plane $\mathbb{A}^2$ if and only if $\mathrm{Pic}(X) = (0)$, $\gamma(X) = 0$ and $\overline{\kappa}(X) = -\infty$ (cf. [159, Chapter 3, Theorem 2.2.1]). This characterization of the affine plane is equivalent to saying that a smooth affine surface is isomorphic to $\mathbb{A}^2$ if and only if X is ML_0, $\mathrm{Pic}(X)_{\mathrm{tor}} = (0)$ and $\rho(X) = 0$. Hence one can ask if a smooth affine surface X contains $\mathbb{A}^2$ as an open set provided X is ML_0 and $\mathrm{Pic}(X)_{\mathrm{tor}} = (0)$. It will be shown later that the answer is negative. In order to construct a counterexample, we shall describe birational transformations of smooth completions of X which do not affect X. For this purpose, we employ the terminology and notation in Section 2.6 after Lemma 2.6.8 and the beginning of Section 2.7.

Let D be a weighted linear chain. We denote by $w(A_i)$ (or simply w_i) the weight of a component D_i of D. Let $Q(D)$ be the intersection form of the components of D. Namely, if D is a linear chain $D_1+D_2+\cdots+D_n$, then $Q(D)$ is an $(n\times n)$-matrix such that its (i,j)-entry is given by $w(D_i)$ (resp., 1 or 0) if $i = j$ (resp., $j = i\pm 1$ or otherwise). Let $d(D) = \det(-Q(D))$. Given a tip D_1 of D, we put $(d, d') = (d(D), d(D'))$, where $D' = D - D_1$ and call it the *pair* of D seen from D_1. We have the following result, which can be verified as an easy exercise.

Lemma 2.8.17. *Let D be a weighted linear chain. The following assertions then hold:*

(1) *Suppose that D is admissible, i. e., $w(D_i) \le -2$ for all i. With the above notations, we have $d' < d$ and $\gcd(d, d') = 1$. Let D_n be the tip of D on the opposite side of D_1 and let $d'' = d(D'')$ with $D'' = D - D_n$. Then $d'd'' \equiv 1 \pmod{d}$. (see [56, Lemma 3.6(2)]).*

(2) *Suppose that D has the form $U + E + V$, where U, V are linear chains meeting the component E at their end components. Let (u, u') (resp., (v, v')) be the pair of U seen from E, where u' (resp., v') is $d(U')$ (resp., $d(V')$) with U' (resp., V') being U (resp., V) minus the end component adjacent to E. Let $w = w(E)$. Then we have*

$$d(D) = -wuv - uv' - u'v.$$

(see [184, Lemma 3.1]).

(3) *Let c, p be relatively prime integers with $1 \le p < c$. Let Q be a point on a (0)-curve ℓ on a smooth projective surface. Apply the Euclidean transformation σ with center Q with respect to the data (c, p) (see [150]) and let D be the linear chain consisting of irreducible components of $\sigma^*(\ell)$,*

$$D = B + E + A,$$

where E is the last (-1)-curve and B contains the proper transform $\sigma'(\ell)$ as the end component. More precisely, let $c/p = [b_1, b_2, \dots, b_n]$ be the expansion of c/p as a continued fraction

$$\frac{c}{p} = b_1 - \cfrac{1}{b_2 - \cfrac{1}{b_3 - \cfrac{1}{\ddots - \cfrac{1}{b_n}}}}$$

and let $c/(c-p) = [a_1, a_2, \dots, a_m]$. Then D is written in the simplified form of the weighted dual graph

$$\begin{array}{ccccccc} (-b_1) & - & (-b_2) & - & \cdots & - & (-b_n) \\ \sigma'(\ell) & & & & & & | \\ & & & & & & (-1) \\ & & & & & & | \\ (-a_1) & - & (-a_2) & - & \cdots & - & (-a_m) \end{array}$$

where B, A have respectively the pairs (c, p) and $(c, c-p)$ seen from the tips of B and A with weights $-b_1$ and $-a_1$.

With the notations of Lemma 2.6.9, we have the following result.

Lemma 2.8.18. *Let $D = \ell + M + A$ be a standard chain with $(M^2) = n_0$. Assume that A has the pair (a, a') seen from M. Suppose we blow up a point $Q \in \ell$ not on M according to the pair (c, p), i. e., the Euclidean transformation with center Q and data (c, p). Let E be the last (-1)-curve and let D' be the resulting weighted chain (containing M and A). Let D^* be the chain obtained from D' by changing the weight of E from -1 to 0. Then we have*

$$d(D^*) = -a + c(n_0 ca + ca' + pa).$$

Proof. The chain D' is of the form $U + E + V$ with $w(E) = -1$. Let U, V have pairs $(u, u'), (v, v')$ seen from E. Then we may assume that V has the form $W + M + A$, where W has the pair (c, p) seen from M and $(M^2) = n_0$. Hence $v = d(V) = -n_0 ca - ca' - pa$ by Lemma 2.8.17. We have also $u = d(U) = c$. Now $d(D') = uv - uv' - u'v$ and $d(D^*) = -uv' - u'v$, whence $d(D^*) = d(D') - uv$. On the other hand, $d(D) = -a$ and $d(D') = d(D)$ since D' is obtained by a sequence of blowing-ups from D. The result then follows readily. □

We shall now fix the setting. *Assume that an ML_0-surface X has $\rho(X) = 1$ and contains an open set U which is isomorphic to $\mathbb{A}^2$.* Then we have the following:

(i) $X - U$ is an irreducible curve C which is isomorphic to $\mathbb{A}^1$ by Lemma 2.6.8.
(ii) Let $f : X \to \mathbb{A}^1$ be an $\mathbb{A}^1$-fibration and $\overline{X}$ be a standard completion of X with boundary $D = \ell + M + A$ and $\mathbb{P}^1$-fibration $\overline{f} : \overline{X} \to \mathbb{P}^1$ induced by ℓ which extends f. Then A has two feathers A_1, A_2 of respective multiplicities m_1, m_2. Since X contains $\mathbb{A}^2$, it is simply connected and hence $\operatorname{Pic}(X)_{\mathrm{tor}} = (0)$. So, $\gcd(m_1, m_2) = 1$ by Lemma 2.6.8. Let (a, a') be the pair of A seen from M.
(iii) We assume that for any standard completion $\overline{X}$ of X as above we have $\min(m_1, m_2) > 1$.
(iv) $\operatorname{Pic}(X - A_1) \cong \mathbb{Z}/m_2\mathbb{Z}$ and $\operatorname{Pic}(X - A_2) \cong \mathbb{Z}/m_1\mathbb{Z}$. This implies that $A_i \neq C$ for $i = 1, 2$.

Our aim is to show that the above assumption is realizable for a special choice of the linear chain A and leads to a contradiction. We proceed in several steps.

Claim. *Let $\overline{C}$ be the closure of C in $\overline{X}$. Then we have:*
(1) *$\overline{C}$ meets ℓ in a single point Q which is a one-place point of $\overline{C}$.*
(2) *After suitable elementary transformations on ℓ, we may assume that $Q \neq \ell \cap M$.*
(3) *Let $\widetilde{c} = (\overline{C} \cdot \ell)$ and let $\widetilde{p}$ be the multiplicity of $\overline{C}$ at Q. It cannot occur that $\widetilde{c} = \widetilde{p} = 1$. We may further assume that $\widetilde{c} > \widetilde{p}$.*

Proof. (1) $\overline{C} - C$ is a one-place point of $\overline{C}$ and hence $\overline{C}$ meets D in one point. If $(\overline{C} \cdot \ell) = 0$, then C is a fiber component of f. Since $C \neq A_i$ for $i = 1, 2$, C is a smooth fiber of f. Hence there is a surjective morphism from $\mathbb{A}^2 \to \mathbb{A}^1_*$. This is a contradiction. So, $\overline{C}$ meets ℓ.

(2) If $Q = \ell \cap M$ then perform elementary transformations with center Q and its infinitely near points lying on M until the proper transform of $\overline{C}$ is separated from the proper transform of M.

(3) Suppose $\widetilde{c} = \widetilde{p} = 1$. Then C cannot meet A_1 and A_2 because $m_1 > 1$ and $m_2 > 1$. Since $\overline{C}$ meets ℓ, $\overline{C}$ does not meet A either, hence the fiber of $\overline{f}$ supported by $A + A_1 + A_2$. This is a contradiction. If $\widetilde{c} = \widetilde{p}$ and $\widetilde{p} > 1$, perform blowing-ups with centers at Q and its infinitely near points lying on $\overline{C}$ until the proper transform of $\overline{C}$ meets the last (-1)-curve, say E, with the intersection number greater than the multiplicity of singularity. Then contract the proper transform of ℓ and all the exceptional curves but E. Then we have $\widetilde{c} > \widetilde{p}$. □

We call such a standard completion $\overline{X}$ *normalized* for C and let $n_0 = (M^2)$. Let $\Gamma = D + \overline{C}$, which is the boundary divisor of $U \cong \mathbb{A}^2$ in $\overline{X}$. Let $c = \widetilde{c}/\gcd(\widetilde{c}, \widetilde{p})$ and $p = \widetilde{p}/\gcd(\widetilde{c}, \widetilde{p})$. Let $\widetilde{\Gamma} = \widetilde{D} + \widetilde{C}$ be the boundary divisor of U which results from the Euclidean transformation with center Q with respect to the pair (c, p) and let E be the last (-1)-curve. Then $\widetilde{C}$ meets E in a point not on the proper transform of ℓ or any other exceptional curves. Further, let $\Gamma^* = D^* + C^*$ be the boundary divisor obtained by the minimal blowing-ups such that the proper transform of Γ together with the exceptional curves has only simple normal crossings. This process involves the above Euclidean transformation. We then have the following two possibilities:

(I) $\widetilde{C}$ is a smooth curve meeting normally E and $(\widetilde{C}^2) = -1$.
(II) Either $\widetilde{C}$ is singular or $\widetilde{C}$ is a smooth curve meeting E normally with $(\widetilde{C}^2) \neq -1$.

Claim. *Suppose the second case above occurs. Then we have:*
(1) *Write $\widetilde{D}$ as $W + E + B_3$, where W and B_3 are subchains and W contains $M + A$. Then $n_0 = (M^2) = -1$ and W is contractible to a smooth point.*
(2) *Suppose that $\widetilde{C}$ is singular. By blowing up, if necessary, the point $E \cap \widetilde{C}$ (resp., $E \cap B_3$) and its infinitely near points lying on $\widetilde{C}$ (resp., B_3) and contracting the proper transform of E and a part of exceptional curves, we have a new standard normalized completion $\overline{X}'$ with a standard boundary $D' = \ell' + M' + A'$ with the image of $\widetilde{C}$ (still possibly singular but the singularity better than $\widetilde{C}$) meeting ℓ' at a point Q' not on M'.*
(3) *In the case (2) above, the linear chain A' is the transpose ${}^tA$ of A, which is, by definition, the same A read from the other tip not meeting M.*
(4) *The case where $\widetilde{C}$ is smooth with $(\widetilde{C}^2) \neq -1$ cannot occur.*

Proof. (1) In the divisor Γ^*, the proper transform of E is a branching component with three branches B_1, B_2, B_3, where B_1 contains the proper transform of ℓ, B_2 contains C^* and B_3 is an admissible chain. In fact, it is always the case that the last (-1)-curve of the Euclidean transformation with data (c, p) is a branching component with the image of C as one branch if $c > p$. Note that since $\Gamma^* + C^*$ is the boundary divisor with simple normal crossings it is minimized to a linear chain by a theorem of Ramanujam [192]. Hence one of B_1 and B_2 is contractible to a point. Suppose B_2 is contractible. Then C^* is a (-1)-curve meeting a component of B_2 normally which becomes a 0-curve after the contraction of C^*. This is a contradiction to the assumption that B_2 is contractible. Hence B_1 is contractible in Γ^* and already contractible in $\widetilde{\Gamma}$. In particular, $n_0 = -1$. In fact, we blow up at least twice on ℓ, so only M can be a (-1)-curve in B_1.

(2) After the contraction of B_1, the image of $\widetilde{\Gamma}$ is of the form $\Gamma' = D'' + \widetilde{C}$, where D'' is a chain with the image E' of E as a tip and $({E'}^2) \geq 0$. If $({E'}^2) > 0$, we blow up the point $E' \cap B_3$ and its infinitely near points lying on E' until the proper transform ℓ' of E' becomes a (0)-curve. Let B_3' be the inverse image of B_3 at this stage and M' the component of B_3' meeting ℓ' and $A' = B_3' - M'$, the rest of B_3'. Then $D' = \ell' - M' - A'$ is the boundary divisor of a standard completion of X. After performing an elementary transformation involving ℓ' if necessary, we may assume that it is normalized for C, i. e., we have $(\overline{C}' \cdot \ell') >$ multiplicity of $\overline{C}'$ at $\overline{C}' \cap \ell'$, where $\overline{C}'$ is the proper transform of $\widetilde{C}$.

(3) We refer to [33, Theorem 3.12].

(4) When $\widetilde{C}$ is smooth with $(\widetilde{C}^2) \neq -1$, the component E is also a branching component of $\widetilde{\Gamma}$ and the branch B_1 contracts to a point. As in the case (2), we have a standard completion $\overline{X}'$ of X with the boundary divisor $D' = \ell' + M' + A'$, where $\widetilde{C}$ meets normally ℓ'. This contradicts (3) of the Claim. □

We call the operation of changing the standard normalized completions $\overline{X} \to \overline{X}'$ a *flip*. After repeating flips several times, we reach to the case (I) described after the Claim. Namely, we may assume that $\widetilde{C}$ is a (-1)-curve meeting E normally. Contract $\widetilde{C}$ to a point and obtain the image of E having intersection number 0. Let $\overline{\Gamma}$ be the image of $\widetilde{\Gamma}$. Then $\overline{\Gamma}$ is a linear chain which is the boundary divisor of $\mathbb{A}^2$ with simple normal crossings. Hence we have $d(\overline{\Gamma}) = \pm 1$. By Lemma 2.8.18, we have one of the following two equalities, where $n_0 = (M^2)$:

$$a \pm 1 = c((n_0 c + p)a + ca'),$$
$$a \pm 1 = c((n_0 c + p)a + ca'').$$

CONSTRUCTION OF A COUNTEREXAMPLE.

We construct an ML_0-surface X as in Lemma 2.6.9 by constructing a chain

$$\underset{\ell_0}{(-3)} \; - \; \underset{E_3}{(-1)} \; - \; \underset{E_2}{(-2)} \; - \; \underset{E_1}{(-2)}$$

and attaching feathers A_1, A_2 to E_3 and E_2 of multiplicities 3 and 2, respectively. The corresponding chain A is now

$$\underset{\ell_0'}{(-3)} \; - \; \underset{E_3'}{(-2)} \; - \; \underset{E_2'}{(-3)} \; - \; \underset{E_1'}{(-2)}$$

with $a = 19, a' = 8$, and $a'' = 12$. It is readily verified that the only way to attach feathers to A or its transpose $A' = {}^tA$ so as to produce a complete fiber in a $\mathbb{P}^1$-fibration is to attach them to E_3 and E_2. Hence the requirements (2) and (3) before step I are satisfied. It is also readily verified that there are no solutions c, n_0, and p satisfying one of the above equalities. So, X does not contain $\mathbb{A}^2$. On the other hand, $\mathrm{Pic}(X) \cong \mathbb{Z}$ is torsion-free. Thus we have constructed an example of an ML_0-surface X with $\mathrm{Pic}(X)_{\mathrm{tor}} = (0)$ and $\rho(X) = 1$ which does not contain an open set isomorphic to $\mathbb{A}^2$. □

Recall that an $\mathbb{A}^1$-fibration on an ML_0-surface with Picard number zero has at most one multiple fiber (which is irreducible) and the multiplicity is the order of the Picard group. Hence this multiplicity is invariant for any $\mathbb{A}^1$-fibration on the surface. This is not the case for the multiplicities of the components of singular fibers of an ML_0-surface with positive Picard number. We exhibit this by giving an example.

Example 2.8.19. Let C be a smooth conic on $\mathbb{P}^2$ and let $Q \in C$ be a point. Let $\sigma : V_0 \to \mathbb{P}^2$ be the blowing-up of Q with $M_0 := \sigma^{-1}(Q)$. The surface V_0 is the Hirzebruch surface of degree 1, M_0 is the minimal section and the standard $\mathbb{P}^1$-fibration $p_0 : V_0 \to \mathbb{P}^1$ is given by the pencil of lines through Q. Let $C' = \sigma'(C)$ and let $X = V_0 - C'$. Then the following assertions hold:

(1) X is an ML_0-surface with $\rho(X) = 1$ and $\mathrm{Pic}(X)_{\mathrm{tor}} = (0)$. Furthermore, X is a half-point attachment of $\mathbb{P}^2 - C$. In particular, $\mathbb{P}^2 - C$ is an open set of X.

(2) $f_0 := p_0|_X : X \to \mathbb{P}^1$ is an $\mathbb{A}^1$-fibration such that every fiber is irreducible and reduced.
(3) Let $P \in C$ be a point other than Q and let T be the tangent line to C at P. The pencil generated by $2T$ and C defines an $\mathbb{A}^1$-fibration on $\mathbb{P}^2 - C$ which extends to an $\mathbb{A}^1$-fibration $f_1 : X \to \mathbb{P}^1$; f_1 has one singular fiber which is irreducible of multiplicity 2 and all other fibers are irreducible and reduced.
(4) There exists an irreducible quintic Y on $\mathbb{P}^2$ which has six consecutive cusps of multiplicity 2 infinitely near to the point P with five of them lying on C [231]. The pencil generated by $2Y$ and $5C$ induces an $\mathbb{A}^1$-fibration on $\mathbb{P}^2 - C$ which extends to an $\mathbb{A}^1$-fibration $f_2 : X \to \mathbb{P}^1$ such that $Y' := \sigma(Y) \cap X$ and $M_0' := M_0 \cap X$ support multiple fibers of f_2 with respective multiplicities 2 and 5 and all other fibers are irreducible and reduced.

Remark 2.8.20. The construction of X in Example 6.3 can be generalized as follows. By a result in [24], a smooth affine surface $\mathbb{F}_s - S$, where $\mathbb{F}_s$ is the Hirzebruch surface of degree s and S is an ample section with $(S^2) = k+1$, contains an open set U which is an ML_0-surface with $\mathrm{Pic}(U) = \mathbb{Z}/k\mathbb{Z}$. Concretely, U can be realized as the complement of the zero locus C of a weighted homogeneous polynomial $f := X_0X_2 - X_1^{k+1}$ in the weighted projective plane $\mathbb{P}$ with weight $(1,1,k)$ and $\mathbb{F}_s - S$ as the complement of the proper transform C' of C in the blowing-up of $\mathbb{P}$ at the (singular) point $Q = (0,0,1)$. Again, there exist $\mathbb{A}^1$-fibrations of $\mathbb{P}$ with C and another curve Y as singular fibers.

2.9 Deformations of $\mathbb{A}^1$-fibrations

2.9.1 Motivation

By Lemma 2.2.7, an $\mathbb{A}^1$-fibration $\rho : X \to B$ on a smooth affine surface X to a smooth curve B is given as the quotient morphism of a G_a-action if the parameter curve B is an affine curve. Meanwhile, it is not so if B is a complete curve. When we deform the surface X over a parameter curve under a suitable setting (*log-deformation*), our question is if the neighboring surfaces still have $\mathbb{A}^1$-fibrations of affine type or of complete type according to the type of the $\mathbb{A}^1$-fibration on X being affine or complete. Assuming that neighboring surfaces have $\mathbb{A}^1$-fibrations, propagation of the type of $\mathbb{A}^1$-fibration is proved in Lemma 2.9.7, whose proof reflects the structure of the boundary divisor at infinity of an affine surface with an $\mathbb{A}^1$-fibration. The stability of the boundary divisor under small deformations, e. g., the stability of the weighted dual graphs has been discussed in topological methods (e. g., by Neumann [180]). Furthermore, if such a property is inherited by the neighboring surfaces, we still ask if the ambient threefold has an $\mathbb{A}^1$-fibration, or equivalently if the generic fiber has an $\mathbb{A}^1$-fibration.

The answer to this question is subtle. We consider first in Subsection 2.9.2 the case where each of fiber surfaces of the deformation has an $\mathbb{A}^1$-fibration of affine type induced by a global vector field on the ambient threefold. This global vector field is in fact given by a locally nilpotent derivation (Theorem 2.9.1). If the $\mathbb{A}^1$-fibrations on the fiber surfaces are of affine type, we can show (Theorem 2.9.13), with the absence of monodromies of boundary components, that there exists an $\mathbb{A}^1$-fibration on the ambient threefold such that the $\mathbb{A}^1$-fibration on each general fiber surface is induced by the global one up to an automorphism of the fiber surface. The proof of Theorem 2.9.13 depends on Lemma 2.9.7 which we prove by observing the behavior of the boundary rational curves. This is done by the use of Hilbert scheme (see [120]) and by killing monodromies by étale finite changes of the base curve.

As a consequence, we can prove the generic triviality of an $\mathbb{A}^2$-fibration over a curve. Namely, if $f : Y \to T$ is a smooth morphism from a smooth affine threefold to a smooth affine curve such that the fiber over every *closed* point of T is isomorphic to the affine plane $\mathbb{A}^2$, then the generic fiber of f is isomorphic to $\mathbb{A}^2$ over the function field $k(T)$ of T and f is an $\mathbb{A}^2$-bundle over an open set of T (see Theorem 2.9.15). This fact, together with a theorem of Sathaye [204], shows that f is an $\mathbb{A}^2$-bundle over T in the Zariski topology.

The question on the generic triviality is also related to a question on the triviality of a k-form of a surface with an $\mathbb{A}^1$-fibration (see Problem 2.2 at the end of Subsection 2.9.3). In the case of an $\mathbb{A}^1$-fibration of complete type, the answer is negative by Dubouloz–Kishimoto [42] (see Theorem 2.9.24).

Theorem 2.9.15 was proved by Kaliman–Zaidenberg [105] in a more comprehensive way and without assuming that the base is a curve. The idea in our first proof of Theorem 2.9.15 is of more algebraic nature and consists of using the existence of a locally nilpotent derivation on the coordinate ring of Y and the second proof of using the Ramanujam–Morrow graph of the normal minimal completion of $\mathbb{A}^2$ was already used in [105]. The related results are also discussed in the article [155, 199]. We cannot still avoid the use of a theorem of Kambayashi [106] on the absence of separable forms of the affine plane.

Some of the algebro-geometric arguments using Hilbert scheme in Subsection 2.9.3 can be replaced by topological arguments using Ehresmann's fibration theorem (Theorem 1.1.5) which might be more appreciated than the use of the Hilbert scheme. But they are restricted to the case of small deformations. This is done in Subsection 2.9.4.

In Subsection 2.9.5, we extend the above result on the generic triviality of an $\mathbb{A}^2$-fibration over a curve by replacing $\mathbb{A}^2$ by an affine pseudoplane of ML_0-type which was discussed in details in Sections 2.6 and 2.8 and has properties similar to $\mathbb{A}^2$, e. g., the boundary divisor for a minimal normal completion is a linear chain of rational curves. But we still need a condition on the monodromy. An affine pseudoplane, not necessarily of ML_0-type, is a $\mathbb{Q}$-homology plane, and we note that Flenner–Zaidenberg [50] made a fairly exhaustive consideration for the log-deformations of $\mathbb{Q}$-homology planes.

In the final Subsection 2.9.6, we observe the case of $\mathbb{A}^1$-fibration of complete type and show by an example of Dubouloz–Kishimoto [42] that the ambient threefold does not have an $\mathbb{A}^1$-fibration. But it is now a fact by Dubouloz–Kishimoto [42] that the ambient threefold is affine-uniruled in the stronger sense that the fiber product of the ambient deformation space by a suitable étale lifting of the base curve has a global $\mathbb{A}^1$-fibration.

We use two notations for the intersection of (not necessarily irreducible) subvarieties A, B of codimension one in an ambient threefold. Namely, $A \cap B$ is the intersection of two subvarieties, and $A \cdot B$ is the intersection of effective divisors. In most cases, both are synonymous, but distinction is necessary in special cases.

2.9.2 Triviality of deformations of locally nilpotent derivations

Let $Y = \operatorname{Spec} B$ be an irreducible affine algebraic variety. We define the tangent sheaf $\mathcal{T}_{Y/k}$ as $\mathcal{H}om_{\mathcal{O}_Y}(\Omega^1_{Y/k}, \mathcal{O}_Y)$. A regular vector field on Y is an element of $\Gamma(Y, \mathcal{T}_{Y/k})$. A regular vector field Θ on Y is identified with a derivation D on B via isomorphisms

$$\Gamma(Y, \mathcal{T}_{Y/k}) \cong \operatorname{Hom}_B(\Omega^1_{B/k}, B) \cong \operatorname{Der}_k(B, B)$$

(see Subsection 1.3.1). We say that Θ is *locally nilpotent* if so is D. In the first place, we are interested in finding a necessary and sufficient condition for D to be locally nilpotent. Suppose that Y has a fibration $f : Y \to T$. A natural question is to ask whether D is locally nilpotent if the restriction of D on each closed fiber of f is locally nilpotent. The following result shows that this is the case.[14]

Theorem 2.9.1. *Let $Y = \operatorname{Spec} B$ and $T = \operatorname{Spec} R$ be irreducible affine varieties defined over k and let $f : Y \to T$ be a dominant morphism such that general fibers are irreducible and reduced. We consider R to be a subalgebra of B. Let D be an R-trivial derivation of B such that, for each closed point $t \in T$, the restriction $D_t = D \otimes_R R/\mathfrak{m}$ is a locally nilpotent derivation of $B \otimes_R R/\mathfrak{m}$, where $\mathfrak{m}$ is the maximal ideal of R corresponding to t. Then D is locally nilpotent.*

We need some preliminary results. We retain the notations and assumptions in the above theorem.

Lemma 2.9.2. *There exist a finitely generated field extension k_0 of the prime field $\mathbb{Q}$ which is a subfield of the ground field k, geometrically integral affine varieties $Y_0 = \operatorname{Spec} B_0$ and $T_0 = \operatorname{Spec} R_0$, a dominant morphism $f_0 : Y_0 \to T_0$ and an R_0-trivial derivation D_0 of B_0 such that the following conditions are satisfied:*
(1) *Y_0, T_0, f_0 and D_0 are defined over k_0.*
(2) *$Y = Y_0 \otimes_{k_0} k, T = T_0 \otimes_{k_0} k, f = f_0 \otimes_{k_0} k$ and $D = D_0 \otimes_{k_0} k$.*
(3) *D_0 is locally nilpotent if and only if so is D.*

14 The result is also remarked in [42, Remark 13].

Proof. Since B and R are integral domains finitely generated over k, write B and R as the residue rings of certain polynomial rings over k modulo the finitely generated ideals. Write $B = k[x_1, \dots, x_r]/I$ and $R = k[t_1, \dots, t_s]/J$. Furthermore, the morphism f is determined by the images $f^*(\eta_j) = \varphi_j(\xi_1, \dots, \xi_r)$ in B, where $\xi_i = x_i \pmod I$ and $\eta_j = t_j \pmod J$. Adjoin to $\mathbb{Q}$ all coefficients of finite sets of generators of I and J as well as the coefficients of the φ_j to obtain a subfield k_0 of k. Let $B_0 = k_0[x_1, \dots, x_r]/I_0$ and $R_0 = k_0[t_1, \dots, t_s]/J_0$, where I_0 and J_0 are respectively the ideals in $k_0[x_1, \dots, x_r]$ and $k_0[t_1, \dots, t_s]$ generated by the same generators of I and J. Furthermore, define the homomorphism f_0^* by the assignment $f_0^*(\eta_j) = \varphi_j(\xi_1, \dots, \xi_r)$. Let $Y_0 = \operatorname{Spec} B_0$, $T_0 = \operatorname{Spec} R_0$ and let $f_0 : Y_0 \to T_0$ be the morphism defined by f_0^*. The derivation D corresponds to a B-module homomorphism $\delta : \Omega^1_{B/R} \to B$. Since $\Omega^1_{B/R} = \Omega^1_{B_0/R_0} \otimes_{k_0} k$, we can enlarge k_0 so that there exists a B_0-homomorphism $\delta_0 : \Omega^1_{B_0/R_0} \to B_0$ satisfying $\delta = \delta_0 \otimes_{k_0} k$. Let $D_0 = \delta_0 \cdot d_0$, where $d_0 : B_0 \to \Omega^1_{B_0/R_0}$ is the universal standard differentiation. Then we have $D = D_0 \otimes_{k_0} k$.

Let $\Phi_0 : B_0 \to B_0[[u]]$ be the R_0-homomorphism into the formal power series ring in t over B_0 defined by

$$\Phi_0(b_0) = \sum_{i \geq 0} \frac{1}{i!} D_0^i(b_0) u^i.$$

Let $\Phi : B \to B[[u]]$ be the R-homomorphism defined in a similar fashion. Then Φ_0 and Φ are determined by the images of the generators of B_0 and B. Since the generators of B_0 and B are the same, we have $\Phi = \Phi_0 \otimes_{k_0} k$. The derivation D_0 is locally nilpotent if and only if Φ_0 splits via the polynomial subring $B_0[u]$ of $B_0[[u]]$. This is the case for D as well. Since Φ_0 splits via $B_0[u]$ if and only if Φ splits via $B[u]$, D_0 is locally nilpotent if and only if so is D. □

Lemma 2.9.3. *Let k_1 be the algebraic closure of k_0 in k. Let $Y_1 = \operatorname{Spec} B_1$ with $B_1 = B_0 \otimes_{k_0} k_1$, $T_1 = \operatorname{Spec} R_1$ with $R_1 = R_0 \otimes_{k_0} k_1$ and $f_1 = f_0 \otimes_{k_0} k_1$. Let $D_1 = D_0 \otimes_{k_0} k_1$. Then the following assertions hold:*

(1) *Let t_1 be a closed point of T_1. Then the restriction of D_1 on the fiber $f_1^{-1}(t_1)$ is locally nilpotent.*
(2) *D_1 is locally nilpotent if and only if so is D.*

Proof. (1) Let t be the unique closed point of T lying over t_1 by the projection morphism $T \to T_1$, where $R = R_1 \otimes_{k_1} k$. (If $\mathfrak{m}_1$ is the maximal ideal of R_1 corresponding to t_1, $\mathfrak{m}_1 \otimes_{k_1} k$ is the maximal ideal of R corresponding to t.) Then $F_t = f^{-1}(t) = f_1^{-1}(t_1) \otimes_{k_1} k$ and the restriction D_t of D onto F_t is given as $D_{1,t_1} \otimes_{k_1} k$, where D_{1,t_1} is the restriction of D_1 onto $f_1^{-1}(t_1)$. We consider also the R-homomorphism $\Phi : B \to B[[u]]$ and the R_1-homomorphism $\Phi_1 : B_1 \to B_1[[u]]$. As above, let $\mathfrak{m}$ and $\mathfrak{m}_1$ be the maximal ideals of R and R_1 corresponding to t and t_1. Then D_t gives rise to the $R/\mathfrak{m}$-homomorphism $\Phi \otimes_R R/\mathfrak{m} : B \otimes_R R/\mathfrak{m} \to (B \otimes_R R/\mathfrak{m})[[u]]$. Similarly, D_{1,t_1} gives rise to the $R_1/\mathfrak{m}_1$-homomorphism $\Phi_1 \otimes_{R_1} R_1/\mathfrak{m}_1 : B_1 \otimes_{R_1} R_1/\mathfrak{m}_1 \to (B_1 \otimes_{R_1} R_1/\mathfrak{m}_1)[[u]]$, where

$R/\mathfrak{m} = k$ and $R_1/\mathfrak{m}_1 = k_1$. Then $\Phi \otimes_R R/\mathfrak{m} = (\Phi_1 \otimes_{R_1} R_1/\mathfrak{m}_1) \otimes_{k_1} k$. Hence $\Phi \otimes_R R/\mathfrak{m}$ splits via $(B \otimes_R R/\mathfrak{m})[u]$ if and only if $\Phi_1 \otimes_{R_1} R_1/\mathfrak{m}_1$ splits via $(B_1 \otimes_{R_1} R_1/\mathfrak{m}_1)[u]$. Hence $D_{1,t}$ is locally nilpotent as so is D_t.

(2) The same argument as above using the homomorphism Φ can be applied. □

The field k_0 can be embedded into the complex field $\mathbb{C}$ because it is a finitely generated field extension of $\mathbb{Q}$. Hence we can extend the embedding $k_0 \hookrightarrow \mathbb{C}$ to the algebraic closure k_1. Thus k_1 is viewed as a subfield of $\mathbb{C}$. Then Lemma 2.9.3 holds if one replaces the extension k/k_1 by the extension $\mathbb{C}/k_1$. Hence it suffices to prove Theorem 2.9.1 with an additional hypothesis $k = \mathbb{C}$.

Lemma 2.9.4. *Theorem 2.9.1 holds if k is the complex field $\mathbb{C}$.*

Proof. Let $Y(\mathbb{C})$ be the set of closed points which we view as a complex analytic space embedded into a complex affine space $\mathbb{C}^N$ as a closed set. Consider the Euclidean metric on $\mathbb{C}^N$ and the induced metric topology on $Y(\mathbb{C})$. Then $Y(\mathbb{C})$ is a complete metric space.

Let b be a nonzero element of B. For a positive integer m, define a Zariski closed subset $Y_m(b)$ of $Y(\mathbb{C})$ by

$$Y_m(b) = \{Q \in Y(\mathbb{C}) \mid D^m(b)(Q) = 0\}.$$

Since Q lies over a closed point t of $T(\mathbb{C})$ and D_t is locally nilpotent on $f^{-1}(t)$ by the hypothesis, we have

$$f^{-1}(t) \subset \bigcup_{m>0} Y_m(b).$$

This implies that $Y(\mathbb{C}) = \bigcup_{m>0} Y_m(b)$. We claim that $Y(\mathbb{C}) = Y_m(b)$ for some $m > 0$. In fact, this follows by Baire category theorem, which states that if the $Y_m(b)$ are all proper closed subsets, its countable union cannot cover the uncountable set $Y(\mathbb{C})$. If $Y(\mathbb{C}) = Y_m(b)$ for some $m > 0$ then $D^m(b) = 0$. This implies that D is locally nilpotent on B.

One can avoid the use of Baire category theorem in the following way. Suppose that $Y_m(b)$ is a proper closed subset for every $m > 0$. Let H be a general hyperplane in $\mathbb{C}^N$ such that the section $Y(\mathbb{C}) \cap H$ is irreducible, $\dim Y(\mathbb{C}) \cap H = \dim Y(\mathbb{C}) - 1$, and $Y(\mathbb{C}) \cap H = \bigcup_{m>0}(Y_m(b) \cap H)$ with $Y_m(b) \cap H$ a proper closed subset of $Y(\mathbb{C}) \cap H$ for every $m > 0$. We can further take hyperplane sections and find a general linear subspace L in $\mathbb{C}^N$ such that $Y(\mathbb{C}) \cap L$ is an irreducible curve and $Y(\mathbb{C}) \cap L = \bigcup_{m>0}(Y_m(b) \cap L)$, where $Y_m(b) \cap L$ is a proper Zariski closed subset. Hence $Y_m(b) \cap L$ is a finite set, and $\bigcup_{m>0}(Y_m(b) \cap L)$ is a countable set, while $Y(\mathbb{C}) \cap L$ is not a countable set. This is a contradiction. Thus $Y(\mathbb{C}) = Y_m(b)$ for some $m > 0$. □

Let D be a k-derivation on a k-algebra B. It is called *surjective* if D is so as a k-linear mapping. The following result is a consequence of Theorem 2.9.1.

Corollary 2.9.5. *Let $Y = \operatorname{Spec} B$, $T = \operatorname{Spec} R$ and $f : Y \to T$ be the same as in Theorem 2.9.1. Let D be an R-derivation of B such that D_t is a surjective k-derivation for every closed point $t \in T$. Assume further that the relative dimension of f is one. Then D is a locally nilpotent derivation and f is an $\mathbb{A}^1$-fibration.*

Proof. Let t be a closed point of T such that the fiber $f^{-1}(t)$ is irreducible and reduced. By [80, Theorem 1.2 and Proposition 1.7], the coordinate ring $B \otimes R/\mathfrak{m}$ of $f^{-1}(t)$ is a polynomial ring $k[x]$ in one variable and $D_t = \partial/\partial x$, where $\mathfrak{m}$ is the maximal ideal of R corresponding to t. Then D_t is locally nilpotent. Taking the base change of $f : Y \to T$ by $U \hookrightarrow T$ if necessary, where U is a small open set of T, we may assume that D_t is locally nilpotent for every closed point t of T. By Theorem 2.9.1, the derivation D is locally nilpotent and hence f is an $\mathbb{A}^1$-fibration. □

2.9.3 Deformations of $\mathbb{A}^1$-fibrations of affine type

In the present subsection, we assume that the ground field k is the complex field $\mathbb{C}$. We consider the following result on deformations. For the complex analytic case, one can refer to [118] and also to [94, p. 269].

Lemma 2.9.6. *Let $\overline{f} : \overline{Y} \to T$ be a smooth projective morphism from a smooth algebraic threefold $\overline{Y}$ to a smooth algebraic curve T. Let C be a smooth rational complete curve contained in $\overline{Y}_0 = \overline{f}^{-1}(t_0)$ for a closed point t_0 of T.*[15] *Then the following assertions hold:*

(1) *The Hilbert scheme $\operatorname{Hilb}(\overline{Y})$ has dimension less than or equal to $h^0(C, \mathrm{N}_{C/\overline{Y}})$ in the point $[C]$. If $h^1(C, \mathrm{N}_{C/\overline{Y}}) = 0$ then the equality holds and $\operatorname{Hilb}(\overline{Y})$ is smooth at $[C]$. Here $\mathrm{N}_{C/\overline{Y}}$ denotes the normal bundle of C in $\overline{Y}$.*
(2) *Let $n = (C^2)$ on $\overline{Y}_0$. Then $\mathrm{N}_{C/\overline{Y}} \cong \mathcal{O}_C \oplus \mathcal{O}_C(n)$ provided $n \ge -1$.*
(3) *Suppose $n = 0$. Then there exists an étale finite covering $\sigma_2 : T' \to T$ such that the morphism $\overline{f}_{T'}$ splits as*

$$\overline{f}_{T'} : \overline{Y} \times_T T' \xrightarrow{\varphi} V \xrightarrow{\sigma_1} T',$$

where φ is a $\mathbb{P}^1$-fibration with C contained as a fiber and σ_1 makes V a smooth T'-scheme of relative dimension one with irreducible fibers. Assume further that every smooth rational complete curve C' in $\overline{Y}_0$ satisfies $(C' \cdot C) = 0$ provided C' is algebraically equivalent to C in $\overline{Y}$. Then the covering $\sigma_2 : T' \to T$ is trivial, i. e., σ_2 is the identity morphism.
(4) *Suppose $n = -1$. Then C does not deform in the fiber $\overline{Y}_0$ but deforms along the morphism $\overline{f}$ after an étale finite base change. Namely, there are an étale finite morphism $\sigma : T' \to T$ and an irreducible subvariety Z of codimension one in $\overline{Y}' := \overline{Y} \times_T T'$ such*

15 When we write $t \in T$, we tacitly assume that t is a closed point of T.

that Z can be contracted along the fibers of $\overline{f}' : \overline{Y}' \to T'$, *where* T' *is an irreducible smooth algebraic curve and* $\overline{f}'$ *is the second projection of* $Y \times_T T'$ *to* T'.

(5) *Assume that there are no* (-1)*-curves* E *and* E' *in* $\overline{Y}_0$ *such that* $E \cap E' \neq \emptyset$ *and* E *is algebraically equivalent to* E' *as 1-cycles on* $\overline{Y}$. *Then, after shrinking* T *to a smaller open set if necessary, we can take* Z *in the assertion (4) above as a subvariety of* $\overline{Y}$. *The contraction of* Z *gives a factorization* $\overline{f}|_Z : Z \xrightarrow{g} T' \xrightarrow{\sigma} T$, *where* g *is a* $\mathbb{P}^1$*-fibration,* C *is a fiber of* g *and* σ *is as above.*

Proof. (1) The assertion follows from [65, Corollary 5.4].

(2) We have an exact sequence

$$0 \to \mathrm{N}_{C/\overline{Y}_0} \to \mathrm{N}_{C/\overline{Y}} \to \mathrm{N}_{\overline{Y}_0/\overline{Y}}|_C \to 0,$$

where $\mathrm{N}_{C/\overline{Y}_0} \cong \mathcal{O}_C(n)$ and $\mathrm{N}_{\overline{Y}_0/\overline{Y}}|_C \cong \mathcal{O}_C$. The obstruction for this exact sequence to split lies in $\mathrm{Ext}^1(\mathcal{O}_C, \mathcal{O}_C(n)) \cong \mathrm{H}^1(C, \mathcal{O}_C(n))$, which is zero if $n \geq -1$.

(3) Suppose $n = 0$. Then $\dim_{[C]} \mathrm{Hilb}(\overline{Y}) = 2$ and $[C]$ is a smooth point of $\mathrm{Hilb}(\overline{Y})$. Let H be a relatively ample divisor on $\overline{Y}/T$ and set $P(n) := P_C(n) = h^0(C, \mathcal{O}_C(nH))$ the Hilbert polynomial in n of C with respect to H. Then $\mathrm{Hilb}^P(\overline{Y})$ is a scheme which is projective over T. Let V be the irreducible component of $\mathrm{Hilb}^P(\overline{Y})$ containing the point $[C]$. Then V is a T-scheme with a morphism $\sigma : V \to T$, $\dim V = 2$ and V has relative dimension one over T. Furthermore, there exists a subvariety Z of $\overline{Y} \times_T V$ such that the fibers of the composite morphism

$$g : Z \hookrightarrow \overline{Y} \times_T V \xrightarrow{p_2} V$$

are curves on $\overline{Y}$ parametrized by V. For a general point $v \in V$, the corresponding curve $C' := C_v$ is a smooth rational complete curve because $P_{C'}(n) = P(n)$, and $(C')^2 = 0$ on $\overline{Y}_t = \overline{f}^{-1}(t)$ with $t = \sigma(v)$ because $\dim_{[C']} \mathrm{Hilb}(\overline{Y}) = 2$. In fact, if $(C')^2 \leq -1$ then the exact sequence of normal bundles in (2) implies $h^0(C', \mathrm{N}_{C'/\overline{Y}}) \leq 1$, which contradicts $\dim_{[C']} \mathrm{Hilb}(\overline{Y}) = 2$. If $(C')^2 > 0$ then $\dim_{[C']} \mathrm{Hilb}(\overline{Y}) > 2$, which is again a contradiction. So, $(C')^2 = 0$. Hence $\overline{Y}_t$ has a $\mathbb{P}^1$-fibration $\varphi_t : \overline{Y}_t \to \overline{B}_t$ such that C' is a fiber and $\overline{B}_t$ is a smooth complete curve. By the universality of the Hilbert scheme, there are an open set U of $\overline{B}_t$ and a morphism $\alpha_t : U \to V_t$ such that $\varphi_t^{-1}(U) = Z \times_V U$. Since V is smooth over T, α_t induces an isomorphism from $\overline{B}_t$ to a connected component of $V_t := \sigma^{-1}(t)$. This is the case if we take $v \in V$ from a different connected component of V_t. Let $\sigma : V \xrightarrow{\sigma_1} T' \xrightarrow{\sigma_2} T$ be the Stein factorization of σ. Then σ_2 is an étale finite morphism and $\sigma_1 : V \to T'$ is a smooth morphism of relative dimension one with irreducible fibers. Furthermore, the morphism g above factors as a composite of T'-morphisms

$$g : Z \hookrightarrow (\overline{Y} \times_T T') \times_{T'} V \xrightarrow{p_2} V,$$

where Z is identified with $\overline{Y} \times_T T'$ by the above construction. Hence g induces a T'-morphism $\varphi : \overline{Y} \times_T T' \to V$ such that the composite $\sigma_1 \cdot \varphi : \overline{Y} \times_T T' \to V \to T'$ is the pull-back $\overline{f}_{T'} : \overline{Y} \times_T T' \to T'$ of the morphism $\overline{f}$.

In the above argument, we take two curves C, C' corresponding to two points v, v' in V_{t_0}. Then C is algebraically equivalent to C' in $\overline{Y}$, and hence $(C \cdot C') = 0$ by the hypothesis. So, $C = C'$ or $C \cap C' = \emptyset$. This implies that C' and C are the fibers of the same $\mathbb{P}^1$-fibration $\varphi_{t_0} : \overline{Y}_{t_0} \to \overline{B}_{t_0}$ and hence that V_{t_0} is irreducible. Namely, $\sigma_2^{-1}(t_0)$ consists of a single point. Hence $\deg \sigma_2 = 1$, i. e., σ_2 is the identity morphism.

(4) Suppose $n = -1$. Then $h^0(C, \mathrm{N}_{C/\overline{Y}}) = 1$ and $h^1(C, \mathrm{N}_{C/\overline{Y}}) = 0$. Hence $\mathrm{Hilb}^P(\overline{Y})$ has dimension one and is smooth at $[C]$, where $P(n) = P_C(n)$ is the Hilbert polynomial of C with respect to H. Let T' be the irreducible component of $\mathrm{Hilb}^P(\overline{Y})$ containing $[C]$. Note that $\dim T' = 1$. Then we find a subvariety Z in $\overline{Y} \times_T T'$ such that C is a fiber of the T-morphism $g = p_2|_Z : Z \to T'$ and every fiber of g is a (-1)-curve in the fiber $\overline{Y}_t$. In fact, the nearby fibers of C are (-1)-curves as a small deformation of C by Kodaira [118]. Hence, by covering T' by small disks, we know that every fiber of g is a (-1)-curve. Further, the projection $\sigma : T' \to T$ is a finite morphism as it is projective and T' is smooth because each fiber is a smooth rational curve in $\overline{Y}$ (see the above argument for $[C]$). Furthermore, σ is étale since $\overline{f}$ is locally a product of the fiber and the base in the Euclidean topology. Hence σ induces a local isomorphism between T' and T. This implies that $\overline{Y} \times_T T'$ is a smooth algebraic threefold and the second projection $\overline{f}' : \overline{Y} \times_T T' \to T'$ is a smooth projective morphism. Now, after an étale finite base change $\sigma : T' \to T$, we may assume that Z is identified with a subvariety of $\overline{Y}$. Since C is a (-1)-curve in $\overline{Y}_0$, it is an extremal ray in the cone $\overline{NE}(\overline{Y}_0)$. Since C is algebraically equivalent to the fibers of $g : Z \to T$, it follows that C is an extremal ray in the relative cone $\overline{NE}(\overline{Y}/T)$. Then it follows from Kollar–Mori theorem [122, Theorem 3.25] that Z is contracted along the fibers of g in $\overline{Y}$ and the threefold obtained by the contraction is smooth and projective over T.

(5) Let $\sigma^{-1}(t_0) = \{u_1, \ldots, u_d\}$ and let $Z_{u_i} = Z \cdot (\overline{Y} \times \{u_i\})$ for $1 \le i \le d$. Then the Z_{u_i} are the (-1)-curves on $\overline{Y}_0$ which are algebraically equivalent to each other as 1-cycles on $\overline{Y}$. By the assumption, $Z_{u_i} \cap Z_{u_j} = \emptyset$ whenever $i \neq j$. This property holds for all $t \in T$ if one shrinks to a smaller open set of t_0. Then we can identify Z with a closed subvariety of $\overline{Y}$. In fact, the projection $p : Z \hookrightarrow Y \times_Y T' \to Y$ is a T-morphism. For the point $t_0 \in T$, the morphism $p \otimes_{\mathcal{O}_{T,t_0}} \widehat{\mathcal{O}}_{T,t_0}$ with the completion $\widehat{\mathcal{O}}_{T,t_0}$ of $\mathcal{O}_{T,t_0}$ is a direct sum of the closed immersions from $Z \otimes_{\mathcal{O}_{T,t_0}} \widehat{\mathcal{O}}_{T',u_i}$ into $Y \otimes_{\mathcal{O}_{T,t_0}} \widehat{\mathcal{O}}_{T',u_i}$ for $1 \le i \le r$. So, $p \otimes_{\mathcal{O}_{T,t_0}} \widehat{\mathcal{O}}_{T,t_0}$ is a closed immersion. Hence p is a closed immersion locally over t_0 because $\widehat{\mathcal{O}}_{T,t_0}$ is faithfully flat over $\mathcal{O}_{T,t_0}$. The rest is the same as in the proof of the assertion (4). □

Let Y_0 be a smooth affine surface and let $\overline{Y}_0$ be a smooth projective surface containing Y_0 as an open set in such a way that the complement $\overline{Y}_0 \setminus Y_0$ supports a reduced effective divisor D_0 with simple normal crossings. We call $\overline{Y}_0$ a *normal comple-*

tion of Y_0 and D_0 the boundary divisor of Y_0. An irreducible component of D_0 is called a (−1)-*component* if it is a smooth rational curve with self-intersection number −1. We say that $\overline{Y}_0$ is a *minimal* normal completion if the contraction of a (−1)-component of D_0 (if any) results the image of D_0 losing the condition of simple normal crossings.

Let $\overline{f} : \overline{Y} \to T$ be a smooth projective morphism from a smooth algebraic threefold $\overline{Y}$ to a smooth algebraic curve T and let $S = \sum_{i=1}^r S_i$ be a reduced effective divisor on $\overline{Y}$ with simple normal crossings. Let $Y = \overline{Y} \setminus S$ and let $f = \overline{f}|_Y$. We assume that for every point $t \in T$, the intersection cycle $D_t = \overline{f}^{-1}(t) \cdot S$ is a reduced effective divisor of $\overline{Y}_t = \overline{f}^{-1}(t)$ with simple normal crossings[16] and $Y_t = Y \cap \overline{Y}_t$ is an affine open set of $\overline{Y}_t$. For a point $t_0 \in T$, we assume that $\overline{Y}_{t_0} = \overline{Y}_0, D_{t_0} = D_0$, and $Y_{t_0} = Y_0$. A collection $(Y, \overline{Y}, S, \overline{f}, t_0)$ is called a *family of logarithmic deformations* of a triple $(Y_0, \overline{Y}_0, D_0)$. We call it simply a *log-deformation* of the triple $(Y_0, \overline{Y}_0, D_0)$. Since f is smooth and S is a divisor with simple normal crossings, $(Y, \overline{Y}, S, \overline{f}, t_0)$ is a family of logarithmic deformations in the sense of Kawamata [115, 114], which is a log version of Kodaira's theory of deformations [119].

From time to time, we have to make a base change by an étale finite morphism $\sigma : T' \to T$ with irreducible T'. Let $\overline{Y}' = \overline{Y} \times_T T', \overline{f}' = \overline{f} \times_T T', S' = S \times_T T'$, and $Y' = Y \times_T T'$. Since the field extension $k(\overline{Y})/k(T)$ is a regular extension, $\overline{Y}'$ is an irreducible smooth projective threefold, and S' is a divisor with simple normal crossings. Hence $(Y', \overline{Y}', S', \overline{f}', t_0')$ is a family of logarithmic deformations of the triple $(Y_0', \overline{Y}_0', D_0') \cong (Y_0, \overline{Y}_0, D_0)$, where $t_0' \in T'$ with $\sigma(t_0') = t_0$.

We have the following result on logarithmic deformations of affine surfaces with $\mathbb{A}^1$-fibrations.

Lemma 2.9.7. *Let $(Y, \overline{Y}, S, \overline{f}, t_0)$ be a log-deformation of the triple $(Y_0, \overline{Y}_0, D_0)$. Then the following assertions hold:*

(1) *Assume that Y_0 has an $\mathbb{A}^1$-fibration. Then Y_t has an $\mathbb{A}^1$-fibration for every $t \in T$.*
(2) *If Y_0 has an $\mathbb{A}^1$-fibration of affine type (resp., of complete type), then Y_t has also an $\mathbb{A}^1$-fibration of affine type (resp., of complete type) for every $t \in T$.*

Proof. (1) Note that $K_{\overline{Y}_t} = (K_{\overline{Y}} + \overline{Y}_t) \cdot \overline{Y}_t = K_{\overline{Y}} \cdot \overline{Y}_t$ because $\overline{Y}_t$ is algebraically equivalent to $\overline{Y}_{t'}$ for $t' \neq t$. Then $K_{\overline{Y}_t} + D_t = (K_{\overline{Y}} + S) \cdot \overline{Y}_t$. By the hypothesis, $h^0(\overline{Y}_0, \mathcal{O}(n(K_{\overline{Y}_0} + D_0))) = 0$ for every $n > 0$. Then the semicontinuity theorem [89, Theorem 12.8] implies that $h^0(\overline{Y}_t, \mathcal{O}(n(K_{\overline{Y}_t} + D_t))) = 0$ for every $n > 0$. Hence $\overline{\kappa}(Y_t) = -\infty$. Since Y_t is affine, this implies that Y_t has an $\mathbb{A}^1$-fibration.

16 In order to avoid the misreading, it is better to specify our definition of simple normal crossings in the case of dimension three. We assume that every irreducible component S_i of S and every fiber $\overline{Y}_t$ are smooth and that analytic-locally at every intersection point P of $S_i \cap S_j$ (resp., $S_i \cap S_j \cap S_k$ or $S_i \cap \overline{Y}_t$), S_i and S_j (resp., S_i, S_j and S_k, or S_i and $\overline{Y}_t$) behave like coordinate hypersurfaces. Hence $S_i \cap S_j$ or $S_i \cap \overline{Y}_t$ are smooth curves at the point P.

(2) Suppose that Y_0 has an $\mathbb{A}^1$-fibration $\rho_0 : Y_0 \to B_0$ which is of affine type. Then ρ_0 defines a pencil Λ_0 on $\overline{Y}_0$.

Suppose first that Λ_0 has no base points and hence defines a $\mathbb{P}^1$-fibration $\overline{\rho}_0$: $\overline{Y}_0 \to \overline{B}_0$ such that $\overline{\rho}_0|_{Y_0} = \rho_0$ and $\overline{B}_0$ is a smooth completion of B_0. If $\overline{\rho}_0$ is not minimal, let E be a (-1)-curve contained in a fiber of $\overline{\rho}_0$, which is necessarily not contained in Y_0. By Lemma 2.9.6, E extends along the morphism $\overline{f}$ if one replaces the base T by a suitable étale finite covering T' and can be contracted simultaneously with other (-1)-curves contained in the fibers $\overline{Y}_t$ $(t \in T)$. Note that this étale finite change of the base curve does not affect the properties of the fiber surfaces. Hence we may assume that all simultaneous blowing-ups and contractions as applied below are achieved over the base T.

The contraction is performed either within the boundary divisor S or the *simultaneous half-point detachments* in the respective fibers Y_t for $t \in T$. (For the definition of half-point detachment (resp., attachment), see, for example, [56]). Hence the contraction does not change the hypothesis on the simple normal crossing of S and the intersection divisor $S \cdot \overline{Y}_t$. Thus we may assume that $\overline{\rho}_0$ is minimal. Since $B_0 \subsetneq \overline{B}_0$, a fiber of $\overline{\rho}_0$ is contained in a boundary component, say S_1. Then the intersection $S_1 \cdot \overline{Y}_0$ as a cycle is a disjoint sum of the fibers of $\overline{\rho}_0$ with multiplicity one. Hence $(S_1^2 \cdot \overline{Y}_0) = ((S_1 \cdot \overline{Y}_0)^2)_{\overline{Y}_0} = 0$. Since $\overline{Y}_t$ and $\overline{Y}_0$ are algebraically equivalent, we have $(S_1^2 \cdot \overline{Y}_t) = 0$ for every $t \in T$. Note that $\overline{Y}_t$ is also a ruled surface by Iitaka [94] and minimal by the same reason as for $\overline{Y}_0$. Considering the deformations of a fiber of $\overline{\rho}_0$ appearing in $S_1 \cdot \overline{Y}_0$, we know by Lemma 2.9.6 that $S_1 \cdot \overline{Y}_t$ is a disjoint sum of smooth rational curves with self-intersection number zero. Namely, $S_1 \cdot \overline{Y}_t$ is a sum of the fibers of a $\mathbb{P}^1$-fibration. This implies that Y_t has an $\mathbb{A}^1$-fibration of affine type. Here we may have to replace the $\mathbb{P}^1$-fibration $\overline{\rho}_t$ by the second one if $\overline{Y}_t \cong \mathbb{P}^1 \times \mathbb{P}^1$. In fact, if a smooth complete surface has two different $\mathbb{P}^1$-fibrations and is minimal with respect to one fibration, then the surface is isomorphic to $\mathbb{P}^1 \times \mathbb{P}^1$ and two $\mathbb{P}^1$-fibrations are the vertical and horizontal fibrations.

Suppose next that Λ_0 has a base point, say P_0, and that the $\mathbb{A}^1$-fibration ρ_0 is of affine type. Then all irreducible components of $D_0 := S \cdot \overline{Y}_0$ are contained in the members of Λ_0. Since the boundary divisor D_0 of $\overline{Y}_0$ is assumed to be a connected divisor with simple normal crossings, there are at most two components of $S \cdot \overline{Y}_0$ passing through P_0, and if there are two of them, they lie on different components of S and P_0 lies on their intersection curve. In particular, if S_1 is a component of S containing P_0, then $S_1 \cdot \overline{Y}_0$ is a disjoint sum of smooth rational curves. Let C_1 be the component of $S_1 \cdot \overline{Y}_0$ passing through P_0 and let F_0 be the member of Λ_0 which contains C_1. We may assume that F_0 is supported by the boundary divisor D_0. If F_0 contains a (-1)-curve E such that $P_0 \notin E$, then E extends along the morphism $\overline{f}$ and can be contracted simultaneously along $\overline{f}$ after the base change by an étale finite covering $T' \to T$. So, we may assume that every irreducible component of F_0 not passing P_0 has self-intersection number ≤ -2 on $\overline{Y}_0$. Then we may assume that $(C_1^2)_{\overline{Y}_0} \ge 0$. In fact, if there are two irreducible components of $S \cdot \overline{Y}_0$ passing through P_0 and belonging to the same member F_0 of Λ_0, one of them must have self-intersection number ≥ 0, for otherwise all the components of the member of Λ_0, after the elimination of base points, would have self-intersection number ≤ -2, which is a contradiction. So, we may assume that the one on S_1, i. e., C_1, has self-intersection number ≥ 0. Then the proper transform of C_1 is the unique (-1)-curve with multiplicity > 1 in the fiber corresponding to F_0 after the elimination of base points of Λ_0.

On the other hand, $S_1 \cdot \overline{Y}_0$ (as well as $S_i \cdot \overline{Y}_0$ if it is nonempty) is a disjoint sum of smooth rational curves, one of which is C_1. Let $n := (C_1^2)_{\overline{Y}_0} \geq 0$. Then $\mathrm{Hilb}^P(\overline{Y})$ has dimension $n+2$ and is smooth at the point $[C_1]$. Since $C_1 \cong \mathbb{P}^1$ and $\mathrm{N}_{C_1/\overline{Y}} \cong \mathcal{O}(n) \oplus \mathcal{O}$, C_1 extends along the morphism $\overline{f}$. Namely, $\overline{f}|_{S_1} : S_1 \to T$ is a composite of a $\mathbb{P}^1$-fibration $\sigma_1 : S_1 \to T'$ and an étale finite morphism $\sigma_2 : T' \to T$, where C_1 is a fiber of σ_1. By the base change by σ_2, we may assume that $S_1 \cdot \overline{Y}_0 = C_1$. In particular, $(C_1^2)_{S_1} = 0$.

Suppose that C_2 is a component of F_0 meeting C_1. Then C_2 is contained in a different boundary component, say S_2, which intersects S_1. Since $(H \cdot S_2 \cdot \overline{Y}_0) > 0$, we have $(H \cdot S_2 \cdot \overline{Y}_t) > 0$ for every $t \in T$, where H is a relatively ample divisor on $\overline{Y}$ over T. Furthermore, $S_2 \cdot \overline{Y}_0$ is algebraically equivalent to $S_2 \cdot \overline{Y}_t$. Note that $S_2 \cdot \overline{Y}_0$ is a disjoint sum of smooth rational curves, one of which is the curve C_2 connected to C_1. By considering the factorization of $\overline{f}|_{S_2} : S_2 \to T$ into a product of a $\mathbb{P}^1$-fibration and an étale finite morphism as in the case for $S_1 \cdot \overline{Y}_0$ and taking the base change by an étale finite morphism, we may assume that $S_2 \cdot \overline{Y}_0 = C_2$. Hence we have

$$(S_1 \cdot S_2 \cdot \overline{Y}_t) = (S_1 \cdot S_2 \cdot \overline{Y}_0) = (C_1 \cdot (S_2 \cdot \overline{Y}_0))_{\overline{Y}_0} = (C_1 \cdot C_2)_{\overline{Y}_0} = 1.$$

This implies that $S_2 \cdot \overline{Y}_t$ is irreducible for a general point $t \in T$. For otherwise, by the Stein factorization of the morphism $\overline{f}|_{S_2} : S_2 \to T$, the fiber $S_2 \cdot \overline{Y}_t$ is a disjoint sum $A_1 + \cdots + A_s$ of distinct irreducible curves which are algebraically equivalent to each other on S_2. Since

$$\begin{aligned} 1 &= (S_1 \cdot S_2 \cdot \overline{Y}_t) = ((S_1 \cdot S_2) \cdot (S_2 \cdot \overline{Y}_t))_{S_2} \\ &= ((S_1 \cdot S_2) \cdot (A_1 + \cdots + A_s))_{S_2} = s((S_1 \cdot S_2) \cdot A_1), \end{aligned}$$

we have $s = 1$ and $(S_1 \cdot S_2 \cdot A_1) = 1$. So, $\overline{f}|_{S_2} : S_2 \to T$ is now a $\mathbb{P}^1$-bundle and $(C_2^2)_{S_2} = 0$. This implies that $\mathrm{N}_{C_2/\overline{Y}} \cong \mathcal{O}(m) \oplus \mathcal{O}$ with $m = (C_2^2)_{\overline{Y}_0} \leq -2$ and that C_2 extends along the morphism $\overline{f}$. We can argue in the same way as above with irreducible components of F_0 other than C_1.

Assume that no members of Λ_0 except F_0 have irreducible components outside of Y_0. If C_i is shown to move on the component S_i along the morphism $\overline{f}$, we consider a component C_{i+1} anew which meets C_i. Each of them is contained in a distinct irreducible boundary component of S and extends along the morphism $\overline{f}$. Let $S_1, S_2, \ldots, S_r$ be all the boundary components which meet $\overline{Y}_0$ along the irreducible components of

F_0. Then $\overline{Y}_t$ intersects $S_1 + S_2 + \cdots + S_r$ in an effective divisor which has the same form as F_0. Furthermore, we have

$$((S_i \cdot \overline{Y}_t)^2)_{\overline{Y}_t} = (S_i^2 \cdot \overline{Y}_t) = (S_i^2 \cdot \overline{Y}_0) = ((S_i \cdot \overline{Y}_0)^2)_{\overline{Y}_0}$$

for $1 \le i \le r$. Namely, the components $S_i \cdot \overline{Y}_t$ $(1 \le i \le r)$ with the same multiplicities as $S_i \cdot \overline{Y}_0$ in F_0 is a member F_t of the pencil Λ_t lying outside of Y_t. This implies that Λ_t has a base point P_t and at least one member of Λ_t lies outside of Y_t. So, the $\mathbb{A}^1$-fibration ρ_t on Y_t is of affine type.

If the pencil Λ_0 contains two members F_0, F_0' such that the components C_1, C_1' of F_0, F_0' lie outside of Y_0 and pass through the point P_0, we may assume that F_0 is supported by the boundary components, while F_0' may not. Then no other members of Λ_0 have irreducible components outside of Y_0 because $\overline{Y}_0 \setminus Y_0$ is connected. We can argue as above to show, after a suitable étale finite base change, that the member F_0 moves along the morphism $\overline{f}$, and further that every boundary component of F_0' moves on a boundary component, say S_j', as a fiber of $f|_{S_j'} : S_j' \to T$. Hence the pencil Λ_t has the member F_t corresponding to F_0 whose all components lie outside of Y_t and the member F_t' corresponding to F_0'. In fact, the part of F_t' lying outside of Y_t is determined as above, but since $Y_t \cap F_t'$ is a disjoint union of the $\mathbb{A}^1$ which correspond to the (-1)-components of F_t' (the *half-point attachments*), the member F_t' is determined up to its weighted graph. This proof also implies that if ρ_0 is of complete type then ρ_t is of complete type for every $t \in T$. □

Remark 2.9.8.

(1) In the above proof of Lemma 2.9.7, the case where the pencil Λ_0 has a base point P_0 on one of the connected components $S_1 \cap \overline{Y}_0$, say C_1, there might exist a monodromy on $\overline{Y}$ which transform Λ_0 to a pencil Λ_0' on $\overline{Y}_0$ having a base point P_0' on a different connected component C_1' of $S_1 \cap \overline{Y}_0$. However, we have $(C_1^2)_{\overline{Y}_0} \ge 0$ as shown in the proof, and $(C_1^2) = (C_1'^2)$. Since C_1' is contained in a member of Λ_0, whence $(C_1'^2) < 0$. This is a contradiction. So, $S_1 \cap \overline{Y}_0$ is irreducible.

(2) In the step of the above proof of Lemma 2.9.7 where we assume that no members of Λ_0 except F_0 have irreducible components outside of Y_0, let P_t' be a point on $C_{1,t} := S_1 \cdot \overline{Y}_t$ other than P_t which is the base point of the given pencil Λ_t. Then there is a pencil Λ_t' on $\overline{Y}_t$ which is similar to Λ_t. In fact, note first that $\overline{Y}_t$ is a rational surface. Perform the same blowing-ups with centers at P_t' and its infinitely near points as those with centers at P_t and its infinitely near points which eliminate the base points of Λ_t. Then we find an effective divisor $\widetilde{F}_t'$ supported by the proper transforms of $S_i \cdot \overline{Y}_t$ $(1 \le i \le r)$ and the exceptional curves of the blowing-ups such that $\widetilde{F}_t'$ has the same form and multiplicities as the corresponding member $\widetilde{F}_t$ in the proper transform $\widetilde{\Lambda}_t$ of Λ_t after the elimination of base points. Then $(\widetilde{F}_t')^2 = 0$ and hence $\widetilde{F}_t'$ is a fiber of an $\mathbb{P}^1$-fibration on the blown-up surface of $\overline{Y}_t$. Then the fibers of the $\mathbb{P}^1$-fibration form the pencil Λ_t' on $\overline{Y}_t$ after the reversed contractions.

In fact, the surface $Y_t = \overline{Y}_t \setminus D_t$ is the affine plane with two systems of coordinate lines given as the fibers of Λ_t and Λ'_t. Hence the $\mathbb{A}^1$-fibrations induced by Λ_t and Λ'_t are transformed by an automorphism of $\overline{Y}_t$. □

The following is one of the simplest examples of our situation.

Example 2.9.9. Let C be a smooth conic in $\mathbb{P}^2$ and let S be the subvariety of codimension one in $\mathbb{P}^2 \times C$ defined by

$$S = \{(P, Q) \mid P \in L_Q,\ Q \in C\},$$

where L_Q is the tangent line of C at Q. Let $Y = (\mathbb{P}^2 \times C) \setminus S$ and let $f : Y \to C$ be the projection onto C. We set $T = C$ to fit the previous notations. Set $\overline{Y} = \mathbb{P}^2 \times C$. Then $\overline{f} : \overline{Y} \to T$ is the second projection and the boundary divisor S is irreducible. For every point $Q \in C$, $Y_Q := \mathbb{P}^2 \setminus L_Q$ has a linear pencil Λ_Q generated by C and $2L_Q$, which induces an $\mathbb{A}^1$-fibration of affine type. The restriction $\overline{f}|_S : S \to T$ is a $\mathbb{P}^1$-bundle. Let C be defined by $X_0X_2 = X_1^2$ with respect to a system of homogeneous coordinates (X_0, X_1, X_2) of $\mathbb{P}^2$ and let $\eta = (1, t, t^2)$ be the generic point of C with t an inhomogeneous coordinate on $C \cong \mathbb{P}^1$. Then L_η is defined by $t^2X_0 - 2tX_1 + X_2 = 0$. The generic fiber Y_η of f has an $\mathbb{A}^1$-fibration induced by the linear pencil Λ_η whose general members are the conics defined by $(X_0X_2 - X_1^2) + u(t^2X_0 - 2tX_1 + X_2)^2 = 0$, where $u \in \mathbb{A}^1$. Indeed, the conics are isomorphic to $\mathbb{P}^1_{k(t)}$ since they have the $k(t)$-rational point $(1, t, t^2)$, and Y_η is isomorphic to $\mathbb{A}^2_{k(t)}$. This implies that the affine threefold Y itself has an $\mathbb{A}^1$-fibration. □

In the course of the proof of Lemma 2.9.7, we frequently used the base change by a finite étale morphism $\sigma : T' \to T$, where T' is taken in such a way that for every $t \in T$, the points $\sigma^{-1}(t)$ correspond bijectively to the connected components of $S_i \cap \overline{Y}_t$, where S_i is an irreducible component of S. Suppose that $\deg \sigma > 1$. Let $\overline{Y}' = \overline{Y} \times_T T'$ and $\overline{f}' = \overline{f} \times_T T'$. Note that $\overline{Y}'$ is smooth because σ is étale. The morphism $\sigma : T' \to T$ gives the Stein factorization $\overline{f}|_{S_i} : S_i \xrightarrow{\varphi} T' \xrightarrow{\sigma} T$. Then the subvariety S_i is considered to be a subvariety of $\overline{Y}'$ via a closed immersion $(\mathrm{id}_{S_i}, \varphi) : S_i \to S_i \times_T T' \hookrightarrow \overline{Y} \times_T T'$. We denote it by S'_i. Let t_1, t_2 be points of T' such that they correspond to the connected components A, B of $S_i \cap \overline{Y}_t$, whence $\sigma(t_1) = \sigma(t_2) = t$. Then A, B are the fibers of S'_i over the points t_1, t_2 of T'. Hence A and B are algebraically equivalent in $\overline{Y}$. Since T' is étale over T, we say more precisely that they are *étale-algebraically equivalent*. We have $(A^2)_{\overline{Y}_t} = (B^2)_{\overline{Y}_t}$. In fact, noting that $\overline{Y}'_{t_1}$ and $\overline{Y}'_{t_2}$ are algebraically equivalent in $\overline{Y}'$ and that $\overline{Y}'_{t_1}$ and $\overline{Y}'_{t_2}$ are isomorphic to $\overline{Y}_t$, we have

$$\begin{aligned}(A^2)_{\overline{Y}_t} &= (A^2)_{\overline{Y}'_{t_1}} = (S'_i \cdot S'_i \cdot \overline{Y}'_{t_1}) \\ &= (S'_i \cdot S'_i \cdot \overline{Y}'_{t_2}) = (B^2)_{\overline{Y}'_{t_2}} = (B^2)_{\overline{Y}_t}.\end{aligned}$$

Let C be an irreducible curve in $\overline{Y}_0 \cap S$, say a connected component of $\overline{Y}_0 \cap S_1$ with an irreducible component S_1 of S. We say that C has *no monodromy* in $\overline{Y}$ if $\overline{f}|_{S_1} : S_1 \to T$

has no splitting $\overline{f}|_{S_1} : S_1 \xrightarrow{\sigma_1} T' \xrightarrow{\sigma_2} T$, where σ_2 is an étale finite morphism with $\deg \sigma_2 > 1$. Note that, after a suitable étale finite base change $\overline{Y} \times_T T'$, this condition is fulfilled. Namely, *the monodromy is killed*. Concerning the extra hypothesis in Lemma 2.9.6(3) and the possibility of achieving the contractions over the base curve T in Lemma 2.9.6(5), we have the following result.

Lemma 2.9.10. *Let $(Y, \overline{Y}, S, \overline{f}, t_0)$ be a family of logarithmic deformation of the triple $(Y_0, \overline{Y}_0, D_0)$. Assume that Y_0 has an $\mathbb{A}^1$-fibration of affine type. Let Λ_0 be the pencil on $\overline{Y}_0$ whose general members are the closures of fibers of the $\mathbb{A}^1$-fibration. Suppose that Λ_0 defines a $\mathbb{P}^1$-fibration $\varphi_0 : \overline{Y}_0 \to \overline{B}_0$. Suppose further that the section of φ_0 in $S \cap \overline{Y}_0$ has no monodromy in $\overline{Y}$.*[17] *Then the following assertions hold:*

(1) *If C is a fiber of φ_0 with $C \cap Y_0 \neq \emptyset$ and C' is a smooth rational complete curve in $\overline{Y}_0$ which is algebraically equivalent to C in $\overline{Y}$, then $(C \cdot C') = 0$.*
(2) *There are no two (-1)-curves E_1 and E_2 on $\overline{Y}_0$ such that they belong to the same connected component of the Hilbert scheme* $\mathrm{Hilb}(\overline{Y})$, *$E_1$ is an irreducible component of a fiber of φ_0 and $E_1 \cap E_2 \neq \emptyset$.*

Proof. (1) Let S_0 be an irreducible component of S such that $(S_0 \cdot F) = 1$ for a general fiber F of φ_0. Then $S_0 \cap \overline{Y}_0$ contains a cross-section of φ_0. The assumption on the absence of the monodromy implies that $S_0 \cap \overline{Y}_0$ is irreducible and is the section of φ_0. Note that φ_0 contains a fiber F_∞ at infinity which is supported by the intersection of $\overline{Y}_0$ with the boundary divisor S in $\overline{Y}$. Such a fiber exists by the assumption that the $\mathbb{A}^1$-fibration on Y_0 is of affine type. Since $S_0 \cap \overline{Y}_0$ gives the cross-section, F_∞ is supported by $S \setminus S_0$. Since $C \cap (S \setminus S_0) = \emptyset$ and C' is algebraically equivalent to C in $\overline{Y}$, C' does not meet the components of $S \setminus S_0$. Hence $C' \cap F_\infty = \emptyset$, and C' is a component of a fiber of φ_0. So, $(C \cdot C') = 0$.[18]

(2) Suppose that such E_1 and E_2 exist. Since E_1 and E_2 are algebraically equivalent 1-cycles on $\overline{Y}$, E_1 and E_2 have the same intersections with subvarieties of codimension one in $\overline{Y}$. We consider possible cases separately:

(i) Suppose that both E_1 and E_2 are contained in the fiber at infinity F_∞. Since $E_1 \cap E_2 \neq \emptyset$, it follows that $F_\infty = E_1 + E_2$ with $(E_1 \cdot E_2) = 1$. If E_1 meets the section $S_0 \cap \overline{Y}_0$, then $(E_1 \cdot S_0) = 1$, whence $(E_2 \cdot S_0) = 1$ because E_1 and E_2 are algebraically equivalent in $\overline{Y}$. This is a contradiction. Hence $E_1 \cap E_2 = \emptyset$.

(ii) Suppose that only E_1 is contained in the fiber at infinity F_∞. Take a smooth fiber F_0 of φ_0 with $F_0 \cap Y_0 \neq \emptyset$ and consider a deformation of F_0 in $\overline{Y}$. Then there exist an étale finite morphism $\sigma_2 : T' \to T$ and a decomposition of $\overline{f}_{T'} : \overline{Y} \times_T T' \xrightarrow{\varphi} V \xrightarrow{\sigma_1} T'$

17 Intuitively speaking, if one moves the fiber $\overline{Y}_t$ along a closed loop γ in the Riemann surface T passing through the point t_0, the section of φ in $S \cap \overline{Y}$ might not come back to the section we started with. So the assumption requires that this occurs indeed.

18 We note here that without the condition on the absence of the monodromy of the cross-section, the assertion fails to hold. See Proposition 2.9.11.

such that F_0 is a fiber of φ (see Lemma 2.9.6(3)). Let B be an irreducible curve on V such that $\varphi(F_\infty) \notin B$ and let $W = \varphi^{-1}(B)$. Note that E_1 and E_2 are also algebraically equivalent in $\overline{Y} \times_T T'$. Since $(E_1 \cdot W) = 0$ by the above construction, it follows that $(E_2 \cdot W) = 0$. This implies that E_2 is contained in a fiber of φ_0. Hence $E_1 \cap E_2 = \emptyset$.

(iii) Suppose that E_1 and E_2 are not contained in the fiber F_∞. Then E_1 and E_2 are the fiber components of φ_0 because $(E_i \cdot F_\infty) = 0$ for $i = 1, 2$. If they belong to the same fiber, we obtain a contradiction by the same argument as in the case (i). If they belong to different fibers, then $E_1 \cap E_2 = \emptyset$. □

The following result shows that Lemma 2.9.10(1) does not hold without the monodromy condition on the section of φ_0.

Proposition 2.9.11. *Let $Q = \mathbb{P}^1 \times \mathbb{P}^1$ and $T' = \mathbb{A}^1_*$ which is the affine line minus one point and hence is the underlying scheme of the multiplicative group G_m. We denote by ℓ (resp., M) a general fiber of the first projection $p_1 : Q \to \mathbb{P}^1$ (resp., the second projection $p_2 : Q \to \mathbb{P}^1$). Let x (resp., y) be an inhomogeneous coordinate on the first (resp., the second) factor of Q. Set $\ell_\infty = p_1^{-1}(\infty)$ and $M_\infty = p_2^{-1}(\infty)$. We consider an involution ι on $Q \times T'$ defined by $(x, y, z) \mapsto (y, x, -z)$, where z is a coordinate of $\mathbb{A}^1_*$. Let Q' be the blowing-up of Q with center $P_\infty := \ell_\infty \cap M_\infty$ and let E be the exceptional curve. Then the involution ι extends to the threefold $Q' \times T'$ in such a way that $E \times T'$ is stable under ι. Let $\overline{Y}$ be the quotient threefold of $Q' \times T'$ by this $\mathbb{Z}_2$-action induced by the involution ι. Since the projection $p_2 : Q' \times T' \to T'$ is $\mathbb{Z}_2$-equivariant, it induces a morphism $\overline{f} : \overline{Y} \to T$, where $T = T'//\mathbb{Z}_2 \cong \mathbb{A}^1_*$. Let $S_1 = ((\ell_\infty \cup M_\infty) \times T')//\mathbb{Z}_2$, $S_2 = (E \times T')//\mathbb{Z}_2$ and $S = S_1 + S_2$. Further, we let $Y = \overline{Y} \setminus S$ and $f : Y \to T$ the restriction of $\overline{f}$ onto Y. Then the following assertions hold:*

1. *The surfaces S_1 and S_2 are smooth irreducible surfaces intersecting normally.*
2. *Fix a point $t_0 \in T$ and denote the fibers over t_0 with the subscript 0. Then the collection $(Y, \overline{Y}, S, \overline{f}, t_0)$ is a family of logarithmic deformations of the triple $(Y_0, \overline{Y}_0, D_0)$, where $D_0 = S \cdot \overline{Y}_0$.*
3. *For every $t \in T$, $S_1 \cap \overline{Y}_t$ is a disjoint union of two smooth curves C_{1t}, C'_{1t} and $S_2 \cap \overline{Y}_t$ is a smooth rational curve C_{2t}, where $(C_{1t} \cdot C_{2t}) = (C'_{1t} \cdot C_{2t}) = 1$ and $(C_{1t}^2) = ({C'_{1t}}^2) = (C_{2t}^2) = -1$. In particular, C_{1t} is étale-algebraically equivalent to C'_{1t}, and hence has a nontrivial monodromy.*
4. *Each fiber $\overline{Y}_t$ is isomorphic to Q' with C_{1t}, C'_{1t} and C_{2t} identified with the proper transforms of M_∞, ℓ_∞ on Q' and E.*
5. *Let $\varphi_t : \overline{Y}_t \to \mathbb{P}_1$ be the $\mathbb{P}^1$-fibration induced by the first projection $p_1 : Q \to \mathbb{P}^1$. Then a general fiber $\ell = p_1^{-1}(x)$ is algebraically equivalent to $M = p_2^{-1}(x)$ for $x \in T$.*
6. *For every $t \in T$, the affine surface Y_t has an $\mathbb{A}^1$-fibration of affine type.*

Proof. (1) Since $(Q \setminus (\ell_\infty \cup M_\infty) = \operatorname{Spec} k[x, y]$, the quotient threefold $V = (Q \times T')//\mathbb{Z}_2$ contains an open set $(\mathbb{A}^2 \times T')//\mathbb{Z}_2$, which has the coordinate ring over k generated by elements $X = x + y$, $U = xy$, $Z = z^2$, and $W = (x - y)z$. Hence the open set is a hypersurface $W^2 = Z(X^2 - 4U)$. The quotient threefold V has a similar open neighborhood of the image of the curve $\{P_\infty\} \times T'$. This can be observed by taking inhomogeneous coordinates x', y' on Q such that $x' = 1/x$ and $y' = 1/y$, where $\ell_\infty \cup M_\infty$ is given by $x'y' = 0$. If we put $W' = x' + y'$, $U' = x'y'$ and $W' = (x' - y')z$, the open neighborhood is defined by a similar equation $W'^2 = Z(X'^2 - 4U')$. Then the image of $(\ell_\infty \cup M_\infty) \times T'$ is given by $U' = 0$. Hence it has an equation $W'^2 = ZX'^2$. So, this is a smooth irreducible surface. The curve E has inhomogeneous coordinate x'/y' (or y'/x'). Hence E is stable under the involution ι. Note that the involution ι has no fixed point because there are no fixed points on the factor T'. The surface S_1 is simultaneously contracted along T, and by the contraction, $\overline{Y}$ becomes a $\mathbb{P}^2$-bundle and the surface S_2 becomes an immersed $\mathbb{P}^1$-bundle. Then assertion (1) follows easily.

(2) The threefold $\overline{Y}$ is smooth and $\overline{f}$ is a smooth morphism. In fact, every closed fiber of $f = \overline{f}|_Y : Y \to T$ is isomorphic to the affine plane.

(3) If $t = z^2$, C_{1t} (resp., C'_{1t}) is identified with M'_∞ (resp., ℓ'_∞) in $Q' \times \{z\}$ and ℓ'_∞ (resp. M'_∞) in $Q' \times \{-z\}$ under the identification $\overline{Y}_t \cong Q' \times \{z\} \cong Q' \times \{-z\}$, where ℓ'_∞ and M'_∞ are the proper transforms of ℓ_∞ and M_∞ on Q'. Now the rest of the assertions are easily verified. □

A sufficient condition on the absence of the monodromy in Lemma 2.9.10 is given by the following result.

Lemma 2.9.12. *Let the notations and the assumptions be the same as in Lemma 2.9.10 and its proof. Let $S_0 \cap \overline{Y}_0 = C_{01} \cup \cdots \cup C_{0m}$. Suppose that C_{01} is a section of the $\mathbb{P}^1$-fibration φ_0. If $(C_{01}^2) \geq 0$, then C_{01} has no monodromy in $\overline{Y}$. Namely, $m = 1$ and $S_0 \cap \overline{Y}_0$ is irreducible.*

Proof. Suppose that $m > 1$. Note that $C_{02}, \ldots, C_{0m}$ are mutually disjoint and do not meet a general fiber of φ_0 because they lie outside Y_0 and a general fiber meets only C_{01} in the boundary at infinity. This implies that $C_{02}, \ldots, C_{0m}$ are rational curves and the fiber components of φ_0. By the remark given before Lemma 2.9.10, we have $(C_{0i}^2) = (C_{01}^2) \geq 0$ for $2 \leq i \leq m$. Then $C_{02}, \ldots, C_{0m}$ are full fibers of φ_0 and hence they meet the section C_{01}. This is a contradiction. □

We prove one of our main theorems.

Theorem 2.9.13. *Let $f : Y \to T$ be a morphism from a smooth affine threefold onto a smooth curve T with irreducible general fibers. Assume that general fibers of f have $\mathbb{A}^1$-fibrations of affine type. Then, after shrinking T if necessary and taking an étale finite morphism $T' \to T$, the fiber product $Y' = Y \times_T T'$ has an $\mathbb{A}^1$-fibration which factors the morphism $f' = f \times_T T'$. Indeed, suppose that there is a relative normal completion $\overline{f} : \overline{Y} \to T$ of $f : Y \to T$ satisfying the following conditions:*

(1) *$(Y, \overline{Y}, S, \overline{f}, t_0)$ with $t_0 \in T$ and $S = \overline{Y} \setminus Y$ is a family of logarithmic deformations of $(Y_0, \overline{Y}_0, D_0)$ as above, where $Y_0 = f^{-1}(t_0)$, $\overline{Y}_0 = \overline{f}^{-1}(t_0)$, and $D_0 = S \cdot \overline{Y}_0$.*

(2) *The given $\mathbb{A}^1$-fibration of affine type on each fiber Y_t extends to a $\mathbb{P}^1$-fibration $\varphi_t : \overline{Y}_t \to \overline{B}_t$.*
(3) *A section of φ_0 in the fiber $\overline{Y}_0$ lying in D_0 has no monodromy in $\overline{Y}$.*

Then the given morphism $f : Y \to T$ is factored by an $\mathbb{A}^1$-fibration.

Proof. Embed Y into a smooth threefold $\overline{Y}$ in such a way that f extends to a projective morphism $\overline{f} : \overline{Y} \to T$. We may assume that the complement $S := \overline{Y} \setminus Y$ is a reduced divisor with simple normal crossings. Let $S = S_0 + S_1 + \cdots + S_r$ be the irreducible decomposition of S. For a general point $t \in T$, let Y_t be the fiber $f^{-1}(t)$ and let $\rho_t : Y_t \to B_t$ be the given $\mathbb{A}^1$-fibration on Y_t. By the assumption, B_t is an affine curve. We may assume that Y_t is smooth and hence B_t is smooth. Let $\overline{Y}_t$ be the closure of Y_t in $\overline{Y}$ which we may assume to be a smooth projective surface with t a general point of T. By replacing T by a smaller Zariski open set, we may assume that $\overline{f}$ is a smooth morphism and that $S \cdot \overline{Y}_t$ is a divisor with simple normal crossings for every $t \in T$. Hence we may assume that condition (1) above is realized.

For each $t \in T$, let Λ_t be the pencil generated by the closures (in $\overline{Y}_t$) of the fibers of the $\mathbb{A}^1$-fibration ρ_t. If Λ_t has a base point, we can eliminate the base points by simultaneous blowing-ups on the boundary at infinity after an étale finite base change of T. In this step, we may have to replace, for some $t \in T$, the pencil Λ_t by another pencil Λ'_t which also induces an $\mathbb{A}^1$-fibration of affine type on Y_t (see the proof of Lemma 2.9.7). So, we may assume that condition (2) above is also satisfied.

If $S_0 \cap \overline{Y}_0$ contains a section of φ_0, we may assume by an étale finite base change that $S_0 \cap \overline{Y}_0$ is irreducible (see the remark before Lemma 2.9.10). So, we may assume that condition (3) is satisfied as well.

Hence, we may assume from the beginning that three conditions are satisfied. The fibration ρ_t extends to a $\mathbb{P}^1$-fibration $\varphi_t : \overline{Y}_t \to \overline{B}_t$ for every $t \in T$, where $\overline{B}_t$ is a smooth completion of B_t. For $t_0 \in T$, we consider the fibration $\varphi_0 : \overline{Y}_0 \to \overline{B}_0$. A general fiber of φ_0 meets one of the irreducible components, say S_0, of S in one point. Then so does every fiber of φ_0 because $S_0 \cdot \overline{Y}_0$ is an irreducible divisor on $\overline{Y}_0$ and the fibers of φ_0 are algebraically equivalent to each other on $\overline{Y}_0$. Hence $S_0 \cdot \overline{Y}_0$ is a section. We claim that:
(1) $\overline{Y}_t$ meets the component S_0 for every $t \in T$.
(2) After possibly switching the $\mathbb{A}^1$-fibrations if some Y_t has two $\mathbb{A}^1$-fibrations, we may assume that for every $t \in T$, the fibers of the $\mathbb{P}^1$-fibration φ_t on $\overline{Y}_t$ meet S_0 along a curve $\overline{A}_t$ such that $\overline{A}_t$ is a cross-section of φ_t and hence φ_t induces an isomorphism between $\overline{A}_t$ and $\overline{B}_t$.

In fact, for a relatively ample divisor H of $\overline{Y}$ over T, we have $(H \cdot S_0 \cdot \overline{Y}_0) > 0$, whence $(H \cdot S_0 \cdot \overline{Y}_t) > 0$ for every $t \in T$ because $\overline{Y}_t$ is algebraically equivalent to $\overline{Y}_0$. This implies assertion (1). To prove assertion (2), we consider the deformation of a smooth fiber C of φ_0 in $\overline{Y}_0$. Since general fibers Y_t of f have $\mathbb{A}^1$-fibrations of affine type, by Lemma 2.9.6(3) and Lemma 2.9.10(1), there is a $\mathbb{P}^1$-fibration $\varphi : \overline{Y} \to V$ such that C

is a fiber of φ. Then the restriction $\varphi|_{\overline{Y}_0}$ is the $\mathbb{P}^1$-fibration φ_0. For every $t \in T$, the restriction $\varphi|_{\overline{Y}_t}$ is a $\mathbb{P}^1$-fibration on $\overline{Y}_t$. If it is different from φ_t, we replace φ_t by $\varphi|_{\overline{Y}_t}$. Then $(S_0 \cdot C') = (S_0 \cdot C) = 1$ for a general fiber C' of φ_t because C' is algebraically equivalent to C. The assertion follows immediately.

With the notations in the proof of Lemma 2.9.6, the isomorphisms $\overline{A}_t \xrightarrow{\sim} V_t := \sigma^{-1}(t) \cong \overline{B}_t$ shows that the morphism

$$S_0 \hookrightarrow \overline{Y} \xrightarrow{\varphi} V \xrightarrow{\sigma} T$$

induces a birational T-morphism $S_0 \to V$ and S_0 is a cross-section of φ. It is clear that the boundary divisor S contains no other components which are horizontal to φ. Hence Y has an $\mathbb{A}^1$-fibration. □

As a consequence of Theorem 2.9.13, we have the following result.

Corollary 2.9.14. *Let $f : Y \to T$ be a smooth morphism from a smooth affine threefold Y to a smooth affine curve T. Assume that f has a relative projective completion $\overline{f} : \overline{Y} \to T$ which satisfies the same conditions on the boundary divisor S and the intersection of each fiber $\overline{Y}_t$ with S as set in Lemma 2.9.7. If a fiber Y_0 has a G_a-action, then there exists an étale finite morphism $T' \to T$ such that the threefold $Y' = Y \times_T T'$ has a G_a-action as a T'-scheme. Furthermore, if the relative completion $\overline{f} : \overline{Y} \to T$ is taken so that the three conditions in Theorem 2.9.13 are satisfied, the threefold Y itself has a G_a-action as a T-scheme.*

Proof. By Lemma 2.9.7, every fiber Y_t has an $\mathbb{A}^1$-fibration of affine type $\rho_t : Y_t \to B_t$, where B_t is an affine curve. As in the proof of Theorem 2.9.13, we may assume that the three conditions therein are satisfied. By the same theorem, Y has an $\mathbb{A}^1$-fibration $\rho : Y \to U$ such that f is factored as

$$f : Y \xrightarrow{\rho} U \xrightarrow{\sigma} T,$$

where $U_t := \sigma^{-1}(t) \cong B_t$ for every $t \in T$. Then U is an affine scheme after restricting T to a Zariski open set. Then Y has a G_a-action by Lemma 2.2.7. □

Given a smooth affine morphism $f : Y \to T$ from a smooth algebraic variety Y to a smooth curve T such that every *closed* fiber is isomorphic to the affine space $\mathbb{A}^n$ of fixed dimension, one can ask if the generic fiber of f is isomorphic to $\mathbb{A}^n$ over the function field $k(T)$. If this is the case with f, we say that *the generic triviality* holds for f. In the case $n = 2$, this holds by the following theorem. If the generic triviality for $n = 2$ holds for $f : Y \to T$ in the setup of Theorem 2.9.15, a theorem of Sathaye [204] shows that f is an $\mathbb{A}^2$-bundle in the sense of Zariski topology. So, this will give a different proof of Corollary 2.1.11.

Theorem 2.9.15. *Let $f : Y \to T$ be a smooth morphism from a smooth affine threefold Y to a smooth affine curve T. Assume that the fiber Y_t is isomorphic to $\mathbb{A}^2$ for every closed point of T. Then the generic fiber Y_η of f is isomorphic to the affine plane over the function field of T. Hence $f : Y \to T$ is an $\mathbb{A}^2$-bundle over T after replacing T by an open set if necessary.*

Before giving a proof, we prepare two lemmas where an integral k-scheme is a reduced and irreducible algebraic k-scheme and where a *separable K-form* of $\mathbb{A}^2$ over a field K is an algebraic variety X defined over K such that $X \otimes_K K'$ is K'-isomorphic to $\mathbb{A}^2$ for a separable algebraic extension K' of K (see Section 2.1).

Lemma 2.9.16. *Let $p : X \to T$ be a dominant morphism from an integral k-scheme X to an integral k-scheme T. Assume that the fiber X_t is an integral k-scheme for every closed point t of T. Then the generic fiber $X_\eta = X \times_T \operatorname{Spec} k(T)$ is geometrically integral $k(T)$-scheme.*

Proof. We have only to show that the extension of the function fields $k(X)/k(T)$ is a regular extension. Namely, $k(X)/k(T)$ is a separable extension, i. e., a separable algebraic extension of a purely transcendental extension of $k(T)$ and $k(T)$ is algebraically closed in $k(X)$. Since the characteristic of k is zero, it suffices to show that $k(T)$ is algebraically closed in $k(X)$. Suppose the contrary. Let K be the algebraic closure of $k(T)$ in $k(X)$, which is a finite algebraic extension of $k(T)$. Let T' be the normalization of T in K. Let $\nu : T' \to T$ be the normalization morphism which is a finite morphism. Then $p : X \to T$ splits as $p : X \xrightarrow{p'} T' \xrightarrow{\nu} T$, which is the Stein factorization. Then the fiber X_t is not irreducible for a general closed point $t \in T$, which is a contradiction to the hypothesis. □

The following result is due to Kambayashi [106].

Lemma 2.9.17. *Let X be a separable K-form of $\mathbb{A}^2$ for a field K. Then X is isomorphic to $\mathbb{A}^2$ over K.*

The proof uses the vanishing of the Galois cohomology $H^1(G, \operatorname{Aut}(\mathbb{A}^2))$, where G is a finite group and $\operatorname{Aut}(\mathbb{A}^2)$ is an amalgamated product of the affine transformation group and the group of de Jonquière transformations. The following proof of Theorem 2.9.15 uses a locally nilpotent derivation and hence is of purely algebraic nature.

Proof of Theorem 2.9.15. Every closed fiber Y_t has an $\mathbb{A}^1$-fibration of affine type and hence a G_a-action. By Corollary 2.9.14, there exists an étale finite morphism $T' \to T$ such that $Y' = Y \times_T T'$ has a G_a-ation as a T'-scheme. Suppose that the generic fiber $Y'_{\eta'}$ of $f_{T'} : Y' \to T'$ is isomorphic to $\mathbb{A}^2$ over the function field $k(T')$. Since $Y'_{\eta'} = Y_\eta \otimes_{k(T)} k(T')$, it follows by Lemma 2.9.17 that Y_η is isomorphic to $\mathbb{A}^2$ over $k(T)$. Hence, we may assume from the beginning that Y has a G_a-action which induces $\mathbb{A}^1$-fibrations on general closed fibers Y_t. The G_a-action on a T-scheme Y is induced by a locally nilpotent derivation δ on the coordinate ring B of Y, i. e., $Y = \operatorname{Spec} B$. Let $T = \operatorname{Spec} R$.

Here δ is an R-trivial derivation on B. Let A be the kernel of δ. Since B is a smooth k-algebra of dimension 3, A is a finitely generated, normal k-algebra of dimension 2. The derivation δ induces a locally nilpotent derivation δ_t on $B_t = B\otimes_R R/\mathfrak{m}_t$, where $\mathfrak{m}_t$ is the maximal ideal of R corresponding to a general point t of T. We assume that $\delta_t \neq 0$. Since B_t is a polynomial k-algebra of dimension 2 by the hypothesis, $A_t := \operatorname{Ker} \delta_t$ is a polynomial ring of dimension 1. In fact, the kernel of a nontrivial locally nilpotent derivation on a polynomial ring of dimension 2 is a polynomial ring of dimension 1.

Claim 1. *$A_t = A \otimes_R R/\mathfrak{m}_t$ if δ_t is nonzero.*

Proof. Let $\varphi : B \to B[u]$ be the k-algebra homomorphism defined by

$$\varphi(b) = \sum_{i\geq 0} \frac{1}{i!}\delta^i(b)u^i.$$

Then $\operatorname{Ker} \delta = \operatorname{Ker}(\varphi - \mathrm{id}\,)$. Hence we have an exact sequence of R-modules

$$0 \to A \to B \overset{\varphi-\mathrm{id}}{\longrightarrow} B[u].$$

Let $\mathcal{O}_t$ be the local ring of T at t, i. e., the localization of R with respect to $\mathfrak{m}_t$, and let $\widehat{\mathcal{O}}_t$ be the $\mathfrak{m}_t$-adic completion of $\mathcal{O}_t$. Since $\widehat{\mathcal{O}}_t$ is a flat R-module, we have an exact sequence

$$0 \to A \otimes_R \widehat{\mathcal{O}}_t \to B \otimes_R \widehat{\mathcal{O}}_t \to (B \otimes_R \widehat{\mathcal{O}}_t)[u]. \tag{2.14}$$

The completion $\widehat{\mathcal{O}}_t$ as a k-module decomposes as $\widehat{\mathcal{O}}_t = k \oplus \widehat{\mathfrak{m}}_t$, where $\widehat{\mathfrak{m}}_t = \mathfrak{m}_t\widehat{\mathcal{O}}_t$, and the above exact sequence splits as a direct sum of exact sequences of k-modules

$$0 \to A \otimes_R k \to B \otimes_R k \to (B \otimes_R k)[u],$$
$$0 \to A \otimes_R \widehat{\mathfrak{m}}_t \to B \otimes_R \widehat{\mathfrak{m}}_t \to (B \otimes_R \widehat{\mathfrak{m}}_t)[u].$$

The first one is, in fact, equal to

$$0 \to A \otimes_R R/\mathfrak{m}_t \to B_t \overset{\varphi_t-\mathrm{id}}{\longrightarrow} B_t[u],$$

where φ_t is defined by δ_t in the same way as φ by δ. Hence $\operatorname{Ker} \delta_t = A\otimes_R R/\mathfrak{m}_t = A_t$. □

Let $X = \operatorname{Spec} A$ and let $p : X \to T$ be the morphism induced by the inclusion $R \hookrightarrow A$. Thus $f : Y \to T$ splits as

$$f : Y \overset{q}{\longrightarrow} X \overset{p}{\longrightarrow} T,$$

where q is the quotient morphism by the induced G_a-action on Y.

Claim 2. *Suppose that $\delta_t \neq 0$ for every $t \in T$. Then X is a smooth surface with $\mathbb{A}^1$-bundle structure over T.*

Proof. Note that R is a Dedekind domain and A is an integral domain. Hence p is a flat morphism. Since f is surjective, p is also surjective. Hence p is a faithfully flat morphism. Further, by Claim 1, $X_t = \operatorname{Spec}(A \otimes_R R/\mathfrak{m}_t)$ is equal to $\operatorname{Spec} A_t$ for every t, which is isomorphic to $\mathbb{A}^1$. The generic fiber of p is geometrically integral by Lemma 2.9.16. Hence, by Lemma 2.1.3, X is an $\mathbb{A}^1$-bundle over T. In particular, X is smooth. □

Let $K = k(T)$ be the function field of T. The generic fiber $X_K = X \times_T \operatorname{Spec} K$ is geometrically integral as shown in the above proof of Claim 2.

Claim 3. *The generic fiber $Y_K = Y \times_T \operatorname{Spec} K$ is isomorphic to $\mathbb{A}^2_K$.*

Proof. We consider $q_K : Y_K \to X_K$, where $X_K \cong \mathbb{A}^1_K$. We prove the following two assertions:

(1) For every closed point x of X_K, the fiber $Y_K \times_{X_K} \operatorname{Spec} K(x)$ is isomorphic to $\mathbb{A}^1_{K(x)}$.
(2) The generic fiber of q_K is geometrically integral.

Note that $K(x)$ is a finite algebraic extension of K. Let T' be the normalization of T in $K' := K(x)$. We consider $Y' := Y \times_T T'$ instead of Y. Then the G_a-action on Y lifts to Y' and the quotient variety is $X' = X \times_T T'$. Indeed, the normalization R' of R in K' is the coordinate ring of T' and is a flat R-module. Then the sequence of R'-modules

$$0 \to A \otimes_R R' \to B \otimes_R R' \xrightarrow{\varphi' - \mathrm{id}} (B \otimes_R R')[u]$$

is exact, where $\varphi' = \varphi \otimes_R R'$. Hence $q_{K'} : Y'_{K'} \to X'_{K'}$, which is the base change of q_K with respect to the field extension K'/K, is the quotient morphism by the G_a-action on $Y'_{K'}$ induced by δ. Since $X'_{K'} = X \times_T \operatorname{Spec} K'$, there exists a K'-rational point x' on $X'_{K'}$ such that x is the image of x' by the projection $X'_{K'} \to X_K$. If the fiber of $q_{K'}$ over x', i. e., $Y'_{K'} \times_{X'_{K'}} (\operatorname{Spec} K', x')$, is isomorphic to $\mathbb{A}^1_{K'}$, then $Y_K \times_{X_K} \operatorname{Spec} K'$, which is the fiber of q_K over the point x, is isomorphic to $\mathbb{A}^1_{K'}$ because $Y'_{K'} \times_{X'_{K'}} \operatorname{Spec} K' = Y_K \times_{X_K} \operatorname{Spec} K'$. Thus we may assume that x is a K-rational point. Let C be the closure of x in X. Then C is a cross-section of $p : X \to T$. Let $Z := Y \times_X C$. Then $q_C : Z \to C$ is a faithfully flat morphism such that the fiber $q_C^{-1}(w)$ is isomorphic to $\mathbb{A}^1$ for every closed point $w \in C$. In fact, $q_C^{-1}(w)$ is the fiber of $Y_t \to X_t$ over the point $w \in C$, where $t = p(w)$, $Y_t \cong \mathbb{A}^2, X_t \cong \mathbb{A}^1$ and $X_t = Y_t /\!/ G_a$. By Lemma 2.9.16 (which is extended to a nonclosed field K), the generic fiber of q_C is geometrically integral, and the generic fiber of q_C, which is $Y_K \times_{X_K} \operatorname{Spec} K(x)$, is isomorphic to $\mathbb{A}^1_{K'}$ by Lemma 2.1.3. This proves the first assertion.

The generic point of X_K corresponds to the quotient field $L := Q(A)$. Then it suffices to show that $B \otimes_A Q(A)$ is geometrically integral over $Q(A)$. Meanwhile, $B \otimes_A Q(A)$ has a locally nilpotent derivation $\delta \otimes_A Q(A)$ such that $\operatorname{Ker}(\delta \otimes_A Q(A)) = Q(A)$. Hence $B \otimes_A Q(A)$ is a polynomial ring $Q(A)[u]$ in one variable over $Q(A)$ because $\delta \otimes_A Q(A)$ has a slice. So, $B \otimes_A Q(A)$ is geometrically integral over $Q(A)$. □

Now, by Lemma 2.1.3, Y_K is an $\mathbb{A}^1$-bundle over $X_K \cong \mathbb{A}^1_K$. Hence Y_K is isomorphic to $\mathbb{A}^2_K$. We have to replace T by an open set $T \setminus F$, where $F = \{t \in T \mid \delta_t = 0\}$. This completes the proof of Theorem 2.9.15.

We can prove Theorem 2.9.15 in a more geometric way by making use of a theorem of Ramanujam–Morrow on the boundary divisor of a minimal normal completion of the affine plane [192, 170] (see Subsection 1.3.10). The proof given below is explained in more precise and explicit terms in [105, Lemma 3.2]. In particular, the step to show that $\overline{Y}_K \cong \mathbb{P}^2_K$ and $Y_K \cong \mathbb{A}^2_K$ is due to [105, Lemma 3.2]. □

The second proof of Theorem 2.9.15. Let $f : Y \to T$ be as in Theorem 2.9.15. Let $\overline{Y}$ be a relative completion such that $\overline{Y}$ is smooth and f extends to a smooth projective morphism $\overline{f} : \overline{Y} \to T$ with the conditions in Lemma 2.9.7 being satisfied together with $S := \overline{Y} \setminus Y$. To obtain this setting, we may have to shrink T to a smaller open set of T. As in the first proof and the proof of Lemma 2.9.7, we can apply an étale finite base change $T' \to T$ by which the intersection $S_i \cap \overline{Y}_t$ is irreducible for every irreducible component S_i of S and every $t \in T$. In particular, we assume that $\overline{Y}_t$ is a smooth normal completion of Y_t for every $t \in T$, where Y_t is isomorphic to $\mathbb{A}^2$. Fix one such completed fiber, say $\overline{Y}_0 = \overline{f}^{-1}(t_0)$, and consider the reduced effective divisor $D_0 := \overline{Y}_0 - Y_0$ with $Y_0 = f^{-1}(t_0) \cong \mathbb{A}^2$. Namely, $(Y, \overline{Y}, S, \overline{f}, t_0)$ is a log-deformation of $(Y_0, \overline{Y}_0, D_0)$. If the dual graph of this divisor is not linear then it contains a (-1)-curve meeting at most two other components of D_0 by a result of Ramanujam [192]. By (4) of Lemma 2.9.6, such a (-1)-curve deforms along the fibers of $\overline{f}$ and we get an irreducible component, say S_1, of $S = \sum_{i=0}^{r} S_i$ which can be contracted. Repeating this argument, we can assume that all the dual graphs for $\overline{Y}_t - Y_t$, as t varies on the set of closed points of T, are linear chains of smooth rational curves. By [170], at least one of these curves is a (0)-curve. Fix such a (0)-curve C_1 in $\overline{Y}_0 - Y_0$. Then C_1 deforms along the fibers of $\overline{f}$ and forms an irreducible component, say S_1, of S by abuse of notation. By the argument in the proof of Lemma 2.9.7, if C_2 is a component of $\overline{Y}_0 - Y_0$ meeting C_1, it deforms along the fibers of $\overline{f}$ on an irreducible component, say S_2, of S. Repeating this argument, we know that all irreducible components of $\overline{Y}_0 - Y_0$ extend along the fibers of $\overline{f}$ to form the irreducible components of S and that the dual graphs of $\overline{Y}_t - Y_t$ are the same for every $t \in T$. Now let K be the function field of T over k. We consider the generic fibers $\overline{Y}_K$ and Y_K of $\overline{f}$ and f. Then the dual graph of $\overline{Y}_K - Y_K$ is the same linear chain of smooth rational curves as the closed fibers $\overline{Y}_t \setminus Y_t$. Write $\overline{Y}_0 - Y_0 = \sum_{i=1}^{r} C_i$. If C_i and C_j meet for $i \neq j$, then the intersection point $C_i \cap C_j$ moves on the intersection curve $S_i \cdot S_j$. Since any minimal normal completion of $\mathbb{A}^2$ can be brought to $\mathbb{P}^2$ by blowing-ups and blowing-downs with centers on the boundary divisor, we can blow up simultaneously the intersection curves and blow down the proper transforms of the S_i according to the blowing-ups and blowing-downs on $\overline{Y}_0$. Here we note that the beginning center of the blowing-up is a point on a (0)-curve C_1. In this case, we choose a suitable cross-section on the irreducible component S_1 which is a $\mathbb{P}^1$-bundle in the Zariski topology because $\dim T = 1$. Note that if T is irrational, then the chosen cross-section may meet the intersection curves on S_1 with other components of S. Then we shrink T so that the cross-section does not meet the intersection curves. If T is rational, S_1 is a trivial $\mathbb{P}^1$-bundle, hence we do not need the procedure of shrinking T. Thus we may assume that, for every $t \in T$, $\overline{Y}_t$ is isomorphic to $\mathbb{P}^2$ and $\overline{Y}_t \setminus Y_t$ is a single curve C_t with $(C_t)^2 = 1$. This implies that $\overline{Y}_K \cong \mathbb{P}^2_K$ and $Y_K \cong \mathbb{A}^2_K$. □

In connection with Theorem 2.9.15, we can pose the following

Problem 2.2. Let K be a field of characteristic zero and let X be a smooth affine surface defined over K. Suppose that $X \otimes_K \overline{K}$ has an $\mathbb{A}^1$-fibration of affine type, where $\overline{K}$ is an algebraic closure of K. Does X then have an $\mathbb{A}^1$-fibration of affine type?

If we consider an $\mathbb{A}^1$-fibration of complete type, an example of Dubouloz–Kishimoto gives a counterexample to a similar problem for the complete type (see Theorem 2.9.24). In view of Proposition 2.9.11 and Theorem 2.9.13, we need perhaps some condition for a positive answer in the case of affine type which guarantees the absence of monodromy of a cross-section of a given $\mathbb{A}^1$-fibration.

2.9.4 Topological arguments instead of Hilbert schemes

In this section we will briefly indicate topological proofs of some of the results in the previous subsections. The use of topological arguments would make the cumbersome geometric arguments more transparent for readers who do not appreciate a heavy machinery like the Hilbert scheme.

We will use the following basic fibration theorem due to Ehresmann (see Theorem 1.1.5 and [227, Chapter V, Proposition 6.4]).

Theorem 2.9.18. *Let M be a connected differentiable manifold, S a closed submanifold, $f : M \to N$ a proper differentiable map such that the tangent maps corresponding to f and $f|_S : S \to N$ are surjective at any point in M and S. Then $f|_{M\setminus S} : M \setminus S \to N$ is a locally trivial fiber bundle with respect to the base N.*

Note that the normal bundle of any fiber of f is trivial. We can give a proof of Ehresmann's theorem using this observation, and the well-known result from differential topology that given a compact submanifold S of a C^∞ manifold X there are arbitrarily small tubular neighborhoods of S in X which are diffeomorphic to neighborhoods of S in the total space of normal bundle of S in X [19, Chapter II, Theorem 11.14].

Now let $\overline{f} : \overline{Y} \to T$ be a smooth projective morphism from a smooth algebraic threefold onto a smooth algebraic curve T. Let $\overline{Y}_t = \overline{f}^{-1}(t)$ be the fiber over $t \in T$. Let S be a simple normal crossing divisor on $\overline{Y}$ such that $D_t := S \cap \overline{Y}_t$ is a simple normal crossing divisor for each $t \in T$ and $Y_t := \overline{Y}_t \setminus D_t$ is affine for each $t \in T$.

We can assume that $\overline{f} : \overline{Y} \to T$ has the property that the tangent map is surjective at each point. It follows from Ehresmann's theorem that all the surfaces $\overline{Y}_t$ are mutually diffeomorphic. In particular, they have the same topological invariants like the fundamental group π_1 and the Betti number b_i. By shrinking T if necessary, we will assume that the restricted map $\overline{f} : S_i \to T$ is smooth for each i. For fixed i and t_0 the intersection $S_i \cap \overline{Y}_{t_0}$ is a disjoint union of smooth, compact, irreducible curves. Let $C_{t_0,i}$ be one of these irreducible curves. Then for each t which is close to t_0, there is an irreducible curve $C_{t,i}$ in $S_i \cap \overline{Y}_t$ and suitable tubular neighborhoods of $C_{t_0,i}$, $C_{t,i}$ in $\overline{Y}_{t_0}$, $\overline{Y}_t$, respectively, are diffeomorphic by Ehresmann's theorem. This implies that $C^2_{t_0,i}$ in $\overline{Y}_{t_0}$ and $C^2_{t,i}$ in $\overline{Y}_t$ are equal. This proves that the weighted dual graphs of the curves D_t in $\overline{Y}_t$ are the same for each $t \in T$.

Recall that if X is a smooth projective surface with a smooth rational curve $C \subset X$ such that $C^2 = 0$ then C is a fiber of a $\mathbb{P}^1$-fibration on X. If the irregularity $q(X) > 0$ then the Albanese morphism $X \to \mathrm{Alb}(X)$ gives a $\mathbb{P}^1$-fibration on X with C as a fiber. By the above discussion the fiber surfaces $\overline{Y}_t$ have the same irregularity.

Suppose that $\overline{Y}_0$ has an $\mathbb{A}^1$-fibration of affine type $f : Y_0 \to B$. If $\overline{f} : \overline{Y}_0 \to \overline{B}$ is an extension of f to a smooth completion of Y_0 then, after simultaneous blowing-ups and blowing-downs along the fibers of $\overline{f}$, we may assume that $D_0 := \overline{Y}_0 - Y_0$ contains at least one (0)-curve which is a tip, i. e., the end component of a maximal twig of D_0. Since D_t and D_0 have the same weighted dual graphs D_t also contains a (0)-curve which is a tip of D_t. Hence, Y_t also has an $\mathbb{A}^1$-fibration of affine type. This proves assertion (2) in Lemma 2.9.7.

We can also shorten the part of showing the invariance of the boundary weighted graphs in the second proof of Theorem 2.9.15. Suppose now that $f : Y \to T$ is a fibration on a smooth affine threefold Y onto a smooth curve T such that every scheme-theoretic fiber of f is isomorphic to $\mathbb{A}^2$. We can embed Y in a smooth projective threefold $\overline{Y}$ such that f extends to a morphism $\overline{f} : \overline{Y} \to T$. By shrinking T we can assume that $\overline{f}$ is smooth, each irreducible component S_i of $\overline{Y} - Y$ intersects each $\overline{Y}_t$ transversally. By the above discussions, each $D_t := \overline{Y}_t - Y_t$ has the same weighted dual graph. Since Y_t is isomorphic to $\mathbb{A}^2$, we can argue as in the second proof of Theorem 2.9.15 using the result of Ramanujam–Morrow to conclude that f is a trivial $\mathbb{A}^2$-bundle on a nonempty Zariski-open subset of T. This observation applies also to the proof of Theorem 2.9.23.

2.9.5 Deformations of ML_0-surfaces

In this subsection, we assume that the ground field k is the complex number field $\mathbb{C}$. Let $\mathcal{F} = (Y, \overline{Y}, S, \overline{f}, t_0)$ be a family satisfying the conditions of Lemma 2.9.7. Let $D_0 = S \cap \overline{Y}_0$.

Lemma 2.9.19. *Let $\mathcal{F} = (Y, \overline{Y}, S, \overline{f}, t_0)$ be a log-deformation of $(Y_0, \overline{Y}_0, D_0)$. Assume that D_0 is a tree of smooth rational curves satisfying one of the following conditions:*
(i) *D_0 contains an irreducible component C_1 such that $(C_1^2) \geq 0$.*
(ii) *D_0 contains a (−1)-curve C_1 which meets more than two other components of D_0.*

Then the following assertions hold after changing T by an étale finite covering of an open set of T if necessary.

(1) *Every irreducible component of D_0 deforms along the fibers of $\overline{f}$. Namely, if $D_0 = \sum_{i=1}^r C_i$ is the irreducible decomposition, then, for every $1 \le i \le r$, there exists an irreducible component S_i of S such that $\overline{f}|_{S_i} : S_i \to T$ has the fiber $(\overline{f}|_{S_i})^{-1}(t_0) = C_i$. Furthermore, $S = \sum_{i=1}^r S_i$.*

(2) *For $t \in T$, let $C_{i,t} = (\overline{f}|_{S_i})^{-1}(t)$. Then $D_t = \sum_{i=1}^r C_{i,t}$ and D_t has the same weighted graph on $\overline{Y}_t$ as D_0 does on $\overline{Y}_0$.*

(3) *For every i, $\overline{f}|_{S_i} : S_i \to T$ is a trivial $\mathbb{P}^1$-bundle over T.*

Proof. By a suitable étale finite base change of T, we may assume that $S_i \cap \overline{Y}_0$ is irreducible for every irreducible component S_i of S. Then the argument is analytic-locally almost the same as in the proof for assertion (2) of Lemma 2.9.7. Consider the deformation of C_1 along the fibers of $\overline{f}$, which moves along the fibers because $(C_1^2) \ge -1$. Then the components of D_0 which are adjacent to C_1 also move along the fibers of $\overline{f}$. Once these components of D_0 move, then the components adjacent to these components move along the fibers of $\overline{f}$. Since D_0 is connected because Y_0 is affine, all the components of D_0 move along the fibers of $\overline{f}$. If S contains an irreducible component which does not intersect $\overline{Y}_0$, it is a fiber component of $\overline{f}$. Then we remove the fiber by shrinking T. This proves assertion (1).

Let $S = \sum_{i=1}^r S_i$ be the irreducible decomposition of S. As shown in (1), $S_i \cap \overline{Y}_0 \neq \emptyset$ for every i. Then $S_i \cap \overline{Y}_t \neq \emptyset$ as well by the argument in the proof of Lemma 2.9.7.

Note that $((S_i \cdot \overline{Y}_t)^2)_{\overline{Y}_t} = (S_i^2 \cdot \overline{Y}_t) = (S_i^2 \cdot \overline{Y}_0) = ((S_i \cdot \overline{Y}_0)^2)_{\overline{Y}_0}$ because $\overline{Y}_t$ is algebraically equivalent to $\overline{Y}_0$. Hence D_0 and D_t have the same dual graphs. □

In order to prove the following result, we use Ehresmann's theorem, which is Theorem 2.9.18.

Lemma 2.9.20. *Let $\mathcal{F} = (Y, \overline{Y}, S, \overline{f}, t_0)$ be a log-deformation of $(Y_0, \overline{Y}_0, D_0)$ which satisfies the same conditions as in Lemma 2.9.19. Assume further that $p_g(\overline{Y}_0) = q(\overline{Y}_0) = 0$. Then the following assertions hold:*

(1) $\mathrm{Pic}(Y_t) \cong \mathrm{Pic}(Y_0)$ *for every $t \in T$.*

(2) $\Gamma(Y_t, \mathcal{O}_{Y_t}^*) \cong \Gamma(Y_0, \mathcal{O}_{Y_0}^*)$ *for every $t \in T$.*

Proof. Since p_g and q are deformation invariants, we have $p_g(\overline{Y}_t) = q(\overline{Y}_t) = 0$ for every $t \in T$. The exact sequence

$$0 \longrightarrow \mathbb{Z} \longrightarrow \mathcal{O}_{\overline{Y}_t} \xrightarrow{\exp} \mathcal{O}_{\overline{Y}_t}^* \longrightarrow 0$$

induces an exact sequence

$$H^1(\overline{Y}_t, \mathcal{O}_{\overline{Y}_t}) \to H^1(\overline{Y}_t, \mathcal{O}_{\overline{Y}_t}^*) \to H^2(\overline{Y}_t; \mathbb{Z}) \to H^2(\overline{Y}_t, \mathcal{O}_{\overline{Y}_t}).$$

Since $p_g(\overline{Y}_t) = q(\overline{Y}_t) = 0$, we have an isomorphism

$$H^1(\overline{Y}_t, \mathcal{O}^*_{\overline{Y}_t}) \cong H^2(\overline{Y}_t; \mathbb{Z}).$$

Now consider the canonical homomorphism $\theta_t : H_2(D_t; \mathbb{Z}) \to H_2(\overline{Y}_t; \mathbb{Z})$, where $H_2(\overline{Y}_t; \mathbb{Z}) \cong H^2(\overline{Y}_t; \mathbb{Z}) = \mathrm{Pic}(\overline{Y}_t)$ by the Poincaré duality. Then $\mathrm{coim}\,\theta_t = \mathrm{Pic}(Y_t)$ and $\mathrm{Ker}\,\theta_t = \Gamma(Y_t, \mathcal{O}^*_{Y_t})/k^*$, where $\mathrm{coim}\,\theta$ is the coimage of θ.

Let N be a nice tubular neighborhood of S with boundary in $\overline{Y}$. The smooth morphism $\overline{f} : \overline{Y} \to T$ together with its restriction on the pair $(N, \partial N)$ gives a proper differential mapping which is surjective and submersive. By Theorem 2.9.18, it is differentiably a locally trivial fibration. Namely, there exists a small disc U of t_0 in T and a diffeomorphism $\varphi_0 : \overline{Y}_0 \times U \xrightarrow{\approx} \overline{f}^{-1}(U)$ such that its restriction induces a diffeomorphism

$$\varphi_0 : (N \cap \overline{Y}_0) \times U \xrightarrow{\approx} (\overline{f}|_N)^{-1}(U).$$

For $t \in U$, noting that U is contractible and hence $H_2(\overline{Y}_0 \times U; \mathbb{Z}) = H_2(\overline{Y}_0; \mathbb{Z})$ and $H_2((N \cap \overline{Y}_0) \times U; \mathbb{Z}) = H_2(N \cap \overline{Y}_0; \mathbb{Z})$, the inclusions $\overline{Y}_t \hookrightarrow \overline{f}^{-1}(U)$ and $N \cap \overline{Y}_0 \hookrightarrow (\overline{f}|_N)^{-1}(U)$ induces compatible isomorphisms

$$p_t : H_2(\overline{Y}_t; \mathbb{Z}) \to H_2(\overline{f}^{-1}(U); \mathbb{Z}) \xrightarrow{(\varphi^{-1})_*} H_2(\overline{Y}_0 \times U; \mathbb{Z}) = H_2(\overline{Y}_0; \mathbb{Z})$$

and its restriction $q_t : H_2(N \cap \overline{Y}_t; \mathbb{Z}) \xrightarrow{\sim} H_2(N \cap \overline{Y}_0; \mathbb{Z})$. Since S and hence D_t are strong deformation retracts of N and $N \cap \overline{Y}_t$, respectively, the isomorphism q_t induces an isomorphism $r_t : H_2(D_t; \mathbb{Z}) \xrightarrow{\sim} H_2(D_0; \mathbb{Z})$ such that the following diagram is true

$$\begin{array}{ccc} H_2(D_t; \mathbb{Z}) & \xrightarrow{\theta_t} & H_2(\overline{Y}_t; \mathbb{Z}) \\ {\scriptstyle r_t}\downarrow & & \downarrow{\scriptstyle p_t} \\ H_2(D_0; \mathbb{Z}) & \xrightarrow[\theta_0]{} & H_2(\overline{Y}_0; \mathbb{Z}) \end{array}$$

This implies that $\mathrm{Pic}(Y_t) \cong \mathrm{Pic}(Y_0)$ and $\Gamma(Y_t, \mathcal{O}^*_{Y_t}) \cong \Gamma(Y_0, \mathcal{O}^*_{Y_0})$. If t is an arbitrary point of T, we choose a finite sequence of points $\{t_0, t_1, \dots, t_n = t\}$ such that t_i is in a small disc U_{i-1} around t_{i-1} $(1 \le i \le n)$ for which we can apply the above argument. □

Remark 2.9.21. By a result of W. Neumann [180, Theorem 5.1], if X is a normal affine surface, D an SNC divisor at infinity of X which does not contain any (-1)-curve meeting at least three other components of D and all whose maximal twigs are smooth rational curves with self-intersections ≤ -2, then the boundary 3-manifold of a nice tubular neighborhood N of D determines the dual graph of D. If we use the local differentiable triviality of a tubular neighborhood N, this result of Neumann shows that the weighted dual graph of D_t is deformation invariant.

We will consider a log deformation of an ML_0-surface.

Theorem 2.9.22. *Let $\mathcal{F} = (Y, \overline{Y}, S, \overline{f}, t_0)$ be a log-deformation of $(Y_0, \overline{Y}_0, D_0)$, where Y_0 is an* ML_0*-surface. Then Y_t is an* ML_0*-surface for every $t \in T$.*

Proof. If $S \cap \overline{Y}_t$ contains a (−1)-curve, then it deforms along the fibers of $\overline{f}$ after an étale finite base change of T, and these (−1)-curves are contracted simultaneously by Lemma 2.9.6. Hence we may assume that $\overline{Y}_t$ is a minimal normal completion of Y_t for every $t \in T$. By Theorem 2.6.2, $D_0 := S \cap \overline{Y}_0$ is a linear chain of smooth rational curves. Hence $D_t := S \cap \overline{Y}_t$ is also a linear chain of smooth rational curves. By Lemma 2.6.6, $\Gamma(Y_t, \mathcal{O}^*_{Y_t}) = k^*$ for every $t \in T$ because $\Gamma(Y_0, \mathcal{O}^*_{Y_0}) = k^*$. So, Y_t is an ML_0-surface by Theorem 2.6.2. □

A smooth affine surface X is, by definition, an *affine pseudoplane* if it has an $\mathbb{A}^1$-fibration of affine type $p : X \to \mathbb{A}^1$ admitting at most one multiple fiber of the form $m\mathbb{A}^1$ as a singular fiber (see [160] for the definition and relevant results). An affine pseudoplane is a $\mathbb{Q}$-homology plane, its Picard group is a cyclic group $\mathbb{Z}/m\mathbb{Z}$ and there are no nonconstant invertible elements. An ML_0-surface is an affine pseudoplane if the Picard number is zero by Lemma 2.6.11.

If $\overline{X}$ is a minimal normal completion of an affine pseudoplane X, the boundary divisor $D = \overline{X} - X$ is a tree of smooth rational curves, which is not necessarily a linear chain. By blowing-ups and blowing-downs with centers on the boundary divisor D, we can make the completion $\overline{X}$ satisfy the following conditions (see Lemma 2.6.8):

(i) There is a $\mathbb{P}^1$-fibration $\overline{p} : \overline{X} \to \mathbb{P}^1$ which extends the $\mathbb{A}^1$-fibration $p : X \to \mathbb{A}^1$.
(ii) The divisor D is written as $D = \ell + M + A$, where ℓ is an irreducible fiber of p, M is a cross-section of p with $(M^2) = 0$ and A is a tree meeting the section M.
(iii) There is a (−1)-curve F_0 (called *feather*) such that $F_0 \cap X \cong \mathbb{A}^1$ and the union $F_0 + A$ is contractible to a smooth rational curve meeting the image of the component M.

Note that X is an ML_0-surface if and only if A is a linear chain. We then call X an *affine pseudoplane of* ML_0*-type*.

If we are given a log-deformation $(Y, \overline{Y}, S, \overline{f}, t_0)$ of the triple $(Y_0, \overline{Y}_0, D_0)$, it follows by Ehresmann's fibration theorem that p_g and the irregularity q of the fiber $\overline{Y}_t$ is independent of t. Furthermore, by Lemma 2.9.7, Y_t has an $\mathbb{A}^1$-fibration if Y_0 has an $\mathbb{A}^1$-fibration. So, we can expect that Y_t is an affine pseudoplane if so is Y_0. Indeed, we have the following result.

Theorem 2.9.23. *Let $\mathcal{F} = (Y, \overline{Y}, S, \overline{f}, t_0)$ be a log-deformation of $(\overline{Y}_0, D_0, Y_0)$. Assume that Y_0 is an affine pseudoplane. Then the following assertions hold:*

(1) *Y_t is an affine pseudoplane for every point $t \in T$.*
(2) *Assume that Y_0 is an affine pseudoplane of* ML_0*-type. Assume further that the boundary divisor D_0 in $\overline{Y}_0$ has the same weighted dual graph as above. Then $f : Y \to T$ is a trivial bundle with fiber Y_0 after shrinking T if necessary.*

Proof. (1) We have only to show that Y_t is an affine pseudoplane for a small deformation of Y_0. After replacing T by an étale finite covering, we may assume that $\overline{Y}_t$ is a minimal normal completion of Y_t for every $t \in T$. Then, by Lemma 2.9.19, the boundary divisor $D_t = S \cap \overline{Y}_t$ has the same weighted dual graph as shown above for D_0. Hence Y_t has an $\mathbb{A}^1$-fibration of affine type. By Lemma 2.9.20, $\mathrm{Pic}(Y_t) \cong \mathrm{Pic}(Y_0)$ which is a finite cyclic group. This implies that Y_t is an affine pseudoplane.

(2) Consider the completion $\overline{Y}_0$ of Y_0. We may assume that $\overline{Y}_0$ is a minimal normal completion of Y_0. In fact, a (-1)-curve contained in the boundary divisor D_0 which meets at most two other components of D_0 deforms to the nearby fibers and contracted simultaneously over the same T by Lemma 2.9.10(2). Note that every fiber Y_t has an $\mathbb{A}^1$-fibration of affine type by Lemma 2.9.7. As in the proof of Lemma 2.9.7(2), by performing simultaneous (i. e., along the fibers of $\overline{f}$) blowing-ups and blowing-downs on the boundary S, we may assume that Y_0 has an $\mathbb{A}^1$-fibration which extends to a $\mathbb{P}^1$-fibration on $\overline{Y}_0$ and that the boundary divisor D_0 has the weighted dual graph $\ell + M + A$ as specified in the condition (ii) above, where A is a linear chain by the hypothesis. To perform a simultaneous blowing-up, we may have to choose as the center a cross-section on an irreducible component S_i which is a $\mathbb{P}^1$-bundle over T. If such a cross-section happens to intersect the curve $S_i \cap S_j$ with another component S_j, we shrink T to avoid this intersection (see the remark in the second proof of Theorem 2.9.15). Note that the interior Y (more precisely, the inverse image of f of the shrunken T) is not affected under these operations. Then the (0)-curve ℓ defines a $\mathbb{P}^1$-fibration $\varphi : \overline{Y} \to V$ (see Lemma 2.9.10(1)). In particular, ℓ moves in an irreducible component, say S_{-1}, of S. The (0)-curve M moves along the fibers of $\overline{f}$ in an irreducible component, say S_0, of S. By Lemma 2.9.19, the curves in A move along the fibers of $\overline{f}$ and fill out the irreducible components $S_1, \ldots, S_r$ of S. Hence $S = S_{-1} \cup S_0 \cup S_1 \cup \cdots \cup S_r$ and $D_t = S \cdot \overline{Y}_t$ has the same weighted dual graph as D_0.

Now consider a (-1)-curve F_0 on $\overline{Y}_0$. By Lemma 2.9.6, F_0 moves along the fibers of $\overline{f}$ and fills out a smooth irreducible divisor F which meets transversally an irreducible component S_i $(1 \le i \le r)$. In fact, the feather F_0 is unique on Y_0 and $(S_i \cdot F \cdot \overline{Y}_t) = (S_i \cdot F \cdot \overline{Y}_0) = 1$. Let S_1 be the component of S meeting S_0. Let $F_t = F \cap \overline{Y}_t$ and $S_{j,t} = S_j \cap \overline{Y}_t$ for every $t \in T$. Then $F_t + \sum_{j=2}^r S_{j,t}$ is contractible to a smooth point P_t lying on $S_{1,t}$. After performing simultaneous elementary transformations on the fiber ℓ which is the fiber at infinity of the $\mathbb{A}^1$-fibration of the affine pseudoplane Y_t, we may assume that P_t is the intersection point $S_{0,t} \cap S_{1,t}$. By applying Lemma 2.9.10(2) repeatedly, we can contract F and the components $S_2, \ldots, S_r$ simultaneously. Let $\overline{Z}$ be the threefold obtained from $\overline{Y}$ by these contractions. Then $\overline{Z}$ has a $\mathbb{P}^1$-fibration $\psi : \overline{Z} \to V$ and the image of S_0 is a cross-section. Let $g = \sigma \cdot \psi : \overline{Z} \xrightarrow{\psi} V \xrightarrow{\sigma} T$ (see Lemma 2.9.6(3) for the notations). For every $t \in T$, $\overline{Z}_t := g^{-1}(t)$ is a minimal $\mathbb{P}^1$-bundle with a cross-section $S_{0,t}$. Since $(S_{0,t})^2 = 0$, $\overline{Z}_t$ is isomorphic to $\mathbb{P}^1 \times \mathbb{P}^1$. Then $\overline{Z}$ is a trivial $\mathbb{P}^1 \times \mathbb{P}^1$-bundle over T after shrinking T if necessary. In fact, $\overline{Z}$ with the images of S_0 and S_{-1} removed is a deformation of $\mathbb{A}^2$, which is locally trivial in the Zariski topology by Theorem 2.9.15. We may assume that $\psi : \overline{Z} \to V$ is the projection of $\mathbb{P}^1 \times \mathbb{P}^1 \times T$ onto the second and the third factors. Choose a section $\overline{S}_0'$ of ψ which is disjoint from the image $\overline{S}_0$ of S_0. Then there is a nontrivial G_m-action on $\overline{Z}$ along the fibers of ψ which has $\overline{S}_0$ and $\overline{S}_0'$ as the fixed point locus.

Now reverse the contractions $\overline{Y} \to \overline{Z}$. The center of the first simultaneous blowing-up with center $S_0 \cap S_1$ and the centers of the consecutive simultaneous blowing-ups except for the blowing-up which produces the component F are G_m-fixed because the blowing-ups are fiberwise subdivisional. Only the center Q_t of the last blowing-up on $\overline{Y}_t$ is non-subdivisional. Let

$$\varphi : \overline{Y} \xrightarrow{\sigma} \overline{Y}_1 \xrightarrow{\sigma_1} \overline{Z}$$

be the factorization of φ where σ is the last non-subdivisional blowing-up. By the construction, the natural T-morphism $\overline{f}_1 : \overline{Y}_1 \to T$ is a trivial fibration with fiber $(\overline{Y}_1)_0 = \overline{f}_1^{-1}(t_0)$. Then there exists an element $\{\rho_t\}_{t\in T}$ of $G_m(T)$ such that $\rho_t(Q_{t_0}) = Q_t$ for every $t \in T$ after shrinking T if necessary.[19] Here note that the G_m-action is nontrivial on the component with the point Q_t thereon, for otherwise the G_m-action is trivial from the beginning. Then these $\{\rho_t\}_{t\in T}$ extends to a T-isomorphism $\widetilde{\rho} : \overline{Y}_0 \times T \xrightarrow{\sim} \overline{Y}$, which induces a T-isomorphism $Y_0 \times T \xrightarrow{\sim} Y$. Hence Y is trivial. □

2.9.6 Deformations of $\mathbb{A}^1$-fibrations of complete type

In the setting of Theorem 2.9.13, if the $\mathbb{A}^1$-fibration of a general fiber Y_t is of complete type, we do not have the same conclusion. This case was treated by Dubouloz and Kishimoto [42]. We consider this case by taking the same example of cubic surfaces in $\mathbb{P}^3$ and explain how it is affine-uniruled.

Taking a cubic hypersurface as an example, we first observe the behavior of the log-Kodaira dimension for a flat family of smooth affine surfaces. Let $\check{\mathbb{P}}^3$ be the dual projective 3-space whose points correspond to the hyperplanes of $\mathbb{P}^3$. We denote it by T. Let S be a smooth cubic hypersurface in $\mathbb{P}^3$ and let $\mathcal{W} = S \times T$ which is a codimension one subvariety of $\mathbb{P}^3 \times T$. Let $\mathcal{H}$ be the universal hyperplane in $\mathbb{P}^3 \times T$, which is defined by $\xi_0 X_0 + \xi_1 X_1 + \xi_2 X_2 + \xi_3 X_3 = 0$, where (X_0, X_1, X_2, X_3) and $(\xi_0, \xi_1, \xi_2, \xi_3)$ are respectively the homogeneous coordinates of $\mathbb{P}^3$ and T. Let $\mathcal{D}$ be the intersection of $\mathcal{W}$ and $\mathcal{H}$ in

19 Q_{t_0} is a point on $\overline{S}_{i1} \cap \overline{f}^{-1}(t_0)$, where $\overline{S}_{i1}$ is a trivial $\mathbb{P}^1$-bundle over T. Hence $\{Q_{t_0}\} \times T$ is a cross-section. We denote the point $(\{Q_{t_0}\} \times T) \cap \overline{f}^{-1}(t)$ by Q_{t_0}. Since the G_m-action on $\overline{f}^{-1}(t)$ is nontrivial and both Q_t and Q_{t_0} are different from the intersection points of $\overline{f}^{-1}(t)$ with adjacent components, there exists $\rho_t \in G_m$ such that $\rho_t(Q_{t_0}) = Q_t$.

$\mathbb{P}^3 \times T$. Let $\pi : \mathcal{W} \to T$ be the projection and let $\pi_{\mathcal{D}} : \mathcal{D} \to T$ be the restriction of π onto $\mathcal{D}$. Then π and $\pi_{\mathcal{D}}$ are flat morphisms. For a closed point $t \in T$, $\mathcal{W}_t = \pi^{-1}(t)$ is identified with S and $\mathcal{D}_t = \pi_{\mathcal{D}}^{-1}(t)$ is the hyperplane section $S \cap \mathcal{H}_t$ in $\mathbb{P}^3$, where $\mathcal{H}_t$ is the hyperplane $\tau_0 X_0 + \tau_1 X_1 + \tau_2 X_2 + \tau_3 X_3 = 0$ with $t = (\tau_0, \tau_1, \tau_2, \tau_3)$. Let $\mathcal{X} = \mathcal{W} \setminus \mathcal{D}$ and $p : \mathcal{X} \to T$ be the restriction of π onto $\mathcal{X}$. Then $\mathcal{X}_t = p^{-1}(t)$ is an affine surface $S \setminus (S \cap \mathcal{H}_t)$.

Since S is smooth, the following types of $S \cap \mathcal{H}_t$ are possible. In the following, $F = 0$ denotes the defining equation of S and $H = 0$ does the equation for $\mathcal{H}_t$:
(1) A smooth irreducible plane curve of degree 3.
(2) An irreducible nodal curve, e. g., $F = X_0(X_1^2 - X_2^2) - X_2^3 + X_0^2 X_3 + X_3^3$ and $H = X_3$.
(3) An irreducible cuspidal curve, e. g., $F = X_0 X_1^2 - X_2^3 + X_3(X_0^2 + X_1^2 + X_2^2 + X_3^2)$ amd $H = X_3$.
(4) An irreducible conic and a line which meets in two points transversally or in one point with multiplicity two. In fact, let ℓ and D be respectively a line and an irreducible conic in $\mathbb{P}^2$ meeting in two points Q_1, Q_2, where Q_1 is possibly equal to Q_2. Let C be a smooth cubic meeting ℓ in three points P_i $(1 \le i \le 3)$ and D in six points P_i $(4 \le i \le 9)$, where the points P_i are all distinct and different from Q_1, Q_2. Choose two points P_1, P_2 on ℓ and four points P_i $(4 \le i \le 7)$ on D. Let $\sigma : S \to \mathbb{P}^2$ be the blowing-up of these six points. Let ℓ', D' and C' be the proper transforms of ℓ, D and C. Then S is a cubic hypersurface in $\mathbb{P}^3$ and $K_S \sim -C'$. Since $\ell' + D' \sim C'$, it is a hyperplane section of S with respect to the embedding $\Phi_{|C'|} : S \hookrightarrow \mathbb{P}^3$.
(5) Three lines which are either meeting in one point or not. Let ℓ_i $(1 \le i \le 3)$ be the lines. Let $Q_1 = \ell_1 \cap \ell_3$ and $Q_2 = \ell_2 \cap \ell_3$. In the setting of (4) above, we consider $\ell = \ell_3$ and $D = \ell_1 + \ell_2$. So, if $Q_1 = Q_2$, three lines meet in one point. Choose a smooth cubic C meeting three lines in nine distinct points P_i $(1 \le i \le 9)$ other than Q_1, Q_2. Choose six points from the P_i, two points lying on each line. Then consider the blowing-up in these six points. The rest of the construction is the same as above.

Note that if S is smooth $S \cap \mathcal{H}_t$ cannot have a nonreduced component. In fact, the nonreduced component is a line in $\mathcal{H}_t$. Hence we may write the defining equation of S as

$$F = X_0^2(aX_1 + X_0) + X_3 G(X_0, X_1, X_2, X_3) = 0,$$

where $G = G(X_0, X_1, X_2, X_3)$ is a quadratic homogeneous polynomial and $a \in k$. We understand that $a = 0$ if the nonreduced component has multiplicity three. By the Jacobian criterion, it follows that S has singularities at the points $G = X_0 = X_3 = 0$.

The affine surface $\mathcal{X}_t$ has log-Kodaira dimension 0 in the cases (1), (2), (4) with the conic and the line meeting in two distinct points and (5) with nonconfluent three lines, and $-\infty$ in the rest of the cases. Although $p : \mathcal{X} \to T$ is a flat family of affine surfaces, the log-Kodaira dimension drops to $-\infty$ exactly at the points $t \in T$ where the boundary divisor $S \cap \mathcal{H}_t$ is not a divisor with normal crossings. This accords with a result of Kawamata concerning the invariance of log-Kodaira dimension under deformations (cf. [115]).

If $\overline{\kappa}(\mathcal{X}_t) = -\infty$, then $\mathcal{X}_t$ has an $\mathbb{A}^1$-fibration. We note that if $\overline{\kappa}(\mathcal{X}_t) = 0$ then $\mathcal{X}_t$ has an $\mathbb{A}^1_*$-fibration. In fact, we consider the case where the boundary divisor $\mathcal{D}_t$ is a smooth cubic curve. Then S is obtained from $\mathbb{P}^2$ by blowing up six points P_i $(1 \le i \le 6)$ on a smooth cubic curve C. Choose four points P_1, P_2, P_3, P_4 and let Λ be a linear pencil of conics passing through these four points. Let $\sigma : S \to \mathbb{P}^2$ be the blowing-up of six points P_i $(1 \le i \le 6)$. The proper transform $\sigma'\Lambda$ defines a $\mathbb{P}^1$-fibration $f : S \to \mathbb{P}^1$ for which the proper transform $C' = \sigma'(C)$ is a 2-section. Since $\mathcal{X}_t$ is isomorphic to $S \setminus C'$, $\mathcal{X}_t$ has an $\mathbb{A}^1_*$-fibration.

Construction of an $\mathbb{A}^1$-fibration in the case $\overline{\kappa}(\mathcal{X}_t) = -\infty$ is not an easy task. Consider, for example, the case where $X = \mathcal{X}_t$ is obtained as $S \setminus (Q \cup \ell)$, where Q is a smooth conic and ℓ is a line in $\mathbb{P}^2$ which meet Q in one point with multiplicity two. As explained in the above, such an X is obtained from $\mathbb{P}^2$ by blowing up six points $P_1, \ldots, P_6$ such that P_1, P_2 lie on a line $\widetilde{\ell}$ and P_3, P_4, P_5, P_6 are points on a conic $\widetilde{Q}$.[20] Then the proper transforms on S of $\widetilde{\ell}, \widetilde{Q}$ are ℓ, Q. Consider the linear pencil $\widetilde{\Lambda}$ on $\mathbb{P}^2$ spanned by $2\widetilde{\ell}$ and $\widetilde{Q}$. Then a general member of Λ is a smooth conic meeting $\widetilde{Q}$ in one point $\widetilde{Q} \cap \widetilde{\ell}$ with multiplicity four. The proper transform Λ of $\widetilde{\Lambda}$ on S defines an $\mathbb{A}^1$-fibration on X.

The following result is due to Dubouloz–Kishimoto [42].

Theorem 2.9.24. *Let S be a cubic hypersurface in $\mathbb{P}^3$ with a hyperplane section $S \cap H$ which consists of a line and a conic meeting in one point with multiplicity two. Let $Y = \mathbb{P}^3 \setminus S$ which is a smooth affine threefold. Then the following assertions hold:*

(1) $\overline{\kappa}(Y) = -\infty$.
(2) *Let $f : Y \to \mathbb{A}^1$ be a fibration induced by the linear pencil on $\mathbb{P}^3$ spanned by S and $3H$. Then a general fiber Y_t of f is a cubic hypersurface S_t minus $Q \cup \ell$, where Q is a conic and ℓ is a line which meet in one point with multiplicity two. Hence $\overline{\kappa}(Y_t) = -\infty$ and Y_t has an $\mathbb{A}^1$-fibration.*
(3) *Y has no $\mathbb{A}^1$-fibration.*

---

20 Note that $-K_S \sim Q + \ell$. Since $p_a(Q) = p_a(\ell) = 0$ and $-K_S$ is ample, it follows that $(Q^2) + 2 = (-K_S \cdot Q) > 0$ and $(\ell^2) + 2 = (-K_S \cdot \ell) > 0$. Further, since $(-K_S{}^2) = 3 = (Q^2) + 4 + (\ell^2)$, either $(Q^2) = 0$ and $(\ell^2) = -1$ or $(Q^2) = -1$ and $(\ell^2) = 0$. In the second case, we have $P_1 \in \widetilde{\ell}$ and $P_2, P_3, P_4, P_5, P_6 \in Q$. Let E_i be the (-1)-curve arising from the blowing-up of P_i. Then, leaving E_2, E_3 for Q, we find three (-1)-curves F_4, F_5, F_6 such that F_j $(j = 4, 5, 6)$ does not meet E_1, E_2, E_3 and $(F_j \cdot \ell) = 1$. In fact, F_j is the proper transform of the line on $\mathbb{P}^2$ passing through two points of P_4, P_5, P_6. Then the set of six (-1)-curves $\{E_1, E_2, E_3, F_4, F_5, F_6\}$ have no mutual intersections. Hence we can blow them down to obtain $\mathbb{P}^2$ as the image of S. Then the images of Q and ℓ are respectively a line and a conic meeting in one point with multiplicity 2. So, we can assume the first case occurs.

(4) *There is a finite covering T' of $\mathbb{A}^1$ such that the normalization of $Y \times_{\mathbb{A}^1} T'$ has an $\mathbb{A}^1$-fibration.*

Proof. (1) Since $K_{\mathbb{P}^3} + S \sim -4H + 3H = -H$, it follows that $\overline{\kappa}(Y) = -\infty$.

(2) The pencil spanned by S and $3H$ has base locus $Q \cup \ell$ and its general member, say S_t, is a cubic hypersurface containing $Q \cup \ell$ as a hyperplane section. It is clear that $S_t \setminus (Q \cup \ell) = Y_t$. Hence, as explained above, Y_t has an $\mathbb{A}^1$-fibration.

(3) Let $\tau : \widetilde{S} \to \mathbb{P}^3$ be the cyclic triple covering of $\mathbb{P}^3$ ramified totally over the cubic hypersurface S. Then $\widetilde{S}$ is a cubic hypersurface in $\mathbb{P}^4$ and $\tau^*(S) = 3\widetilde{H}$, where $\widetilde{H}$ is a hyperplane in $\mathbb{P}^4$. The restriction of τ onto $Z := \widetilde{S} \setminus \widetilde{S} \cap \widetilde{H}$ induces a finite étale covering $\tau_Z : Z \to Y$. Suppose that Y has an $\mathbb{A}^1$-fibration $\varphi : Y \to T$. Then T is a rational surface. Since τ_Z is finite étale, this $\mathbb{A}^1$-fibration φ lifts up to an $\mathbb{A}^1$-fibration $\widetilde{\varphi} : Z \to \widetilde{T}$. By [28], $\widetilde{S}$ is unirational and irrational. Hence $\widetilde{T}$ is a rational surface. This implies that Z is a rational threefold. This is a contradiction because $\widetilde{S}$ is irrational.

(4) There is an open set T of $\mathbb{A}^1$ such that the restriction of f onto $f^{-1}(T)$ is a smooth morphism onto T. By abuse of the notations, we denote $f^{-1}(T)$ by Y anew and the restriction of f onto $f^{-1}(T)$ by f. Hence $f : Y \to T$ is a smooth morphism. Let $K = k(t)$ be the function field of T and let Y_K be the generic fiber. Let $\overline{K}$ be an algebraic closure of K. Then $Y_{\overline{K}} := Y_K \otimes_K \overline{K}$ is identified with $S_{\overline{K}} \setminus (Q \cup \ell)$, where $S_{\overline{K}}$ is a cubic hypersurface in $\mathbb{P}^3_{\overline{K}}$ defined by $F_K = F_0 + tX_3^3 = 0$. Here t is a coordinate of $\mathbb{A}^1$ and (X_0, X_1, X_2, X_3) is a system of homogeneous coordinates of $\mathbb{P}^3$ such that $F_0(X_0, X_1, X_2, X_3) = 0$ is the defining equation of the cubic hypersurface S and the hyperplane H is defined by $X_3 = 0$. Then $Y_{\overline{K}}$ is obtained from $\mathbb{P}^2_{\overline{K}}$ by blowing up six $\overline{K}$-rational points in general position (two points on the image of ℓ and four points on the image of Q). As explained earlier, there is an $\mathbb{A}^1$-fibration on $Y_{\overline{K}}$ which is obtained from conics on $\mathbb{P}^2_{\overline{K}}$ belonging to the pencil spanned by Q and 2ℓ. This construction involves six points on $\mathbb{P}^2_{\overline{K}}$ to be blown up to obtain the cubic hypersurface $S_{\overline{K}}$ and four points (the point $Q \cap \ell$ and its three infinitely near points). Hence there exists a finite algebraic extension K'/K such that all these points are rational over K'. Let T' be the normalization of T in K'. Let $Y' = Y \otimes_K K'$. Then Y' has an $\mathbb{A}^1$-fibration. □

The assertion (4) above holds in more general settings. In fact, Dubouloz–Kishimoto [43] proved the following result.

Theorem 2.9.25. *Let $f : X \to S$ be a dominant morphism between normal complex algebraic varieties whose general fibers are smooth $\mathbb{A}^1$-ruled affine surfaces. Then there exists a dense open subset $S^* \subset S$, a finite étale morphism $T \to S^*$ and a normal T-scheme $h : Y \to T$ such that the induced morphism $f_T = \mathrm{pr}_T : X_T = X \times_{S^*} T \to T$ factors as*

$$f_T = h \cdot \rho : X_T \xrightarrow{\rho} Y \xrightarrow{h} T,$$

where $\rho : X_T \to Y$ is an $\mathbb{A}^1$-fibration.

2.10 Problems for Chapter 2

1. Let A be an affine k-domain, where k is an algebraically closed field of characteristic zero and let $X = \operatorname{Spec} A$. Suppose that X has a G_a-action σ defined by an lnd δ on A. Show that the following assertions hold:
   (1) $X(k) := \operatorname{Hom}_{\operatorname{Spec} k}(\operatorname{Spec} k, X) \cong \operatorname{Hom}_{k\text{-alg}}(A, k)$ is the set of closed points of the affine variety X.
   (2) For a closed point $x \in X(k)$, let θ_x be the corresponding k-homomorphism $A \to k$. Then every element $a \in A$ is considered as a k-valued function on $X(k)$ by $a(x) = \theta_x(a)$.
   (3) For $\lambda \in k$ and $a \in A$, define ${}^{\lambda}a$ by ${}^{\lambda}a = \sum_{i\geq 0} \frac{1}{i!}\delta^i(a)\lambda^i$. Then this defines an action of k^+ on a k-algebra A. Namely, it satisfies the conditions (i) ${}^{\lambda+\lambda'}a = {}^{\lambda'}({}^{\lambda}a)$, (ii) ${}^{0}a = a$, (iii) ${}^{\lambda}(a + a') = {}^{\lambda}a + {}^{\lambda}a'$, (iv) ${}^{\lambda}(aa') = {}^{\lambda}a \cdot {}^{\lambda}a'$, and (v) ${}^{\lambda}c = c$ for $c \in k$.
   (4) For $x \in X(k)$, define a k^+-action ${}^{\lambda}x$ by $\theta_{{}^{\lambda}x} = {}^{\lambda}\theta_x = \sum_{i\geq 0} \frac{1}{i!}\lambda^i\theta_x\delta^i$. Then we have $({}^{\lambda}a)(x) = a({}^{\lambda}x) := \theta_{{}^{\lambda}x}(a)$.
   (5) ${}^{\lambda}a = a$ for every $\lambda \in k$ if and only if $a \in \operatorname{Ker} \delta$.

   Answer. Straightforward. □

2. Let K be a field of characteristic zero and let D be a derivation of K. Set

$$k = \{a \in K \mid D(a) = 0\},$$
$$R = \{a \in K \mid D^n(a) = 0 \ \exists n > 0\}.$$

   Prove the following assertions:
   (1) Show that k is a subfield of K, which is algebraically closed in K. The derivation D is thus a k-derivation.
   (2) Suppose that $k \subsetneq R$. Let A be a finitely generated k-subalgebra such that $k \subsetneq A \subseteq R$ and A is generated over k by $a_1, \dots, a_r$ and $D^i(a_j)$ for $i > 0$ and $1 \leq j \leq r$. Then D is considered as an lnd of A, and hence G_a acts on $X = \operatorname{Spec} A$. Show that $X \cong \mathbb{A}^1$, and hence A is a polynomial ring in one variable over k.
   (3) Suppose that $k \subsetneq R$. Then R is a polynomial ring in one variable over k.

   Answer. (1) Let $a \in k \setminus (0)$. Then $a\alpha = 1$ for $\alpha \in K$. Since $D(a) = D(1) = 0$, we have $D(\alpha) = 0$. This implies that a has the inverse in k. Suppose that $\alpha \in K$ is algebraic over k. Let $f(x)$ be a monic minimal polynomial of α over k and write it as

$$f(x) = x^n + a_1x^{n-1} + \cdots + a_{n-1}x + a_n.$$

   Since $f(\alpha) = \alpha^n + a_1\alpha^{n-1} + \cdots + a_{n-1}\alpha + a_n = 0$, we obtain

$$D(f(\alpha)) = (n\alpha^{n-1} + (n-1)a_1\alpha^{n-2} + \cdots + a_{n-1})D(\alpha) = 0.$$

Since $n\alpha^{n-1} + (n-1)a_1\alpha^{n-2} + \cdots + a_{n-1} \neq 0$ by the minimality of n, it follows that $D(\alpha) = 0$. Hence $\alpha \in k$.
(2) There exists an element $u \in R$ such that $b := D(u) \in k \setminus (0)$. Replacing u by $b^{-1}u$, we may assume that $b = 1$. Since D restricted on A is an lnd, we have $A = (\operatorname{Ker} D)[u]$. Since $\operatorname{Ker} D = k$, it follows that $A = k[u]$.
(3) If $A \subsetneq R$, take $a' \in R \setminus A$ and let $A' = A[a', D^j(a') \mid j > 0]$. Then $A' = k[u']$ with $D(u') = 1$ by (2). Then $D(u - u') = 0$, whence $u' = u + b$ with $b \in k$. Then $A = A'$. This is a contradiction. Hence $A = R$. □

3. Let G be a finite group and let G act faithfully on an affine k-domain A. Let B be the ring of G-invariants of A. Then A is a finite B-module. Let K and L be respectively the quotient fields of A and B. Then L is the G-invariant subfield of K. For a k-algebra R, we denote by $\operatorname{Der}_k(R, R)$ or simply $\operatorname{Der}_k(R)$ the R-module of k-derivations of R into R.
Prove the following assertions:
   (1) G acts on $\operatorname{Der}_k(R, R)$ by $g(\Delta)(x) = g(\Delta(g^{-1}(x)))$, where $g \in G, \Delta \in \operatorname{Der}_k(R, R)$ and $x \in R$.
   (2) Taking the above K as R, $\Delta \in \operatorname{Der}_k(K, K)$ is a lifting of an element $\delta \in \operatorname{Der}_k(L, L)$ if and only if $g(\Delta) = \Delta$ for every $g \in G$.
   (3) Taking the above A as R, $\Delta \in \operatorname{Der}_k(A, A)$ is a lifting of an element $\delta \in \operatorname{Der}_k(B, B)$ if and only if $g(\Delta) = \Delta$ for every $g \in G$.

Answer. (1) It is straightforward to verify that $g(\Delta) \in \operatorname{Der}_k(R, R)$.
(2) Suppose that Δ is a lifting of $\delta \in \operatorname{Der}_k(L, L)$. Then, for $z \in L$, we compute as $g(\Delta)(z) = g(\Delta(g^{-1}(z))) = g(\delta(z)) = \delta(z)$. This implies that $g(\Delta)$ is also a lifting of δ. Since the lifting of δ is unique, we have $g(\Delta) = \Delta$ for every $g \in G$. Conversely, if $g(\Delta) = \Delta$, then for $z \in L$, we have $g(\Delta(z)) = g(\Delta(g^{-1}(z)) = (g(\Delta))(z) = \Delta(z)$ as $g^{-1}z = z$, whence $\Delta(z) \in L$. So, Δ induces an element $\delta \in \operatorname{Der}_k(L, L)$. Hence Δ is a lifting of δ.
(3) In the above proof, if $\Delta \in \operatorname{Der}_k(A, A)$, then $g(\Delta) \in \operatorname{Der}_k(A, A)$. Then the same proof applies to the present case. □

4. Let $A = k[x, y]$ be a polynomial ring in two variables and let $\iota : A \to A$ be a k-algebra automorphism defined by $\iota(x) = y$ and $\iota(y) = x$. The automorphism ι has order 2 and is called an *involution*. Let B be the invariant subring $A^{\langle\iota\rangle}$. Prove the following facts:
   (1) $B = k[t, u]$, where $t = x + y$ and $u = xy$. The finite covering $p : X = \operatorname{Spec} A \to Y = \operatorname{Spec} B$ is branched over the discriminant locus $V(t^2 - 4u)$ of Y.
   (2) Let $f(X) = X^2 - tX + u$, where $f(X) = 0$ is a minimal equation of both x and y over $L = Q(B)$. Let $\delta \in \operatorname{Der}_k(B, B)$. Then δ lifts to $\Delta \in \operatorname{Der}_k(A, A)$ if and only if $f^\delta(X) = -\delta(t)X + \delta(u)$ is divisible by $f'(X) = 2X - t$ in $B[X]$, i. e., $2\delta(u) = t\delta(t)$.

Answer. Both assertions are straightforward. □

5. Let $p : V \to C$ be a $\mathbb{P}^1$-bundle over a smooth projective curve C and let H be an ample cross-section of p. Let $X = V \setminus H$ and $f = p\,|_X : X \to C$. Prove the following assertions:
   (1) The surface X is affine and $f : X \to C$ is an $\mathbb{A}^1$-bundle over C, which is not trivial.
   (2) The morphism $f : X \to C$ is not the quotient morphism of a G_a-action on X.
   (3) If C is not isomorphic to $\mathbb{P}^1$, there are no G_a-actions on X.

   Answer. (1) The proof is standard, and we omit it.
   (2) Write $X = \operatorname{Spec} A$. Suppose that f is the quotient morphism of a G_a-action on X. Let δ be the corresponding lnd of the G_a-action. Then $B = \operatorname{Ker} \delta$ is a k-subalgebra of A of dimension one, and the quotient morphism is given by the inclusion $B \hookrightarrow A$. Then $C = \operatorname{Spec} B$. This is clearly a contradiction because C is a complete curve.
   (3) Suppose that there exists a nontrivial G_a-action on X. Since X is a surface, there exists the quotient morphism $q : X \to Y$. Let F be a general fiber of q. Since $F \cong \mathbb{A}^1$ and C is not rational, $f(F)$ is a point of C. Namely, general fibers of q are fibers of f. Hence q coincides with the morphism f. But this is impossible by assertion (2). □

6. Let $V = \mathbb{P}^1 \times \mathbb{P}^1$ and let D be the diagonal, i. e., $D = \{(P, P) \mid P \in \mathbb{P}^1\}$. Set $X = V \setminus D$. Prove the following assertions.
   (1) X is an affine surface isomorphic to the hypersurface in $\mathbb{A}^3$ defined by $yz = x^2 - 1$. The affine surface X is the simplest case of what is called a *Danielewski surface.*
   (2) Let $p_1 : V \to \mathbb{P}^1$ be the first projection $(P, Q) \mapsto P$. Then $(p_1)|_X : X \to \mathbb{P}^1$ is an $\mathbb{A}^1$-bundle over $\mathbb{P}^1$. Hence it is not the quotient morphism of a G_a-action on X by the previous problem.
   (3) Let M and ℓ be a fiber of the first (resp., the second) projection p_1 (resp., p_2) of V. Then $D \sim \ell + M$ (linear equivalence). Let Λ be the linear pencil on V generated by D and $\ell + M$. Suppose that ℓ and M meets at a point P_0 on D. Then other members of Λ are irreducible smooth rational curves which touch the diagonal D at the point P_0 with multiplicity 2.
   (4) The pencil Λ defines an $\mathbb{A}^1$-fibration $f_{P_0} : X \to \mathbb{A}^1$ such that $(\ell + M) \cap X$ is a unique reducible fiber. If P_0 moves on the diagonal, the linear pencils Λ_{P_0}, which is the above Λ defined by P_0, gives distinct $\mathbb{A}^1$-fibrations, hence distinct G_a-actions on X.

   Answer. (1) Consider the Segre embedding of V into $\mathbb{P}^3$ defined by $((x_0, x_1), (y_0, y_1)) \mapsto (x_0y_0, x_0y_1, x_1y_0, x_1y_1)$, whose image is a quadratic hypersurface $X_0X_3 = X_1X_2$, where $X_0 = x_0y_0, X_1 = x_0y_1, X_2 = x_1y_0, X_3 = x_1y_1$. The diagonal D of V is the hyperplane section by $X_1 = X_2$. Set $X_2 - X_1 = T$. Then the equation of V is given by $X_0X_3 = X_1(X_1 + T)$. Since the divisor D is defined by $T = 0$, the affine surface X is defined in $\mathbb{A}^3 = \operatorname{Spec} k[x, y, z]$ by $yz = x(x - 1)$, where $x = X_1/T, y = X_0/T$ and

$z = X_3/T$. It is clear that the equation $yz = x(x-1)$ is changed to $yz = x^2 - 1$ by a suitable change of coordinates of $\mathbb{A}^3$.
(2)–(4) The rest of the assertions are straightforward to show. □

7. Let C be a smooth conic in $\mathbb{P}^2$ defined by $Z_1^2 = 4Z_0Z_2$ with respect to a system of homogeneous coordinates (Z_0, Z_1, Z_2) of $\mathbb{P}^2$. Prove the following assertions:
(1) Let $p : V = \mathbb{P}^1 \times \mathbb{P}^1 \to \mathbb{P}^2$ be the morphism given by

$$((x_0, x_1), (y_0, y_1)) \mapsto (x_0y_0, x_0y_1 + x_1y_0, x_1y_1).$$

Then p satisfies $p = p \cdot \iota$, where ι is an involution $((x_0, x_1), (y_0, y_1)) \mapsto ((y_0, y_1), (x_0, x_1))$ and $C = \iota_*(D)$, where D is the diagonal in V.
(2) Let $Y = \mathbb{P}^2 \setminus C$. Then $p^{-1}(Y)$ is isomorphic to X in Problem 6. Furthermore, $p|_X : X \to Y$ is an étale finite morphism of degree 2.
(3) Let Q be a point of C and let ℓ_Q be the tangent line of C at Q. Let L_Q be the linear pencil on $\mathbb{P}^2$ generated by C and $2\ell_Q$. The other members C' of L_Q are smooth conics passing through Q and with the intersection multiplicities $(\ell_Q \cdot C') = i(C', \ell_Q; Q) = 2$ and $(C \cdot C') = i(C, C'; Q) = 4$.
(4) Let P be the unique point of D lying over Q. Then the linear pencil p^*L_Q consisting of the pull-backs of the members of L_Q is the linear pencil $\{C' + \iota(C') \mid C' \in \Lambda_P\}$. The linear pencil L_Q defines an $\mathbb{A}^1$-fibration $g : Y \to \mathbb{A}^1$, which is the quotient morphism of a G_a-action. The $\mathbb{A}^1$-fibration g has a unique singular fiber $2(\ell_Q \cap Y)$, which is a multiple irreducible fiber. The pull-back $p^*(g)$ is composed by the $\mathbb{A}^1$-fibration f_P in Problem 6. Namely, $p^*(G)$ of a fiber G of g is the sum $F + \iota(F)$, where F is a fiber of f such that $p(F) = G$.

Answer. (1) We have

$$\begin{aligned} Z_1^2 - 4Z_0Z_2 &= (x_0y_1 + x_1y_0)^2 - 4x_0y_0x_1y_1 \\ &= (x_0y_1 - x_1y_0)^2 = (X_1 - X_2)^2 \end{aligned}$$

and

$$\iota : (X_0, X_1, X_2, X_3) \mapsto (X_0, X_2, X_1, X_3).$$

Since $p : (X_0, X_1, X_2, X_3) \mapsto (X_0, X_1 + X_2, X_3)$, it is clear that $p \cdot \iota = p$ and $C = \iota_* D$.
(2) Since $p^*(C) = 2D$ (because $Z_1^2 - 4Z_0Z_2 = (X_1 - X_2)^2$) and since $P \neq \iota(P)$ for $P \in \mathbb{P}^1 \times \mathbb{P}^1$ if and only if $p(P) \notin C$, p is a finite covering of degree 2 which is totally ramified over C and unramified elsewhere.
(3) We have $C \sim 2\ell_Q$. Hence $2D = p^*(C) \sim 2p^*(\ell_Q)$. Since $\mathrm{Pic}(\mathbb{P}^1 \times \mathbb{P}^1) \cong \mathbb{Z} \times \mathbb{Z}$ has no torsion, we have $D \sim p^*(\ell_Q)$. Since $\ell_Q \cap C = \{Q\}$, we have $p^*(\ell_Q) = E + E'$ with $E' = \iota(E)$. Hence $(p^*(\ell_Q)^2) = 2 = (E^2) + (E'^2) + 2(E \cdot E')$, where $(E^2) = (E'^2) \geq 0$ and $(E \cdot E') > 0$. Hence $(E^2) = (E'^2) = 0$ and $(E \cdot E') = 1$. Then $p^*(\ell_Q) \sim M + \ell$ on $\mathbb{P}^1 \times \mathbb{P}^1$. The assertion follows from this observation.

(4) The assertion also follows from the observation in the proof of (3). In fact, the linear pencil L_Q gives an $\mathbb{A}^1$-fibration $g : Y \to \mathbb{A}^1$, where $Y = \mathbb{P}^2 \setminus C$. The member C of L_Q is the boundary divisor $\mathbb{P}^2 \setminus Y$, and the member $2\ell_Q$ gives a multiple fiber of g with multiplicity 2. It is clear that $X = (\mathbb{P}^1 \times \mathbb{P}^1) \setminus D = p^*(\mathbb{P}^2 \setminus C) = p^*(Y)$. Since $p|_X : X \to Y$ is a finite étale covering of degree 2, $p^*(\ell_Q \setminus \{Q\}) = (\ell \setminus \{P\}) \coprod (M \setminus \{P\})$, where $p^*(\ell_Q) = \ell + M$ and $Q = p(P)$ with $P \in \ell \cap M$. □

8. Let $X = \{xy = z^2 - 1\}$ be a Danielewski surface in Problem 6. Let $\rho : X \to \mathbb{A}^1$ be an $\mathbb{A}^1$-fibration $(x, y, z) \mapsto x$. Prove the following assertions:
   (1) The $\mathbb{A}^1$-fibration ρ is the quotient morphism of X by a G_a-action defined by an lnd δ such that

$$\delta(x) = 0, \quad \delta(y) = 2z, \quad \delta(z) = x.$$

   (2) Let U be the open set of $\mathbb{A}^1$ such that $x \neq 0$, whence $U \cong \mathbb{A}^1_*$. Then $\rho^{-1}(U) \cong U \times \mathbb{A}^1$ and $\rho^{-1}(O) = \mathbb{A}^1(z = 1) \coprod \mathbb{A}^1(z = -1)$, where O is the origin of $\mathbb{A}^1$ defined by $x = 0$, and $\mathbb{A}^1(z = 1)$ (resp., $\mathbb{A}^1(z = -1)$) signifies the closed set $\{x = z - 1 = 0\}$ (resp., $\{x = z + 1 = 0\}$).
   (3) Let $\mathrm{SL}(2, k)$ be the special linear group $\{xv - yu = 1\}$. Consider a torus action $\mathrm{SL}(2, k) \times T \to \mathrm{SL}(2, k)$ defined by $(x, y, u, v; t) \mapsto (xt, yt^{-1}, ut, vt^{-1})$ (the right action of T), where $t \in T \cong G_m$. Then $X \cong \mathrm{SL}(2, k)/T$.

Answer. (1) Let A be the coordinate ring of X. Then $A^{G_a} = \operatorname{Ker} \delta = k[x]$. The assertion follows from this.
(2) $A[x^{-1}] = k[x, x^{-1}][z/x]$, where z/x is a slice of δ. Hence $\rho^{-1}(U) \cong U \times \mathbb{A}^1$. The rest is clear.
(3) Let $B = k[x, y, u, v]$ be the coordinate ring of $\mathrm{SL}(2, k)$, where $xv - yu = 1$. Then $B^T = k[xy, uv, xv]$. Then we have

$$4xyuv = 4(xv)(yu) = 4(yu + 1)(yu) = (2yu + 1)^2 - 1.$$

Let $X = 2xy$, $Y = 2uv$ and $Z = 2yu$. Then $XY = Z^2 - 1$, where $B^T = k[X, Y, Z]$. Hence $\mathrm{SL}(2, k)/T$ is isomorphic to the Danielewski surface X. □

9. Let $X(2, 1)$ be the complex hypersurface $xy - z^2u = 1$ in $\mathbb{A}^4 = \operatorname{Spec} \mathbb{C}[x, y, z, u]$. Prove the following assertions:
   (1) $X(2, 1)$ is factorial and simply-connected.
   (2) $X(2, 1)$ has a G_a-action defined by an lnd δ such that

$$\delta(x) = \delta(u) = 0, \quad \delta(y) = 2zu, \quad \delta(z) = x.$$

   (3) $X(2, 1)$ has a G_m-action defined by

$${}^t(x, y, z, u) = (tx, t^{-1}y, tz, t^{-2}u).$$

Let R be the coordinate ring of $X(2,1)$, i. e., $R = \mathbb{C}[x,y,z,u]$ with $xy - z^2u = 1$. Then

$$A := R^{G_m} = \mathbb{C}[xy, yz, x^2u, z^2u, xzu].$$

Set $X = yz$, $Y = x^2u$, $Z = z^2u$ and $U = xzu$. Then $X(2,1)//G_m$ is a closed subvariety in $\mathbb{A}^4 = \operatorname{Spec} \mathbb{C}[X,Y,Z,U]$ defined by

$$XY = U(Z+1), \quad YZ = U^2, \quad XU = Z(Z+1).$$

(4) Let $V := X(2,1)//G_m$. Then V is a smooth affine surface with two $\mathbb{A}^1$-fibrations $\rho_1, \rho_2 : V \to \mathbb{A}^1$, where ρ_1 (resp., ρ_2) is given by the inclusion $\mathbb{C}[X] \hookrightarrow A$ (resp., $\mathbb{C}[Y] \hookrightarrow A$).

(5) The lnd δ induces an lnd $\overline{\delta}$ on A such that

$$\overline{\delta}(X) = 3Z + 1, \quad \overline{\delta}(Y) = 0, \quad \overline{\delta}(Z) = 2U, \quad \overline{\delta}(U) = Y.$$

The $\mathbb{A}^1$-fibration ρ_2 is the quotient morphism of the G_a-action associated to $\overline{\delta}$.

(6) $\rho_1^{-1}(\mathbb{A}^1_*) \cong \mathbb{A}^1_* \times \mathbb{A}^1$ and $\rho_1^{-1}(O) = \mathbb{A}^1 \coprod \mathbb{A}^1$. The same happens for ρ_2. But, V is not a hypersurface.

Answer. (1) A closed set $\{x = 0\}$ of $X(2,1)$ is a trivial $\mathbb{A}^1$-bundle over an integral curve $z^2u + 1 = 0$ in $\mathbb{A}^2$. Hence x is a prime element of R. Further, $R[x^{-1}] = \mathbb{C}[x,x^{-1}][z,u]$ is a polynomial ring over $\mathbb{C}[x,x^{-1}]$. Hence $R[x^{-1}]$ is a UFD. By Nagata's lemma (see Lemma 2.4.22), R is a UFD. So $X(2,1)$ is factorial. Apply Nori's exact sequence in Lemma 1.1.6 to the projection $p : X(2,1) \to \mathbb{A}^1$ given by $(x,y,z,u) \mapsto x$. We have an exact sequence

$$\pi_1(\mathbb{A}^2) \to \pi_1(X(2,1)) \to \pi_1(\mathbb{A}^1) \to (1).$$

This shows that $\pi_1(X(2,1)) = (1)$. Hence $X(2,1)$ is simply-connected.
(2) Straightforward.
(3) We show that $A = \mathbb{C}[xy, yz, x^2u, xzu, z^2u]$. Suppose that $x^a y^b z^c u^d$ is G_m-invariant. Then $a + c = b + 2d$. Write $b = (a-d) + (c-d) \geq 0$.
(i) Suppose $a \geq d$ and $c \geq d$. Then

$$x^a y^b z^c u^d = (xy)^{a-d}(yz)^{c-d}(xzu)^d.$$

(ii) Suppose $a \geq d$ and $c < d$. Then

$$x^a y^b z^c u^d = (xzu)^c (xy)^b (x^2u)^{d-c}.$$

(iii) Suppose $a < d$ and $c \geq d$. Then

$$x^a y^b z^c u^d = (xzu)^a (yz)^b (z^2u)^{d-a}.$$

Set $X = yz, Y = x^2u, Z = z^2u$ and $U = xzu$. It is then easy to see that

$$XY = U(Z+1), YZ = U^2, XU = Z(Z+1).$$

Let W be the closed subvariety of $\mathbb{A}^4 = \operatorname{Spec}\mathbb{C}[X, Y, Z, U]$ defined by these equations. Then W is smooth by the Jacobian criterion. Since

$$\mathbb{C}[X,Y,Z,U]/\begin{pmatrix} XY = U(Z+1) \\ YZ = U^2 \\ XU = Z(Z+1) \end{pmatrix} \to A = \mathbb{C}[xy, yz, x^2u, xzu, z^2u]$$

is a surjective homomorphism of $\mathbb{C}$-affine domains of dimension 2, it is an isomorphism. Namely, $V \cong W$.

(4)–(6) Let $\rho_1 : V \to \mathbb{A}^1$ be defined by the inclusion $\mathbb{C}[X] \hookrightarrow A$. Then, if $X \neq 0$, we have

$$U = \frac{1}{X}Z(Z+1), \quad Y = \frac{1}{X^2}Z(Z+1)^2.$$

Hence $\rho_1^{-1}(\mathbb{A}^1_*) \cong \mathbb{A}^1_* \times \mathbb{A}^1$. The fiber $V_0 = \rho_1^{-1}(O)$ is a disjoint union of two lines defined by $(X = Z = U = 0)$ and $(X = 0, Z = -1, Y + U^2 = 0)$. Define $\rho_2 : V \to \mathbb{A}^1$ by the inclusion $\mathbb{C}[Y] \hookrightarrow A$. If $Y \neq 0$, then

$$X = \frac{1}{Y^2}(U^3 + YU), \quad Z = \frac{U^2}{Y}.$$

In fact, $\operatorname{Ker}\overline{\delta} = \mathbb{C}[Y]$ and U/Y is a slice of $\overline{\delta}$ extended onto $A[1/Y]$. We have $\rho_2^{-1}(\mathbb{A}^1_*) \cong \mathbb{A}^1_* \times \mathbb{A}^1$ and $\rho_2^{-1}(O) \cong \mathbb{A}^1 \coprod \mathbb{A}^1$. Since $A = \mathbb{C}[yz, x^2u, xzu, z^2u]$ and $\{yz, x^2u, xzu, z^2u\}$ is a minimal set of generators of A, the $\mathbb{C}$-algebra A is not generated by three elements. Hence V is not realized as a hypersurface. □

10. Generalizing $X(2,1)$, define $X(m,1)$ as a hypersurface $\{xy - z^m u = 1\}$ in $\mathbb{A}^4 = \operatorname{Spec}\mathbb{C}[x, y, z, u]$, where $m \geq 2$. Prove the following assertions:
    (1) $X(m,1)$ is factorial and simply-connected.
    (2) The coordinate ring $R = \mathbb{C}[x,y,z,u]/(xy - z^m u = 1)$ of $X(m,1)$ admits an lnd δ defined by

$$\delta(x) = \delta(u) = 0, \quad \delta(y) = mz^{m-1}u, \quad \delta(z) = x.$$

    (3) $X(m,1)$ has a G_m-action defined by

$$^t(x, y, z, u) = (tx, t^{-1}y, tz, t^{-m}u).$$

    Let $A = R^{G_m}$. Then we have

$$A = \mathbb{C}[xy, yz, x^{m-i}z^i u \quad (0 \leq i \leq m)].$$

Define a $\mathbb{C}$-algebra B by

$$B = \mathbb{C}[X, Y_0, Y_1, \dots, Y_m] / \left(\begin{array}{ll} XY_{i-1} = Y_i(1 + Y_m), & 1 \le i \le m \\ Y_i Y_j = Y_{i'} Y_{j'} & \text{if } i + j = i' + j' \\ Y_i Y_j = Y_k^2 & \text{if } i + j = 2k \end{array} \right).$$

Then there exists an isomorphism $\theta : B \to A$ defined by $\theta(X) = yz, \theta(Y_i) = x^{m-i} z^i u \ (0 \le i \le m)$.

(4) Let $V(m,1) = \operatorname{Spec} A = X(m,1)//G_m$. Then $V(m,1)$ is a smooth affine surface. The lnd δ on R induces an lnd $\overline{\delta}$ on A such that

$$\overline{\delta}(X) = (m+1)Y_m + 1, \quad \overline{\delta}(Y_i) = iY_{i-1} \quad (0 \le i \le m).$$

Hence $\operatorname{Ker} \overline{\delta} = \mathbb{C}[Y_0]$. Let $\rho_2 : V(m,1) \to \mathbb{A}^1$ be the quotient morphism by the G_a-action associated to $\overline{\delta}$, where $\mathbb{A}^1 = \operatorname{Spec} \mathbb{C}[Y_0]$. If $Y_0 \ne 0$, then

$$X = \frac{1}{Y_0^m} Y_1 (Y_0^{m-1} + Y_1^m), \quad Y_i = \frac{Y_1^i}{Y_0^{i-1}} \quad (2 \le i \le m).$$

Hence $\rho_2^{-1}(\mathbb{A}^1_*) \cong \mathbb{A}^1_* \times \mathbb{A}^1$ and $\rho_2^{-1}(O)$ is a disjoint union of two affine lines

$$(Y_0 = \cdots = Y_{m-1} = 0, Y_m = -1) \coprod (Y_0 = \cdots = Y_{m-1} = Y_m = 0).$$

(5) Let $\rho_1 : V(m,1) \to \mathbb{A}^1 = \operatorname{Spec} \mathbb{C}[X]$ be defined by the inclusion $\mathbb{C}[X] \hookrightarrow A$. If $X \ne 0$ then

$$Y_i = \frac{Y_m}{X^{m-i}} (1 + Y_m)^{m-i} \quad (0 \le i \le m), \tag{2.15}$$

whence $\rho_1^{-1}(\mathbb{A}^1_*) \cong \mathbb{A}^1_* \times \mathbb{A}^1$, and $\rho_1^{-1}(O)$ is a disjoint union of two reduced $\mathbb{A}^1$s:

$$\begin{aligned} &\{Y_1 = \cdots = Y_m = 0, \ Y_0 \text{ is a variable}\} \\ &\coprod \{Y_m = -1, Y_{m-i} = (-1)^{i+1} Y_{m-1}^i \ (0 \le i \le m)\}. \end{aligned} \tag{2.16}$$

(6) A minimal closed embedding $V(m,1) \hookrightarrow \mathbb{A}^{m+2}$ is given by

$$(x, y, z, u) \mapsto (X, Y_0, \dots, Y_m).$$

Hence, if $m \ne m'$, $V(m,1)$ is not isomorphic to $V(m',1)$. In particular, $V(m,1) \not\cong V(1,1)$ if $m \ge 2$, where $V(1,1)$ is a hypersurface Danielewski surface $xy = z^2 - 1$.

(7) For $m \ne m'$, we have $V(m,1) \times \mathbb{A}^1 \cong V(m',1) \times \mathbb{A}^1$.

Answer. (1)–(2) The same as in the case of $X(2,1)$.
(3) We show that $R^{G_m} = \mathbb{C}[xy, yz, x^{m-i} z^i u \ (0 \le i \le m)]$. Suppose that a monomial $x^a y^b z^c u^d$ is G_m-invariant, equivalently, $a + c = b + md$, where a, b, c, d are nonnegative integers. We consider three cases separately

(i) Case $0 \le c < d$. Then

$$x^a y^b z^c u^d = (xy)^b (x^{m-1}zu)^c (x^m u)^{d-c}.$$

(ii) Case $id \le c < (i+1)d$ for $1 \le i < m$. Then

$$x^a y^b z^c u^d = (xy)^b (x^{m-i}z^i u)^{(i+1)d-c} (x^{m-i-1}z^{i+1}u)^{c-id}.$$

(iii) Case $md \le c$. Then

$$x^a y^b z^c u^d = (xy)^a (yz)^{b-a} (z^m u)^d.$$

Hence R^{G_m} is generated by xy, yz and $x^{m-i}z^i u$ for $0 \le i \le m$.
It is clear that the homomorphism $\theta : B \to A$ is well-defined and surjective. Furthermore, if $X \ne 0$ then

$$Y_i = \frac{Y_m}{X^{m-i}}(1 + Y_m)^{m-i} \quad (0 \le i \le m).$$

Hence $\dim B = 2$, and θ is birational. Since $XY_{m-1} = Y_m(1+Y_m)$, $Y_m = 0$ or $Y_m = -1$ if $X = 0$. Since $XY_i = Y_{i+1}(1 + Y_m)$ for $0 \le i < m$, $Y_1 = \cdots = Y_{m-1} = 0$ if $X = Y_m = 0$. Hence θ induces an isomorphism between $B/(X, Y_m) \cong \mathbb{C}[Y_0]$ and $\mathbb{C}[x^m u]$. Suppose that $Y_m = -1$. We show by induction on i that if $Y_{m-1} \ne 0$ then Y_{m-i} is a monomial in Y_{m-1} of positive degree for $0 \le i < m$. In fact, since $Y_m Y_{m-2} = (Y_{m-1})^2$, we have $Y_{m-2} = -(Y_{m-1})^2$. Suppose that $Y_{m-i+2} \sim (Y_{m-1})^{\alpha_{i-2}}$ and $Y_{m-i+1} \sim (Y_{m-1})^{\alpha_{i-1}}$ with $0 < \alpha_{i-2} < \alpha_{i-1}$ for $i \ge 2$, where $\sim$ means that two terms differ by a nonzero constant. Since $Y_{m-i+2}Y_{m-i} = (Y_{m-i+1})^2$, we have $Y_{m-i} \sim (Y_{m-1})^{\alpha_i}$ with $\alpha_i = 2\alpha_{i-1} - \alpha_{i-2} > 0$. Since $Y_{m-i-1}Y_{m-i+1} = (Y_{m-i})^2$, we have $Y_{m-i-1} \sim Y_{m-1}^{\alpha_{i+1}}$ with

$$\alpha_{i+1} = 2\alpha_i - \alpha_{i-1} = (4\alpha_{i-1} - 2\alpha_{i-2}) - \alpha_{i-1} = 3\alpha_{i-1} - 2\alpha_{i-2},$$

whence $\alpha_{i+1} - \alpha_i = 2(\alpha_{i-1} - \alpha_{i-2}) > 0$ by induction. So, we are done.
Suppose that $X = Y_{m-1} = 0$ and $Y_m = -1$. Since $Y_{m-i} \sim (Y_{m-1})^{\alpha_i}$, we have $Y_i = 0$ for $0 \le i < m$. This implies that $\operatorname{Spec} B/(X, Y_m + 1) \cong \{\text{one point}\} \cup \{Y_{m-1} \ne 0\}$, where $\{Y_{m-1} \ne 0\} \cong \mathbb{A}^1_*$ and θ induces an isomorphism between $B[Y_{m-1}^{-1}]/(X, 1 + Y_m) \cong \mathbb{C}[Y_{m-1}, Y_{m-1}^{-1}]$ and $\mathbb{C}[xz^{-1}, x^{-1}z]$. By the Serre criterion of normality, B is a normal domain, and θ is an isomorphism since A is regular by the assertion (4) below.
(4) Since the G_m-action on $X(m,1)$ has trivial isotropy group everywhere, the quotient surface $V(m,1)$ is smooth by Luna's étale slice theorem [135]. The rest of the assertion (4) is straightforwardly verified.
(5) Equality (2.15) follows from the relations $XY_{i-1} = Y_i(1 + Y_m)$ for $1 \le i \le m$. Let $X = 0$. Then $Y_m = 0$ or $Y_m = -1$ because $XY_{m-1} = Y_m(1+Y_m)$. If $Y_m = 0$ then $XY_{i-1} = Y_i(1 + Y_m)$ implies $Y_i = 0$ for $1 \le i \le m$. Then Y_0 is a variable of the irreducible component of $\rho_1^{-1}(O)$ with $Y_m = 0$. Suppose $Y_m = -1$. By (3), one can show that

$Y_m^{i-1} Y_{m-i} = Y_{m-1}^i$ for $0 \le i \le m$, whence follows the equality (2.16). So, the other irreducible component of $\rho_1^{-1}(O)$ with $Y_m = -1$ is isomorphic to $\mathbb{A}^1 = \operatorname{Spec} \mathbb{C}[Y_{m-1}]$.
(6) It is clear that $\{xy, yz, x^{m-i}z^i u \ (0 \le i \le m)\}$ is a minimal set of generators of A. Hence follows the rest of the assertion.
(7) Note that by assertion (4) $V(m,1)$ has a G_a-bundle structure over a nonseparated scheme $\mathbb{A}^1$ with two points lying over the origin. The assertion can be proved by the same argument as in Dubouloz [40, Example 0.1]. □

11. Let $X_m := X(m,1)$ be an affine hypersurface $xy - z^m u = 1$ as in Problem 10. Prove the following assertions:
    (1) $H_1(X_m;\mathbb{Z}) = H_2(X_m;\mathbb{Z}) = 0$ and $H_3(X_m;\mathbb{Z}) = \mathbb{Z}$.
    (2) Let N_m be the boundary 5-manifold at infinity, i. e., the boundary of a tubular neighborhood of the boundary divisor at infinity with respect to a suitable smooth normal compactification of X_m. Then the homology groups $H_i(N_m;\mathbb{Z})$ is independent of m, where $0 \le i \le 5$.

Answer. (1) As in the previous problem, X_m has a fixed-point free G_m-action. Hence X_m is a $\mathbb{C}^*$-fiber bundle over $V_m := X_m/\!/G_m$ in the sense of C^∞-topology. Hence we have a homotopy exact sequence

$$\pi_2(\mathbb{C}^*) \to \pi_2(X_m) \to \pi_2(V_m) \to \pi_1(\mathbb{C}^*) \to \pi_1(X_m) \to \pi_1(V_m),$$

where $\pi_2(X_m) \cong H_2(X_m;\mathbb{Z})$ and $\pi_2(V_m) \cong H_2(V_m;\mathbb{Z})$ by Hurewicz's isomorphism theorem since $\pi_1(X_m) = \pi_1(V_m) = (1)$. Note that $H_2(V_m;\mathbb{Z}) \cong \mathbb{Z}$ because $e(V_m) = 2$ and $H_2(V_m;\mathbb{Z})$ has no torsion by Kaup–Narasimhan–Hamm theorem. Hence $H_2(X_m;\mathbb{Z}) = 0$ because $\pi_2(\mathbb{C}^*) \cong \pi_2(S^1) = (1)$.
(2) By a change of coordinates, we write the defining equation of X_m as $w^2 + z^2 + x^m y = 1$. Let N_m be as stated as in the assertion. It is also the boundary of a big closed ball taken inside X_m. Then N_m is a ramified covering of N_1 which is totally ramified over $F_1 := N_1 \cap \{x = 0\}$ and unramified outside F_1.
Note that X_1 has a natural compactification $\overline{X}_1 = \{W^2 + Z^2 + XY = U^2\}$ which is a quadric hypersurface in $\mathbb{P}^4$, and $\overline{X}_1 \setminus X_1$ is the quadric surface $Q = \{W^2+Z^2+XY = 0\}$ in $\mathbb{P}^3$. Since N_1 is the boundary of a tubular neighborhood of Q which is smooth, N_1 is an S^1-bundle over Q. The complement $N_1 \setminus F_1$ is a trivial S^1-bundle over $\mathbb{A}^2 = Q \setminus \{X = 0\}$, whence $N_1 \setminus F_1$ is homeomorphic to $\mathbb{A}^2 \times S^1$.
Now F_1 is embedded into N_m since the mapping $N_m \to N_1, (w,z,x,y) \mapsto (w,z,x^m,y)$, is totally ramified over F_1. Since $N_m \setminus F_1 \to N_1 \setminus F_1$ is a cyclic étale covering of order m, we deduce that $N_m \setminus F_1$ is also homeomorphic to $\mathbb{A}^2 \times S^1$. Using the relative cohomology sequence for the pair (N_m, F_1), we have an exact sequence of integral cohomology groups

$$\begin{aligned} 0 &\to H^0(N_m, F_1) \to H^0(N_m) \to H^0(F_1) \\ &\to H^1(N_m, F_1) \to H^1(N_m) \to H^1(F_1) \end{aligned}$$

$$\begin{aligned}&\to H^2(N_m, F_1) \to H^2(N_m) \to H^2(F_1)\\&\to H^3(N_m, F_1) \to H^3(N_m) \to H^3(F_1)\\&\to H^4(N_m, F_1) \to H^4(N_m),\end{aligned}$$

where

$$\begin{aligned}&H^i(N_m, F_1) \cong H_{5-i}(\mathbb{A}^2 \times S^1; \mathbb{Z}) \cong H_{5-i}(S^1; \mathbb{Z}) = 0, \quad i = 0, 1, 2, 3,\\&H^4(N_m, F_1) \cong H_1(\mathbb{A}^2 \times S^1; \mathbb{Z}) \cong H_1(S^1; \mathbb{Z}) \cong \mathbb{Z}\end{aligned}$$

by the Lefschetz duality and by noting that homology groups of $\mathbb{A}^2 \times S^1$ are the same as those of S^1 since $\mathbb{A}^2$ is contractible to a point. Hence it follows that

$$\begin{aligned}&H^i(N_m; \mathbb{Z}) \cong H_i(F_1; \mathbb{Z}), \quad i = 0, 1, 2,\\&H^3(N_m; \mathbb{Z}) \cong \mathrm{Ker}(H^3(F_1; \mathbb{Z}) \to \mathbb{Z}),\\&H^4(N_m; \mathbb{Z}) \cong H_1(N_m; \mathbb{Z}) \cong \mathbb{Z},\\&H^5(N_m; \mathbb{Z}) \cong H_0((N_m; \mathbb{Z}) \cong \mathbb{Z}.\end{aligned}$$

For the computation of $H_1(N_m; \mathbb{Z})$, Nori's exact sequence (see Lemma 1.1.6) is available. Hence it follows that $H_i(N_m; \mathbb{Z})$ is independent of m. □

12. Let $V = \mathbb{F}_n$ with $n > 0$ and let $p : V \to \mathbb{P}^1$ be the $\mathbb{P}^1$-fibration. Let M be the minimal section, i. e., $(M^2) = -n$ and ℓ be a fiber of p. Verify the following assertions:
   (1) Let φ be an automorphism of V. Then φ preserves the $\mathbb{P}^1$-fibration p. Namely, $\varphi(\ell)$ is a fiber of p. Hence there is an exact sequence (see Maruyama [140])
   $$0 \longrightarrow \mathrm{Aut}_{\mathbb{P}^1}(V) \longrightarrow \mathrm{Aut}(V) \xrightarrow{\theta} \mathrm{Aut}(\mathbb{P}^1).$$
   (2) Let $\varphi \in \mathrm{Aut}_{\mathbb{P}^1}(V)$. Assume that φ stabilizes a cross-section $D \sim M + m\ell$ with $m > n$. Then φ is the identity. So, θ is injective.
   (3) Let G be a finite subgroup of $\mathrm{Aut}(V)$ such that each automorphism in G stabilizes the cross-section D as in (2) above. Then G is a cyclic group.

Answer. (1) Since M is the unique negative section of p, φ stabilizes M, i. e., $\varphi(M) = M$. Write $\varphi(\ell) \sim aM + b\ell$. If $a = 0$ then $b = 1$ because $\varphi(\ell)$ is irreducible. Hence $\varphi(\ell) \sim \ell$. Assume that $a \neq 0$. Since $(\varphi(\ell)^2) = 0$, we have $(aM + b\ell)^2 = a(2b - an) = 0$, whence $2b = an$. Since $(\varphi(\ell) \cdot M) \geq 0$, we have $b \geq an$. Hence $an \geq 2an$, which is a contradiction. It is clear that we then have the above exact sequence.
(2) Since $V = \mathrm{Proj}(\mathcal{O}_{\mathbb{P}^1} \oplus \mathcal{O}_{\mathbb{P}^1}(n))$, we can choose a system of local fiber coordinates $\{\xi_i\}_{i \in I}$ over an affine open covering $\mathcal{U} = \{U_i\}_{i \in I}$ of $\mathbb{P}^1$ in such a way that the minimal section M is locally defined by $\xi_i = 0$ over U_i. In fact, we can take the index set I consisting of two elements $\{1, 2\}$ such that $p^{-1}(U_i) \cong \mathbb{A}^1 \times \mathbb{P}^1$, where $U_i \cong \mathbb{A}^1$ for $i = 1, 2$. Then $\varphi^*(\xi_i) = c_i\xi_i$ with $c_i \in k^*$. Then the cross-section M' defined by

$\xi_i^{-1} = 0$ is also φ-stable. Hence φ stabilizes three points $\ell \cap M, \ell \cap M', \ell \cap D$, where $\ell = p^{-1}(Q)$ for $Q \in U_i$. Then φ is the identity on ℓ and the nearby fibers of ℓ. Hence φ is the identity.

(3) In the proof of (2), we take $I = \{1, 2\}$ and write $\xi = \xi_1$. Then $p^{-1}(U_1) \setminus (M' \cap p^{-1}(U_1)) \cong \mathbb{A}^2$, where we can take variables $\{t, \xi\}$ of $\mathbb{A}^2$ with t a coordinate of $U_1 \cong \mathbb{A}^1$. Then $\varphi^*(\xi) = c\xi$ with $c \in k^*$. The defining equation of $D \cap p^{-1}(U_1)$ is given in the form $a(t)\xi + b(t) = 0$ for $a(t), b(t) \in k[t]$. Since $(M \cdot D) = m - n$ and $(M' \cdot D) = m$ with the section M' defined by $\xi^{-1} = 0$, it follows that $\deg a(t) \le m$ and $\deg b(t) \le m - n$. In fact, we can choose the covering $U_1 \cup U_2 = \mathbb{P}^1$ so that these inequalities are exact equalities. Since $\varphi^*(a(t)\xi + b(t))$ is an irreducible polynomial in t, ξ, we have

$$a(\varphi^*(t))(c\xi) + b(\varphi^*(t)) = d(a(t)\xi + b(t)), \quad d \in k^*. \tag{2.17}$$

This implies that $\varphi^*(t) = bt$ with $b \in k^*$ after a necessary linear change of variable t. Assume that $(M \cdot D) = m - n$ and $(M' \cdot D) = m$. Write $a(t) = a_0 t^m + \cdots + a_m$ and $b(t) = b_0 t^{m-n} + \cdots + b_{m-n}$ with $a_0 b_0 \ne 0$. Plug in $\varphi^*(\xi) = c\xi$ and $\varphi^*(t) = bt$ into equation (2.17). Then we obtain $a_0 b^m c = da_0$ and $b_0 b^{m-n} = db_0$, whence $c = b^{-n}$. This implies that G is a finite group in G_m, hence G is a cyclic group. □

13. Let X be a smooth affine surface satisfying the conditions:
    (i) There exists a smooth projective surface V such that $X \subset V$ as an open set and $D := V - X$ is isomorphic to $\mathbb{P}^1$.
    (ii) X is an ML_0-surface.[21]

    Prove the following assertions:
    (1) V is isomorphic to $\mathbb{P}^2$ or $\mathbb{F}_n$ with $n \ge 0$ and that D is either a line or a conic if $V \cong \mathbb{P}^2$ and D is a curve of type $(1, m)$ with $m > n$ if $V \cong \mathbb{F}_n$.
    (2) Let $\sigma : \mathbb{F}_0 \to \mathbb{F}_0$ be the exchange of coordinates $\sigma(P, Q) = (Q, P)$ for $P, Q \in \mathbb{P}^1$ and let $G = \langle \sigma \rangle$. Then the diagonal Δ of $\mathbb{F}_0$ is the fixed point locus of G and the geometric quotient $\mathbb{F}_0/G \cong \mathbb{P}^2$ with $\pi(\Delta) = D$ a conic, where $\pi : \mathbb{F}_0 \to \mathbb{P}^2$ is the quotient morphism.
    (3) If $V = \mathbb{F}_n$ and D is a curve of type $(1, m)$ then a finite automorphism group of V stabilizing D is at most a cyclic group.
    (4) X has no finite automorphism group acting freely except for the case treated in (2) and (3).

Answer. (1) We use Lemma 2.6.8. We denote the divisor $\overline{X} - X$ by Δ to avoid confusion. Then $\Delta = \ell + M + A$, where A is a linear chain of smooth rational curves if $A \ne \emptyset$ and $|\ell|$ defines a $\mathbb{P}^1$-fibration on $\overline{X}$ with a section M. By elementary transformations on ℓ we may assume that $(M^2) = -1$. Then the surface V is obtained by

21 Condition (ii) is redundant. In fact, Problem 13 is based on an article by M. H. Gizatullin [59].

contracting M and all components of A so that the image of ℓ after the contraction is the boundary divisor D. If $A = \emptyset$, then $(D^2) = 1$ and $V \cong \mathbb{P}^2$. Hence D is a line. Suppose that $A \neq \emptyset$. Then A is a linear chain of (-2)-curves. Meanwhile, the components of A together with feathers $A'_1, \ldots, A'_r$ meeting A form a reducible fiber of the $\mathbb{P}^1$-fibration. Then there are three possibilities:

(a) $A = A_1 + A_2 + A_3$ with $(A_i^2) = -2$, $(M \cdot A_1) = 1$, $r = 1$ and A'_1 is a (-1)-curve meeting A_2.

(b) $A = A_1$, $r = 2$ and $(A_i'^2) = -1$ for $i = 1, 2$.

(c) $A = A_1 + A_2 + \cdots + A_{2m-n-1}$ with $(A_i^2) = -2$, $r = 2$ and two (-1)-curves A'_1, A'_2 meeting respectively A_1, A_{2m-n-1}.

In the case (a), the contraction of M, A_1, A_2, and A_3 makes the image of ℓ a conic D on $\mathbb{P}^2$ and the image of A'_1 a line touching D with order 2. In the case (b), the contraction of M, A_1 makes the image $\overline{\ell}$ of ℓ a smooth rational curve with $(\overline{\ell}^2) = 2$ and the images $\overline{A}'_1, \overline{A}'_2$ of A'_1, A'_2 two smooth curves such that $(\overline{A}_1'^2) = (\overline{A}_2'^2) = 0$ and $\overline{\ell}, \overline{A}'_1, \overline{A}'_2$ meet transversally in one point. Then $V \cong \mathbb{F}_0$ and $\overline{\ell}$ is the diagonal. The curves $\overline{A}'_1$ and $\overline{A}'_2$ define two transversal $\mathbb{P}^1$-fibrations on $\mathbb{F}_0$. In this case, X is a Danielewski surface $xy = z^2 - 1$. In the case (c), the contraction of $M, A_1, \ldots, A_{2m-n-1}$ makes ℓ the curves D with $(D^2) = 2m - n$. The curve A'_2 becomes a curve L such that $(L^2) = 0$ and $(D \cdot L) = 1$. Hence the surface V obtained from $\overline{X}$ by these contractions is a minimal ruled surface. The curve A'_1 has the image C such that $(C^2) = 2m - n - 2$ and $(D \cdot C) = i(D, C; C \cap D) = 2m - n - 1$. The existence of such curves C, D is guaranteed by Gizatullin [59].

(2) In fact, let V be a double covering of $\mathbb{P}^2$ ramified over a conic D. Since $D \sim 2\ell$ with a line on $\mathbb{P}^2$, we have $K_V \sim \pi^*(K_{\mathbb{P}^2} + \ell) = \pi^*(-2\ell) \sim -2L - 2M$, where $\pi^*(\ell) = L + M$ if ℓ touches D. Since the covering involution exchanges L and M and since $(L \cdot M) = 1$, it follows that $(L^2) = (M^2) = 0$. Then $|L|$ and $|M|$ define transversal $\mathbb{P}^1$-fibrations. Hence $V \cong \mathbb{P}^1 \times \mathbb{P}^1$ and $\pi^{-1}(D)$ is the diagonal Δ.

(3) The assertion is proved in Problem 12.

(4) Note that $\pi_1(\mathbb{F}_0 \setminus \Delta) = 1$, $\pi_1(\mathbb{P}^2 \setminus D) \cong \mathbb{Z}/2\mathbb{Z}$ and $\pi_1(\mathbb{P}^2 \setminus \ell) = 1$. The assertion follows from this remark. □

14. As in the settings in the proof of Theorem 2.8.2, let $\varphi : X \to Y$ be a (possibly ramified) Galois covering of smooth affine surfaces with group G. Let $X \hookrightarrow V$ be a G-equivariant completion into a smooth projective surface V with an extended G-action such that $D := V - X$ is a linear chain of smooth rational curves. Prove the following assertions:

(1) D is G-stable and a tip (i. e., an end component of the linear chain) of D is either stabilized or mapped to the other tip by the G-action.

(2) The quotient surface V/G is a normal projective surface, and D/G is a simply-connected divisor on V/G.

(3) V/G has at worst cyclic quotient singularities which lie on D/G.

Answer. (1) Since X is G-stable the divisor D is clearly G-stable. As an automorphism of V, any element g of G maps an irreducible component D_i of D to an irreducible component $g(D_i)$ of D in such a way that $n(D_i) = n(g(D_i))$, where $n(D_i)$ is the number of irreducible components of $D - D_i$ meeting D_i. Hence $n(D_i) = 1$ if D is reducible and D_i is a tip.
(2) It is known (see [172]) that if G is a finite group the geometric quotient V/G exists and $\pi^{-1}(z)$ is a G-orbit for any $z \in V/G$, where $\pi : V \to V/G$ is the quotient morphism. Since V is smooth, V/G is normal. Let D_i be an irreducible component of D and let S_i be the stabilizer subgroup $\{g \in G \mid g(D_i) = D_i\}$. Then $\pi(D_i) \cong D_i/S_i$ by the above observation. This implies that $\pi(D_i)$ is isomorphic to $\mathbb{P}^1$ as the quotient of $\mathbb{P}^1$ by a finite automorphism group.
(3) We show that possible singularities on V/G are cyclic quotient singularities. Let Q be a singular point on V/G. Then it necessarily lies on D/G. Let $P \in V$ such that $\pi(P) = Q$. There are two cases.

Case (i) P is the intersection point of two irreducible components of D, say D_1 and D_2. Let S be the stabilizer group of the point P. Take a system of local coordinates $\{u_1, u_2\}$ at P such that D_1 (resp. D_2) is defined by $u_2 = 0$ (resp., $u_1 = 0$). Since the tangential representation $\rho : S \to \mathrm{GL}(T_{V,P})$ is faithful, we may assume that u_1 (resp. u_2) is the tangential direction of D_1 (resp., D_2). Then we have

$$\rho(g) = \begin{pmatrix} a(g) & 0 \\ 0 & b(g) \end{pmatrix}, \quad g \in S.$$

Since $a(g), b(g) \in k^*$, S is a cyclic group. Hence $\widehat{\mathcal{O}}_{V/G,Q} \cong k[[u, v]]^S$ has a cyclic singularity, and so does $\mathcal{O}_{V/G,Q}$.

Case (ii) P lies on an irreducible component D_1 of D (not necessarily a tip of D) and is not the intersection point of D. Let S be the stabilizer group of the point P. Let u, v be the tangential directions of the irreducible component D_1 and a curve C transversal to D at P. Hence $C \cap X \neq \emptyset$. Then the tangential representation $\rho : S \to \mathrm{GL}(T_{V,P})$ with respect to a basis $\{u, v\}$ is

$$\rho(g) = \begin{pmatrix} a(g) & c(g) \\ 0 & b(g) \end{pmatrix}, \quad g \in S,$$

where $g(u) = a(g)u + c(g)v$ and $g(v) = b(g)v$. If $a(g) = b(g)$ then $c(g) = 0$. In fact,

$$\rho(g^n) = \begin{pmatrix} a^n(g) & na^{n-1}(g)c(g) \\ 0 & a^n(g) \end{pmatrix}.$$

Since $g^n = 1$ for some $n > 0$, we have $na^{n-1}(g)c(g) = 0$. Since $a(g) \neq 0$, we have $c(g) = 0$. If $a(g) \neq b(g)$, set $u' = u + d(g)$, where $d(g) = c(g)/(a(g) - b(g))$. Then the representation with respect to $\{u', v\}$ is given by

$$\rho'(g) = \begin{pmatrix} a(g) & 0 \\ 0 & b(g) \end{pmatrix}.$$

Hence, including the case $a(g) = b(g)$, the diagonal representation ρ' of S is faithful. Hence S is a cyclic group, and the singularity is a cyclic singularity. □

15. Let X be a smooth affine surface with $\rho(X) = 0$. Show that there is no surjective $\mathbb{A}^1$-fibration $f : X \to \mathbb{P}^1$.

Answer. Suppose that there exists an $\mathbb{A}^1$-fibration $f : X \to \mathbb{P}^1$. Then there exists a smooth normal completion $X \hookrightarrow V$ such that $D := V - X$ is an SNC-divisor and f extends to a $\mathbb{P}^1$-fibration $p : V \to \mathbb{P}^1$. Then one irreducible component, say D_1, of D is a cross-section of p and other components are fiber components which are algebraically independent. We may assume that every fiber component of D is not a (-1)-curve. Since $\rho(X) = 0$, every fiber of f is irreducible. Let $P_0 \in \mathbb{P}^1$, let $m_0L_0 = f^*(P_0)$ be the scheme-theoretic inverse image and let $F_0 = p^*(P_0)$, where L_0 is reduced. Since $\rho(X) = 0$, some multiple am_0L_0 is linearly equivalent to an integral combination of irreducible components of D. Write this relation as

$$aF_0 \sim a_1D_1 + \sum_{D_i \neq D_1} a_iD_i.$$

Since D_i is a fiber component of p if $D_i \neq D_1$, the intersection of this equivalence relation with a general fiber F of p shows that $a_1 = 0$. The intersection of this equivalence relation with an ample divisor on V shows that some $a_i \neq 0$. Since the intersection form on $D - D_1$ is negative-definite and $a_i \neq 0$ for some i, we have

$$\left(\sum_{D_i \neq D_1} a_iD_i\right)^2 < 0.$$

But this contradicts $(aF_0)^2 = 0$. □

16. Let $X = \operatorname{Spec} A$ be a normal affine surface with a nontrivial G_a-action. Show that the quotient morphism $q : X \to Y = \operatorname{Spec} B$ is surjective.

Answer. Let δ be a nonzero lnd on A corresponding to the given G_a-action on X. Then $B = \operatorname{Ker}\delta$ and q is induced by the inclusion $B \hookrightarrow A$. Note that Y is a smooth affine curve. Suppose that $q(X) \subsetneq Y$. Let $Q \in Y$ be a closed point lying outside of $q(X)$ and let $\mathfrak{n}$ be the maximal ideal of B corresponding to Q. Hence the local ring $\mathcal{O}_{Y,Q} = B_\mathfrak{n}$. Let $A_\mathfrak{n} = A \otimes_B B_\mathfrak{n}$. Then the lnd δ lifts to an lnd $\delta_\mathfrak{n}$ on $A_\mathfrak{n}$ and $B_\mathfrak{n} = \operatorname{Ker}\delta_\mathfrak{n}$. Since $B_\mathfrak{n}$ is a DVR, let t be a uniformizant of $B_\mathfrak{n}$. Since $\mathfrak{n}A_\mathfrak{n} = A_\mathfrak{n}$, we have $ta' = 1$ for $a' \in A_\mathfrak{n}$. Then $a' \in B_\mathfrak{n}$. This is a contradiction. Hence $Y = q(X)$. □

17. Show that the two conditions in Definition 2.4.24 are equivalent.

Answer. (2) ⇒ (1). This is obvious.
(1) ⇒ (2). Let $x_i \in A$ be a local slice with $\delta_i(x_i) = c_i$ for δ_i for $i = 1, 2$. Write $B = \operatorname{Ker}\delta_1 = \operatorname{Ker}\delta_2$. Then $A[c_i^{-1}] = B[c_i^{-1}][x_i]$ and $\delta_i = c_i(\partial/\partial x_i)$ in $A[c_i^{-1}]$. Let $c = c_1c_2$.

Then $A[c^{-1}] = B[c^{-1}][x_1] = B[c^{-1}][x_2]$. Switching x_1 and x_2 if necessary, we may assume that $c^m x_2 = b_1 x_1 + b_2$ with $m \geq 0$ and $b_1, b_2 \in B$. Then we have

$$\frac{\partial}{\partial x_2}(c^m x_2) = c^m = \frac{\partial}{\partial x_2}(b_1 x_1 + b_2) = b_1 \frac{\partial x_1}{\partial x_2},$$

where $c^\ell(\partial x_1/\partial x_2) \in A$ for some $\ell \geq 0$. This implies that $c^\ell(\partial x_1/\partial x_2) \in B$ because $c^{\ell+m} \in B$. Since $(\partial/\partial x_2) = d(\partial/\partial x_1)$ with $d = (\partial x_1/\partial x_2)$, we have

$$c^\ell c_1 \delta_2 = c^\ell c_1 c_2 \frac{\partial}{\partial x_2} = c_1 c_2 (c^\ell d)\frac{\partial}{\partial x_1} = c_2(c^\ell d)\delta_1. \qquad \square$$

3 Fibrations in higher dimension

The notion of $\mathbb{A}^1$-fibration plays an important role in the theory of affine algebraic surfaces, especially for those with logarithmic Kodaira dimension $-\infty$. The same role is expected in the study of affine algebraic varieties of higher dimension.

In the previous chapter, we observed that an $\mathbb{A}^1$-fibration of affine type $f : X \to Y$ is the quotient morphism of a certain G_a-action on X. As a measure of abundance of G_a-actions, we introduced the notion of ML-invariant and accordingly ML_i-surface. In particular, important is the class of ML_0-surfaces which enjoy properties similar to the affine plane.

In the present chapter, we try to generalize these observations in dimension three or higher. Distinction between an $\mathbb{A}^1$-fibration and the quotient morphism of a G_a-action appears even in the case of an $\mathbb{A}^1$-fibration of affine type, i. e., the one with affine base variety. Let $f : X \to Y$ be an $\mathbb{A}^1$-fibration from a normal affine variety $X = \operatorname{Spec} A$ to a normal affine variety $Y = \operatorname{Spec} B$. If $\dim X \geq 3$, then $\dim Y = \dim X - 1 \geq 2$ and the morphism f is factored by the quotient morphism of a G_a-action on X by Lemma 2.2.7 as

$$f : X \xrightarrow{q} \widetilde{Y} \xrightarrow{\nu} Y,$$

where $\widetilde{Y} = \operatorname{Spec} \widetilde{B}$ for the factorial closure $\widetilde{B}$ of B in A, which is defined as $\operatorname{Ker} \delta$ for a locally nilpotent B-derivation δ corresponding to the G_a-action on X, or equivalently $\widetilde{B} = A \cap Q(B)$. Hence $q : X \to \widetilde{Y}$ is the quotient morphism and $\nu : \widetilde{Y} \to Y$ is determined by the inclusion $B \hookrightarrow \widetilde{B}$. The following major obstacles in higher dimension already appear at this step:

(i) The k-algebra $\widetilde{B}$ is an integrally closed domain, birational to B, but it might not be an affine domain if $\dim A \geq 4$. This obstacle is caused by the existence of counterexamples to the Hilbert's 14th problem. But $\widetilde{B}$ is an affine domain if $\dim A = 3$ by Zariski's finiteness theorem.

(ii) Suppose that $\widetilde{B}$ is an affine domain and hence $\widetilde{Y}$ is a normal affine variety. The morphism $\nu : \widetilde{Y} \to Y$ is then birational but not necessarily quasifinite. If ν is quasifinite then ν is an open immersion by Zariski's main theorem. Hence, if f is equidimensional and surjective, $\widetilde{Y}$ is isomorphic to Y. So, f is the quotient morphism by a G_a-action. In general, ν is not quasifinite, and f is not equidimensional or surjective. Hence, in dimension greater than two, a clear distinction exists between an $\mathbb{A}^1$-fibration of affine type and a G_a-action.

(iii) Even if X is smooth, Y may acquire singularities as in Winkelman's example (see Example 1.3.5). But singularity is mild if the fiber lying over the singular point is one-dimensional (see Lemma 3.1.30).

(iv) If G_a acts on a smooth affine variety, it happens that the geometric quotient X/G_a exists as an open set of the algebraic quotient $X//G_a$. A typical example is $X =$

https://doi.org/10.1515/9783110577563-003

$\mathrm{SL}(2,k)$ with a G_a-action given by the left multiplication of lower unipotent subgroup. In this case, X/G_a is the punctured affine plane $\mathbb{A}^2_*$ and $X//G_a$ is the affine plane $\mathbb{A}^2$.

(v) Singular fibers of an $\mathbb{A}^1$-fibration $f : X \to Y$ are not easily described. We can give some description in the case when f is the quotient morphism under a G_a-action on a smooth affine threefold X with some restriction on $\dim X^{G_a}$ and the singular fiber is pure one-dimensional. If f contains a fiber component S of dimension greater than one, it is hard to describe the structure of S.

But, as $\dim X$ (hence $\dim Y$) grows, there arise many obstacles, among which we can list the following ones. None of them appear in the case $\dim X = 2$.

(1) The morphism f will have closed singular fibers of dimension greater than one, and hence f fails to be a flat morphism.
(2) Y may acquire singular points although X is $\mathbb{A}^4$ (see Example 3.4.18 for Weitzenböck derivation) as pointed already in the item (iii) above.
(3) Some of fiber components of f might not be rational as in the case when X is a hypersurface $x^m y = f(x, z_1, \ldots, z_{n-1})$ in Section 3.4, where the hypersurface $f(0, z_1, \ldots, z_{n-1}) = 0$ in $\mathbb{A}^{n-1}$ is not rational.

In Section 3.1, we consider properties of an $\mathbb{A}^1$-fibration and the quotient morphism by G_a-action which can be treated by algebraic methods. We observe when an $\mathbb{A}^1$-fibration $f : X \to Y$ becomes the quotient morphism by a G_a-action on X. The condition that X is factorial makes the situation easier to treat. The G_a-quotient morphism is not necessarily surjective. We classify principal G_a-bundles, i. e., G_a-torsors, over the punctured affine plane $\mathbb{A}^2_*$ up to isomorphisms. For the quotient morphism by a G_a-action, we consider the fixed point locus X^{G_a} by making use of cohomology groups of X with compact supports, singular fibers, and singularities of the algebraic quotient surface when $\dim X = 3$. As the last topic in Section 3.1, we apply our arguments for G_a-action to vector fields on $\mathbb{A}^3$. This section is based on [79].

In Section 3.2, Miyanishi theorem is treated from a topological point of view. For this purpose, homology and contractible n-spaces are defined in analogy to homology and contractible planes. When $n = 3$, the structures of these spaces are studied via $\mathbb{A}^1$-fibrations and $\mathbb{A}^1_*$-fibrations in the next Section 3.3.

Section 3.3 is devoted to a study of affine varieties with $\mathbb{A}^1_*$-fibrations which are or not induced by G_m-actions. Singular fibers or singular points of the base variety behave like in the case of $\mathbb{A}^1$-fibration and are handled in more details because the quotient by a G_m-action is well understood. The Koras–Russell threefold is a contractible 3-space which draws particular attention and leads to hypersurfaces to be studied in Section 3.4.

In Section 3.4, we consider $\mathbb{A}^n$-fibrations on algebraic varieties and affine varieties of dimension $n+1$ containing $\mathbb{A}^n$-cylinders. The existence of an $\mathbb{A}^n$-cylinder on an affine variety X implies the existence of independent, mutually commutative

G_a-actions $\sigma_1, \ldots, \sigma_n$ on X. When $n = 2$, we define an affine pseudo-3-space as one subclass of homology 3-spaces. It has properties more close to the affine 3-space. Theorem 3.4.11, which is proved with the help of affine modification by Kaliman–Zaidenberg [104] will exhibit the significance of an affine pseudo-3-space.

In Section 3.5, we push forward the study of G_a-actions on algebraic varieties developed in Section 3.1 further in the case of an algebraic threefold.

3.1 $\mathbb{A}^1$-fibrations in higher dimension

3.1.1 Motivation and summary of principal results

A major example of an $\mathbb{A}^1$-fibration is given by the quotient morphism $q : X \to Y$ when G_a acts algebraically on an affine normal variety X and the ring of invariants $\Gamma(X, \mathcal{O}_X)^{G_a}$ is finitely generated over k. In the case $\dim X = 2$, any $\mathbb{A}^1$-fibration of affine type is the quotient morphism by a G_a-action. But this will not be the case if $\dim X \geq 3$. Hence we need to know the conditions with which one can distinguish quotient morphisms from $\mathbb{A}^1$-fibrations. It is shown in Lemma 3.1.2 that any $\mathbb{A}^1$-fibration of affine type $f : X \to Y$ is split as a composite $q : X \xrightarrow{\widetilde{q}} \widetilde{Y} \to Y$, where $\widetilde{q}$ is the quotient morphism by a G_a-action on X in the generalized sense and $\widetilde{Y}$ is the *factorial closure* of Y in X. The factorial closure $\widetilde{Y}$ is not necessarily an algebraic variety if $\dim X > 3$, but $\widetilde{Y}$ is affine algebraic in the case $\dim X = 3$.

The quotient morphism by G_a-action itself is not easy to study. Notwithstanding, there are a number of articles which have studied the quotient morphisms under G_a-actions in terms of fiber components of bigger dimension and singularities of the quotient variety (see [18, 37, 38, 39, 44, 226]). We consider in the present book mostly the quotient morphisms by G_a-actions in the case $\dim X = 3$.

In Subsection 3.1.2, we consider the properties of $\mathbb{A}^1$-fibrations and the quotient morphisms by G_a-actions which can be treated by algebraic methods. With the factoriality condition in the case $\dim X = 3$, an $\mathbb{A}^1$-fibration $f : X \to Y$ being the quotient morphism by a G_a-action is equivalent to the morphism f being equidimensional (see Theorem 3.1.8). We also consider one-dimensional, reducible, singular fibers of f (see Lemma 3.1.7). If there are no singular fibers, the $\mathbb{A}^1$-fibration $f : X \to Y$ is an $\mathbb{A}^1$-bundle over the image $f(X)$. We also discuss the $\mathbb{A}^1$-fibrations by replacing the factoriality condition by the $\mathbb{Q}$-factoriality condition (see Lemmas 3.1.7 and 3.1.13).

There is an important well-known example of an $\mathbb{A}^1$-bundle over $\mathbb{A}^2_* := \mathbb{A}^2 \setminus \{0\}$ (see Example 3.1.9). In Subsection 3.1.3, we extend this example and classify all the $\mathbb{A}^1$-bundles over $\mathbb{A}^2_*$. As one of such surfaces, we find a G_a-action on a nonfactorial smooth affine threefold such that the quotient morphism $q : X \to Y$ has the properties that $Y \cong \mathbb{A}^2$, $q^{-1}(0) \cong \mathbb{A}^2$ and q is an $\mathbb{A}^1$-bundle over $\mathbb{A}^2_*$ (see Example 3.1.22). This makes a sharp contrast with the case where X is factorial (see Theorem 3.1.8(1)). We have a conjecture that if $q : X \to Y$ is the quotient morphism by a G_a-action on a smooth affine threefold X, then the singularities on Y are at most cyclic quotient singularities. Our trial toward solving this conjecture is to classify all the $\mathbb{A}^1$-bundles over the Platonic $\mathbb{A}^1_*$-surfaces $\overline{U} := \mathbb{A}^2/G - \{\overline{O}\}$ (see [154, 159] for the relevant results), where G is a cyclic group, and extend some of such $\mathbb{A}^1$-bundles to smooth affine threefolds with G_a-actions such that the quotient morphism coincides with the given $\mathbb{A}^1$-fibration of the $\mathbb{A}^1$-bundle and the quotient space is $\mathbb{A}^2/G$. The equation defining such an $\mathbb{A}^1$-bundle over $\overline{U}$ is rather involved, and we are not successful so far in this direction. The same object has been taken up independently and from a different viewpoint in [44]. Notably, the interpretation of the Čech cohomology $H^1(\mathbb{A}^2_*, \mathcal{O}_{\mathbb{A}^2_*})$ is the same as ours.

In Subsection 3.1.5, we come back again to the study of degenerate fibers and singularities of the quotient variety Y when the $\mathbb{A}^1$-fibration $f : X \to Y$ is given by a G_a-action. The difference between Subsections 3.1.2 and 3.1.5 is that we assume the ground field k to be the complex number field $\mathbb{C}$ in Subsection 3.1.5 in order to use freely the topological and geometric arguments. The variety X is a smooth affine threefold in most results. If X is not complete, the fixed point locus X^{G_a} is either the empty set or is positive-dimensional by the argument of Białynicki–Birula [17] (see also [55, Theorem 1.30 and related references therein]). If the fiber $f^{-1}(Q)$ with a point Q of Y has an irreducible component of dimension one, Q has at most a quotient singularity in Y (see Lemma 3.1.30). It is shown in Lemma 3.1.31 that if F is a one-dimensional fiber over a point $Q \in Y$ of the morphism f, then F is a disjoint union of contractible curves and in fact a disjoint union of the affine lines provided F is a reduced fiber (cf. Corollary 3.1.39). We expect that the last result holds without F being reduced. Although G_a-action is not exploited fully in the higher-dimensional case, Example 3.1.40 may suggest that the existence of G_a-stable embedded components in the fiber $f^{-1}(Q)$ has something to do with unpleasant complicated behaviors of the fibers. In the last part of Subsection 3.1.4, it is observed that for the quotient morphism $q : X \to Y$ by a G_a-action, the triviality of the tangent bundle T_Y implies the triviality of T_X (see Lemma 3.1.41). Even if the ambient variety X with a G_a-action is smooth, the singularities on the quotient variety seem to depend on the way how G_a acts on X. In [38, 39], Deveney and Finston gave some interesting results.

In Subsection 3.1.6, we consider a polynomial ring $A = k[x, y, z]$ and a nonzero k-derivation D of A, which is not necessarily a locally nilpotent derivation. Let $B = \operatorname{Ker} D = \{a \in A \mid D(a) = 0\}$. We set $X = \operatorname{Spec} A$, $Y = \operatorname{Spec} B$, and $f : X \to Y$ the morphism induced by the inclusion $B \hookrightarrow A$. So, we have a situation similar to that considered in the previous subsections. The k-algebra B is an affine k-domain and has dimension at most two. The general fibers of f are irreducible and reduced, although they might not be isomorphic to the affine space. We describe a necessary and sufficient condition for D to be locally nilpotent in terms of the morphism f and the variety Y (see Lemma 3.1.46). Theorem 3.1.47 is a slight generalization of Theorem 3.2.5 saying that the G_a-invariant subalgebra of A is a polynomial ring in two variables over k when G_a acts on $X = \mathbb{A}^3$.

3.1.2 $\mathbb{A}^1$-fibrations and G_a-actions in higher dimension

In Sections 2.1 and 2.2 we described general properties of $\mathbb{A}^1$-fibrations and relationships between $\mathbb{A}^1$-fibrations and quotient morphisms by G_a-actions. We intend to generalize such descriptions to the higher-dimensional case. We begin with supplementing these results.

Let $f : X \to Y$ be an $\mathbb{A}^1$-fibration for algebraic varieties X and Y. By the generic triviality of $\mathbb{A}^1$-fibration, the generic fiber of f is isomorphic to $\mathbb{A}^1$ (see Lemmas 2.1.3 and 2.1.4). By the same results, we know the following result.

Lemma 3.1.1. *Let $f : X \to Y$ be an $\mathbb{A}^1$-fibration of algebraic varieties. Then there exists a nonempty Zariski open subset U of Y such that*
(1) *U is contained in the smooth locus of Y.*
(2) *The restriction $f_U : f^{-1}(U) \to U$ is an $\mathbb{A}^1$-bundle in the Zariski topology of Y.*

An integral domain R is said to be *factorial* if R is a UFD. Similarly, an affine scheme $\operatorname{Spec} R$ is *factorial* if so is R. This use of the term *factoriality* is different from its use in projective algebraic geometry where factoriality means the coincidence of Weil and Cartier divisors. We say that an algebraic k-scheme X is *locally factorial* if the local ring $\mathcal{O}_{X,x}$ is a UFD for every point x of X. In the setting of Lemma 3.1.1, we assume that X and Y are affine varieties. Write $X = \operatorname{Spec} A$ and $Y = \operatorname{Spec} B$. Then the morphism f identifies B with a k-subalgebra of A. In the statement of Lemma 2.2.1, B is *factorially closed* in A in the sense that $b = a_1a_2$ for an element $b \in B \setminus \{0\}$ with $a_1, a_2 \in A$ implies $a_1, a_2 \in B$. In Lemma 2.2.7, we defined the *factorial closure* $\widetilde{B}$ of B in A. The factorial closure $\widetilde{B}$ is not necessarily finitely generated over k. The following lemma is a restatement of Lemma 2.2.7.

Lemma 3.1.2. *Let $f : X \to Y$ be an $\mathbb{A}^1$-fibration of algebraic varieties. Assume that X and Y are affine varieties. Then there exists a G_a-action on X such that the factorial closure $\widetilde{B}$ is the ring of G_a-invariants of A. The morphism f is factored by the* quotient morphism in the generalized sense *$q : X \to \widetilde{Y}$ as*

$$f : X \xrightarrow{q} \widetilde{Y} \xrightarrow{\nu} Y,$$

where $\widetilde{Y} = \operatorname{Spec} \widetilde{B}$.[1] *If B is factorially closed in A, then f is the quotient morphism of a G_a-action on X.*

A quotient morphism in the generalized sense is often abbreviated as a quotient morphism if there is no fear of confusion.

If B is not factorially closed in A, an $\mathbb{A}^1$-fibration is not given by a G_a-action. The following example shows that f is not necessarily equidimensional in such a case.

1 The definition of quotient morphism requires B to be finitely generated over k. If $\widetilde{B}$ is not finitely generated over k, we treat $q : X \to \widetilde{Y}$ as the quotient morphism in a broad sense.

Example 3.1.3. Let $A = k[x,y,z]$ be a polynomial ring and $B = k[x,xy]$. It is clear that B is not factorially closed in A, where $\widetilde{B} = k[x,y]$. Let $X = \operatorname{Spec} A \cong \mathbb{A}^3$ and let $Y = \operatorname{Spec} B \cong \mathbb{A}^2$. The inclusion $B \hookrightarrow A$ defines an $\mathbb{A}^1$-fibration $f : X \to Y$. In fact, for a point $P \in Y$ defined by $x = \alpha, xy = \beta$ with $\alpha, \beta \in k$, the fiber $f^{-1}(P)$ is an affine line $x = \alpha, y = \beta/\alpha$ provided $\alpha \neq 0$, while it is an affine plane $x = 0$ provided $\alpha = 0$ (hence $\beta = 0$). The morphism p has a fiber of dimension 2, and is not surjective since the image of f misses the points on the xy-axis except for the origin.

By Lemma 2.2.9, an $\mathbb{A}^1$-fibration $f : X \to Y$ of normal affine varieties might not be surjective, but the image $f(X)$ contains all codimension one points of Y if f is the quotient morphism of a G_a-action on X.

Concerning the (possibly closed) set $Y - f(X)$ in the case $\dim X = 3$, we give the following example which shows that $\#(Y - f(X))$ is arbitrarily big.

Example 3.1.4. Let $\varphi(x) \in k[x]$ (resp., $\psi(y) \in k[y]$) be a polynomial such that $\gcd(\varphi(x), \varphi'(x)) = 1$ (resp., $\gcd(\psi(y), \psi'(y)) = 1$). Namely, $\varphi(x) = 0$ and $\psi(y) = 0$ have distinct simple roots. Let X be the hypersurface $\varphi(x)z = \psi(y)u + 1$. Then X is smooth and has a G_a-action defined by $\delta(x) = \delta(y) = 0, \delta(z) = \psi(y)$ and $\delta(u) = \varphi(x)$ on the coordinate ring A. Then $B := \operatorname{Ker} \delta = k[x,y]$ and the quotient morphism $q : X \to Y$ has the image $Y - S$, where $S = \{(\alpha, \beta) \mid \varphi(\alpha) = \psi(\beta) = 0\}$.

We remark the following result which gives a sufficient condition for a given $\mathbb{A}^1$-fibration to descend down under a birational morphism.

Lemma 3.1.5. *Let A_i $(i = 1,2)$ be a normal affine k-domain such that A_1 is a subalgebra of A_2 and $Q(A_1) = Q(A_2)$. Let B_i $(i = 1,2)$ be an affine k-subalgebra of A_i such that $B_1 = B_2 \cap A_1$. Let $X_i = \operatorname{Spec} A_i$ and $Y_i = \operatorname{Spec} B_i$. Let $f_i : X_i \to Y_i$ be the morphism induced by the inclusion $B_i \hookrightarrow A_i$. Assume that B_2 is integrally closed in A_2 and $\dim B_1 = \dim A_2 - 1$. Then the following assertions hold:*

(1) *If f_2 is an $\mathbb{A}^1$-fibration, so is f_1. In particular, $Q(B_1) = Q(B_2)$.*
(2) *If B_2 is factorially closed in A_2, so is B_1 in A_1. Hence if f_2 is the quotient morphism of a certain G_a-action on X_2, there exists a G_a-action on X_1 and f_1 is the quotient morphism under this G_a-action.*

Proof. (1) Let $\varphi : X_2 \to X_1$ be a birational morphism induced by the inclusion $A_1 \hookrightarrow A_2$. Suppose that $f_2|_E : E \to Y_2$ is not dominating for every irreducible exceptional subvariety E of φ, i. e., $\dim \varphi(E) < \dim E$. Then a general fiber F_2 of f_2 is isomorphic to the image $\varphi(F_2)$. Hence $f_1 : X_1 \to Y_1$ is factored by an $\mathbb{A}^1$-fibration. Suppose now that $f_2(E)$ is dense in Y_2 for an irreducible exceptional subvariety E of φ. Let $\mathfrak{P}_2$ be a prime ideal of A_2 corresponding to E. Then $\mathfrak{P}_2 \cap B_2 = (0)$ and $\dim A_2/\mathfrak{P}_2 = \dim B_2$ because $\dim A_2 > \dim A_2/\mathfrak{P}_2 \geq \dim B_2 = \dim A_2 - 1$. Set $\mathfrak{P}_1 = \mathfrak{P}_2 \cap A_1$. Then $A_1/\mathfrak{P}_1 \subseteq A_2/\mathfrak{P}_2$ and $\dim A_1/\mathfrak{P}_1 < \dim A_2/\mathfrak{P}_2$ because E is an exceptional subvariety for φ. Since $\mathfrak{P}_1 \cap B_1 = (\mathfrak{P}_2 \cap B_2) \cap B_1 = (0)$, it follows that $B_1 \subseteq A_1/\mathfrak{P}_1$. This is a contradiction since $\dim B_1 = \dim A_2 - 1 = \dim A_2/\mathfrak{P}_2$ by the assumption. So, f_1 is factored by an

$\mathbb{A}^1$-fibration $q_1 : X_1 \to Z = \operatorname{Spec} R$, where R is the normalization of B_1 in A_1. Since B_2 is integrally closed in A_2 and $B_1 = A_1 \cap B_2$, it is then easy to show that $R = B_1$. Hence f_1 is an $\mathbb{A}^1$-fibration.

(2) Since $B_1 = A_1 \cap B_2$ and B_2 is factorially closed in A_2, it is easy to show that B_1 is factorially closed in A_1. Then, by Lemma 3.1.2, f_1 is the quotient morphism under a certain G_a-action on X_1. □

The following example will exhibit the situation treated in the above lemma.

Example 3.1.6. Let $A_1 = k[x, y, z]$ be a polynomial ring and let $A_2 = k[x, \frac{y}{x}, z]$. Define a locally nilpotent derivation δ_2 on A_2 by $\delta_2(z) = \frac{y}{x}, \delta_2(\frac{y}{x}) = x, \delta_2(x) = 0$. Then $B_2 := \operatorname{Ker} \delta_2 = k[x, 2xz - (\frac{y}{x})^2]$. Let $B_1 = A_1 \cap B_2$. Then $B_1 = k[x, 2x^3z - y^2]$. The derivation δ_2 defines a G_a-action on $X_2 = \operatorname{Spec} A_2$, whose quotient morphism $f_2 : X_2 \to Y_2 := \operatorname{Spec} B_2$ is given by the inclusion $B_2 \hookrightarrow A_2$. Let $X_1 = \operatorname{Spec} A_1$ and let $Y_1 = \operatorname{Spec} B_1$. Then the morphism f_2 induces an $\mathbb{A}^1$-fibration $f_1 : X_1 \to Y_1$. Indeed, for a point P of Y_1 given by $x = \alpha, 2x^3z - y^2 = \beta$ with $\alpha, \beta \in k$, the fiber $f_1^{-1}(P)$ is an affine line $2\alpha^3 z - y^2 = \beta$ if $\alpha \neq 0$, a disjoint union of two affine lines if $\alpha = 0$ and $\beta \neq 0$ and a multiple affine line with multiplicity 2 if $\alpha = \beta = 0$. If $\alpha \neq 0$, this fiber is the image of the fiber $\{x = \alpha, 2xz - (\frac{y}{x})^2 = \frac{\beta}{\alpha^2}\}$, but if $\alpha = 0$, the $(\frac{y}{x}, z)$-plane is mapped surjectively to the fiber $\alpha = \beta = 0$. Although δ_2 does not induce a locally nilpotent derivation on A_1, the subalgebra B_1 is still the kernel of a locally nilpotent derivation δ_1 on A_1 that is defined by $\delta_1(x) = 0, \delta_1(y) = x^3$ and $\delta_1(z) = y$.

Let us consider the existence of an irreducible component in a fiber of the morphism $f : X \to Y$ which has dimension greater than one.

Lemma 3.1.7. *With the same notations and assumptions as in Lemma 3.1.2, we assume that $\dim X \geq 3$, X is factorial and the morphism f is the quotient morphism under a G_a-action on X. Then the following assertions hold:*

(1) *There is no irreducible component of codimension one contained in a fiber of f. Hence, if $\dim X = 3$, then every irreducible component contained in a fiber of f is the surjective image of a contractible curve.*
(2) *If $\dim X \geq 4$, then an irreducible component of dimension greater than 1 can exist in a fiber of f.*

Proof. (1) Suppose that a fiber $F = f^{-1}(P)$ contains an irreducible component Z of codimension one. Since X is factorial, Z is defined by an element $a \in A$. Since Z is G_a-stable, it follows that $\delta(a)$ is divisible by a. Then $\delta(a) = 0$, where δ is the locally nilpotent derivation of A corresponding to the given G_a-action. Hence the defining ideal aA of Z is the extension of the ideal aB in B. This contradicts the assumption that $f(Z)$ is the point P. Suppose $\dim X = 3$. Then $\dim Y = 2$. Let F_0 be a possibly reducible fiber of f lying over a point $P \in Y$. Take a sufficiently general irreducible curve C in Y passing through the point P and let $W = f^{-1}(C)$. Then W is an irreducible affine surface having the induced $\mathbb{A}^1$-fibration $f_W : W \to C$, and F_0 is a fiber of f_W. Note that for a small open subset U_0 of C, $f_W^{-1}(U_0) \cong U_0 \times \mathbb{A}^1$ and $f_W^{-1}(U_0)$ is smooth. Let $\widetilde{W}$ be the normalization of W and let $\widetilde{C}$ be the normalization of C. Then $\widetilde{W} \xrightarrow{\nu_W} W \xrightarrow{f_W} C$ splits as $\widetilde{W} \xrightarrow{f_{\widetilde{W}}} \widetilde{C} \xrightarrow{\nu} C$, where ν_W and ν are the normalization morphisms of W and C, respectively. It is clear that $f_{\widetilde{W}}$ is an $\mathbb{A}^1$-fibration, and that the fiber F_0 is a surjective image of a fiber $\widetilde{F}_0$ of $f_{\widetilde{W}}$. Since $\widetilde{F}_0$ is a disjoint union of the affine lines by Lemma 1.4.3, it follows that every irreducible component of F_0 is the image of the affine line. (There is a related result in Lemma 3.1.26 below about a degenerate fiber of f.)

(2) We consider Example 1.3.5(1) by Bonnet, where $X = \mathbb{A}^4 = \operatorname{Spec} k[x, y, u, v]$ with the G_a-action given by

$$^t(x, y, u, v) \mapsto (x, y, u - ty, v + tx).$$

Then the invariant subalgebra A is given by $k[x, y, xu + yv]$ and $Y = \mathbb{A}^3$. The fiber F_0 of q lying over the point of origin $x = y = xu + yv = 0$ is the affine plane $A^2 = \operatorname{Spec} k[u, v]$. See also Example 1.3.5(2) by Winkelmann. □

Concerning the coincidence between $\mathbb{A}^1$-fibrations and the quotient morphisms by G_a-actions, we obtain the following result by summarizing partly the above observations.

Theorem 3.1.8. *With the same notations as above, let $f : X \to Y$ be an $\mathbb{A}^1$-fibration of normal affine algebraic varieties such that the factorial closure $\widetilde{B}$ of B in A is an affine k-domain. Then the following assertions hold:*
1. *If there are no irreducible components of dimension greater than one in the fibers of f and the closed set $\overline{Y - f(X)}$ has codimension greater than one in Y, then f is the quotient morphism by a G_a-action on X.*
2. *Assume that X is factorial and has dimension three. Then f is the quotient morphism by a G_a-action if and only if there is no irreducible component of codimension one in the fibers of f and* $\operatorname{codim}_Y(\overline{Y - f(X)}) > 1$.
3. *Assume that* $\dim X = 2$. *Then f is the quotient morphism by a G_a-action on X if and only if f is surjective.*

Proof. (1) By Lemma 3.1.2, there exists a locally nilpotent derivation δ on A such that the associated quotient morphism coincides with f for general fibers. Then $\operatorname{Ker}\delta$ is the factorial closure $\widetilde{B}$ of B in A by Lemma 2.2.7. Then $\widetilde{B}$ is a normal, affine k-domain by the assumption, B is a subalgebra of $\widetilde{B}$ and $Q(B) = Q(\widetilde{B})$. We shall show that $\widetilde{B} = B$. The $\mathbb{A}^1$-fibration f splits as $f : X \xrightarrow{\widetilde{f}} \widetilde{Y} \xrightarrow{\nu} Y$, where $\widetilde{Y} = \operatorname{Spec}\widetilde{B}$ and $\widetilde{f}$ is the quotient morphism by a G_a-action. Then $\operatorname{codim}_{\widetilde{Y}}(\widetilde{Y} - \widetilde{f}(X)) \geq 2$ by Lemma 2.2.9. Since all the fibers of f (if not empty) are one-dimensional, so are the fibers of $\widetilde{f}$. Let $\widetilde{Z} = \overline{\widetilde{Y} - \widetilde{f}(X)}$ and $Z = \overline{Y - f(X)}$. Then $\operatorname{codim}_{\widetilde{Y}}\widetilde{Z} \geq 2$ as remarked above, $\operatorname{codim}_Y Z \geq 2$ by the hypothesis and hence $\nu(\widetilde{Y} - \widetilde{Z})$ contains all codimension one points of Y. Furthermore, ν

induces a quasifinite morphism from $\widetilde{Y} - \widetilde{Z}$ to $Y - Z$ because over any closed point of $Y - Z$ there are finitely many fibers which have dimension one by the hypothesis and hence are parametrized by finitely many points of $\widetilde{Y} - \widetilde{Z}$. Since $\widetilde{Y}$ and Y are normal, Zariski's main theorem implies that $\widetilde{Y} - \widetilde{Z}$ is an open subset of $Y - Z$. Then $\widetilde{B} = \Gamma(\widetilde{Y} - \widetilde{Z}, \mathcal{O}_{\widetilde{Y}}) = \Gamma(Y - Z, \mathcal{O}_Y) = B$.

(2) The *"if"* part follows from the assertion (1) above. The *"only if"* part follows from the assertion (1) of Lemma 3.1.7.

(3) Suppose f is surjective. Let δ be a locally nilpotent derivation of A as defined in Lemma 3.1.2 and let $A_0 = \operatorname{Ker}\delta$. Then we have the inclusions $B \subseteq A_0 \subseteq A$. Clearly, $Z := \operatorname{Spec} A_0$ is a normal affine curve which is birational to Y. Since f is surjective, the morphism $q : Z \to Y$ induced by the inclusion $B \subseteq A_0$ is surjective. Hence q is an isomorphism. Namely, f is the quotient morphism associated to δ. Suppose that f is the quotient morphism by a certain G_a-action on X. We show that f is surjective. Let $Z := f(X)$. Then Z is an open set of the affine curve Y, whence Z is an affine curve. Let $B' = \Gamma(Z, \mathcal{O}_Z)$. Then we have the inclusions $B \subseteq B' \subseteq A$. We show that $B = B'$. In fact, let $b' \in B'$. Since $Q(B') = Q(B)$, there exist $b_1, b_2 \in B$ such that $b_1 b' = b_2$. Since $B = A^{G_a}$, B is factorially closed in A by Lemma 3.1.2, whence $b' \in B$. So, $Z = Y$, and f is surjective. □

In view of the assertion of Theorem 3.1.8, we say that an $\mathbb{A}^1$-fibration $f : X \to Y$ is *good* if it has equidimensional fibers and $\operatorname{codim}_X(\overline{Y - f(X)}) \geq 2$. A good $\mathbb{A}^1$-fibration is the quotient morphism of a G_a-action on X if the factorial closure of B in A is an affine k-domain, while the converse does not hold by Lemma 3.1.7(3).

The following well-known example shows that the quotient morphism of a G_a-action on a factorial smooth affine threefold is not surjective in general.

Example 3.1.9. Let $X = \mathrm{SL}(2,k)$ and let $A = k[x,y,u,v]/(xv - yu - 1)$ be the coordinate ring of X. Let δ be a locally nilpotent derivation on A defined by $\delta(x) = \delta(y) = 0, \delta(u) = x, \delta(v) = y$. Then $B = \operatorname{Ker}\delta = k[x,y]$ and the quotient morphism $f : X \to Y$ is given by $(x,y,u,v) \mapsto (x,y)$. In fact, G_a is identified with $U = \{\left(\begin{smallmatrix} 1 & 0 \\ t & 1 \end{smallmatrix}\right) \mid t \in k\}$ and the G_a-action with the left multiplication of U to X. Then the image $f(Y)$ is the punctured affine plane $\mathbb{A}^2_* = Y \setminus \{x = 0, y = 0\}$. Hence X is an $\mathbb{A}^1$-bundle over $\mathbb{A}^2_*$ and f is not surjective. Furthermore, A is factorial by Nagata's lemma (see Lemma 2.4.22). In fact, x is a prime element in A and $A[x^{-1}] = k[x,x^{-1}][y,u]$ is factorial.

The first assertion of Lemma 3.1.7 states that there are no codimension one fiber components in the quotient morphism $q : X \to Y$ if X is factorial. We generalize slightly this result to the case where X is a $\mathbb{Q}$-factorial affine variety. Let X be an affine normal variety. We say that X is *$\mathbb{Q}$-factorial* if the Weil divisor class group $\mathrm{C}\ell(X)$ is a torsion group. We also say that a normal affine k-domain A is $\mathbb{Q}$-factorial if so is $\operatorname{Spec} A$.

We need the following result.

Lemma 3.1.10. *Let A be a $\mathbb{Q}$-factorial affine k-domain and let δ be a nontrivial locally nilpotent derivation. Suppose that $B = \operatorname{Ker}\delta$ is an affine k-domain. Assume that $\mathfrak{p}A \neq A$*

for a height one prime ideal $\mathfrak{p}$ *of* B *and* $\sqrt{\mathfrak{p}A} = \sqrt{fA}$ *for an element* $f \in A$. *Then* $f \in B$ *and* $\mathfrak{p} = \sqrt{fB}$.

Proof. Define k-algebra homomorphisms $\varphi : A \to A[t]$ by

$$\varphi(a) = \sum_{i=0}^{\infty} \frac{\delta^i(a)}{i!} t^i,$$

where t is a variable over A, and $\Phi : A[t] \to A[t]$ by setting $\Phi|_A = \varphi$ and $\Phi(t) = t$. Then Φ is a $B[t]$-automorphism of $A[t]$ with the inverse Φ^{-1} defined by $\Phi^{-1}(a) = \sum_{i=0}^{\infty}(\delta^i(a)/i!)(-t)^i$ and $\Phi^{-1}(t) = t$. For $\lambda \in k$ and $a \in A$, we denote ${}^{\lambda}a = \varphi(a)|_{t=\lambda}$. We define a k-automorphism φ_λ of A by $\varphi_\lambda(a) = {}^{\lambda}a$ for $a \in A$.

Since $\mathfrak{p}$ is an ideal of B, $\mathfrak{p}A$ and $\sqrt{\mathfrak{p}A}$ are *δ-ideals* in the sense that $\delta(\mathfrak{p}A) \subseteq \mathfrak{p}A$ and $\delta(\sqrt{\mathfrak{p}A}) \subseteq \sqrt{\mathfrak{p}A}$. An ideal J of A is a δ-ideal if and only if $\varphi_\lambda(J) = J$ for every $\lambda \in k$, and also if and only if $\Phi(JA[t]) = JA[t]$. Note that $\varphi_\lambda(\sqrt{fA}) = \sqrt{{}^{\lambda}fA}$ for $\lambda \in k$, or equivalently $\Phi(\sqrt{fA[t]}) = \sqrt{\varphi(f)A[t]}$. Hence $\varphi(f)^n \in fA[t]$ for some integer $n > 0$. This implies that $({}^{\lambda}f)^n \in fA$ for $\lambda \in k$. Let $m = \deg_t \varphi(f)$. Then the coefficient of the top term in $\varphi(f)$ is $\delta^m(f)/m!$ and it is an element of B. Write

$$\varphi(f)^n = a_0 + a_1 t + \cdots + a_N t^N,$$

where $N = mn, a_0 = f^n$, and $a_N = (\delta^m(f)/m!)^n \in B$. Choose mutually distinct $N+1$ elements λ_i $(0 \le i \le N)$ of k. Then we have

$$\varphi_{\lambda_i}(f)^n = a_0 + a_1\lambda_i + \cdots + a_N\lambda_i^N = fh_i, \quad h_i \in A.$$

Then by solving this system of linear equations in $a_0, a_1, \ldots, a_N$ and noting that the Vandermonde determinant with respect to the λ_i is nonzero, we see that $a_i \in fA$ for every $0 \le i \le N$. In particular, $a_N \in fA$. Hence $f \in B$ because $a_N \in B$.

Since $\sqrt{\mathfrak{p}A} = \sqrt{fA}$, we have $f^r \in \mathfrak{p}A$. Then $f^r \in \mathfrak{p}$ since $\mathfrak{p}A \cap B = \mathfrak{p}$ and hence $f \in \mathfrak{p}$. In fact, since there exists a height one prime ideal $\mathfrak{P}$ of A such that $\mathfrak{P} \cap B = \mathfrak{p}$ (cf. Lemma 2.2.9), we have $\mathfrak{p} = \mathfrak{P} \cap B \supseteq \mathfrak{p}A \cap B \supseteq \mathfrak{p}$. Since $\mathfrak{p} \subseteq \sqrt{fA}$, an element $b \in \mathfrak{p}$ is written as $b^n = fa$ with $a \in A$. Then $a \in B$ and hence $\mathfrak{p} = \sqrt{fB}$. □

The following result shows that the quotient morphism of a G_a-action behaves nicely if one replaces the factoriality by the $\mathbb{Q}$-factoriality.

Lemma 3.1.11. *Let* $X = \operatorname{Spec} A$ *be a* $\mathbb{Q}$*-factorial normal affine variety with a* G_a*-action, let* $B = A^{G_a}$ *and* $Y = \operatorname{Spec} B$. *Suppose that* B *is an affine domain. Then the following assertions hold:*

(1) *Y is also $\mathbb{Q}$-factorial.*
(2) *If* $\dim X \ge 3$, *the quotient morphism* $q : X \to Y$ *has no codimension one fiber component.*

Proof. (1) Let δ be a locally nilpotent derivation of A corresponding to the G_a-action on X. Then $B = \operatorname{Ker}\delta$. Let $\mathfrak{p}$ be a height one prime ideal of B so that $V(\mathfrak{p})$ is an irreducible Weil divisor on Y. Then $V(\mathfrak{p}A)$ is the scheme-theoretic inverse image. Since X is $\mathbb{Q}$-factorial, $\sqrt{\mathfrak{p}A} = \sqrt{fA}$. By Lemma 3.1.10, we have $f \in B$ and $\mathfrak{p} = \sqrt{fB}$. Since the image $q(X)$ contains all codimension one points of Y, it follows that Y is $\mathbb{Q}$-factorial.

(2) Suppose that the fiber $q^{-1}(P)$ contains an irreducible component Z of codimension one. Then the defining prime ideal $\mathfrak{P}$ is equal to $\sqrt{fA}$. Since Z is G_a-stable, it follows that $\mathfrak{P}$ is a δ-ideal. This implies that $\sqrt{fA} = \sqrt{{}^{\lambda}fA}$ for every $\lambda \in k$. Hence there exists a positive integer n such that $\varphi_\lambda(f)^n \in fA$. As in the proof of Lemma 3.1.10, it follows that $f \in B$. Let $\mathfrak{p} = \sqrt{fB}$. Then $\mathfrak{P} = \sqrt{\mathfrak{p}A}$. If $b_1b_2 \in \mathfrak{p}$ for $b_1, b_2 \in B$, then $b_1b_2 \in \mathfrak{P}$. Since $\mathfrak{P}$ is a prime ideal, either $b_1 \in \mathfrak{P}$ or $b_2 \in \mathfrak{P}$. Suppose $b_1 \in \mathfrak{P}$. Then $b_1^m \in fA$ for some $m > 0$. Since $b_1^m, f \in B$, it follows that $b_1^m \in fB$. Hence $b_1 \in \sqrt{fB} = \mathfrak{p}$. So, $\mathfrak{p}$ is a prime ideal. Since B is normal, $\mathfrak{p}$ has height one and defines an irreducible divisor W of Y. Since $q^{-1}(W)_{\mathrm{red}} = Z$, this is a contradiction. □

Given an $\mathbb{A}^1$-fibration $f : X \to Y$, we say that a fiber $F = f^{-1}(P)$ is a *singular fiber* if it is not isomorphic to the affine line $\mathbb{A}^1$ defined over $k(P)$, where $k(P)$ is the residue field of Y at P. A multiple fiber as given in Example 3.1.6 is a singular fiber. Let $\operatorname{Sing}(f)$ be the set of points $P \in Y$ such that the fiber $f^{-1}(P)$ is singular. We call it the *singular locus* or the *degeneracy locus* of f. We do not yet know if the set $\operatorname{Sing}(f)$ is a closed set. So, whenever we speak of an irreducible component of $\operatorname{Sing}(f)$, it is the one of the closure $\overline{\operatorname{Sing}(f)}$ in Y.

Lemma 3.1.12. *Let $f : X \to Y$ be an $\mathbb{A}^1$-fibration of normal affine varieties. Assume that f is a flat morphism and $\overline{Y - f(X)}$ has codimension greater than one in Y. Then the degeneracy locus* $\operatorname{Sing}(f)$ *is either empty or has pure codimension one. If* $\operatorname{Sing}(f) = \emptyset$, *then $f : X \to f(X)$ is an $\mathbb{A}^1$-bundle.*

Proof. Suppose that the degeneracy locus $\operatorname{Sing}(f)$ has an irreducible component D of codimension greater than one. Let P be the generic point of D. Let $R = \mathcal{O}_{Y,P}$ and $S = A \otimes_B R$. By the hypotheses, the restriction $f \otimes_B R : \operatorname{Spec} S \to \operatorname{Spec} R$ satisfies all the necessary conditions in Lemma 2.1.6 for $\operatorname{Spec} S$ to be an $\mathbb{A}^1$-bundle over $\operatorname{Spec} R$. Hence the fiber $f^{-1}(P)$ is not a singular fiber. □

The following result shows that the converse of Lemma 3.1.11, the assertion (1) holds if the morphism f has no reducible fibers or fiber components of dimension greater than one.

Lemma 3.1.13. *Let $X = \operatorname{Spec} A$ be a normal affine variety with a nontrivial G_a-action. Suppose that $Y := X/\!/G_a$ exists as a $\mathbb{Q}$-factorial normal affine variety. Suppose further that* $\operatorname{codim}_Y(\overline{Y - q(X)}) > 1$ *for the quotient morphism $q : X \to Y$ and that every fiber of q over a point of $q(X)$ is irreducible and one-dimensional, i. e., only irreducible multiple fibers of dimension one are admitted as singular fibers. Then X is $\mathbb{Q}$-factorial.*

Proof. With the same notations as in Lemma 3.1.11, let $\mathfrak{P}$ be a prime ideal of height one of A. If $\mathfrak{P} \cap B = (0)$, then $\mathfrak{P} = (\mathfrak{P}S^{-1}A) \cap A$, where $S = B \setminus \{0\}$. Let $K = Q(B) = S^{-1}B$. Then $A \otimes_B K$ is isomorphic to a polynomial ring $K[t]$ over the field K, where we may take t to be an element of A. Then there exists an element $g \in B[t]$ such that $\mathfrak{P}S^{-1}A = gS^{-1}A$ and g is irreducible in $K[t]$. Consider the minimal prime decomposition $\sqrt{gA} = \mathfrak{P}_1 \cap \cdots \cap \mathfrak{P}_r$, where we put $\mathfrak{P}_1 = \mathfrak{P}$. Note that $\mathfrak{P}_i$ $(2 \le i \le r)$ has height one. Consider $\mathfrak{P}_i$ for $2 \le i \le r$. By the choice of the element g, we have $\mathfrak{P}_i \cap B = \mathfrak{p}_i$, where $\mathfrak{p}_i$ is a height 1 prime ideal of B because q has relative equidimension one. Furthermore, since every fiber of q is irreducible, we have $\mathfrak{P}_i = \sqrt{\mathfrak{p}_i A}$. Since Y is $\mathbb{Q}$-factorial, $\mathfrak{p}_i = \sqrt{f_i B}$ for $f_i \in B$. Then $\mathfrak{P}_i = \sqrt{f_i A}$. It follows that some multiple of the irreducible divisor defined by $\mathfrak{P}$ is linearly equivalent to the zero divisor. If $\mathfrak{P} \cap B \neq (0)$, then $\mathfrak{P} = \sqrt{fA}$ for some $f \in B$ by the above argument. □

An example of a normal affine variety which fits to the setting in Lemma 3.1.13 is an affine pseudoplane which is an affine smooth surface X with an $\mathbb{A}^1$-fibration $f : X \to \mathbb{A}^1$ such that f is surjective and has a single irreducible multiple fiber of multiplicity $m > 1$ as singular fibers (see the definition after Theorem 2.9.22). The universal covering $\widetilde{X}$ of X has also an $\mathbb{A}^1$-fibration $\widetilde{f} : \widetilde{X} \to \mathbb{A}^1$ which has only one singular fiber consisting of m reduced irreducible components isomorphic to $\mathbb{A}^1$. In other words, $\widetilde{X}$ is obtained by patching m copies of $\mathbb{A}^2$ along $\mathbb{A}^2 \setminus \{\text{a line}\}$. This is proved by the ramified covering trick in Subsection 1.1.6.

As a generalization of affine pseudoplane to the case of dimension $n \ge 2$, we consider a smooth affine variety X of dimension n endowed with an $\mathbb{A}^1$-fibration $f : X \to \mathbb{A}^{n-1}$ satisfying the following conditions:
(i) f is faithfully flat and all fibers of f are irreducible.
(ii) $\mathrm{Sing}(f)$ consists of a hyperplane H_0 of $\mathbb{A}^{n-1}$ and $f^{-1}(P)$ is a multiple fiber of multiplicity $m > 1$ for general points P of H_0.

As an intuitive image, write the base $\mathbb{A}^{n-1}$ as a direct product $H_0 \times \mathbb{A}^1$. Let $\pi : \mathbb{A}^{n-1} \to H_0$ be the projection and let $p = \pi \cdot f : X \to H_0$. Then a general fiber of p is an affine pseudoplane with a multiple fiber $m\mathbb{A}^1$.

Question 3.1.14. What is the universal covering of the above affine variety X? Is it obtained by patching together m copies of the affine n-space $\mathbb{A}^n$ along $\mathbb{A}^n \setminus H'$, where H' is a hyperplane of $\mathbb{A}^n$ of the form $H_0 \times \mathbb{A}^1$?

3.1.3 $\mathbb{A}^1$-fibrations over surfaces

When we say that an $\mathbb{A}^1$-fibration $f : X \to Y$ is an $\mathbb{A}^1$-bundle over Y, we tacitly assume that the morphism f is surjective. The above Lemma 3.1.12 implies a rather strong result.

Lemma 3.1.15. *Let $f : X \to Y$ be an $\mathbb{A}^1$-fibration of normal affine varieties. Assume that X is factorial, f is flat, the factorial closure of B in A is an affine domain, and $Z := \overline{Y - f(X)}$ has codimension greater than one in Y. Let δ be a locally nilpotent derivation of A giving rise to the morphism f* (see Lemma 2.2.7). *We can take δ in such a way that $b^{-1}\delta(A) \not\subset A$ for every nonzero element $b \in B$ which is not a unit. Then the following assertions hold:*

(1) *Every prime element p of B is a prime element of A.*
(2) *For every prime element $p \in B$, we denote by δ/p the nonzero, locally nilpotent derivation of A/pA induced by δ. Then* $\mathrm{Sing}(f) = \emptyset$ *if and only if* $Q(B/pB) = Q(\mathrm{Ker}(\delta/p))$ *for every prime element p of B.*
(3) *Suppose that $f : X \to f(X)$ is an $\mathbb{A}^1$-bundle. If $H^1(f(X), \mathcal{O}_Y) = (0)$, then $X \cong Y \times \mathbb{A}^1$ and the G_a-action on X is induced by a translation by an element of B on the factor $\mathbb{A}^1$.*
(4) *Suppose that Y is smooth and* $\mathrm{codim}_Y(Z) \geq 3$. *Then $H^1(f(X), \mathcal{O}_Y) = 0$.*

Proof. (1) Since f is flat, f is equidimensional and the image $f(X)$ is an open set of Y. By Theorem 3.1.8, f is the quotient morphism by a G_a-action on X. Since B is factorially closed in A and A is factorial by the assumption, B is factorial as well. Note that all units in A are elements of B. In order to show the assertion (1), it suffices to show that p is an irreducible element of A. But this is clear because B is factorially closed in A.

(2) Suppose that $\mathrm{Sing}(f) \neq \emptyset$. Let D be an irreducible component of $\mathrm{Sing}(f)$. Then $D = V(p)$ for a prime element p. Then $f^{-1}(D)$ is a prime divisor of X isomorphic to $\mathrm{Spec}\, A/pA$. Then the morphism $\overline{f} : \mathrm{Spec}(A/pA) \to \mathrm{Spec}(B/pB)$ induced by f factors through the quotient morphism $\overline{q} : \mathrm{Spec}(A/pA) \to \mathrm{Spec}(\mathrm{Ker}\, \delta/p)$ by the induced G_a-action and a morphism $\overline{p} : \mathrm{Spec}(\mathrm{Ker}\, \delta/p) \to \mathrm{Spec}(B/pB)$, where $\overline{q}$ is an $\mathbb{A}^1$-fibration and $\overline{p}$ is a generically finite morphism. Since D is a component of $\mathrm{Sing}(f)$, $\overline{p}$ is not birational. So, this proves the "*if*" part of the assertion. Conversely, if $\overline{p}$ is not birational for a prime element p of B, then the general fibers of the induced morphism $\overline{f}$ are not irreducible. Hence $\mathrm{Sing}(f) \neq \emptyset$. This proves the "*only if*" part.

(3) Assuming that $H^1(f(X), \mathcal{O}_Y) = (0)$, we show that f is surjective, $X \cong Y \times \mathbb{A}^1$ and the G_a-action on X giving rise to the quotient morphism f is a translation on the factor $\mathbb{A}^1$. We consider f as an $\mathbb{A}^1$-bundle over $f(X)$. Then there exists an affine open covering $\mathfrak{U} = \{U_\lambda\}_{\lambda \in \Lambda}$ of $f(X)$ such that $f^{-1}(U_\lambda) = \mathrm{Spec}\, B_\lambda[t_\lambda]$ for every $\lambda \in \Lambda$, where $B_\lambda = \Gamma(U_\lambda, \mathcal{O}_Y)$. Then we have

$$t_\mu = f_{\mu\lambda} t_\lambda + g_{\mu\lambda} \quad \text{with } f_{\mu\lambda} \in B^*_{\mu\lambda},\ g_{\mu\lambda} \in B_{\mu\lambda},$$

where $B_{\mu\lambda} = \Gamma(U_\mu \cap U_\lambda, \mathcal{O}_Y)$. Then it holds on $U_\lambda \cap U_\mu \cap U_\nu$ that

$$f_{\nu\mu} f_{\mu\lambda} = f_{\nu\lambda} \quad \text{and} \quad g_{\nu\lambda} = f_{\nu\mu} g_{\mu\lambda} + g_{\nu\mu}.$$

Hence there exists an invertible sheaf $\mathcal{L}$ on $f(X)$ whose transition functions with respect to the covering $\mathfrak{U}$ are given by $\{f_{\mu\lambda} \mid \lambda, \mu \in \Lambda\}$. Note that $H^1(f(X), \mathcal{O}_Y^*) = \mathrm{Pic}(f(X)) = \mathrm{C}\ell((f(X)) = \mathrm{C}\ell(Y)$ because $Y - f(X)$ has codimension greater than one in Y and $\mathrm{C}\ell(Y) =$

(0) by the hypothesis that $C\ell(X) = (0)$. Hence we find, after passing to a finer affine open covering of $f(X)$ if necessary, a family $\{h_\lambda \in A_\lambda^* \mid \lambda \in \Lambda\}$ such that $f_{\mu\lambda} = h_\mu^{-1}h_\lambda$. Then, by replacing t_λ by $h_\lambda t_\lambda$, we may assume that $f_{\mu\lambda} = 1$ for every pair $(\lambda, \mu) \in \Lambda^2$. Then $g_{\nu\lambda} = g_{\mu\lambda} + g_{\nu\mu}$. Since $H^1(f(X), \mathcal{O}_Y) = (0)$ by the hypothesis, again after passing to a finer affine open covering of $f(X)$ if necessary, there exists $a_\lambda \in B_\lambda$ for every $\lambda \in \Lambda$ such that $g_{\mu\lambda} = a_\lambda - a_\mu$. Then $t_\mu + a_\mu = t_\lambda + a_\lambda$. Hence there exists an element $t \in \Gamma(X, \mathcal{O}_X) = A$ such that $t_\lambda + a_\lambda$ is the restriction of t onto $f^{-1}(U_\lambda)$. Namely, t is a common parameter for $f^{-1}(U_\lambda)$ for every $\lambda \in \Lambda$. Now consider a k-subalgebra $B[t]$ of A. Then the associated morphism induces an open immersion $X \hookrightarrow \operatorname{Spec} B[t] = Y \times \mathbb{A}^1$ which is a morphism over Y. In fact, it induces an isomorphism $X \xrightarrow{\sim} f(X) \times \mathbb{A}^1$. Since $(\operatorname{Spec} B[t]) \setminus X$ then has codimension greater than one, we have $A = \Gamma(X, \mathcal{O}_X) = B[t]$. Let δ be a locally nilpotent derivation giving rise to the G_a-action. Then $\delta(t) = b \in B$. Hence the G_a-action is a translation by b.

(4) We have an exact sequence of local cohomologies (see [64, Exposés II, III]),

$$H_Z^1(Y, \mathcal{O}_Y) \longrightarrow H^1(Y, \mathcal{O}_Y) \longrightarrow H^1(f(X), \mathcal{O}_Y) \longrightarrow H_Z^2(Y, \mathcal{O}_Y),$$

where $H^1(Y, \mathcal{O}_Y) = 0$ because Y is affine. If Y is smooth and $\operatorname{codim}_Y(Z) \geq 3$ then $H_Z^2(Y, \mathcal{O}_Y) = 0$ because $\operatorname{depth}(\mathcal{O}_{Y,z}) \geq 3$ for any point $z \in Z$. This implies $H^1(f(X), \mathcal{O}_Y) = 0$. □

Remark 3.1.16. (1) In terms of group cohomologies, the argument for the assertion (3) of Lemma 3.1.15 proceeds as follows. Let GA be a k-group scheme such that, for a k-algebra R, $GA(R)$ is the group of affine transformations of the affine line $\mathbb{A}^1$ over R. The group $GA(R)$ is given by a matrix group with entries in R,

$$GA(R) = \left\{ \begin{pmatrix} a & b \\ 0 & 1 \end{pmatrix} \mid a \in R^*,\ b \in R \right\}.$$

Hence there exists an exact sequence of k-group schemes

$$0 \longrightarrow G_a \longrightarrow GA \longrightarrow G_m \longrightarrow 0.$$

Then we have an exact sequence of cohomology groups over $U := f(X)$ in the Zariski topology

$$\begin{aligned} 0 \longrightarrow H^0(U, G_a) &\longrightarrow H^0(U, GA) \longrightarrow H^0(U, G_m) \\ \longrightarrow H^1(U, G_a) &\longrightarrow H^1(U, GA) \longrightarrow H^1(U, G_m), \end{aligned}$$

where $H^0(U, G_a) = \Gamma(U, \mathcal{O}_Y) = B$, $H^0(U, GA) = GA(B)$, $H^0(U, G_m) = B^*$, $H^1(U, G_a) = H^1(U, \mathcal{O}_Y)$, and $H^1(U, G_m) = H^1(U, \mathcal{O}_Y^*) = (0)$. Since GA is not abelian, $H^1(U, GA)$ has no group structure and the exactness of the second row means that the image of an element $\xi \in H^1(U, GA)$ in $H^1(U, G_m)$ is zero if and only if ξ is the image of an element of $H^1(U, G_a)$. The $\mathbb{A}^1$-bundle $f : X \to U$ has an obstruction to be trivial in the group $H^1(U, GA)$. Note that $H^0(U, GA) \to H^0(U, G_m)$ is surjective and $H^1(U, G_m) = (0)$ by the factoriality of B. Hence we have an isomorphism $H^1(U, \mathcal{O}_Y) \cong H^1(U, GA)$. If $H^1(U, \mathcal{O}_Y) = (0)$, then the bundle $f : X \to U$ is trivial, whence $X = U \times \mathbb{A}^1$. Since X is affine, U must be affine as well. Hence $U = Y$ as $\operatorname{codim}_Y(Z) \geq 2$.

(2) There is a related result in [129, Lemma 1] when $\dim X = 3$. In the case of Example 3.1.9, X is an $\mathbb{A}^1$-bundle over $f(X)$, where $Z = Y - f(X)$ is the closed set $Z = \{(0,0)\}$ and $U = f(X)$ is an open set. We have

$$H^1(U, \mathcal{O}_Y) \cong H^2_Z(Y, \mathcal{O}_Y) \cong \varinjlim_n \operatorname{Ext}^2_B(B/\mathfrak{m}^n, B) \neq 0,$$

where $\mathfrak{m}$ is the maximal ideal (x, y) of $B = k[x, y]$. In fact, by the syzygy sequence

$$0 \longrightarrow Be \xrightarrow{\psi} Be_1 \oplus Be_2 \xrightarrow{\varphi} B \longrightarrow k \longrightarrow 0,$$

where $\psi(e) = ye_1 - xe_2$, $\varphi(a_1e_1 + a_2e_2) = a_1x + a_2y$ and $\operatorname{Im}\varphi = \mathfrak{m}$, we conclude that $\operatorname{Ext}^1_B(k, B) = 0$, $\operatorname{Ext}^2_B(k, B) = k$ and $\operatorname{Ext}^2_B(B/\mathfrak{m}^n, B) \hookrightarrow \operatorname{Ext}^2_B(B/\mathfrak{m}^{n+1}, B)$. So, $\mathrm{SL}(2, k)$ is not a trivial $\mathbb{A}^1$-bundle over $\mathbb{A}^2_*$. □

The argument used in the proof of Lemma 3.1.15 proves the following result because $H^1(Y, \mathcal{O}_Y) = (0)$. It also follows from Lemma 2.1.7.

Lemma 3.1.17. *Let Y be a factorial affine variety. Then every $\mathbb{A}^1$-bundle over Y is trivial.*

By a *threefold* we mean an algebraic variety of dimension three. We are interested in an $\mathbb{A}^1$-fibration on a threefold $f : X \to Y$. Then the base variety Y is a surface. If X and Y are affine varieties, by Lemma 2.2.7, f is factored as $f : X \xrightarrow{\widetilde{f}} \widetilde{Y} \xrightarrow{\nu} Y$, where ν is a birational morphism. If $X = \operatorname{Spec} A$ and $Y = \operatorname{Spec} B$, let $\widetilde{B} = A \cap K$, where $K = Q(B)$ and the intersection of A and K are considered in $Q(A)$. Then $\widetilde{B}$ is an affine domain by Zariski's finiteness theorem (see Theorem 2.2.4), which is called the *factorial closure* of B in A. By Lemma 2.2.7, $\widetilde{f} : X \to \widetilde{Y} = \operatorname{Spec} \widetilde{B}$ is the quotient morphism of a G_a-action. If f is equidimensional then ν is an isomorphism under a mild condition of normality on X and Y. We give here the following result.

Lemma 3.1.18. *Let X be a normal affine threefold with a nontrivial G_a-action and let $q : X \to Y$ be the quotient morphism. Let C be an irreducible curve on Y and let Z be the scheme-theoretic inverse image of C, i. e., $Z = X \times_Y C$. If Z is irreducible and reduced, then Z has a nontrivial $\mathbb{A}^1$-fibration. If X is factorial, the assumption on Z being irreducible and reduced is automatically satisfied.*

Proof. As above, we write $A = \Gamma(X, \mathcal{O}_X)$, $B = \Gamma(Y, \mathcal{O}_Y)$ and δ the locally nilpotent derivation on A associated to the given G_a-action. Let $\mathcal{O} = \mathcal{O}_{Y,C}$ be the local ring at the generic point of C. Since Y is normal, $\mathcal{O}$ is a discrete valuation ring (DVR, for short). Let t be a uniformizant of $\mathcal{O}$. Let $R = A \otimes_B \mathcal{O}$. Then δ extends to R in a natural way and $\mathcal{O}$ is the kernel of the extended δ. Furthermore, δ induces a locally nilpotent derivation $\overline{\delta}$

on R/tR. If it is trivial, we have $\delta(R) \subset tR$. Then we replace δ by $t^{-1}\delta$ which is a locally nilpotent derivation on R. Thus, by replacing δ by $t^{-r}\delta$, we may assume that $\overline{\delta}$ is nontrivial. Then R/tR is a polynomial ring in one variable over K, where K is a field containing $\mathcal{O}/t\mathcal{O}$. In fact, R/tR is a finitely generated integral domain of dimension one containing the field $\mathcal{O}/t\mathcal{O}$. Since $\overline{\delta}$ is a nonzero, locally nilpotent derivation, the kernel K of $\overline{\delta}$ has dimension zero and contains $\mathcal{O}/t\mathcal{O}$. So, K is a field and R/tR is a polynomial ring over K because $\overline{\delta}$ then has a slice. Hence Z has an $\mathbb{A}^1$-fibration. If A is factorial, then B is factorial. An element f defining the curve C is a prime element of A. This implies that Z is irreducible and reduced. □

Besides the hypersurface $\{xv = yu + 1\}$ in Example 3.1.9, there are many factorial $\mathbb{A}^1$-bundles over the punctured affine plane $\mathbb{A}^2_* := \mathbb{A}^2 - \{(0,0)\}$. The hypersurface $\{x^m v = y^n u + f(x,y)\}$ is such an example for arbitrary positive integers m, n and a polynomial $f(x,y) \in k[x,y]$ with $f(0,0) \neq 0$. Note that the hypersurface is factorial by Nagata's lemma (see Lemma 2.4.22) with some restriction on $f(x,y)$. Related to the following results, we note that there is an interesting work by Isac Hedén [91] on principal G_a-bundles (equivalently G_a-torsors) over a punctured surface.

Remark 3.1.19. Let $f : X \to Y$ be an $\mathbb{A}^1$-bundle. Then it is a G_a-torsor if there exists a G_a-action $\sigma : G_a \times X \to X$ such that the graph morphism $\Phi : G_a \times X \to X \times_Y X$ is an isomorphism (see Subsection 2.2.2). We say that an $\mathbb{A}^1$-bundle $f : X \to Y$ is *affine* if X is affine. Even if an $\mathbb{A}^1$-bundle is affine, Y is not necessarily affine. By the assertion (1) of Theorem 3.1.8, an affine $\mathbb{A}^1$-bundle is a G_a-torsor if X is normal and Y is embedded as an open set into an affine variety $\widetilde{Y} = \operatorname{Spec} \widetilde{B}$ in such a way that $Z = \widetilde{Y} \setminus Y$ has codimension greater than one in $\widetilde{Y}$. In fact, if X is normal, then Y is normal because f is faithfully flat. Hence, by replacing $\widetilde{Y}$ by the normalization of $\widetilde{Y}$, we may assume that $\widetilde{Y}$ is normal. Then the composite of f and the open immersion $\iota : Y \hookrightarrow \widetilde{Y}$ is the quotient morphism of a G_a-action. So, $f : X \to Y$ is a G_a-torsor. □

For a polynomial $f(x,y) \in k[x,y]$ and positive integers m, n, we set

$$X(m,n,f) = \{x^m v = y^n u + f(x,y)\}.$$

Clearly, $X(m,n,f)$ is irreducible. The mapping $(x,y,u,v) \mapsto (x,y)$ defines an $\mathbb{A}^1$-fibration $f : X(m,n,f) \to \mathbb{A}^2$. Then f is the quotient morphism of a G_a-action on $X(m,n,f)$ defined by a locally nilpotent derivation δ such that

$$\delta(x) = \delta(y) = 0,\ \delta(u) = x^m,\ \delta(v) = y^n.$$

If $f(x,y)$ has a nonzero monomial cx^iy^j with either $i \geq m$ or $j \geq n$, then we can replace v or u by $v - cx^{i-m}y^j$ or $u + cx^iy^{j-n}$. Hence we may assume that any nonzero monomial cx^iy^j of $f(x,y)$ satisfies $i < m$ and $j < n$. We then say that $X(m,n,f)$ is *preslimmed*. Let $U_1 = D(x)$ and $U_2 = D(y)$ be the open sets of $Y := \operatorname{Spec} k[x,y]$. Then $U_1 \cup U_2 = U :=$

$Y - \{(0,0)\}$. Let $t_2 = v/y^n$ and $t_1 = u/x^m$. Suppose that $f(x,y) \neq 0$ and $X(m,n,f)$ is preslimmed. Then we have

$$t_2 = t_1 + \frac{f(x,y)}{x^m y^n} \quad \text{and} \quad \frac{f(x,y)}{x^m y^n} = \sum_{\substack{0<i\le m\\ 0<j\le n}} \frac{c_{ij}}{x^i y^j},$$

where $f(x,y) = \sum_{i,j} c_{ij} x^{m-i} y^{n-j}$. Let $m_0 = \max\{i \mid c_{ij} \neq 0\}$ and $n_0 = \max\{j \mid c_{ij} \neq 0\}$. We define an element $\widetilde{f}(x,y)$ of $k[x,y]$ by

$$\widetilde{f}(x,y) = x^{m_0} y^{n_0} \sum_{\substack{0<i\le m\\ 0<j\le n}} \frac{c_{ij}}{x^i y^j}.$$

Let $\widetilde{v} = y^{n_0} t_2$ and $\widetilde{u} = x^{m_0} t_1$. Then $x^{m_0}\widetilde{v} = y^{n_0}\widetilde{u} + \widetilde{f}(x,y)$. Since $\widetilde{f}(x,y) \in k[x,y]$ and $\widetilde{v}$ (resp., $\widetilde{u}$) is defined over $f^{-1}(U_2)$ (resp., $f^{-1}(U_1)$), the functions $\widetilde{v}$ and $\widetilde{u}$ are defined on $f^{-1}(U_1 \cup U_2)$. Let $\widetilde{A}$ be the coordinate ring of $X(m_0, n_0, \widetilde{f})$, and let $\theta : X(m_0, n_0, \widetilde{f}) \to X(m,n,f)$ be the birational morphism defined by an $k[x,y]$-homomorphism $A \hookrightarrow \widetilde{A}$ such that $v = y^{n-n_0}\widetilde{v}$ and $u = x^{m-m_0}\widetilde{u}$. Note that θ induces an isomorphism over $U_1 \cap U_2$. We say that $X(m_0, n_0, \widetilde{f})$ is the *slimmed form* of $X(m,n,f)$ and $X(m,n,f)$ is *slimmed* if $X(m,n,f) = X(m_0, n_0, \widetilde{f})$. Note that $\widetilde{f}(x,y)/x^{m_0}y^{n_0} = f(x,y)/x^m y^n$ and $f(x,y)/x^m y^n$ determines uniquely a slimmed hypersurface.

Suppose $f(x,y) = 0$. Then $t_2 = t_1$, which we set t. Then t is defined over $f^{-1}(U)$. Hence $f^{-1}(U)$ is a trivial $\mathbb{A}^1$-bundle with a vertical coordinate t. We define the slimmed form of $X(m,n,0)$ to be $X(1,1,0)$, which is defined by $xv = yu$. In the case $f(x,y) = 0$, the fiber $f^{-1}(0,0)$ is isomorphic to the affine plane $\mathbb{A}^2 = \operatorname{Spec} k[u,v]$.

We leave a proof of the following result to the readers.

Lemma 3.1.20. *With the above notations, assume that $X(m,n,f)$ is preslimmed and $f(x,y) \neq 0$. Then the following assertions hold:*

(1) *$X(m,n,f)$ is slimmed if and only if $m = m_0$ and $n = n_0$. In other words, $f(x,y)$ contains monomials x^i and y^j with nonzero coefficients.*

(2) *Assume that $X(m,n,f)$ is slimmed. Then $f^{-1}(0,0) = \emptyset$ if and only if $f(0,0) \neq 0$. Furthermore, $X(m,n,f)$ is smooth if and only if one of $f(0,0), f_x(0,0)$ and $f_y(0,0)$ is nonzero.*

(3) *Suppose that $X(m,n,f)$ is not smooth. If $m > 1$ and $n > 1$, $X(m,n,f)$ is nonnormal along the fiber $f^{-1}(0,0)$ which is isomorphic to $\mathbb{A}^2 = \operatorname{Spec} k[u,v]$. If $m = 1$ and $n > 1$, then $f(x,y) \in k[y]$ with $f(0) = f_y'(0) = 0$ and the singular locus is the line $\{x = y = v = 0\}$ in $f^{-1}(0,0)$. If $m > 1$ and $n = 1$, then $f(x,y) \in k[x]$ with $f(0) = f_x'(0) = 0$ and the singular locus is the line $\{x = y = u = 0\}$. If $m = n = 1$, then $X(1,1,f)$ is defined by $xv - yu = 0$. Hence the point $(x,y,u,v) = (0,0,0,0)$ is the unique singular point. Thus $X(m,n,f)$ is normal except for the case $m > 1$ and $n > 1$.*

Let $\mathcal{H}$ be the set of slimmed hypersurfaces $X(m,n,f)$ with $f \in k[x,y]$ and positive integers m,n. Let $X_i(m_i,n_i,f_i)$ be slimmed hypersurfaces for $i = 1,2$. Let $m_3 =$

$\max(m_1, m_2)$ and $n_3 = \max(n_1, n_2)$. For $c_1, c_2 \in k$, define a polynomial $f_3(x, y)$ by

$$\frac{f_3(x,y)}{x^{m_3}y^{n_3}} = \frac{c_1 f_1(x,y)}{x^{m_1}y^{n_1}} + \frac{c_2 f_2(x,y)}{x^{m_2}y^{n_2}}.$$

Define the sum $c_1X_1 + c_2X_2$ by the slimmed form of the hypersurface $X(m_3, n_3, f_3)$. With this sum and scalar product, $\mathcal{H}$ is a k-vector space. By Nagata's lemma, all hypersurfaces belonging to $\mathcal{H}$ are factorial if $f(0,0) \neq 0$.

We have the following result.

Theorem 3.1.21. *Let $Y = \mathbb{A}^2$ and $U = \mathbb{A}^2 \setminus \{(0,0\}$. Then $H^1(U, \mathcal{O}_Y)$, which is the set of isomorphism classes of G_a-torsors over U, is an infinite-dimensional k-vector space with a basis $\{x^i y^j \mid i < 0, j < 0\}$. Furthermore, the correspondence*

$$X(m, n, f(x,y)) \in \mathcal{H} \mapsto \frac{f(x,y)}{x^m y^n} \in H^1(U, \mathcal{O}_Y)$$

is an isomorphism of k-vector space.

Proof. The following argument is an exercise in [90, Exercise 4.3]. Since $U = U_1 \cup U_2$, we have an exact sequence (the Mayer–Vietoris sequence)

$$\begin{aligned} 0 &\to H^0(U, \mathcal{O}_Y) \to H^0(U_1, \mathcal{O}_Y) \oplus H^0(U_2, \mathcal{O}_Y) \to H^0(U_1 \cap U_2, \mathcal{O}_Y) \\ &\to H^1(U, \mathcal{O}_Y) \to H^1(U_1, \mathcal{O}_Y) \oplus H^1(U_2, \mathcal{O}_Y) = 0, \end{aligned}$$

where $H^1(U_1, \mathcal{O}_Y) \oplus H^1(U_2, \mathcal{O}_Y) = 0$ because U_1 and U_2 are affine open sets. The exact sequence gives an exact sequence

$$0 \longrightarrow B \longrightarrow B[x^{-1}] \oplus B[y^{-1}] \longrightarrow B[(xy)^{-1}] \longrightarrow H^1(U, \mathcal{O}_X) \longrightarrow 0,$$

where $B = k[x, y]$. This gives the isomorphism

$$H^1(U, \mathcal{O}_Y) \cong \sum_{i<0,\, j<0} kx^i y^j.$$

The isomorphism between $\mathcal{H}$ and $H^1(U, \mathcal{O}_X)$ is clear. □

Example 3.1.22. Let $X = X(3, 3, f)$ with $f(x, y) = x + y$. Then X is a smooth, affine 3-fold. Let $f : X \to Y = \operatorname{Spec} k[x, y]$ be the quotient morphism by the G_a-action given by a locally nilpotent derivation δ on $A = \Gamma(\mathcal{O}_X)$ such that $\delta(x) = \delta(y) = 0, \delta(v) = y^3$ and $\delta(u) = x^3$. Then the fiber $f^{-1}(0,0)$ is the affine plane $\mathbb{A}^2 = \operatorname{Spec} k[u, v]$. Note that A is not factorial because the closed set $f^{-1}(0,0)$ is defined by the ideal (x, y), not by a principal ideal. So, this makes a clear distinction between the case of $\Gamma(X, \mathcal{O}_X)$ being factorial (see Theorem 3.1.8) and the case of X being nonfactorial.

Let $\overline{U}$ be a Platonic $\mathbb{A}^1_*$-surface over k (see Section 1.2). Namely, $\overline{U}$ is isomorphic to $(\mathbb{A}^2//G) \setminus \{\overline{O}\}$ for a small finite subgroup G of $\mathrm{GL}(2,k)$ (i. e., G contains no reflections) and the singular point $\overline{O}$ of $\mathbb{A}^2//G$. Note that the group G acts freely on the set $U := \mathbb{A}^2 \setminus \{(0,0)\}$. In fact, U is the universal covering space of $\overline{U}$. We are interested in the existence of nontrivial $\mathbb{A}^1$-bundles over $\overline{U}$. We consider the case G is cyclic. The noncyclic case is not yet settled.

With a slight change of notations, let $\overline{f} : \overline{X} \to \overline{U}$ be an $\mathbb{A}^1$-bundle over $\overline{U}$ and let $X = \overline{X} \times_{\overline{U}} U$. Then the projection $f : X \to U$ is an $\mathbb{A}^1$-bundle. Furthermore, the group G acts on X freely in such a way that the G-action preserves the $\mathbb{A}^1$-bundle structure f and the quotient $X//G$ is equal to $\overline{X}$. Let $U = U_1 \cup U_2$ be the open covering of U as above and let t_i be a fiber coordinate of $f^{-1}(U_i)$ as chosen above for $i = 1, 2$. By Theorem 3.1.21, we may assume that

$$t_2 = t_1 + \sum_{i>0,\, j>0} \frac{c_{ij}}{x^i y^j}. \tag{3.1}$$

Furthermore, since G is cyclic, we identify G with the group of rth roots of unity, i. e., $G = \{\zeta^i \mid 0 \le i < r\}$, where r is the order of G and ζ is a primitive rth root of unity. We may (and shall) assume that the G-action on $\mathbb{A}^2$ is given by ${}^\zeta x = \zeta x$ and ${}^\zeta y = \zeta^d y$, where $\gcd(d, r) = 1$. Since U_i is now G-stable, $f^{-1}(U_i)$ is G-stable for $i = 1, 2$. Note that $f^{-1}(U_1) = \operatorname{Spec} k[x, y, x^{-1}, t_1]$ and $f^{-1}(U_2) = \operatorname{Spec} k[x, y, y^{-1}, t_2]$, where

$${}^\zeta t_1 = \zeta^a t_1 + F_1(x, y, x^{-1}), \quad {}^\zeta t_2 = \zeta^b t_2 + F_2(x, y, y^{-1}) \tag{3.2}$$

with $a, b \ge 0$ and $F_1(x, y, x^{-1})$ (resp., $F_2(x, y, y^{-1})$) is a polynomial in x, y, x^{-1} (resp., x, y, y^{-1}). Applying ζ to the terms in (3.1) and using the equality (3.2), we have

$$\begin{aligned} &\zeta^b\Big(t_1 + \sum_{i>0,\, j>0} \frac{c_{ij}}{x^i y^j}\Big) + F_2(x, y, y^{-1}) \\ &\quad = \zeta^a t_1 + F_1(x, y, x^{-1}) + \sum_{i>0,\, j>0} \frac{\zeta^{-(i+dj)} c_{ij}}{x^i y^j}. \end{aligned}$$

Thence we have $a \equiv b \pmod r$ and

$$\sum_{i>0,\, j>0} \frac{(\zeta^{-(i+dj)} - \zeta^b) c_{ij}}{x^i y^j} = F_2(x, y, y^{-1}) - F_1(x, y, x^{-1}). \tag{3.3}$$

But, the latter equality (3.3) holds if and only if both sides are zero separately because the left-hand side is a sum of purely fractional parts, i. e., each term has a constant numerator and $x^i y^j$ with $i, j > 0$ as a denominator. Hence $c_{ij} = 0$ whenever $a + i + dj \not\equiv 0 \pmod r$ and $F_1(x, y, x^{-1}) = F_2(x, y, y^{-1})$. For the latter equality, the fractional parts of $F_1(x, y, x^{-1})$ and $F_2(x, y, y^{-1})$ vanish, hence $F_1(x, y, x^{-1}) = F_2(x, y, y^{-1}) = g(x, y) \in k[x, y]$.

Lemma 3.1.23. *With the above notation, we may assume $g(x, y) = 0$.*

Proof. Write the G-action on $k[x,y,t_i]$ by σ. Hence $\sigma(x) = \zeta x, \sigma(y) = {}^\zeta y$ and $\sigma(t_i) = {}^\zeta t_i$. It is well-known that the first G-cohomology group $H^1(G,M)$ with value in a G-module M is given as

$$H^1(G,M) = \{m \in M \mid N(m) = 0\}/(\sigma - 1)M, \tag{3.4}$$

where $N = 1 + \sigma + \cdots + \sigma^{r-1}$. If $\zeta^a = 1$, then $g(x,y)$ is a 1-cocycle of G with value in a G-module $k[x,y]$. If $\zeta^a \neq 1$, we can derive the following equality from (3.1):

$$(\zeta^a - 1)(t_i + \sigma(t_i) + \cdots + \sigma^{r-1}(t_i)) = -(g + \sigma(g) + \cdots + \sigma^{r-1}(g)),$$

where

$$t_i + \sigma(t_i) + \cdots + \sigma^{r-1}(t_i) = (1 + \zeta^a + \cdots + \zeta^{(r-1)a})(t_i) = \frac{1-\zeta^{ra}}{1-\zeta^a}(t_i) = 0.$$

Hence $g + \sigma(g) + \cdots + \sigma^{r-1}(g) = 0$. This implies that $g(x,y)$ is a 1-cocycle of G. We show that $H^1(G,k[x,y]) = 0$. Since $\sigma(x) = \zeta x$ and $\sigma(y) = \zeta^d y$, it follows that $\sigma(x^iy^j) = \zeta^{i+dj}x^iy^j$. Hence it suffices to show that $H^1(G,M) = 0$ for a G-module $M = ke$ with $\sigma(e) = \zeta^s e$, where $0 \leq s < r$. If $s = 0$, it is clear by the above expression of $H^1(G,M)$ because $N(\lambda e) = r\lambda e = 0$ implies $\lambda = 0$. If $s > 0$, then $\zeta^s - 1 \neq 0$. Hence $e = (\sigma - 1)(e/(\zeta^s - 1))$. This implies that $H^1(G,M) = H^1(G,k[x,y]) = 0$.

Suppose that $\zeta^a = 1$. Since $g(x,y)$ is a 1-cocycle, there exists $h(x,y) \in k[x,y]$ such that $g = \sigma(h) - h$. Then we have $\sigma(t_i - h) = (t_i - h)$ for $i = 1, 2$. So, we make a change of variables $t_i \mapsto t_i - h$. Suppose that $\zeta^a \neq 1$. Consider a new G-action on $k[x,y]$ defined by $\sigma' = \zeta^{-a}\sigma$, hence defined by $\sigma'(x) = \zeta^{1-a}x$ and $\sigma'(y) = \zeta^{d-a}y$. Then $\sigma'(t_i) = t_i + \zeta^{-a}g(x,y)$ for $i = 1,2$. By the same argument as above, $\zeta^{-a}g(x,y)$ is a 1-cocycle. Hence we can write $\zeta^{-a}g = \zeta^{-a}\sigma(h) - h$ for $h \in k[x,y]$. Namely, we have $g = \sigma(h) - \zeta^a h$. Then make a change of variables $t_i \mapsto t_i - h$ for $i = 1,2$. Then, with a new variable $t_i' = t_i - h$, we have $\sigma(t_i') = \zeta^a(t_i')$. Hence we may assume that $g(x,y) = 0$. □

Hence ${}^\zeta t_i = \zeta^a t_i$ for $i = 1,2$. With the above action of G, the quotient $f^{-1}(U_1)//G$ is given by $\operatorname{Spec} k[x^r, x^{-r}, y/x^d, t_1/x^a]$, which is equal to $\overline{f}^{-1}(\overline{U}_1)$. Similarly, we have $\overline{f}^{-1}(\overline{U}_2) = \operatorname{Spec} k[y^r, y^{-r}, x/y^e, t_2/y^{b'}]$, where e and b' are positive integers such that $de \equiv 1 \pmod r$ and $a \equiv db' \pmod r$. By (3.1), we have

$$\frac{t_2}{y^{b'}} = \left(\frac{x^a}{y^{b'}}\right)\frac{t_1}{x^a} + \sum_{i>0,\, j>0} \frac{c_{ij}}{x^iy^{j+b'}}, \tag{3.5}$$

where $x^iy^{j+b'}$ is G-invariant if $c_{ij} \neq 0$.

Theorem 3.1.24. *With the above notations, we have an exact sequence*

$$0 \longrightarrow H^1(U,\mathcal{O}_U)^G \xrightarrow{\alpha} H^1(\overline{U},GA) \xrightarrow{\beta} C\ell(\mathbb{A}^2//G) \longrightarrow 0,$$

where $H^1(U,\mathcal{O}_U)^G = H^1(\overline{U},\mathcal{O}_{\overline{U}})$.

Proof. Note that $H^1(\overline{U}, GA)$ denotes the set of isomorphism classes of $\mathbb{A}^1$-bundles over $\overline{U}$. Consider an $\mathbb{A}^1$-bundle $\overline{f} : \overline{X} \to \overline{U}$ which is represented by (3.5) with respect to the covering $\overline{U} = \overline{U}_1 \cup \overline{U}_2$. Then the image of $\overline{X}$ by the mapping β is an invertible sheaf whose transition function is $x^a/y^{b'}$. Suppose that it is trivial. Namely, $x^a/y^{b'} = u/v$ with $u \in k[x^r, x^{-r}, y/x^d]^*$ and $v \in k[y^r, y^{-r}, x/y^e]^*$. Then $a \equiv b' \equiv 0 \pmod r$. Then the $\mathbb{A}^1$-bundle determined by (3.5) gives rise to a class of $H^1(U, \mathcal{O}_U)^G$. The surjectivity of the homomorphism β follows from an observation that any invertible sheaf on $\overline{U}$ is trivialized over $\overline{U}_1$ and $\overline{U}_2$ and its transition function is used to define an $\mathbb{A}^1$-bundle over $\overline{U}$ whose transition matrix is a diagonal matrix without the unipotent part. The exactness of the sequence and the equality $H^1(\overline{U}, \mathcal{O}_{\overline{U}}) = H^1(U, \mathcal{O}_U)^G$ follows from the exact sequence of cohomology groups used in Remark 3.1.16. □

The above argument shows the following general result.

Proposition 3.1.25. *Let Y be a normal algebraic variety. Then we have an exact sequence*

$$0 \longrightarrow H^1(Y, \mathcal{O}_Y) \longrightarrow H^1(Y, GA) \longrightarrow \operatorname{Pic} Y \longrightarrow 0,$$

where $H^1(Y, GA)$ represents the set of isomorphism classes of $\mathbb{A}^1$-bundles over Y.

Proof. We have an exact sequence as in Remark 3.1.16

$$\begin{aligned} 0 \longrightarrow \Gamma(Y, \mathcal{O}_Y) \longrightarrow GA(\Gamma(Y, \mathcal{O}_Y)) \longrightarrow \Gamma(Y, \mathcal{O}_Y)^* \\ \longrightarrow H^1(Y, \mathcal{O}_Y) \longrightarrow H^1(Y, GA) \longrightarrow H^1(Y, \mathcal{O}_Y^*), \end{aligned}$$

where $GA(\Gamma(Y, \mathcal{O}_Y)) \to \Gamma(Y, \mathcal{O}_Y^*)$ is surjective by the definition of the group scheme GA and $H^1(Y, GA) \longrightarrow H^1(Y, \mathcal{O}_Y^*) = \operatorname{Pic} Y$ is surjective as well because the scheme Y defined by the symmetric algebra sheaf $\mathcal{S}_X(\mathcal{O}_Y \oplus \mathcal{I})$ is an $\mathbb{A}^1$-bundle over Y for any invertible sheaf $\mathcal{I}$. □

3.1.4 Fixed point locus of G_a-actions

In this subsection, we assume that the ground field k is the complex number field $\mathbb{C}$ in order to use some topological arguments. However, many of the results are easily verified over k by the Lefschetz principle. Recall that the algebraic quotient of $X = \operatorname{Spec} A$ by a G_a-action is defined if the ring of invariants $B := A^{G_a} = \operatorname{Ker} \delta$ is finitely generated over $\mathbb{C}$. In this case, we let Y (or $X//G_a$) be the normal affine variety corresponding to this ring of invariants, and we write $q : X \to Y$ for the natural quotient morphism. Furthermore, X^{G_a} denotes the fixed point locus in X under the G_a-action, which is a closed subset defined by the ideal of A generated by $\delta(A)$. Note that $\delta(A) \cap B$ is an ideal of B. Let $\overline{\Sigma}$ be the closed set $V(\delta(A) \cap B)$ in Y. If $X^{G_a} \neq \emptyset$, then the set X^{G_a} is contained in $q^{-1}(\overline{\Sigma})$. The ideal $\delta(A) \cap B$ of B is called the *plinth ideal*.

For any topological space T, let $H_c^i(T)$ denote the ith cohomology group with compact support and with integer coefficients. Other homology and cohomology groups have also integer coefficients unless otherwise specified.

Lemma 3.1.26. *The following general results hold:*
(1) *For a smooth affine irreducible variety X of dimension $d > 1$, the group $H_c^i(X) = (0)$ for $i = 0, 1$.*
(2) *An affine irreducible curve C is contractible if and only if $H_1(C) = (0)$. Any singular point of such a curve is a unibranch singularity.*
(3) *An affine irreducible curve C has $H_c^0(C) = (0)$. If further $H_c^1(C) = (0)$, then C is contractible.*

Proof. (1) By duality, $H_c^i(X) \cong H_{2d-i}(X)$ (see Section 1.1.3). By Theorem 1.1.4, for an affine irreducible variety X of $\dim X = d$ the homology groups $H_j(X)$ are trivial for $j > d$. The assertion follows from this.

(2) $H_2(C) = 0$ by Theorem 1.1.4. Hence $e(C) = 1$. Let $\widetilde{C}$ be the normalization of C. Then $e(\widetilde{C}) \geq e(C)$ and $e(\widetilde{C}) = 2 - 2g - r$ with $r > 0$, where g is the geometric genus of $\widetilde{C}$. Hence $g = 0$ and $r = 1$. It follows that $\widetilde{C} \cong \mathbb{A}^1$ and $e(\widetilde{C}) = e(C)$. This implies the assertion.

(3) The first assertion follows from Lemma 1.1.3. For the proof of the second assertion, let S be the singular locus of C. We use the long exact sequence for cohomology with compact support for the pair (C, S).

$$\begin{aligned} 0 \to H_c^0(C,S) &\to H_c^0(C) \to H_c^0(S) \\ &\to H_c^1(C,S) \to H_c^1(C) \to H_c^1(S) \to \cdots . \end{aligned}$$

Suppose $H_c^0(C) = (0)$ and $H_c^1(C) = (0)$. Then we get $H_c^0(S) \cong H_c^1(C,S) \cong H_1(C - S)$. Since S is compact, the rank of $H_c^0(S)$ is equal to the cardinality of S. If either C is not rational, or C has a singular point which is not unibranch, or the normalization of C in its function field is isomorphic to $\mathbb{A}^1$ minus at least one point, then we see easily that the rank of $H_1(C - S)$ is strictly bigger than the cardinality of S. It follows that C is rational, it has at most unibranch singular points and its normalization is isomorphic to $\mathbb{A}^1$. This means that C is contractible. □

The following result due to Białynicki–Birula [17, Corollaries 1 and 4] will be crucial in subsequent arguments.

Lemma 3.1.27. *Let G_a act regularly on a reduced complex algebraic variety X and let Z be a G_a-stable closed subvariety of X which contains X^{G_a}. Then $H_c^i(X) \cong H_c^i(Z)$ for $i = 0, 1$. In particular, we have $H_c^i(X) \cong H_c^i(X^{G_a})$ for $i = 0, 1$. If X is irreducible, incomplete and $\dim X > 0$, then X^{G_a} cannot have an isolated fixed point.*

Proof. This follows immediately from the result in [17] stated in terms of étale cohomologies with compact supports and converted easily into $H_c^i(X)$ for a complex algebraic variety X. Namely, we have $H_c^i(X) \cong H_c^i(X^{G_a}) \cong H_c^i(Z)$ for $i = 0, 1$.

We indicate how to prove this result. Let X be an irreducible, algebraic variety on which G_a acts. Let Z be a reduced, possibly reducible, G_a-stable closed subvariety of X which contains X^{G_a}. Then we can find a stratification

$$X = X_0 \supset X_1 \supset \cdots \supset X_j \supset \cdots \supset X_r = Z$$

by closed G_a-stable (possibly reducible) subvarieties X_j of X such that $X_j \setminus X_{j+1}$ is a smooth variety on which the G_a-action is *split* in the sense that $X_j \setminus X_{j+1} \cong Y_j \times \mathbb{A}^1$ for some variety Y_j and the action of G_a on the right-hand side of this isomorphism is by translation on $\mathbb{A}^1$. For the existence of such a filtration, we use a result of Rosenlicht [196] which says that there is a nonempty G_a-stable Zariski-open subset U of X for which the geometric quotient U/G_a exists. Since X cannot contain any positive-dimensional complete subvariety, each Y_j is incomplete, except when some Y_j contains a finite set of isolated points.

We will prove that $H_c^i(X) \cong H_c^i(Z)$ for $i = 0, 1$ by induction on r. It suffices to prove this when $r = 1$, i. e., $X \setminus Z \cong Y_0 \times \mathbb{A}^1$. First assume that each irreducible component of Y_0 has positive dimension. Since $X \setminus Z$ is smooth by assumption, Y_0 is a disjoint union of smooth irreducible varieties, each of which is incomplete. Let Y_{0j} be an irreducible component of Y_0 of dimension, say $d_{0j} > 0$. Then $H_{2d_{0j}}(Y_{0j}) = (0)$ because Y_{0j} is incomplete. It follows from the Poincaré duality that

$$H_c^k(Y_{0j} \times \mathbb{A}^1) \cong H_{2d_{0j}+2-k}(Y_{0j} \times \mathbb{A}^1) \cong H_{2d_{0j}+2-k}(Y_{0j}) = (0)$$

for $k = 0, 1, 2$. Hence $H_c^k(Y_0 \times \mathbb{A}^1) = (0)$ for $k = 0, 1, 2$. Now the long exact sequence for cohomology with compact support for the pair (X, Z) shows that $H_c^i(X) \cong H_c^i(Z)$ for $i = 0, 1$. If some Y_{0j} consists of finitely many points then $X \setminus Z$ has finitely many connected components which are nontrivial G_a-orbits. Each such G_a-orbit Gx satisfies $H_c^i(G_a x) = (0)$ for $i = 0, 1$. Since Z is a closed affine subvariety, we see easily that in this case $G_a x$ is actually a connected component of X. Therefore, without loss of generality, we can assume that X has no nontrivial orbits as connected components. Combining these observations, we obtain the result.

For the last assertion, note that $H_c^0(X) = (0)$ since X is incomplete and irreducible. Meanwhile, if X^{G_a} contains an isolated fixed point, then $H_c^0(X^{G_a}) \neq (0)$. □

Corollary 3.1.28. *With the notation of Lemma 3.1.27, suppose that X is affine and $X/\!/G_a$ is defined as an algebraic variety. If a fiber F of the quotient morphism q contains a fixed point then F contains a positive-dimensional fixed point subset.*

Remark 3.1.29. By a similar argument, it is proved in [17, Corollary 4] that if X is complete and connected then X^{G_a} is connected.

3.1.5 Degenerate fibers of the quotient morphism and singularities of the quotient space

Now let X be an irreducible affine variety such that $Y := X//G_a$ is defined as an algebraic variety. A general fiber of the quotient morphism $q : X \to Y$ is isomorphic to $\mathbb{A}^1$. First we study the singularities of $X//G_a$ by restricting the type of a fiber of q when $\dim X = 3$.

Lemma 3.1.30. *Assume that X is a smooth affine threefold. If F is a fiber of q which contains an irreducible component of dimension one, say C, then the point $Q := q(F)$ is at worst a quotient singular point of Y.*

Proof. Let S be a general hyperplane section of X and let $P \in S \cap C$. Then the quotient morphism q restricted to S has an isolated point in the fiber at P. This implies that the analytic local ring of S at P is an integral extension of the analytic local ring of Y at Q. Since P is a smooth point of S, it follows from Mumford's well-known result on the topology of a normal surface singular point (Theorem 1.3.17 and its remark) that the local fundamental group $\pi_1^Q(Y)$ is a finite group. Hence Y has at worst a quotient singularity at Q (see the proof of Lemma 3.5.4). □

If $\dim X = 3$, then there are only finitely many fibers of q each of which contain a 2-dimensional irreducible component. By Theorem 3.1.8, there are no 2-dimensional fiber component if X is factorial. We shall consider degenerate fibers of the quotient morphism $q : X \to Y = X//G_a$. Note that a special case was treated in Lemma 3.1.7.

Lemma 3.1.31. *Assume that X is a smooth affine threefold with a G_a-action such that $\dim X^{G_a} \le 1$. Let F be a one-dimensional fiber of q. Then F_{red} is a disjoint union of contractible irreducible curves.*

Proof. We can assume that F is a singular fiber of q. We claim that $q(X)$ is an open subset of Y after shrinking Y to avoid the fibers containing dimension two fiber components. To see this, we argue as in the proof of Lemma 3.1.30. For a general hyperplane section Z of X the induced morphism $q_Z : Z \to Y$ is quasifinite. It follows that $q_Z(Z)$ is a Zariski-open subset of Y. In fact, let Y° be the smooth locus of Y. Then q_Z restricted to $(q_Z)^{-1}(Y^\circ)$ is a flat morphism because Y° and Z are smooth and q_Z is quasifinite. If a singular point Q is in the image of q_Z, the proof of Lemma 3.1.30 shows that the point Q is a quotient singular point and Q is contained in the interior of $q_Z(Z)$. By shrinking X, we can assume that q is surjective and every fiber of q is one-dimensional. Note that F_{red} consists of a finite number of G_a orbits and components of X^{G_a}. Since G_a orbits are contractible we can, by shrinking Y, assume that $X^{G_a} \subset F$. To complete the proof, it suffices to show that $F^{G_a} = X^{G_a}$ is a disjoint union of contractible irreducible curves. Using Lemma 3.1.26, we see that $H_c^i(X) \cong H_c^i(F^{G_a}) = (0)$ for $i = 0, 1$. For, since X is smooth and affine of dimension three, the groups $H_c^i(X)$ are trivial for $i = 0, 1$. If Δ is a connected component of F^{G_a}, then it follows that $H_c^i(\Delta) = (0)$ for $i = 0, 1$. We write

$\Delta = C \cup C'$, where C is some irreducible component of Δ. Then $C \cap C'$ is a nonempty finite set if $C' \neq \emptyset$. We claim that $\Delta = C$. The Mayer–Vietoris sequence for cohomology with compact support for the union $C \cup C'$ is

$$\cdots \to H_c^0(C) \oplus H_c^0(C') \to H_c^0(C \cap C') \to H_c^1(C \cup C') \to \cdots,$$

where $H_c^0(C \cap C') \neq (0)$ as $C \cap C'$ is a nonempty finite set, $H_c^0(C) = H_c^0(C') = (0)$ as C and C' are incomplete and $H_c^1(C \cup C') = (0)$ by the hypothesis. But this is a contradiction and thus proves the claim. It follows that $H_c^1(C) = (0)$, so that C is contractible by Lemma 3.1.26. □

Remark 3.1.32. With the notations and assumptions of Lemma 3.1.31, if X is factorial, the quotient morphism $q : X \to Y$ has pure equidimension one by Theorem 3.1.8 and any two-dimensional component S of X^{G_a} has an $\mathbb{A}^1$-fibration induced by q by Lemma 3.1.37 below, hence $\dim q(S) = 1$. If a two-dimensional component S of X^{G_a} is defined by $a = 0$ for a prime element $a \in A = \Gamma(X, \mathcal{O}_X)$, then $S \subset X^{G_a}$ implies that $\delta(A) \subset aA$, where δ is an lnd associated to the G_a-action. Then $a \in \operatorname{Ker}\delta$ and $a^{-1}\delta$ is an lnd of A. Repeating this process, we can get rid of the two-dimensional component S of X^{G_a} and assume that $\dim X^{G_a} \leq 1$. Even with this process, the assumption that $\dim X^{G_a} \leq 1$ is meaningful. It is important to ask in Lemma 3.1.31 if F_{red} is a disjoint union of the affine lines. When compared with the case of a surface, it is also significant if the normality assumption on X is enough to conclude the assertion.

Following Abhyankar [1], we call a normal surface singularity *quasirational* if the exceptional divisor in a suitable resolution of singularity is a divisor with simple normal crossings, all whose irreducible components are rational curves and the dual graph is a tree.

Let X be a factorial, smooth, affine threefold with a G_a-action. Every fiber of the quotient morphism $q : X \to Y$ is either one-dimensional or the empty set. Let Q be a point of Y. If $F_Q := q^{-1}(Q)$ is nonempty, then Q is at worst a quotient singularity by Lemma 3.1.30. So we assume that F_Q is empty. By [176], we can find an embedding $X \subset X'$, where X' is a smooth threefold such that q extends to a proper morphism $q' : X' \to Y$ and that the G_a-action extends to X'. Since $\dim X' = 3$, there are only finitely many fibers F'_Q of q' which contain a two-dimensional component. By shrinking Y we will assume that except possibly for F'_Q every other fiber of q' is one-dimensional. With these notations and assumptions, we have the following result.

Lemma 3.1.33. *Assume that Q is a smooth point of Y. Then the fiber F'_Q is simply-connected.*

Proof. Let $C \subset Y$ be the curve such that $q : X - q^{-1}(C) \to Y - C$ is an $\mathbb{A}^1$-bundle. For any irreducible component C_i of C, the surface $S_i := q^{-1}(C_i)$ is an irreducible and reduced surface by Lemma 3.1.15, (1). Hence a general fiber of the morphism $S_i \to C_i$ is reduced. It follows that except for finitely many points Q' of Y the fiber $F_{Q'}$ is reduced.

Hence we may assume that for $Q' \neq Q$ the fiber $F'_{Q'}$ is one-dimensional and contains a reduced irreducible component. A general fiber of q' is isomorphic to $\mathbb{P}^1$. Let N be a small contractible Euclidean neighborhood of Q in Y. By Lemma 1.1.6, we have a short exact sequence

$$\pi_1(\mathbb{P}^1) \longrightarrow \pi_1(q'^{-1}(N)) \longrightarrow \pi_1(N).$$

Thus, $q'^{-1}(N)$ is simply-connected. Since q' is proper we can assume that F'_Q is a strong deformation retract of $q'^{-1}(N)$. This proves that F'_Q is simply-connected. □

Remark 3.1.34. Assume now that Q is not necessarily a smooth point of Y. As above, F'_Q can be assumed to be a strong deformation retract of X'. If F'_Q is simply-connected then so is X'. Let $\tau : \widetilde{Y} \to Y$ be a resolution of singularities of Y. There exists a smooth threefold $\widetilde{X'}$ obtained by blowing up X' such that there is a proper morphism $\widetilde{q} : \widetilde{X'} \to \widetilde{Y}$ extending q'. It is easy to see that since $\widetilde{X'} \to X'$ is proper and birational, we have a natural isomorphism $\pi_1(\widetilde{X'}) \xrightarrow{\sim} \pi_1(X')$. The natural map $\pi_1(\widetilde{X'}) \to \pi_1(\widetilde{Y})$ is surjective since the fibers of $\widetilde{q}$ are connected. It follows that $\widetilde{Y}$ is simply-connected if F'_Q is simply-connected. We can assume that Y is so small that the exceptional divisor E for τ is a strong deformation retract of $\widetilde{Y}$. So we conclude that if F'_Q is simply-connected then so is E, i. e., Q is a quasirational singular point of Y.

Let $q : X \to Y$ be the quotient morphism under a G_a-action. It is an interesting problem to ask whether a fiber component F of dimension greater than one of the morphism q is rational (see Remark 3.1.32). The following example suggests that it might not be rational if q is assumed to be only an $\mathbb{A}^1$-fibration, though the example is incomplete since X is not affine.

Example 3.1.35. Let S be the affine cone over a smooth irrational curve C embedded in some projective space $\mathbb{P}^n$. Let $\widetilde{S} \to S$ be the blow-up of the singular point of S. Let $W := \widetilde{S} \times \mathbb{P}^1$ and consider the composite morphism $W \to \widetilde{S} \to S$. A general fiber of the morphism $W \to S$ is $\mathbb{P}^1$ but the fiber over the singular point of S is isomorphic to $C \times \mathbb{P}^1$ and has an infinite fundamental group. If we take $X = \widetilde{S} \times \mathbb{A}^1$ instead of W and the composite morphism $q : X \to S$, then the fiber of q lying over the singular point is not rational, though X is not affine.

We have the following result.

Lemma 3.1.36. *Assume that X is a smooth affine threefold with a G_a-action such that $\dim X^{G_a} \leq 1$. Let F be a fiber of the quotient morphism $q : X \to Y$. Let S be an irreducible component of the fiber F such that $\dim S = 2$. Then any one-dimensional irreducible component C of F is disjoint from S.*

Proof. Since X is smooth and affine, $X' := X - \cup S_i$ is again a smooth affine threefold, where S_i are the irreducible divisors in F. By Lemma 3.1.31, the fiber of the morphism $q' = q|_{Y'} : X' \to Y$ containing $C - \cup S_i$ is a disjoint union of contractible irreducible curves. If $C \cap S \neq \emptyset$, then C will be a complete curve, which is a contradiction. □

In order to investigate the properties of a two-dimensional fiber component of $q: X \to Y = X//G_a$, we need the following result.

Lemma 3.1.37. *Assume that X is a smooth factorial affine threefold. Let S be a two-dimensional irreducible component of the fixed point locus X^{G_a}. Then S has an $\mathbb{A}^1$-fibration.*

Proof. By the hypothesis, using Theorem 3.1.8, we know that every fiber of q is one-dimensional and $A := \Gamma(X, \mathcal{O}_X)$ is factorial. By factoriality of X, we can find a prime element f in A whose zero locus is S. Since S is G_a-stable and G_a has no nontrivial characters, f is in the kernel of δ. Since X^{G_a} is the zero locus of $\delta(A)$ we also have $\delta(A) \subset (f)$. Since S is contained in X^{G_a}, the induced locally nilpotent derivation $\overline{\delta}$ on A/fA is trivial. As in Lemma 3.1.15, we may replace δ by $\Delta := f^{-r}\delta$ with $r > 0$ such that $\Delta(A) \not\subset fA$. Then Δ (= δ/f with the notation therein) gives rise to an $\mathbb{A}^1$-fibration on S. □

We consider a singular fiber of dimension one in the case when it is reduced. We need the following result for an algebraic surface with G_a-action.

Lemma 3.1.38. *Let Z be a (not necessarily normal) affine algebraic surface with a G_a-action and let $q: Z \to C$ be the associated quotient morphism. Let F be a singular fiber of q and let $F = \sum_{i=1}^n F_i$ be the decomposition into irreducible components. Assume that F is reduced. Then each irreducible component F_i is a connected component and isomorphic to $\mathbb{A}^1$.*

Proof. Our proof consists of several steps.

STEP 1. We may assume that C is a normal algebraic curve. In fact, let $\widetilde{C} \to C$ be the normalization, let $Z' = Z \times_C \widetilde{C}$ and let $q': Z' \to \widetilde{C}$ be the second projection. Then q' is the quotient morphism by the associated G_a-action. Let $Q = q(F)$ and let $\widetilde{Q}$ be a point of $\widetilde{C}$ lying over Q. Since Q and $\widetilde{Q}$ are k-rational points, we have ${q'}^{-1}(\widetilde{Q}) = q^{-1}(Q) = F$. Hence we may assume without loss of generality that C is normal.

STEP 2. Assume that F is reduced. Let $\mathcal{O}$ be the local ring of C at Q. It is a discrete valuation ring with uniformizant t. Set $A = \Gamma(Z, \mathcal{O}_Z), B = \Gamma(C, \mathcal{O}_C)$ and $R = A \otimes_B \mathcal{O}$. Then tR is a radical ideal and $tR = \mathfrak{P}_1 \cap \cdots \cap \mathfrak{P}_n$, where $\mathfrak{P}_i$ is the defining prime ideal of F_i in Spec R. Note that every component F_i is G_a-stable. If the induced G_a-action on F_i is nontrivial, then F_i is isomorphic to $\mathbb{A}^1$ and disjoint from other components. In fact, if $F_i \cap F_j \neq \emptyset$ for $i \neq j$, then $F_i \cap F_j$ consists of G_a-stable points, while there are no G_a-stable points on F_i provided the G_a-action on F_i is nontrivial.

We claim that if F_i is disjoint from $\sum_{j \neq i} F_j$ then $U := Z \setminus F_i$ is affine. In fact, let $\nu: \widetilde{Z} \to Z$ be the normalization. Then $\widetilde{Z}$ has a G_a-action such that the associated quotient morphism is the composite $q \cdot \nu : \widetilde{Z} \to C$ (see Lemmas 2.2.16 and 2.2.19). Then $\widetilde{U} := \widetilde{Z} \setminus \nu^{-1}(F_i)$ is an affine open set. Meanwhile, $\nu_{\widetilde{U}} : \widetilde{U} \to U$ is a finite surjective morphism. Hence, by Chevalley's theorem [90, Exercise 4.2, p. 222], U is affine if $\widetilde{U}$ is affine. Note that this argument applies also to the case where the induced G_a-action on F_i is trivial.

Hence we may assume that the G_a-action on every irreducible component is trivial. Namely, if δ is the associated locally nilpotent derivation on R, then $\delta(R) \subset \mathfrak{P}_i$ for every i. Then $\delta(R) \subset tR$ so that $t^{-1}\delta$ is a locally nilpotent derivation on R. We may replace δ by $t^{-1}\delta$ and look for an irreducible component F_i on which the induced G_a-action is nontrivial. Since δ is not divisible by t infinitely many times, we can find an irreducible component F_i on which the induced (new) G_a-action is nontrivial. Then we can remove F_i while keeping the affineness condition on Z. So, we are done by induction on the number n of the irreducible components. □

Corollary 3.1.39. *Assume that $q : X \to Y$ is an $\mathbb{A}^1$-fibration for affine varieties X and Y. Assume further that q is the quotient morphism of a G_a-action on X. Let $F = q^{-1}(Q)$ be a singular fiber of pure dimension one. If F is reduced, then F is a disjoint union of the affine lines.*

Proof. We may choose a general irreducible curve C on Y through Q such that $Z := q^{-1}(C)$ is an algebraic surface with the induced G_a-action. A general fiber of the morphism $Z \to C$ is smooth and irreducible. Let $q' : Z \to C'$ be the quotient morphism. Then the morphism q is a composition $\mu \cdot q'$, where $\mu : C' \to C$ is a birational morphism since a general fiber of $Z \to C$ is smooth and irreducible. Let $\mu^{-1}(Q) = \{Q'_1, \dots, Q'_r\}$. Then the fiber of q' over Q'_i is reduced for any $1 \le i \le r$ and $F = \coprod_{i=1}^r q'^{-1}(Q'_i)$. By Lemma 3.1.38, $q'^{-1}(Q'_i)$ is a reduced disjoint union of the affine lines, hence so is F. □

In order to exhibit the situation where the one-dimensional fiber F of the quotient morphism $q : X \to Y$ under a G_a-action is not reduced, we consider the following example.

Example 3.1.40. Let $\widetilde{A} = k[x, y, t]$ be a polynomial ring of dimension three with a locally nilpotent derivation δ defined by $\delta(x) = y, \delta(y) = t$ and $\delta(t) = 0$. (It is called a *Weitzenböck derivation* in dimension three.) Then the kernel of δ is a subalgebra $B = k[t, y^2 - 2xt]$. Consider a subalgebra A of $\widetilde{A}$ defined by

$$A = k[t, y, x^2, x^3, xy, xt],$$

which is a nonnormal algebra of dimension three and birational to $\widetilde{A}$. Then A contains B and δ induces a locally nilpotent derivation on A, i. e., $\delta(A) \subseteq A$. Let $\widetilde{X} = \operatorname{Spec} \widetilde{A}$, $Y = \operatorname{Spec} B$, and $X = \operatorname{Spec} A$. Let $\widetilde{q} : \widetilde{X} \to Y$ and $q : X \to Y$ be respectively the quotient morphism of the G_a-actions on $\widetilde{X}$ and X. Let $\widetilde{F}$ and F be the fibers of $\widetilde{q}$ and q over the point Q of Y defined by $t = y^2 - 2xt = 0$. Then the following assertions hold:

(1) $\widetilde{F} = \operatorname{Spec} \widetilde{R}$ and $F = \operatorname{Spec} R$, where

$$\widetilde{R} = k[x, y]/(y^2) \quad \text{and} \quad R = k[t, y, x^2, x^3, xy, xt]/(t, y^2 - 2xt).$$

The nilradicals of $\widetilde{R}$ and R are respectively

$$\widetilde{\mathfrak{n}} = (y)/(y^2) \quad \text{and} \quad \mathfrak{n} = (t, y, xy, xt)/(t, y^2 - 2xt).$$

Hence $\widetilde{F}_{\mathrm{red}} \cong \mathbb{A}^1$ and $F_{\mathrm{red}} = \operatorname{Spec} k[x^2, x^3]$. In fact, $y^2 = 2xt$ and $t = 0$ in R. Hence $y^3 = 2xt \cdot y = 2xy \cdot t = 0, (xy)^2 = x^2 \cdot y^2 = 2x^3 \cdot t = 0$, and $(xt)^2 = x^2 \cdot t^2 = 0$.

(2) $\widetilde{\mathfrak{n}} = (0 : y)$ and $\mathfrak{n} = (0 : xy)$. In fact, $(xy) \cdot y = xy^2 = 2x^2 \cdot t = 0, (xy) \cdot (xy) = x^2y^2 = 0$ and $(xy) \cdot (xt) = x^2 \cdot y \cdot t = 0$.

(3) The maximal ideal $\mathfrak{m} = (t, y, x^2, x^3, xy, xt)/(t, y^2 - 2xt)$ of R is an embedded prime ideal. In fact, $\mathfrak{m} = (0 : xt)$.

In the next result we study the tangent bundles of X, Y assuming that they are both smooth varieties and that the G_a-action is fixed point free. There is a related result in [38, Theorem 2.1].

Lemma 3.1.41. *Assume that X, Y are smooth affine varieties and there is a smooth, dominant morphism $f : X \to Y$ such that general fibers are isomorphic to $\mathbb{A}^1$, $\operatorname{codim}_Y(\overline{Y - f(X)}) > 1$ and the factorial closure of $\Gamma(Y, \mathcal{O}_Y)$ in $\Gamma(X, \mathcal{O}_X)$ is an affine domain over $\mathbb{C}$, whence f is the quotient morphism by a G_a-action by Theorem 3.1.8(1). Assume that the tangent bundle T_Y of Y is trivial and that either X is factorial or the G_a-action is fixed point free. Then the tangent bundle of X is also trivial.*

Proof. Since X, Y are smooth, the coherent sheaves of modules Ω^1_X, Ω^1_Y are locally free on X, Y, respectively. We have a short exact sequence of coherent sheaves on Y

$$(0) \to f^*\Omega^1_Y \to \Omega^1_X \to \Omega^1_{X/Y} \to (0).$$

Since f is a smooth morphism, the sheaf $\Omega^1_{X/Y}$ is locally free of rank one on X [90, Chapter III, Proposition 10.4]. Since all the three sheaves in the exact sequence are locally free and X is affine, the exact sequence splits, so that $\Omega^1_X \cong f^*\Omega^1_Y \oplus \Omega^1_{X/Y}$, where Ω^1_Y is trivial because T_Y is trivial by the assumption. We show that the invertible sheaf $\Omega^1_{X/Y}$ is trivial. This is so if X is factorial. Suppose that the G_a-action is fixed point free. Let δ be the associated locally nilpotent derivation. Then δ defines an element of $T_{X/Y} := \operatorname{Hom}_{\mathcal{O}_X}(\Omega^1_{X/Y}, \mathcal{O}_X)$, which is nowhere zero. Hence $T_{X/Y} = \mathcal{O}_X\delta$ and it is free. Hence $\Omega^1_{X/Y}$ is trivial. It follows that Ω^1_X is trivial, and so is T_X. □

Corollary 3.1.42. *Let X be a smooth affine variety with a G_a-action which is fixed point free. Assume that $Y := X//G_a$ is defined as a normal affine variety. Suppose further that the quotient morphism q is surjective and $T_X \cong T_{X/Y} \oplus q^*T_Y$. If T_Y is trivial then so is T_X.*

Proof. By [37, Proposition 2.1(1)], since the G_a-action is fixed point free, $T_{X/Y} \cong \mathcal{O}_X$ Now the result follows immediately by Lemma 3.1.41. □

The following result is an application of Corollary 3.1.42 to the universal covering of an affine pseudoplane.

Corollary 3.1.43. *Let X be an affine pseudo-plane with an $\mathbb{A}^1$-fibration $f : X \to C$, where $C \cong \mathbb{A}^1$ and $f^{-1}(Q) = mF$ is a unique multiple fiber with $m > 1$ and $Q \in C$. Let $\widetilde{X}$ be the universal covering of X. Then the tangent bundle $T_{\widetilde{X}}$ is trivial.*

Proof. By Theorem 3.1.8(3), the morphism f is the quotient morphism by a G_a-action. Choose a coordinate t of C so that $t = 0$ at Q. Since $f^{-1}(C - \{Q\}) \cong (C - \{Q\}) \times \mathbb{A}^1 = \operatorname{Spec} \mathbb{C}[t, t^{-1}, x]$, we may choose the corresponding locally nilpotent derivation δ on $\Gamma(X, \mathcal{O}_X)$ in the form $t^r(\partial/\partial x)$ with $r \geq 0$. Hence the G_a-action is free on $f^{-1}(C - \{Q\})$. The universal covering $\widetilde{X}$ is obtained as the normalization of $X \times_C \widetilde{C}$, where $\widetilde{C} \to C$ is the cyclic covering of degree m ramifying totally over the point Q. Then $\widetilde{X}$ has an $\mathbb{A}^1$-fibration $\widetilde{f} : \widetilde{X} \to \widetilde{C}$ which is the composite of the normalization morphism $\widetilde{X} \to X \times_C \widetilde{C}$ and the second projection from $X \times_C \widetilde{C}$ onto $\widetilde{C}$. Furthermore, the G_a-action on X lifts uniquely to $\widetilde{X}$ and $\widetilde{f}$ is the quotient morphism by this G_a-action. Let Γ be the cyclic group of degree m. Then Γ acts on $\widetilde{X}$, and the Γ-action commutes with the G_a-action. Let $\widetilde{Q}$ be the unique point of $\widetilde{C}$ lying over Q. Then $\widetilde{f}^{-1}(\widetilde{Q})$ consists of a disjoint union of m components $\widetilde{F}_1, \ldots, \widetilde{F}_m$ with multiplicity one, and Γ acts transitively on the set $\{\widetilde{F}_1, \ldots, \widetilde{F}_m\}$. So, if the G_a-action is trivial on some component, say $\widetilde{F}_i$, then it is trivial on every component. We replace $\widetilde{\delta}$ by $\tau^{-1}\widetilde{\delta}$, where $\widetilde{\delta}$ is the locally nilpotent derivation on $\Gamma(\widetilde{X}, \mathcal{O}_{\widetilde{X}})$ corresponding to the G_a-action and τ is a coordinate of $\widetilde{C}$ such that $\tau = 0$ at $\widetilde{Q}$. After this replacement repeated several times, we may assume that the corresponding G_a-action is not trivial on each component of $\widetilde{f}^{-1}(\widetilde{Q})$. By the construction of δ, the G_a-action on $\widetilde{X}$, which corresponds to the unique lift $\widetilde{\delta}$ of δ, is free on $\widetilde{f}^{-1}(\widetilde{C} - \{\widetilde{Q}\})$. Hence we may assume that the G_a-action on $\widetilde{X}$ is fixed point free. Then $T_{\widetilde{X}}$ is trivial by Corollary 3.1.42. □

Question 3.1.44. Is the tangent bundle of the universal covering of the smooth affine variety X defined before Question 3.1.14 or an affine pseudo-3-space (see Definition 3.4.9 below) trivial?

3.1.6 Kernel of a vector field on $k[x, y, z]$

In the present subsection, we consider the case where A is a polynomial ring $k[x, y, z]$ in three variables. Let D be a nonzero regular vector field on the affine 3-space $\mathbb{A}^3 = \operatorname{Spec} A$, which we identify with a k-derivation on A. Then D has the following expression:

$$D = a_1 \frac{\partial}{\partial x} + a_2 \frac{\partial}{\partial y} + a_3 \frac{\partial}{\partial z} \tag{3.6}$$

with $a_1, a_2, a_3 \in A$. We say that D is *reduced* if $\gcd(a_1, a_2, a_3)$ is a nonzero constant. In general, the *reduced form* of D is $D_{\mathrm{red}} = d^{-1}D$, where $d = \gcd(a_1, a_2, a_3)$. The derivation

D extends in a natural way to a k-derivation on the quotient field $K = Q(A)$, which we denote by the same letter D or D_K. Let B be the kernel of D, i. e.,

$$\operatorname{Ker} D = \{a \in A \mid D(a) = 0\}.$$

We denote $\operatorname{Ker} D$ by B. Then B is a k-subalgebra of A, and $B = A \cap \operatorname{Ker} D_K$. By Zariski's finiteness theorem, B is an affine k-domain. Furthermore, it follows that B is integrally closed in K and hence B is normal, and that $B^* = k^*$ because $A^* = k^*$. It is clear that $\dim B \le 2$. For these results, see also [183]. Let $X = \operatorname{Spec} A$, $Y = \operatorname{Spec} B$, and let $f : X \to Y$ be the morphism induced by the inclusion $B \hookrightarrow A$. We are interested in the structure of B (or Y) and the morphism f.

Example 3.1.45.
(1) If D is the *Euler derivation* $x\frac{\partial}{\partial x} + y\frac{\partial}{\partial y} + z\frac{\partial}{\partial z}$, then $B = k$.
(2) If $D = x\frac{\partial}{\partial x} - y\frac{\partial}{\partial y} - z\frac{\partial}{\partial z}$, then $B = k[xy, xz]$ and $f : X \to Y$ is an $\mathbb{A}^1_*$-fibration. The morphism f has the plane $\{x = 0\}$ as a fiber component.
(3) If $D = x\frac{\partial}{\partial x} + y\frac{\partial}{\partial y}$, then $B = k[z]$ and $f : X \to Y$ is an $\mathbb{A}^2$-fibration.

We say that two derivations D_1, D_2 are *equivalent* if $D_2 = cD_1$ for some $c \in k^*$.

Lemma 3.1.46. *With the above notations, assume that* $\dim B = 2$ *and the morphism* $f : X \to Y$ *is an* $\mathbb{A}^1$*-fibration. Then there exists a locally nilpotent derivation* δ *of* A *such that* $B = \operatorname{Ker} \delta$. *Hence* B *is a polynomial ring in two variables over* k. *If* D *is reduced, it is equivalent to the reduced form of*

$$\begin{vmatrix} g_y & g_z \\ h_y & h_z \end{vmatrix} \frac{\partial}{\partial x} + \begin{vmatrix} g_z & g_x \\ h_z & h_x \end{vmatrix} \frac{\partial}{\partial y} + \begin{vmatrix} g_x & g_y \\ h_x & h_y \end{vmatrix} \frac{\partial}{\partial z},$$

where g, h *are generators of* B *over* k, *i. e.,* $B = k[g, h]$. *Hence* D *is locally nilpotent.*

Proof. By Lemma 3.1.2, there exists a splitting $f : X \xrightarrow{\widetilde{q}} \widetilde{Y} \xrightarrow{\nu} Y$, where $\widetilde{q}$ is the quotient morphism by a G_a-action on X and $\widetilde{Y} = \operatorname{Spec} \widetilde{B}$ with $\widetilde{B}$ the factorial closure of B in A. Let δ be the locally nilpotent derivation associated to the G_a-action. Then $\operatorname{Ker} \delta = \widetilde{B}$. Since $Q(\widetilde{B}) = Q(B)$ and $B = A \cap \operatorname{Ker} D_K$, it follows that $B = \widetilde{B}$. By Theorem 3.2.5 below, $\widetilde{B}$ is a polynomial ring in two variables over k. For the last assertion, we can show that the coefficient row vector (a_1, a_2, a_3) of the derivation D written in the form (3.6) is equivalent to

$$\left(\begin{vmatrix} g_y & g_z \\ h_y & h_z \end{vmatrix}, \begin{vmatrix} g_z & g_x \\ h_z & h_x \end{vmatrix}, \begin{vmatrix} g_x & g_y \\ h_x & h_y \end{vmatrix} \right), \tag{3.7}$$

where the equivalence means that the row vector in the displayed form (3.7) divided by the gcd of the entries (*reduced form*) is equal to a constant multiple of (a_1, a_2, a_3). For the locally nilpotent derivation δ, the coefficient row vector of δ put into the reduced form is equivalent to the above displayed row vector put into the reduced form. The reduced form of a locally nilpotent derivation is also a locally nilpotent derivation. In fact, write $\delta = d\overline{\delta}$, where $d \in A$ and $\overline{\delta}$ is the reduced form of δ. Since δ is a locally nilpotent derivation, there exists an element $a \in A$ such that $\delta(a)$ is a nonzero element of $B = \operatorname{Ker}\delta$. Then $\delta(a) = d\overline{\delta}(a) \in B$ implies that $d \in B$. Then $\overline{\delta}$ is a locally nilpotent derivation. The assertion follows from these observations. □

If $\dim B = 1$, then B is a polynomial ring in one variable over k. In fact, B is rational by Lüroth's theorem and B is a normal algebra with $B^* = k^*$. Let $\dim B = 2$. In Theorem 3.2.5 below, it is shown that if D is locally nilpotent then B is a polynomial ring in two variables. We will show in this subsection a slight generalization of this result. We consider the following two conditions:
(H1) Every fiber of $f : X \to Y$ is one-dimensional.
(H2) Y is $\mathbb{Q}$-factorial.

Then we have the following result.

Theorem 3.1.47. *Let the notations be as above. Suppose that* $\dim B = 2$ *and that the conditions* (H1) *and* (H2) *are satisfied. Then B is a polynomial ring in two variables.*

Let $\operatorname{Sing} Y$ be the singular locus of Y and let $Y^\circ = Y - \operatorname{Sing} Y$. Our proof follows essentially the same arguments in [154] and consists of several lemmas below. So, the readers are referred to [154] for the assertions without proofs. In the following lemmas, we tacitly assume that the conditions $(H1)$ and $(H2)$ are satisfied.

Lemma 3.1.48. *The following assertions hold:*
1. *Y° has logarithmic Kodaira dimension $-\infty$.*
2. *$X - f^{-1}(\operatorname{Sing} Y)$ is simply-connected.*

Proof. For the proof of the assertion (1), see [154, Lemma 4 and a part of the proof of Theorem 3]. For the assertion (2), note that $\operatorname{Sing} Y$ is the empty set or a finite set. Hence $f^{-1}(\operatorname{Sing} Y)$ has codimension ≥ 2 by the assumption $(H1)$. So, $X - f^{-1}(\operatorname{Sing} Y)$ is simply-connected. □

Lemma 3.1.49. *The surface Y satisfies one of the following two conditions:*
1. *Y has an $\mathbb{A}^1$-fibration $\rho : Y \to C$ such that $C \cong \mathbb{A}^1$ and every fiber of ρ is irreducible.*
2. *Y is isomorphic to $\mathbb{A}^2/\Gamma$ with Γ a finite subgroup of* $\mathrm{GL}(2, k)$.

Proof. Since Y° has logarithmic Kodaira dimension $-\infty$, either Y° is affine-ruled or Y° contains a Platonic $\mathbb{A}^1_*$-fiber space $\mathbb{A}^2/\Gamma - \{\overline{O}\}$ as an open set U so that $Y^\circ - U$ is a disjoint union of the affine lines. We consider these two cases separately. In both cases, it is crucial that Y being $\mathbb{Q}$-factorial implies that $\operatorname{Pic}(Y)$ is a torsion group.

CASE 1. Suppose that Y° is affine-ruled. Then there exists an $\mathbb{A}^1$-fibration $\rho^\circ : Y^\circ \to C$, where C is a smooth curve and ρ° is surjective. Since Y is affine, ρ° extends to an $\mathbb{A}^1$-fibration $\rho : Y \to C$. Since the composite of the morphisms $\rho \cdot f : X \to C$ is a dominant morphism, C is rational. If $C \cong \mathbb{P}^1$, Y is not $\mathbb{Q}$-factorial [152, Proof of Lemma 4.4, Chapter I]. Hence C is affine. Meanwhile, since $B^* = k^*$, it follows that $C \cong \mathbb{A}^1$. If a fiber $\rho^{-1}(Q)$ over a point $Q \in C$ has r irreducible components, the Picard number rank $\operatorname{Pic}(Y) \otimes \mathbb{Q}$ is greater than or equal to $r-1$. Hence each irreducible component of ρ is irreducible.

CASE 2. Suppose that Y° contains an open set U isomorphic to $\mathbb{A}^2/\Gamma - \{\overline{O}\}$ and $Y^\circ - U$ is a disjoint union $\coprod_{i=1}^s C_i$. Then rank $\operatorname{Pic}(X) \otimes \mathbb{Q} \geq s$. Hence $s = 0$ and $Y^\circ = U$. Since Y is affine and normal, we have

$$B = \Gamma(Y^\circ, \mathcal{O}_Y) \cong \Gamma(U, \mathcal{O}_U) = \Gamma(\mathbb{A}^2/\Gamma, \mathcal{O}_{\mathbb{A}^2/\Gamma}).$$

This implies that $Y \cong \mathbb{A}^2/\Gamma$. □

Lemma 3.1.50. *Suppose that Y has an $\mathbb{A}^1$-fibration $\rho : Y \to C$ as in Lemma* 3.1.49. *Then the following assertions hold:*

(1) *If $Q \in Y$ is a singular point of Y, then Y has a cyclic quotient singularity at Q and the fiber $\rho^{-1}(\rho(Q))$ is a multiple fiber.*
(2) *If $m_1F_1, \ldots, m_rF_r$ exhaust all the multiple fibers of ρ, then $\pi_1(Y^\circ) \cong \operatorname{Pic}(Y^\circ) \cong \prod_{i=1}^r \mathbb{Z}/m_i\mathbb{Z}$.*

Proof. See [153] for the assertion (1) and [74, Lemma 4.3] for the assertion (2). □

Proof of Theorem 3.1.47. In the case (1) of Lemma 3.1.49, if ρ has no multiple fibers, then $Y = Y^\circ$ by the proof of Lemma 3.1.50(1) and $\rho : Y \to C$ is an $\mathbb{A}^1$-bundle over $C \cong \mathbb{A}^1$. Hence $Y \cong \mathbb{A}^2$. So, we assume that ρ has a multiple fiber. In both cases of Lemma 3.1.50, let $\mu : Z \to Y^\circ$ be the universal covering. Then $\deg \mu > 1$ by the assumption. Let $W = f^{-1}(Y^\circ)$ and let $\widetilde{W} = W \times_{Y^\circ} Z$. Then the first projection $p_1 : \widetilde{W} \to W$ is a finite étale morphism. Since $W = X - f^{-1}(\operatorname{Sing} Y)$ is simply connected by Lemma 3.1.48, $\widetilde{W}$ is a disjoint sum of the components isomorphic to W. Hence there is a splitting

$$f_W : W \xrightarrow{\widehat{f}} Z \xrightarrow{\mu} Y^\circ.$$

This yields the ring homomorphism

$$B = \Gamma(Y^\circ, \mathcal{O}_{Y^\circ}) \xrightarrow{\mu^*} \Gamma(Z, \mathcal{O}_Z) \xrightarrow{\widehat{f}^*} \Gamma(X, \mathcal{O}_X) = A,$$

where $\Gamma(Z, \mathcal{O}_Z)$ is a subalgebra of (in fact, equal to) the integral closure of B in A. Since B is integrally closed in A, this is a contradiction. So, only the case $Y \cong \mathbb{A}^2$ takes place. □

3.2 Homology and contractible spaces

3.2.1 A topological proof of $\mathbb{A}^3//G_a \cong \mathbb{A}^2$

We summarize various results which we make use of in the subsequent arguments. The ground field is always the complex number field $\mathbb{C}$. A *homology n-space* is a smooth affine algebraic variety X of dimension n such that $H_i(X;\mathbb{Z}) = 0$ for every $i > 0$. If $H_i(X;\mathbb{Q}) = 0$ for every $i > 0$ instead, we call X a $\mathbb{Q}$-*homology n-space*. If $n = 2$ it is called *homology plane* and $\mathbb{Q}$-*homology plane*, respectively. A homology n-space is a *contractible n-space* if it is topologically contractible.

Lemma 3.2.1. *Let* $X = \operatorname{Spec} A$ *be a homology n-space. Let* (V, D) *be a pair of a smooth projective variety* V *and a reduced effective divisor* D *on* V *such that* X *is a Zariski open set of* V *with* $D = V - X$ *and* D *is a divisor with simple normal crossings. Then the following assertions hold:*
(1) X *is factorial and* $A^* = \mathbb{C}^*$.
(2) $H^1(V, \mathcal{O}_V) = H^2(V, \mathcal{O}_V) = 0$.

Proof. To simplify the arguments, we treat the case $n = 3$. Consider the exact sequence of $\mathbb{Z}$-cohomology groups for the pair (V, D),

$$\begin{array}{cccccccc}
0 & \to & H^0(V,D) & \to & H^0(V) & \to & H^0(D) & \to & H^1(V,D) \\
 & \to & H^1(V) & \to & H^1(D) & \to & H^2(V,D) & \to & H^2(V) \\
 & \to & H^2(D) & \to & H^3(V,D) & \to & H^3(V) & \to & H^3(D) \\
 & \to & H^4(V,D) & \to & H^4(V) & \to & H^4(D) & \to & H^5(V,D) \\
 & \to & H^5(V) & \to & H^5(D). & & & &
\end{array}$$

By Lefschetz duality, we have $H^i(V, D) \cong H_{6-i}(X)$ for $0 \le i \le 6$. Since X is a homology 3-space, $H_i(X) = 0$ for $1 \le i \le 3$, and since X is affine, $H_i(X) = 0$ for $4 \le i \le 6$ by Theorem 1.1.4. Hence we obtain the isomorphism $H^i(V) \cong H^i(D)$ for $0 \le i < 6$. Since $H^5(D) = 0$ as $\dim D = 2$, we have $H^5(V;\mathbb{Z}) = 0$. By Poincaré duality, we have $H_1(V;\mathbb{Z}) = 0$. Then the universal coefficient theorem [213, Theorem 3, p. 243] implies that $H^1(V;\mathbb{Z}) = 0$, and hence $H^1(V;\mathbb{C}) = 0$. The Hodge decomposition then implies that $H^1(V, \mathcal{O}_V) = 0$. Now consider an exact sequence

$$0 \longrightarrow \mathbb{Z} \longrightarrow \mathcal{O}_V \overset{\exp(2\pi i\ )}{\longrightarrow} \mathcal{O}_V^* \longrightarrow 0$$

and the associated exact sequence

$$H^1(V, \mathcal{O}_V) \to H^1(V, \mathcal{O}_V^*) \to H^2(V;\mathbb{Z}) \to H^2(V, \mathcal{O}_V). \tag{3.8}$$

On the other hand, consider the isomorphism $H^4(V;\mathbb{Z}) \cong H^4(D;\mathbb{Z})$. By the Mayer–Vietoris exact sequence, it follows that $H^4(D;\mathbb{Z})$ is a free abelian group generated by the classes of the irreducible components of D. The Poincaré duality and the universal coefficients theorem implies

$$H^4(V;\mathbb{Z}) \cong H_2(V;\mathbb{Z}) \cong H^2(V;\mathbb{Z}),$$

where we note that $H_1(V;\mathbb{Z}) = 0$. Hence we know that $H^2(V;\mathbb{Z})$ is a free abelian group generated by the first Chern classes of the irreducible components of D. Since all these classes are algebraic, we conclude that $H^2(V;\mathbb{Z}) = H^{1,1}(V) \cap H^2(V;\mathbb{C})$. Hence $H^2(V,\mathcal{O}_V) = 0$.

Now, by (3.8) above, we know that

$$\operatorname{Pic}(V) \cong H^2(V;\mathbb{Z}) \cong H^4(D;\mathbb{Z}).$$

Here $\operatorname{Pic}(X)$ is isomorphic to the residue group of $\operatorname{Pic}(V)$ modulo the subgroup generated by the classes of irreducible components of D. Hence $\operatorname{Pic}(X) = 0$, and X is factorial. Since there are no linear equivalence relations among the irreducible components of D, it follows that $A^* = \mathbb{C}^*$. □

The same argument with the $\mathbb{Z}$-coefficients replaced by $\mathbb{Q}$-coefficients in the proof of Lemma 3.2.1 has the following result.

Corollary 3.2.2. *Let X be a $\mathbb{Q}$-homology n-space X. Then the following assertions hold:*
(1) *X is $\mathbb{Q}$-factorial, i. e., $\operatorname{Pic}(X)$ is a finite abelian group, and $A^* = \mathbb{C}^*$.*
(2) *$H^1(V,\mathcal{O}_V) = H^2(V,\mathcal{O}_V) = 0$.*

The following result is an important consequence of a result of Hamm [88] that an affine variety of dimension n defined over $\mathbb{C}$ has the homotopy type of a CW complex of real dimension n (see Theorem 1.1.4).

Lemma 3.2.3. *Let X be an affine variety of dimension n. Then $H_n(X;\mathbb{Z})$ is a free abelian group of finite rank. Furthermore, $H^i(X;\mathbb{Z}) = 0$ for $i > n$.*

Proof. The vanishing of $H^i(X;\mathbb{Z})$ follows from the first assertion and the vanishing of $H_i(X;\mathbb{Z})$ for $i > n$. By the universal coefficient theorem, we have

$$H^i(X;\mathbb{Z}) = \operatorname{Hom}(H_i(X;\mathbb{Z}),\mathbb{Z}) \oplus \operatorname{Ext}^1(H_{i-1}(X;\mathbb{Z}),\mathbb{Z}).$$

If $i = n+1$, $H^{n+1}(X;\mathbb{Z}) = 0$ because $H_n(X;\mathbb{Z})$ is torsion free. For $i > n+1$, the result is clear because $H_i(X;\mathbb{Z}) = H_{i-1}(X;\mathbb{Z}) = 0$. □

The following result (see [182, Lemma 1.5]) is frequently used below (see also Lemma 1.1.6).

Lemma 3.2.4. *Let $f : X \to Y$ be a dominant morphism of algebraic varieties such that the general fibers are irreducible. Then the natural homomorphism $\pi_1(X) \to \pi_1(Y)$ is surjective.*

Proof. Let U be a nonempty open set of Y such that $f^{-1}(U) \to U$ is a smooth morphism. Then $\pi_1(f^{-1}(U)) \to \pi_1(U)$ is surjective. Then the assertion follows from Lemma 1.1.6 and the commutative diagram

$$\begin{array}{ccc} \pi_1(f^{-1}(U)) & \longrightarrow & \pi_1(U) \\ \downarrow & & \downarrow \\ \pi_1(X) & \longrightarrow & \pi_1(Y) \end{array}$$

where the vertical arrows are surjective. □

As an extension of the above argument, one can give a different proof of the following result in [154], which is called *Miyanishi theorem*. In the original proof, one has to use, in a crucial step of the proof, a rather difficult result that a smooth, quasiaffine surface X of log-Kodaira dimension $-\infty$ is either affine-ruled or contains a Platonic $\mathbb{A}^1_*$-fiber space $U := (\mathbb{A}^2/\Gamma) \setminus \{\overline{O}\}$ as an open set so that $X - U$ is a disjoint union of affine lines (see Theorem 1.2.4 and the proof of Theorem 3.1.47).

Theorem 3.2.5. *Let the additive group scheme G_a act nontrivially on the affine 3-space $\mathbb{A}^3$. Then $\mathbb{A}^3//G_a \cong \mathbb{A}^2$.*

For a topological proof different from the original, we need two subsequent lemmas.

Lemma 3.2.6. *Let X be a contractible 3-space with a nontrivial G_a-action and let $Y = X//G_a$. Assume that the quotient morphism $q : X \to Y$ is surjective.*[2] *Let V be a normal projective surface containing Y as an open set such that V is smooth along $D := V - Y$ and D is a divisor of simple normal crossings. Let $\rho : V' \to V$ be a minimal resolution of singularities of Y. Let $p_g(V')$ be the geometric genus of the surface V' which is a birational invariant independent of the choice of V'. Then the following assertions hold:*
1. *Both Y and Y° are simply connected, where Y° is the smooth part of Y.*
2. *$H_1(Y;\mathbb{Z}) = 0$. Further, if $p_g(V') = 0$ and $H^1(D;\mathbb{Z}) = 0$, then $H_2(Y;\mathbb{Z}) = 0$ and hence Y is contractible.*
3. *Y is smooth under the additional assumptions in the assertion* (2).

Proof. (1) Write $X = \operatorname{Spec} A$ and $Y = \operatorname{Spec} B$. Then A is factorial by Lemma 3.2.1 and hence B is factorial, too. So, the singular locus of Y is a finite set of quotient singular points by Theorem 3.1.8 and Lemma 3.1.30. Hence the quotient morphism $q : X \to Y$ has no fiber components of dimension two and $X - q^{-1}(Y^\circ)$ has dimension ≤ 1. Hence $\pi_1(q^{-1}(Y^\circ)) = \pi_1(X) = (1)$. By Lemma 3.2.4, both Y and Y° are simply-connected.

(2) Let Y' be the inverse image $\rho^{-1}(Y)$ and let E be the exceptional locus of ρ which is a divisor of simple normal crossings. Let $F = D + E$. Then we have an exact sequence

2 This assumption that $q : X \to Y$ is surjective is not necessary by [99, Remark 3.3].

of cohomology groups

$$H^3(V',F;\mathbb{Z}) \to H^3(V';\mathbb{Z}) \to H^3(F:\mathbb{Z}),$$

where $H^3(F;\mathbb{Z}) = 0$ and $H^3(V',F;\mathbb{Z}) \cong H_1(Y^\circ;\mathbb{Z})$ by the Lefschetz duality and $H_1(Y^\circ;\mathbb{Z}) = 0$ because $\pi_1(Y^\circ) = 1$. Hence $H_1(V';\mathbb{Z}) \cong H^3(V';\mathbb{Z}) = 0$. As in the proof of Lemma 3.2.1, this implies that $H^1(V';\mathbb{C}) = 0$ and $H^1(V',\mathcal{O}_{V'}) = 0$. Hence we have an exact sequence

$$0 \to H^1(V',\mathcal{O}^*_{V'}) \to H^2(V';\mathbb{Z}) \to H^2(V',\mathcal{O}_{V'}).$$

Since $H^2(V',\mathcal{O}_{V'}) = 0$ by the hypothesis, we have $\mathrm{Pic}(V') \cong H^1(V',\mathcal{O}^*_{V'}) \cong H^2(V';\mathbb{Z})$. Namely the group $H^2(V';\mathbb{Z})$ of topological 2-cocycles is generated by algebraic classes of divisors. Since Y is factorial, $\mathrm{Pic}(V')$ is generated by the classes of irreducible components of F. Since $H^2(F;\mathbb{Z})$ is a free abelian group generated by the classes of irreducible components of F, the natural homomorphism $H^2(V';\mathbb{Z}) \to H^2(F;\mathbb{Z})$ is an isomorphism.

Consider the long exact sequence associated to the pair (V',F),

$$H^1(V';\mathbb{Z}) \to H^1(F;\mathbb{Z}) \to H^2(V',F;\mathbb{Z}) \to H^2(V';\mathbb{Z}) \to H^2(F;\mathbb{Z}),$$

where $H^1(V';\mathbb{Z}) = 0$ as $H_1(V';\mathbb{Z}) = 0$ and $H^2(V';\mathbb{Z}) \to H^2(F;\mathbb{Z})$ is an isomorphism. Hence $H^1(F;\mathbb{Z}) \cong H^2(V',F;\mathbb{Z})$. Since the divisors D and E are disjoint from each other, $H^1(F;\mathbb{Z}) \cong H^1(D;\mathbb{Z}) \oplus H^1(E;\mathbb{Z})$, where $H^1(D;\mathbb{Z}) = 0$ by the hypothesis and $H^1(E;\mathbb{Z}) = 0$ because Y has at worst quotient singular points. So, $H^1(F;\mathbb{Z}) = 0$. By the Lefschetz duality, $H^2(V',F;\mathbb{Z}) \cong H_2(Y^\circ;\mathbb{Z})$. It follows that $H_2(Y^\circ;\mathbb{Z}) = 0$.

Now let $\{Q_1,\dots,Q_r\}$ be the set of singular points of Y. Choose a closed neighborhood T_i of Q_i for every i so that $T_i \cap T_j = \emptyset$ if $i \neq j$. Let ∂T_i be the boundary of T_i and let $\partial T = \partial T_1 \cup \cdots \cup \partial T_r$. By [166, Proof of Lemma 2.2], we have an exact sequence

$$0 \to H_2(Y^\circ;\mathbb{Z}) \to H_2(Y;\mathbb{Z}) \to H_1(\partial T;\mathbb{Z}) \to H_1(Y^\circ;\mathbb{Z}),$$

where $H_1(Y^\circ;\mathbb{Z}) = 0$ as $\pi_1(Y^\circ) = 1$ and $H_1(\partial T;\mathbb{Z})$ is a finite abelian group because $\pi_1(\partial T_i)$ is a finite group as the local fundamental group $\pi_1^{Q_i}(X)$ of the quotient singular point Q_i. Since $H_2(Y^\circ;\mathbb{Z}) = 0$ as shown above, $H_2(Y;\mathbb{Z})$ is a finite abelian group. Since $H_2(Y;\mathbb{Z})$ is torsion free by Lemma 3.2.4, it follows that $H_2(Y;\mathbb{Z}) = 0$.

(3) Since $\pi_1(Y) = \pi_1(Y^\circ) = 1$ and Y is contractible, Y is smooth by the so-called *affine Mumford theorem* [85, Theorem 3.6] which we state below. □

Theorem 3.2.7. *Let Y be a normal affine surface such that Y is contractible and the smooth part Y° of Y is simply-connected. Then Y is smooth.*

Now a different proof of Theorem 3.2.5 is given as follows. With the notations in Lemma 3.2.6, let $X = \mathbb{A}^3$. Then the quotient morphism $q : X \to X/\!/G_a$ is surjective by Bonnet [18]. Choose a general linear hyperplane L in such a way that the quotient morphism q restricted on L is a dominant morphism to $Y := \mathbb{A}^3//G_a$ and L meets the inverse image $q^{-1}(\operatorname{Sing} Y)$ in finitely many points. Then it follows that Y is rational and $\overline{\kappa}(Y^\circ) = -\infty$. This implies that $p_g(V') = 0$ and $H^1(D; \mathbb{Z}) = 0$ with the notations in Lemma 3.2.6. Hence Y is smooth. It is clear that B is factorial, $B^* = \mathbb{C}^*$ and $\overline{\kappa}(Y) = -\infty$. By a characterization of the affine plane, Y is isomorphic to $\mathbb{A}^2$.

Remark 3.2.8. To be accurate, we have to use the result in [154] mentioned before Theorem 3.2.5 in the course of the proof of Theorem 3.2.7. Hence it is very desirable to give a new proof of Theorem 3.2.7 which is more topological and not depending on the result in [154].

In [103, Theorem 2.7], Kaliman and Saveliev state a result stronger than Lemma 3.2.6. We call it *Kaliman–Saveliev theorem*. By the above argument, it shows that if $X \cong \mathbb{A}^3$ then $Y \cong \mathbb{A}^2$.

Theorem 3.2.9. *For every nontrivial G_a-action on a contractible 3-space X, the algebraic quotient $Y = X//G_a$ is a smooth contractible affine surface.*

3.2.2 Homology 3-spaces with $\mathbb{A}^1$-fibrations

In Theorem 3.1.8, it is shown that an $\mathbb{A}^1$-fibration $f : X \to Y$ from a smooth affine threefold to a normal surface Y has a factorization

$$f : X \xrightarrow{q} \widetilde{Y} \xrightarrow{\sigma} Y,$$

where $q : X \to \widetilde{Y}$ is the quotient morphism by a G_a-action and $\sigma : \widetilde{Y} \to Y$ is the birational morphism such that $\widetilde{B} = \Gamma(\widetilde{Y}, \mathcal{O}_{\widetilde{Y}})$ is the factorial closure of $B = \Gamma(Y, \mathcal{O}_Y)$ in $A = \Gamma(X, \mathcal{O}_X)$. Namely,

$$\widetilde{B} = \{a \in A \mid a \text{ is a factor of an element } b \in B\}.$$

In fact, we have $\widetilde{B} = A \cap Q(B)$. Thus we may look into the quotient morphism $q : X \to Y$ by a G_a-action. We consider the singular locus $\operatorname{Sing}(q) = \{Q \in Y \mid F_Q \not\cong \mathbb{A}^1\}$, where F_Q is the scheme-theoretic fiber of q over $Q \in Y$, i. e., $F_Q = X \times_Y \operatorname{Spec} k(Q)$ with the residue field $k(Q)$.

We prove a preparatory result.

Lemma 3.2.10. *Let $f : X \to Y$ be a dominant morphism from an algebraic variety X of dimension $n + r$ to an algebraic variety Y of dimension n. Let P be a point of X and let Q be a point of Y such that $f(P) = Q$. Suppose that X is smooth at P and the fiber $F := F_Q$ is reduced, r-dimensional and smooth at P. Then Q is smooth at Y.*

Proof. Let $(R, \mathfrak{m})$ (resp., $(S, \mathfrak{M})$) be the local ring of Y (resp., X) at Q (resp., P). Since the fiber F is defined by $\mathfrak{m}S$ at P and since $S/\mathfrak{m}S$ is a regular local ring as P is a smooth point of F, it follows that $\mathfrak{M} = \mathfrak{m}S + (z_{n+1}, \ldots, z_{n+r})S$ for an element $z_{n+1}, \ldots, z_{n+r} \in \mathfrak{M}$. Since $(S, \mathfrak{M})$ is a regular local ring of dimension $n + r$, we find n elements $z_1', \ldots, z_n'$ of $\mathfrak{m}$ with images $z_1, \ldots, z_n$ in S such that $\{z_1, \ldots, z_n, z_{n+1}, \ldots, z_{n+r}\}$ is a regular system of parameters of $(S, \mathfrak{M})$. Since the completion $(\widehat{S}, \widehat{\mathfrak{M}})$ is isomorphic to $\mathbb{C}[[z_1, \ldots, z_n, z_{n+1}, \ldots, z_{n+r}]]$, we can express any $h \in \widehat{S}$ as a formal power series in $z_{n+1}, \ldots, z_{n+r}$ with coefficients in $\mathbb{C}[[z_1, \ldots, z_n]]$,

$$h = \sum_{\vec{i}=0}^{\infty} \alpha_{\vec{i}} z^{\vec{i}}, \quad \alpha_{\vec{i}} \in \mathbb{C}[[z_1, \ldots, z_n]],$$

where $z^{\vec{i}} = z_{n+1}^{i_{n+1}} \cdots z_{n+r}^{i_{n+r}}$ for $\vec{i} = (i_{n+1}, \ldots, i_{n+r})$. We shall show that the completion $(\widehat{R}, \widehat{\mathfrak{m}})$ is isomorphic to the formal power series ring $\mathbb{C}[[z_1, \ldots, z_n]]$. We consider the local complete intersection

$$H = \{z_j = 0 \mid n + 1 \le j \le n + r\}$$

in X near the point P as a section transversal to the fiber F at the point P. We may assume that the restriction $f|_H : H \to Y$ is quasifinite. Hence the injective local ring homomorphism $R \to S \to S/zS$ induces an injective local homomorphism with z signifying $(z_{n+1}, \ldots, z_{n+r})$

$$(\widehat{R}, \widehat{\mathfrak{m}}) \to (\widehat{S}, \widehat{\mathfrak{M}}) \to (\widehat{S}/z\widehat{S}, \widehat{\mathfrak{M}}/z\widehat{S}),$$

because $(\widehat{S}/(z_{n+1}, \ldots, z_{n+r})\widehat{S}, \widehat{\mathfrak{M}}/(z_{n+1}, \ldots, z_{n+r})\widehat{S})$ is viewed as the completion of the local ring $\mathcal{O}_{H,P}$. Sending z_i to z_i', $i = 1, \ldots, n$, we obtain a homomorphism

$$\mathbb{C}[[z_1, \ldots, z_n]] \longrightarrow \widehat{R},$$

which gives an isomorphism when composed with the mapping $\widehat{R} \to (\widehat{S}/z\widehat{S})$. Let h' be an element of $\widehat{R}$ and h its image in $\widehat{S}$. With h expressed as a power series as above, we find that h' and the image in $\widehat{R}$ of $\alpha_{\vec{0}} \in \mathbb{C}[[z_1, \ldots, z_n]]$ have the same image in $\widehat{S}/z\widehat{S}$ and hence coincide. Hence $\widehat{R} \cong \mathbb{C}[[z_1, \ldots, z_n]]$. This implies that Y is smooth at Q. □

We now prove the following result.

Lemma 3.2.11. *Let $q : X \to Y$ be the quotient morphism of a smooth affine threefold X with respect to a G_a-action. Suppose that q is equidimensional. Then the following assertions hold:*

(1) *If a fiber F_Q has an irreducible component which is reduced in F_Q, then the point Q is smooth in Y.*
(2) *$\mathrm{Sing}(q)$ is a closed set.*

Proof. (1) The assertion follows from Lemma 3.2.10.

(2) Since q has equidimension one, the fixed point locus X^{G_a} consists of fiber components by Corollary 3.1.28. If $\dim X^{G_a} \leq 1$ then every fiber of q is a disjoint union of contractible curves by Lemma 3.1.31. Further, a contractible irreducible fiber component is isomorphic to $\mathbb{A}^1$ if it contains a nonfixed point.

For the assertion it suffices to show that given a fiber F_Q isomorphic to $\mathbb{A}^1$ there exists an open neighborhood U of Q in Y with $F_{Q'} \cong \mathbb{A}^1$ for all $Q' \in U$. Let C be an irreducible curve on Y through Q and let Z be the normalization of $X \times_Y C$. Suppose that the fiber of q over a general point of C consists of m copies of $\mathbb{A}^1$, where $m > 1$ (see Lemma 1.4.4). Then Z has an induced G_a-action such that

$$\begin{array}{ccc} Z & \xrightarrow{\mu} & X \times_Y C \\ \downarrow{\scriptstyle \widetilde{q}} & & \downarrow{\scriptstyle q} \\ \widetilde{C} & \xrightarrow{\nu} & C \end{array}$$

where $\widetilde{q}$ is the quotient morphism and $\widetilde{C}$ is the normalization of C in Z. The morphism $\nu : \widetilde{C} \to C$ is a finite covering of degree m (Stein factorization), which is ramified over the point Q. It then follows that the fiber F_Q is nonreduced, and this is a contradiction.

Consider the closure $\overline{\mathrm{Sing}(q)}$ and remove from Y all the irreducible components of dimension one of $\overline{\mathrm{Sing}(q)}$ over which a general fiber of q consists of more than one irreducible component. By the above argument, such a component has no points over which the fibers of q are isomorphic to $\mathbb{A}^1$, hence contained in $\mathrm{Sing}(q)$. Thus we may assume that all singular fibers of q are of the form $m\mathbb{A}^1$ with $m > 1$. Note that Q is a smooth point by the assertion (1). Replacing Y by an affine open neighborhood U of Q and accordingly X by the inverse image $q^{-1}(U)$, we may assume that Y is smooth, and hence that q is faithfully flat. Consider the tangential homomorphism of the tangent bundles

$$dq : \mathcal{T}_X \to q^*\mathcal{T}_Y$$

and let $\mathcal{C}$ be the cokernel of dq. Then $\mathcal{C}$ is a coherent $\mathcal{O}_X$-module. The support $T = \mathrm{Supp}(\mathcal{C})$ is a closed set such that $T = q^{-1}(q(T))$. If $Q \in \overline{\mathrm{Sing}(q)}$, then $F_Q \subset T$, but $F_Q \cap T = \emptyset$ since q is smooth over Q as $F_Q \cong \mathbb{A}^1$. This is a contradiction. Hence $\mathrm{Sing}(q)$ contains no codimension one points of Y. Then, by Lemma 2.1.6, there exists an open neighborhood U of Q such that $q^{-1}(U)$ is an $\mathbb{A}^1$-bundle over U. Hence $\mathrm{Sing}(q)$ is a closed set. □

Concerning a G_a-action, we ask the following

Question 3.2.12. Let X be an affine variety with a G_a-action. Suppose that the algebraic quotient $Y = X/\!/G_a$ exists as an affine variety, i. e., the G_a-invariant subring of $\Gamma(X, \mathcal{O}_X)$ is an affine domain. Let P be a point of X with trivial isotropy group. Is then the fiber F_Q of the quotient morphism $q : X \to Y$ passing through the point P reduced, where $Q = q(P)$?

The answer is negative and an example is given by an affine pseudoplane. For the following example, see Problems 6 and 7 in Section 2.10.

Example 3.2.13. Let C be a smooth conic in $\mathbb{P}^2 = \operatorname{Proj} \mathbb{C}[X_0, X_1, X_2]$ defined by $X_2^2 = X_0X_1$. Let L be the tangent line of C at the point $Q : (X_0, X_1, X_2) = (0, 1, 0)$. Hence L is defined by $X_0 = 0$. Let Λ be the linear pencil generated by C and $2L$. Then the G_a-action on $\mathbb{P}^2$ defined by

$$^tX_0 = X_0, \quad ^tX_1 = X_1 + 2X_2t + X_0t^2, \quad ^tX_2 = X_2 + X_0t$$

has a unique fixed point Q and preserves the members of the pencil Λ. Let X be the complement of C in $\mathbb{P}^2$. Then Λ defines an $\mathbb{A}^1$-fibration $q : X \to \mathbb{A}^1$ which turns out to be the quotient morphism of the induced G_a-action on X. Although the G_a-action has no fixed points, q has a multiple fiber 2ℓ which comes from the member $2L$ of Λ.

This is the case for every affine pseudoplane $q : X \to B$ of type (m, r) with an integer $r \geq 1$ such that $r \equiv 1 \pmod{m}$. An *affine pseudoplane of type* (m, r) is, by definition, an affine pseudoplane having an $\mathbb{A}^1$-fibration $q : X \to B \cong \mathbb{A}^1$ and a unique multiple fiber mF with $m > 1$ and $F \cong \mathbb{A}^1$ such that there exists a normal completion (V, D) for which the dual graph of D is given as below

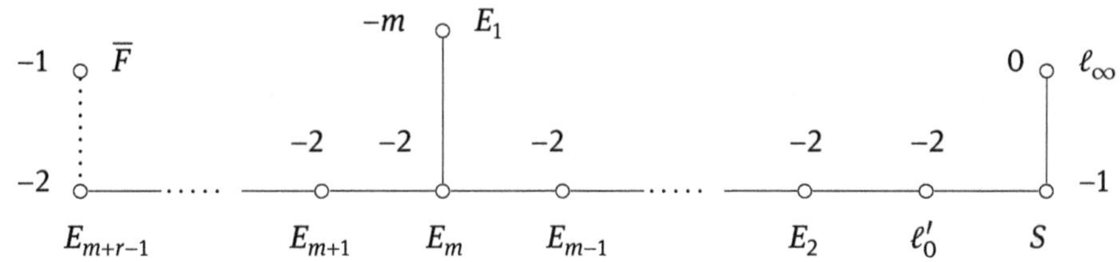

In the above figure, ℓ_∞ is the fiber at infinity of a $\mathbb{P}^1$-fibration $\overline{q} : V \to \overline{B}$ extending q and S is the section at infinity of $\overline{q}$ (see [144] for the details). Also X is denoted by $X(m, r)$.

Lemma 3.2.14. *Let $q : X \to B$ be an affine pseudoplane of type (m, r) with $r \equiv 1 \pmod{m}$. Then q is given by a fixed-point free G_a-action.*

Proof. Let $\widetilde{X}(m, r)$ be the universal covering of $X(m, r)$. Let $H(m) = \mathbb{Z}/m\mathbb{Z}$ be the covering Galois group which is identified with the mth roots of unity. By [144], $\widetilde{X}(m, r)$ is isomorphic to a hypersurface in $\mathbb{A}^3 = \operatorname{Spec} \mathbb{C}[x, y, z]$ defined by

$$x^rz + (y^m + a_1xy^{m-1} + \cdots + a_{m-1}x^{m-1}y + a_mx^m) = 1, \quad a_i \in \mathbb{C}.$$

The Galois group $H(m)$ acts as

$$^\zeta(x, y, z) = (\zeta x, \zeta y, \zeta^{-r}z), \quad \zeta \in H(m).$$

The projection $(x,y,z) \mapsto x$ defines an $\mathbb{A}^1$-fibration $\widetilde{q} : \widetilde{X}(m,r) \to \widetilde{B}$, where $\widetilde{B} \cong \mathbb{A}^1$. Further, there is a G_a-action on $\widetilde{X}(m,r)$ defined by

$$\begin{aligned} &{}^t(x,y,z) = \\ &(x, y + tx^r, z - x^{-r}\{((y+tx^r)^m + a_1x(y+tx^r)^{m-1} + \cdots + a_mx^m) \\ &\quad -(y^m + a_1xy^{m-1} + \cdots + a_{m-1}x^{m-1}y + a_mx^m)\}). \end{aligned}$$

Then it follows that:
(1) the fiber $\widetilde{F}_0$ of $\widetilde{q}$ over the point $x = 0$ is a disjoint union of m-copies of the affine line,
(2) the G_a-action preserves the fibration $\widetilde{q}$,
(3) the G_a-action preserves and acts nontrivially on each connected component of $\widetilde{F}_0$,
(4) if $r \equiv 1 \pmod m$ then the $H(m)$-action commutes with the G_a-action.

Hence the G_a-action descends to a G_a-action on $X(m,r)$ which has no fixed points. □

We consider, however, a result implied by the assumptions in Question 3.2.12. We need some auxiliary results. Let $A = \Gamma(X, \mathcal{O}_X)$ and $B = \Gamma(Y, \mathcal{O}_Y)$. Let δ be a locally nilpotent derivation on A which corresponds to the given G_a-action. Further, let $\mathfrak{n}$ be the maximal ideal of B corresponding to the point Q. First we recall the following result in [162, Theorem 3.3].

Lemma 3.2.15. *Let A be an affine domain over $\mathbb{C}$ and let δ be a locally nilpotent derivation on A. Let $B = \operatorname{Ker}\delta$. Suppose that $\operatorname{Spec} A/\mathfrak{n}A$ is an irreducible scheme of dimension one and that the associated G_a-action on $\operatorname{Spec} A/\mathfrak{n}A$ has no fixed points. Then the following assertions hold:*
(1) *For any integer $n > 0$, there exists an element $z_n \in A$ such that $A/\mathfrak{n}^nA = R_n[z_n]$, where R_n is an Artin local ring and we denote the residue class of z_n in $A/\mathfrak{n}^nA$ by the same letter.*
(2) *For $m > n$, we have a natural exact sequence*

$$0 \to \mathfrak{n}^n R_m \to R_m \xrightarrow{\theta_{nm}} R_n \to 0,$$

where θ_{nm} is a local homomorphism. Let $\widehat{R} = \varprojlim_n R_n$. Then $\widehat{R}$ is a complete local ring.

Let $\mathcal{O}$ be the local ring $\mathcal{O}_{Y,Q} = B_{\mathfrak{n}}$ and denote the ideal $\mathfrak{n}\mathcal{O}$ by the same letter $\mathfrak{n}$. Let $\widehat{\mathcal{O}}$ be the $\mathfrak{n}$-adic completion of $\mathcal{O}$. Then $A \otimes_B \mathcal{O}$ is an affine domain over $\mathcal{O}$ with the associated locally nilpotent derivation δ such that δ is nonzero and $\operatorname{Ker}\delta = \mathcal{O}$. Further we assume the following condition:

$$\text{(H)} \quad (\mathfrak{n}^n(A \otimes_B \mathcal{O})) \cap B = \mathfrak{n}^n \quad \text{for every } n > 0.$$

This condition is satisfied if $A \otimes_B \mathcal{O}$ is $\mathcal{O}$-flat (see [25, Lemma 1.4]).

Proposition 3.2.16. *With the above notations and assumptions, the following assertions hold:*

(1) $(A \otimes_B \mathcal{O})/\mathfrak{n}^n(A \otimes_B \mathcal{O}) \cong A/\mathfrak{n}^n A = R_n[z_n]$.

(2) *The projective limit* $\widehat{A} = \varprojlim_n (A \otimes_B \mathcal{O})/\mathfrak{n}^n(A \otimes_B \mathcal{O})$ *contains* $A \otimes_B \mathcal{O}$ *as a subring. Further,* $\widehat{A}$ *has a derivation* $\widehat{\delta}$ *and an element* $\widehat{z}$ *such that* $\widehat{\delta}|_{A\otimes_B\mathcal{O}} = \delta$, $\operatorname{Ker}\widehat{\delta} = \widehat{R}$, *and* $\widehat{\delta}(\widehat{z}) = 1$. *The ring* $\widehat{A}$ *itself is a subring of a formal power series ring* $\widehat{R}[[\widehat{z}]]$.

(3) *Let* $\widehat{A}_{\mathrm{ln}}$ *be the subring of* $\widehat{A}$ *consisting of elements for which* $\widehat{\delta}$ *is locally nilpotent. Then* $\widehat{A}_{\mathrm{ln}} = \widehat{R}[\widehat{z}]$ *and* $A \otimes_B \mathcal{O}$ *is a subring of* $\widehat{A}_{\mathrm{ln}}$.

Proof. (1) The assertion follows from Lemma 3.2.15.

(2) It suffices to show that if a is an element of $\bigcap_n \mathfrak{n}^n(A \otimes_B \mathcal{O})$ then $a = 0$. Suppose that $a \neq 0$. Then there exists an integer $r \geq 0$ such that $\delta^r(a) \neq 0$ and $\delta^{r+1}(a) = 0$. Then $\delta^r(a) \in \operatorname{Ker}\delta = \mathcal{O}$. Since $\delta(\mathfrak{n}^n(A \otimes_B \mathcal{O})) \subseteq \mathfrak{n}^n(A \otimes_B \mathcal{O})$ for every $n > 0$, by the condition (H), we have $\delta^r(a) \in (\bigcap_n \mathfrak{n}^n(A \otimes_B \mathcal{O})) \cap \mathcal{O} = \bigcap_n \mathfrak{n}^n$, which is zero by Krull's intersection theorem. Hence $\delta^r(a) = 0$. This is a contradiction. Hence $A \otimes_B \mathcal{O}$ is a subring of $\widehat{A}$. The sequence $\{z_n\}_{n\geq 1}$ is a Cauchy sequence because $z_m - z_n \in \mathfrak{n}^n(A \otimes_B \mathcal{O})$ for $m > n$. Hence it converges to an element $\widehat{z}$ of $\widehat{A}$. The derivation δ extends to a derivation $\widehat{\delta}$ of $\widehat{A}$. Since we may assume that $\delta(z_n) \equiv 1 \pmod{\mathfrak{n}^n(A \otimes_B \mathcal{O})}$, it follows that $\widehat{\delta}(\widehat{z}) = 1$ in $\widehat{A}$. Let a_n be an element of $A \otimes_B \mathcal{O}$. Since $(A \otimes_B \mathcal{O})/\mathfrak{n}^n(A \otimes_B \mathcal{O}) \cong R_n[z_n]$ and $\widehat{z} - z_n \in \mathfrak{n}^n\widehat{A}$, there exists a polynomial $f_n(\widehat{z}) \in \widehat{R}[\widehat{z}]$ such that $a_n - f_n(\widehat{z}) \in \mathfrak{n}^n\widehat{A}$. This implies that a Cauchy sequence $\{a_n\}_{n\geq 0}$ in $A \otimes_B \mathcal{O}$ is approximated by a Cauchy sequence $\{f_n(\widehat{z})\}_{n\geq 0}$ in $\widehat{A}$. So, $\widehat{A}$ is a subring of $\widehat{R}[[\widehat{z}]]$.

(3) Let a be an element of $A \otimes_B \mathcal{O}$. Write

$$a = \alpha_0 + \alpha_1\widehat{z} + \cdots + \alpha_i\widehat{z}^i + \cdots = \sum_{i=0}^{\infty} \alpha_i\widehat{z}^i$$

as an element of $\widehat{R}[[\widehat{z}]]$. If this is an infinite series, the module $M(a)$ generated over $\widehat{\mathcal{O}}$ by $\widehat{\delta}^i(a)$ for $i \geq 0$ is not finitely generated over $\widehat{\mathcal{O}}$ because it is not finitely generated over $\widehat{R}$. Meanwhile, $\widehat{\delta}^i(a) = \delta^i(a)$ for every $i \geq 0$ and $M(a)$ is finitely generated over $\widehat{\mathcal{O}}$ because δ is locally nilpotent. Hence $a \in \widehat{R}[\widehat{z}]$. This implies that $A \otimes_B \mathcal{O}$ is a subring of $\widehat{R}[\widehat{z}]$. It is clear that $\widehat{A}_{\mathrm{ln}} = \widehat{R}[\widehat{z}]$. □

Finally we raise the following challenging problem inspired by Kaliman–Saveliev theorem (see Theorem 3.2.9).

Question 3.2.17. Let X be a homology 3-space with a fixed-point free G_a-action. Is X G_a-equivariantly isomorphic to a product $Y \times \mathbb{A}^1$ with G_a acting on the second factor, where $Y = X/\!/G_a$?

If X is replaced by a $\mathbb{Q}$-homology 3-space, the answer is negative.

Example 3.2.18. Let $q : X \to B$ be an affine pseudoplane as in Lemma 3.2.14. Then X is a $\mathbb{Q}$-homology plane. Let $\widetilde{X} = X \times \mathbb{A}^1$ with G_a acting on X and trivially on the second factor. Then $\widetilde{X}$ is a $\mathbb{Q}$-homology 3-space, $\widetilde{X}//G_a = B \times \mathbb{A}^1$ and the quotient morphism is $q_{\widetilde{X}} = q \times \mathrm{id}_{\mathbb{A}^1} : X \times \mathbb{A}^1 \to B \times \mathbb{A}^1$. Hence $\mathrm{Sing}(q_X) \cong \mathbb{A}^1$. This implies that the G_a-action on $\widetilde{X}$ is never like in Question 3.2.17.

3.3 $\mathbb{A}^1_*$-fibrations in higher dimension

A main theme of the present section is $\mathbb{A}^1_*$-fibration. We consider differences between $\mathbb{A}^1_*$-fibrations and quotient morphisms by G_m. Contrary to the case of an $\mathbb{A}^1$-fibration, an $\mathbb{A}^1_*$-fibration is not necessarily factored by a quotient morphism by G_m. Possibility of factorization depends on the nature of singular fibers of the $\mathbb{A}^1_*$-fibration (see Theorem 3.3.21). The singular fibers of an $\mathbb{A}^1_*$-fibration are fully understood in the surface case (see Lemma 3.3.2) but not completely in the case of higher dimension. The locus of singular fibers is shown to be a closed set with the unmixedness condition on singular fibers (cf. Lemma 3.3.11).

Our primary purpose in this section is to develop a theory of $\mathbb{A}^1_*$-fibrations by combining algebraic geometry with algebraic topology and commutative algebra and to apply it to elucidate the structure of such objects as homology (or contractible) 3-spaces with $\mathbb{A}^1_*$-fibrations. The ground field is also the complex number field $\mathbb{C}$.

3.3.1 $\mathbb{A}^1_*$-fibration

Let $f : X \to Y$ be a dominant morphism of algebraic varieties. We call f an $\mathbb{A}^1_*$-fibration if general fibers of f are isomorphic to $\mathbb{A}^1_*$. A *singular fiber* of f is a fiber which is scheme-theoretically not isomorphic to $\mathbb{A}^1_*$. We say that an $\mathbb{A}^1_*$-fibration is *untwisted* (resp *twisted*) if the generic fiber $X_K := X \times_Y \mathrm{Spec}\, K$ has two K-rational places at infinity (resp., if X_K has one non-K-rational place at infinity), where K is the function field of Y over $\mathbb{C}$. In the untwisted case, X_K is isomorphic to $\mathbb{A}^1_{*,K} := \mathrm{Spec}\, K[t, t^{-1}]$, while X_K is a nontrivial K-form of $\mathbb{A}^1_{*,K}$ and $\mathrm{Pic}(X_K) \cong \mathbb{Z}/2\mathbb{Z}$ in the twisted case. If X is a factorial affine variety, then any $\mathbb{A}^1_*$-fibration on X is untwisted because X_K is then factorial. In fact, if $X = \mathrm{Spec}\, A$ and $Y = \mathrm{Spec}\, B$ then A is factorial and hence the quotient ring $A \otimes_B K$ is factorial. Since $X_K = \mathrm{Spec}\, A \otimes_B K$, X_K is factorial. In the present section, an $\mathbb{A}^1_*$-fibration is *always assumed to be untwisted.*

If the multiplicative group scheme G_m acts on an affine variety $X = \mathrm{Spec}\, A$ then the algebraic quotient $X//G_m$ exists. Namely, the invariant subring $B = A^{G_m}$ is an affine domain over $\mathbb{C}$ by the well-known result on reductive group action and $Y = \mathrm{Spec}\, B$ is the algebraic quotient variety. The quotient morphism $q : X \to Y$ is given by the inclusion $B \hookrightarrow A$. In the case of G_m-actions, we have only inequalities

$$0 \le \dim Y < \dim X$$

in general, not necessarily $\dim Y = \dim X - 1$, but we assume *unless otherwise mentioned* that $\dim Y = \dim X - 1$. Hence q has relative dimension one. In the case of a G_m-action, the quotient morphism q is surjective and either an $\mathbb{A}^1$-fibration or an $\mathbb{A}^1_*$-fibration depending on whether general orbits admit fixed points or not. If the quotient morphism $q : X \to Y$ is an $\mathbb{A}^1_*$-fibration, it is untwisted by a theorem of Rosenlicht [196]. We recall the following fundamental result on the quotients under reductive algebraic group actions.

Lemma 3.3.1. *Let G be a connected reductive algebraic group acting on a smooth affine algebraic variety X and let $q : X \to Y$ be the quotient morphism. Then q is surjective, and for two points P_1, P_2 of X, we have $q(P_1) = q(P_2)$ if and only if $\overline{GP_1} \cap \overline{GP_2} \neq \emptyset$. Hence for any point Q of Y, the fiber $q^{-1}(Q)$ contains a unique closed orbit.*

In the surface case, singular fibers of an $\mathbb{A}^1_*$-fibration are classified as follows.

Lemma 3.3.2. *Let X be a normal affine surface and let $f : X \to C$ be an $\mathbb{A}^1_*$-fibration. Then the following assertions hold:*

(1) *A singular fiber F of f is written in the form $F = \Gamma + \Delta$, where $\Gamma = m\mathbb{A}^1_*, \emptyset$ or $m\mathbb{A}^1 + n\mathbb{A}^1$ with two $\mathbb{A}^1$s meeting in one point which may be a cyclic quotient singular point of X, $\Gamma \cap \Delta = \emptyset$ and Δ is a disjoint union of affine lines with multiplicities, each of which may have one cyclic quotient singular point of X. Each type of a singular fiber is realizable.*
(2) *If f is the quotient morphism of a G_m-action on X and f is an $\mathbb{A}^1_*$-fibration, the part Δ is absent in a singular fiber F. Hence F is either mA^1_* or $m\mathbb{A}^1 + n\mathbb{A}^1$ in the above list.*

Proof. (1) In [166, Lemma 2.9], singular fibers of an $\mathbb{A}^1_*$-fibration are classified, where singularities of the surface X are assumed to be quotient singularities, but one can drop this condition and make the same argument by assuming simply that X is normal.

(2) By the assertion (1) and Lemma 3.3.1, the only possible cases of a singular fiber of the quotient morphism f are the two cases listed in the statement and the case where F_{red} is irreducible and isomorphic to $\mathbb{A}^1$ which may contain a unique cyclic quotient singularity. We show that the last case does not occur. Suppose that $F_{\text{red}} \cong \mathbb{A}^1$. Then F contains a unique fixed point P. Suppose first that X is smooth at P. Look at the induced tangential representation of G_m on $T_{X,P}$ which is diagonalizable and has weights $-a, b$ with $a, b > 0$.[3] Then it follows that F is locally near P a union of two irreducible components meeting in P. This is a contradiction. Suppose that X is singular at P. Let $\sigma : \widetilde{X} \to X$ be a minimal resolution of singularity at P. Then the inverse image $\sigma^{-1}(F_{\text{red}})$ is the proper transform G of F_{red} and a linear chain of $\mathbb{P}^1$'s with G meeting one of the

3 Let $\{x, y\}$ be a system of local coordinates of X at P such that the G_m-action is given by ${}^t(x, y) = (t^{-a}x, t^b y)$ for $t \in G_m$. If either a or b is zero, say $a = 0$, then the local curve $y = 0$ is a section of f. Hence f is not an $\mathbb{A}^1_*$-fibration. Since $\dim B = \dim X - 1$, we must have $ab < 0$.

terminal components of the linear chain. Since the G_m-action on X lifts to $\widetilde{X}$, each irreducible component of $\sigma^{-1}(F_{\mathrm{red}})$ is G_m-stable and the other terminal component, say $\widetilde{F}$, of the linear chain has an isolated fixed point, say $\widetilde{P}$. Note that we obtain an affine surface from $\widetilde{X}$ by removing all components of $\sigma^{-1}(F_{\mathrm{red}})$ except for the component $\widetilde{F}$. Now looking at the induced tangential representation of G_m on $T_{\widetilde{X},\widetilde{P}}$, we have a contradiction as in the above smooth case. So, the case $F_{\mathrm{red}} \cong \mathbb{A}^1$ does not occur. □

Special attention has to be paid in the case where the G_m-quotient morphism $q : X \to Y$ is an $\mathbb{A}^1$-fibration. We consider the surface case first and then treat the case of threefolds. There are related references [52, 200, 219].

Lemma 3.3.3. *Let X be a normal affine surface with a G_m-action. Suppose that the quotient morphism $q : X \to C$ is an $\mathbb{A}^1$-fibration, where C is a normal affine curve. Then the following assertions hold:*

(1) *There exists a closed curve S on X such that the restriction of q onto S induces an isomorphism between S and C, i. e., S is a cross-section of q.*
(2) *Suppose further that X is smooth. Then every fiber of q is reduced and isomorphic to $\mathbb{A}^1$. Hence X is a line bundle over C.*
(3) *Suppose that X is smooth and C is rational. Then X is isomorphic to a direct product $C \times \mathbb{A}^1$.*

Proof. (1) Let $\overline{q} : \overline{X} \to \overline{C}$ be a G_m-equivariant completion of $q : X \to C$. Namely, $\overline{X}$ is a normal projective surface containing X as an open set and $\overline{C}$ is a smooth completion of C. We may assume that $\overline{X}$ is smooth along $\overline{X} \setminus X$, q extends to a morphism $\overline{q}$ and the G_m-action extends to $\overline{X}$ so that $\overline{q}$ is a $\mathbb{P}^1$-fibration. Since q is an $\mathbb{A}^1$-fibration, there exists an open set U of C such that $q^{-1}(U) \cong U \times \mathbb{A}^1$ with G_m acting on $\mathbb{A}^1$ in a natural fashion. Let S_0 be the fixed point locus in $q^{-1}(U)$ and let S (resp., $\overline{S}$) be the closure of S_0 in X (resp., $\overline{X}$). Then the restriction of $\overline{q}$ onto $\overline{S}$ is a birational morphism onto $\overline{C}$, hence an isomorphism. Thus $\overline{S}$ is a cross-section of $\overline{q}$. On the other hand, $\overline{q}$ has another cross-section $\overline{S}_\infty$ lying outside X. Since q is an $\mathbb{A}^1$-fibration, every fiber of q is a disjoint union of the affine lines, hence it consists of a single affine line by Lemma 3.3.1. Suppose that for a point $\alpha \in C$, the fiber F_α does not meet S. Then $\overline{S}$ meets the fiber $\overline{F}_\alpha := \overline{q}^{-1}(\alpha)$ outside F_α, and F_α has an isolated G_m-fixed point P. This leads to a contradiction if we argue as in the proof of Lemma 3.3.2. Namely, if X is smooth at P, then consider the induced tangential representation at P. Otherwise, consider a minimal resolution of singularity at P and a G_m-fixed point appearing in one of the terminal components of the exceptional locus which does not meet the proper transform of F_α. Thus S meets every fiber of q and is smooth.

(2) With the above notations, suppose that X has a singular point on the fiber F_α. Since G_m acts transitively on $F_\alpha - F_\alpha \cap S$, X has cyclic singularity at the point $P_\alpha := F_\alpha \cap S$. Let $\sigma : \widetilde{X} \to \overline{X}$ be a minimal resolution of singularities and let $\Delta = \sigma^{-1}(P_\alpha)$. The $\mathbb{P}^1$-fibration $\overline{q} : \overline{X} \to \overline{C}$ lifts to a $\mathbb{P}^1$-fibration $\widetilde{q} : \widetilde{X} \to \overline{C}$. The proper transform $\widetilde{S}$ of $\overline{S}$ is a cross-section of $\widetilde{q}$. The fiber $\widetilde{F}_\alpha$ of $\widetilde{q}$ corresponding to $\overline{F}_\alpha$ contains the linear chain

Δ in such a way that one terminal component, say G, meets $\widetilde{S}$ and the other terminal component meets the proper transform of F_α. Further, all the components in the fiber $\widetilde{F}_\alpha$ except for the terminal component G meeting $\widetilde{S}$ can be contracted smoothly. In fact, we can replace the fiber F_α by $G - \{Q\}$ without losing the affineness of X, where Q is the point where G meets the adjacent component of $\widetilde{F}_\alpha$. If X is smooth along F_α, it is reduced. If X is smooth, the $\mathbb{A}^1$-fibration q has no singular fibers. Hence X is an $\mathbb{A}^1$-bundle over C by Lemma 2.1.6. Since the two cross-sections $\overline{S}$ and $\overline{S}_\infty$ are disjoint from each other over C, it follows that X is a line bundle over C.

(3) Since $\mathrm{Pic}(C) = 0$, every line bundle is trivial. □

The proof of the assertion (2) implies that every normal affine surface with a G_m-action and an $\mathbb{A}^1$-fibration as the quotient morphism is constructed from a line bundle over C by a succession of blow-ups with centers on the zero section and contractions of linear chains. The following result is well-known.

Lemma 3.3.4. *Let $X = \mathrm{Spec}\,A$ be a smooth affine variety of dimension greater than one with a G_m-action and let $q : X \to Y := \mathrm{Spec}\,B$ be the quotient morphism. Assume that q is an $\mathbb{A}^1$-fibration and that q is equidimensional. Then the following assertions hold:*

1. *We may assume that A is positively graded, that is, A is a graded B-algebra indexed by $\mathbb{Z}_{\geq 0}$.*
2. *B is normal and factorially closed in A.*
3. *$q : X \to Y$ has a cross-section S. Namely, each fiber of q has dimension one and meets S in one point transversally.*
4. *Y is smooth, and X is a line bundle over Y.*
5. *If X is factorial and has dimension three, then the equidimensionality condition is automatically satisfied and X is isomorphic to a direct product $Y \times \mathbb{A}^1$ with G_m acting on the factor $\mathbb{A}^1$ in a natural fashion.*

Proof. (1) By Lemma 2.1.8, there exists a nonzero element $b \in B$ such that $A[b^{-1}] = B[b^{-1}][u]$, where u is an element of A algebraically independent over the quotient field $Q(B)$. In fact, u is a variable on a general fiber of q. We may assume that the G_m-action on the fiber is given by ${}^t u = tu$. Note that the G_m-action gives a $\mathbb{Z}$-graded ring structure on A (see Problem 9, Section 3.6). Let a be a homogeneous element of A. We can write $b^r a = f(u)$, where $f(u) = b_0 u^m + b_1 u^{m-1} + \cdots + b_m \in B[u]$ with $b_0 \neq 0$. Then we have

$$b^r({}^t a) = f(tu) = b_0 t^m u^m + b_1 t^{m-1} u^{m-1} + \cdots + b_m.$$

Hence $b^r a = b_0 u^m$ and a has degree m. This implies that every homogeneous element of A has degree ≥ 0.

(2) The normality of B is well-known [128, p. 100], and the factorial closedness follows from the assertion (1). In fact, B being normal follows from the condition that A is normal and B is factorially closed in A.

(3) Let S_0 be the closed set in the open set $q^{-1}(D(b))$ defined by $u = 0$. Hence S_0 is the locus of the fixed points in the fibers $q^{-1}(Q)$ when $Q \in D(a)$. Let S be the closure of S_0 in X. Then S is a rational cross-section of q. We shall show that S meets every fiber of q in one point. Let Q be a point of Y, let C be a general irreducible curve on Y passing through Q and let $Z = X \times_Y C$. Then $q_C : Z \to C$, the projection onto C, is an $\mathbb{A}^1$-fibration. Let $\nu_Z : \widetilde{Z} \to Z$ and $\nu_C : \widetilde{C} \to C$ be the normalizations of Z and C, respectively. Then there exists an $\mathbb{A}^1$-fibration $q_{\widetilde{C}} : \widetilde{Z} \to \widetilde{C}$ such that $\nu_C \cdot q_{\widetilde{C}} = q_C \cdot \nu_Z$. The normal affine surface $\widetilde{Z}$ has the induced G_m-action and $q_{\widetilde{C}}$ is the quotient morphism. Choosing the open set $D(b)$ small enough, we may assume that $C \cap D(b)$ is a nonempty set contained in the smooth part of C. Let $\Sigma_0 = S_0 \cap Z$ and let $\widetilde{\Sigma}$ be the closure of Σ_0 in $\widetilde{Z}$. Let $\widetilde{Q}$ be a point of $\widetilde{C}$ lying over Q. Let $\widetilde{F} = q_{\widetilde{C}}^{-1}(\widetilde{Q})$ and $F = q^{-1}(Q)$. By Lemma 3.3.3, $\widetilde{\Sigma}$ meets $\widetilde{F}$ in a single point. This implies that S meets F in a point, say R. By Lemmas 3.1.12, 3.1.31, and 3.3.1, F consists of one irreducible component which is contractible and smooth outside a point R because it admits a nontrivial G_m-action with the fixed point R. Since X is smooth, the local intersection multiplicity $i(S, F; R) = 1$. This implies that F is reduced and smooth also at R. Hence F is isomorphic to $\mathbb{A}^1$.

(4) The morphism $q : X \to Y$ induces a quasifinite birational morphism $q|_S : S \to Y$. Then it is an isomorphism by Zariski's main theorem. Since S is smooth by the argument in the proof of the assertion (3), Y is also smooth. Furthermore, every fiber of q is an affine line. Hence X is an $\mathbb{A}^1$-bundle over Y. Since a smooth completion $\overline{q} : \overline{X} \to \overline{Y}$ is a $\mathbb{P}^1$-fibration with two cross-sections (the zero section and the infinity section) which are disjoint over Y, X is a line bundle over Y.

(5) Since B is factorially closed in A, by Lemma 3.1.2, q is also the quotient morphism of a G_a-action on X. By Lemma 3.1.7, q does not have codimension one fiber components. As $\dim X = 3$, this implies that q has equidimension one. Since Y is factorial, by (4), X is isomorphic to $Y \times \mathbb{A}^1$. This fact follows also from the proof of the assertion (1). In fact, one can take the element u to be a prime element of A. Then, the relation $b^r a = b_0 u^m$ for a homogeneous element a of degree m implies that $a = cu^m$ for a certain element $c \in B$. Hence $A = B[u]$. □

In Lemma 3.1.2, it is shown that any $\mathbb{A}^1$-fibration $f : X \to Y$ on an affine threefold X is factored by the quotient morphism $q : X \to \widetilde{Y}$ by a certain G_a-action on X because the factorial closure $\widetilde{B}$ is an affine domain by Zariski's finiteness theorem. Meanwhile, this is not the case with an $\mathbb{A}^1_*$-fibration. Examples can be easily produced by using Lemma 3.3.2(1).

A homology 3-space X is a *contractible 3-space* if it is topologically contractible. A homology 3-space is contractible if and only if $\pi_1(X) = 1$. We consider a G_m-action on such a threefold.

Lemma 3.3.5. *Let G_m act nontrivially on a $\mathbb{Q}$-homology 3-space X. Then the fixed point locus X^{G_m} is a nonempty connected closed subset of X. Furthermore, $H_i(X^{G_m}; \mathbb{Z}/p\mathbb{Z}) = 0$ for every $i > 0$ and for infinitely many prime numbers p.*

Proof. Since $H_i(X;\mathbb{Z})$ is a finite group, by the universal coefficient theorem, there exist infinitely many primes p such that X is $\mathbb{Z}/p\mathbb{Z}$-acyclic, i. e., $H_i(X;\mathbb{Z}/p\mathbb{Z}) = 0$ for every $i > 0$. We choose such a prime p with the additional property that p does not divide any weight of the induced tangential G_m-action on $T_{X,Q}$ for every fixed point Q. We take a point $Q \in X^{G_m}$. By [123, Theorem 1], there exists a G_m-stable open neighborhood U of Q in the Euclidean topology and algebraic functions x, y, z around the point Q which form a system of analytic coordinates on U. The G_m-action is given by ${}^t(x,y,z) = (t^a x, t^b y, t^c z)$ with integers a, b, c, where $p \nmid \alpha$ for any $\alpha \in \{a,b,c\} \setminus \{0\}$. Then $U \cap X^{G_m} = U \cap X^{\mathbb{Z}/p\mathbb{Z}}$. Since $X^{G_m} \subseteq X^{\mathbb{Z}/p\mathbb{Z}}$, it follows that X^{G_m} is a connected component of $X^{\mathbb{Z}/p\mathbb{Z}}$. In fact, by the Smith theory of group actions (see [53]), we have an inequality

$$\sum_i \dim H_i(X^{\mathbb{Z}/p\mathbb{Z}};\mathbb{Z}/p\mathbb{Z}) \le \sum_i \dim H_i(X;\mathbb{Z}/p\mathbb{Z}) = 1.$$

This implies that $X^{\mathbb{Z}/p\mathbb{Z}}$ is connected and equal to X^{G_m}. Hence X^{G_m} is connected and $H_i(X^{G_m};\mathbb{Z}/p\mathbb{Z}) = 0$ for every $i > 0$. □

Before going further, we recall some of necessary definitions on G_m-actions on affine varieties. Giving a G_m-action on an affine variety $X = \operatorname{Spec} A$ is equivalent to endowing the coordinate ring A with a $\mathbb{Z}$-graded ring structure. Write $A = \bigoplus_{n\in\mathbb{Z}} A_n$ as a graded ring over the ground field k, which corresponds to the given G_m-action. Let $A_{<0} = \bigoplus_{n<0} A_n$, $A_0 = A^{G_m}$ and $A_{>0} = \bigoplus_{n>0} A_n$. The fixed point locus X^{G_m} is the closed set defined by the ideal generated by $A_{<0}$ and $A_{>0}$. The given G_m-action is called *elliptic* (*parabolic*, or *hyperbolic*, resp.) if $A_{<0} = 0$ and $A_0 = k$ ($A_{<0} = 0$ and $\operatorname{tr.deg}(A_0) > 0$, or $A_{<0} \ne 0$, resp.). Hence an elliptic G_m-action has a unique fixed point defined by the ideal $A_{>0}$ and a parabolic G_m-action has the quotient morphism $q : X \to Y = \operatorname{Spec} A_0$ for which X^{G_m} is isomorphic to Y via q because $A/A_{>0} \cong A_0$. A fixed point free G_m-action is included in the class of hyperbolic G_m-action according to the above definition. Suppose that X is smooth and there exists a fixed point $P \in X$. Then the induced linear representation of G_m on the tangent space $T_{X,P}$ is diagonalizable. Namely, there exists a basis $\{\xi_1, \xi_2, \dots, \xi_d\}$ of $T_{X,P}$ such that

$${}^t(\xi_1, \xi_2, \dots, \xi_d) = (t^{a_1}\xi_1, t^{a_2}\xi_2, \dots, t^{a_d}\xi_d),$$

where $d = \dim X$. The G_m-action is elliptic (parabolic, or hyperbolic, resp.) if and only if $a_1 > 0, \dots, a_d > 0$ ($a_1 = \cdots = a_r = 0, a_{r+1} > 0, \dots, a_d > 0$, or $a_1 < 0$, resp.) after a suitable permutation of $\xi_1, \dots, \xi_d$, where $r = \operatorname{tr.deg} A_0$.

We have the following result.

Theorem 3.3.6. *Let G_m act nontrivially on a $\mathbb{Q}$-homology 3-space X and let $q : X \to Y$ be the quotient morphism. Suppose that q has relative dimension one. Then the following assertions hold:*

(1) *X^{G_m} is $\mathbb{Q}$-acyclic. If X is a homology 3-space, then X^{G_m} is $\mathbb{Z}$-acyclic.*
(2) *If $\dim X^{G_m} = 2$ then $X^{G_m} \cong Y$ and X is a line bundle over Y. If X is a homology 3-space, then $X \cong Y \times \mathbb{A}^1$.*
(3) *If $\dim X^{G_m} = 1$ then $X^{G_m} \cong \mathbb{A}^1$ and $q|_{X^{G_m}} : X^{G_m} \to Y$ is a closed embedding. If X is contractible and q is equidimensional, then Y is a smooth contractible surface of log-Kodaira dimension $-\infty$ or 1.*
(4) *If $\dim X^{G_m} = 0$ then the G_m-action on X is hyperbolic, i. e., the tangential representation on $T_{X,P}$ for the unique fixed point P has mixed weights, e. g., either $a_1 < 0, a_2 > 0$ and $a_3 > 0$, or $a_1 > 0, a_2 < 0$ and $a_3 < 0$. In this case, there is an irreducible component of codimension one contained in the fiber of q passing through the point P.*

Proof. (1) The $\mathbb{Q}$-acyclicity is verified below case by case according to $\dim X^{G_m}$. When X is a homology 3-space, the acyclicity of X^{G_m} follows from the Smith theory [198, §22.3].

(2) By Lemma 3.3.1, X^{G_m} lies horizontally to the morphism q and the restriction $q|_{X^{G_m}} : X^{G_m} \to Y$ is a bijection (hence a birational morphism). In particular, X^{G_m} is irreducible. Further, $q : X \to Y$ is an $\mathbb{A}^1$-fibration, hence X has a G_a-action. Since X is $\mathbb{Q}$-factorial by Corollary 3.2.2, q is equidimensional by Lemma 3.1.11. Then Lemma 3.3.4 implies that Y is smooth and $q : X \to Y$ is a line bundle. By Zariski's main theorem, we conclude that $X^{G_m} \cong Y$. Furthermore, since X is contractible to Y, it follows that X^{G_m} is $\mathbb{Q}$-acyclic (resp., $\mathbb{Z}$-acyclic) if X is a $\mathbb{Q}$-homology 3-space (resp., $\mathbb{Z}$-homology 3-space). If X is a homology 3-space, then Y is factorial and a line bundle over Y is trivial. Hence $X \cong Y \times \mathbb{A}^1$.

(3) X^{G_m} is a connected curve by Lemma 3.3.5. Further, X^{G_m} is smooth. In fact, the smoothness of the fixed point locus is a well-known fact for a reductive algebraic group action on a smooth affine variety (cf. [198]).[4] Hence X^{G_m} is irreducible. Then X^{G_m} is an affine line because $H_1(X^{G_m}; \mathbb{Z}/p\mathbb{Z}) = 0$. By Lemma 3.3.1, q induces a closed embedding of X^{G_m} into Y.[5] If X is contractible, Y is contractible by [131]. By Lemma 3.2.4, $\pi_1(Y^\circ) = (1)$, where Y° is the smooth part of Y. Then Y is smooth by Theorem 3.2.7. So, Y is a smooth contractible surface containing a curve isomorphic to $\mathbb{A}^1$. Hence Y has log-Kodaira dimension $-\infty$ or 1 (cf. [232, 72]).

---

4 In the case where G is a connected reductive algebraic group acting on a smooth affine variety X, the G-action near a fixed point P is locally analytically G-equivalent to a linear representation [123]. In the case of a linear representation of G on the affine space $\mathbb{C}^n$, the fixed point locus near the origin is a linear subspace and it is G-equivariantly a direct summand of $\mathbb{C}^n$ by the complete reducibility of G. The smoothness of the fixed point locus X^G near the point P follows from this observation.

5 Let $P \in X^{G_m}$. In view of [123], there exists a system of local (analytic) coordinates $\{x, y, z\}$ at P such that G_m acts as ${}^t(x, y, z) = (x, t^{-a}y, t^b z)$ with $ab > 0$ and $y = z = 0$ defining the curve X^{G_m} near P. Then the quotient surface Y at the point $Q = q(P)$ has a system of local analytic coordinates $\{x, x^{b'}y^{a'}\}$ with $a' = a/d, b' = b/d$ and $d = \gcd(a, b)$. Then the curve $q(X^{G_m})$ is defined by $x^{a'}y^{b'} = 0$ near Q and Y is smooth near the curve $q(X^{G_m})$. Hence X^{G_m} and $q(X^{G_m})$ are locally isomorphic at P and Q. Since $q|_{X^{G_m}} : X^{G_m} \to Y$ is injective, it is a closed embedding. Furthermore, the fiber of q through the point P is a *cross* $a'\mathbb{A}^1 + b'\mathbb{A}^1$. See the definition after Corollary 3.3.8.

(4) The induced G_m-action on the tangent space $T_{X,P}$ of the unique fixed point P must have mixed weights. Otherwise, either the relative dimension of q is greater than one or $\dim X^{G_m} > 0$ near the point P. This is a contradiction. To prove the last assertion, we have only to show it when X is identified with the tangent linear 3-space $T_{X,P}$. Then the G_m-action is diagonalized and ${}^t(x,y,z) = (t^{a_1}x, t^{a_2}y, t^{a_3}z)$ with respect to a suitable system of coordinates (x,y,z). Then the invariant elements, viewed as elements in $\Gamma(X, \mathcal{O}_X)$, are divisible by x. Hence the fiber F of q passing through P contains an irreducible component $\{x = 0\}$ which is a hypersurface of X. □

Question 3.3.7. In the case (3) above, is X isomorphic G_m-equivariantly to a direct product $Z \times X^{G_m}$, where Z is a smooth affine surface with G_m acting on it? If this is the case, q is the direct product $q = q' \times X^{G_m}$ of the quotient morphism $q' : Z \to C$, where $C = Z//G_m$.

We discuss this question in the next Subsection 3.3.2 in detail. Theorem 3.3.6 has the following consequence.

Corollary 3.3.8. *Let the notations and the assumptions be the same as in Theorem* 3.3.6. *If the quotient morphism $q : X \to Y$ has equidimension one, then the fixed point locus X^{G_m} has positive dimension.*

Now we consider an $\mathbb{A}^1_*$-fibration $f : X \to Y$, where X is a smooth affine threefold[6] and Y is a normal affine surface. A fiber F_Q of f is called *singular* if F_Q is not isomorphic to $\mathbb{A}^1_*$ over $k(Q)$. The set of points $Q \in Y$ such that F_Q is singular is denoted by $\operatorname{Sing}^{st}(p)$ and called the *strict singular locus* or the *degeneracy locus* of f. For a technical reason (cf. Lemmas 3.3.13 and 3.3.14), we define the *singular locus* $\operatorname{Sing}(f)$ as the union of $\operatorname{Sing}^{st}(p)$ and the set $\operatorname{Sing}(Y)$ of singular points. A singular fiber F_Q is called a *cross* (resp., *tube*)[7] if it has the form $F_Q \cong m\mathbb{A}^1 + n\mathbb{A}^1$ (resp., $m\mathbb{A}^1_*$). When we speak of a cross on a threefold or a surface which is smooth at the intersection point P of two lines, we assume that two affine lines meet each other transversally at P. By abuse of the terminology, we call a singular fiber F on a normal surface a cross even when the surface has a cyclic quotient singularity at the intersection point P and the proper transforms of the two lines on the minimal resolution of singularity form a linear chain together with the exceptional locus.

We recall a result of Bhatwadekar–Dutta [15, Theorem 3.11] on $\mathbb{A}^1_*$-fibrations.

Lemma 3.3.9. *Let R be a noetherian normal domain and let A be a finitely generated flat R-algebra such that:*

(1) *The generic fiber $A \otimes_R K$ is a Laurent polynomial ring $K[T, T^{-1}]$ in one variable over $K = Q(R)$.*

---

6 We can generalize the definitions to the case where $\dim X \geq 3$.

7 The naming of cross is apparent. The fiber $m\mathbb{A}^1_*$ is a projective line with two end points missing and thickened with multiplicity. It looks like a tube.

(2) *For each prime ideal* $\mathfrak{q}$ *of* R *of height one, the fiber* $\operatorname{Spec} A \otimes_R k(\mathfrak{q})$ *is geometrically integral but is not* $\mathbb{A}^1$ *over* $k(\mathfrak{q})$.

Then there exists an invertible ideal I *in* R *such that* A *is a* $\mathbb{Z}$*-graded* R*-algebra isomorphic to the* R*-subalgebra* $R[IT, I^{-1}T^{-1}]$ *of* $K[T, T^{-1}]$. *In particular,* $\operatorname{Spec} A$ *is locally* $\mathbb{A}^1_*$ *and hence an untwisted* $\mathbb{A}^1_*$*-fibration over* $\operatorname{Spec} R$.

In geometric terms, a weaker version of Lemma 3.3.9 in the case of dimension three is stated as follows.

Lemma 3.3.10. *Let* $f : X \to Y$ *be an* $\mathbb{A}^1_*$*-fibration with a smooth affine threefold* X *and let* Q *be a closed point of* Y. *Suppose that the following conditions are satisfied:*
(1) *There is an open neighborhood* U *of* Q *such that* U *is smooth and* f *is equidimensional over* U.
(2) *Every fiber* $F_{Q'}$ *of* f *for* $Q' \in U \setminus \{Q\}$ *is isomorphic to* $\mathbb{A}^1_*$.

Then the fiber F_Q *is isomorphic to* $\mathbb{A}^1_*$ *and* $f : f^{-1}(U) \to U$ *is an* $\mathbb{A}^1_*$*-bundle.*

Proof. By replacing Y with a smaller affine open neighborhood of Q contained in U, we may assume that Y is smooth, f is equidimensional and $\operatorname{Sing}(f) \cap (Y \setminus \{Q\}) = \emptyset$. Then f is flat. Further, there exists a $\mathbb{P}^1$-fibration $\overline{f} : \overline{X} \to Y$ such that X is an open set of $\overline{X}$ and $\overline{f}|_X = f$. Since the $\mathbb{A}^1_*$-fibration f is assumed to be untwisted, the generic fiber $\overline{X}_K$ has two K-rational points ξ_1, ξ_2, where K is the function field of Y over $\mathbb{C}$. Let S_i be the closure of ξ_i in $\overline{X}$ for $i = 1, 2$. By the assumption (2) above, S_1, S_2 are cross-sections over $Y \setminus \{Q\}$. Namely, $f^{-1}(Y \setminus \{Q\})$ is an $\mathbb{A}^1_*$-bundle. Hence $f^{-1}(Y \setminus \{Q\}) \cong \mathcal{S}pec\,(\mathcal{O}_{Y\setminus\{Q\}}[\mathcal{L}', \mathcal{L}'^{-1}])$, where $\mathcal{L}'$ is an invertible sheaf on $Y \setminus \{Q\}$. Since Y is smooth and Q is a point of codimension two, $\mathcal{L}'$ is extended to an invertible sheaf $\mathcal{L}$ on Y. Since X is affine and $\operatorname{codim}_X f^{-1}(Q) = 2$, it follows that $X \cong \mathcal{S}pec\,(\mathcal{O}_Y[\mathcal{L}, \mathcal{L}^{-1}])$. So, $f : X \to Y$ is an $\mathbb{A}^1_*$-bundle and $F_Q \cong \mathbb{A}^1_*$. □

Let F_Q be a fiber of an $\mathbb{A}^1_*$-fibration $f : X \to Y$ with $Q \in Y$. Let C be a general smooth curve on Y passing through Q. Let Z be the normalization of $X \times_Y C$. Then f induces an $\mathbb{A}^1_*$-fibration $f_C : Z \to C$ and the fiber $\widetilde{F}_Q$ over the point Q is a finite covering of the fiber F_Q in the sense that the normalization morphism induces a finite morphism $\widetilde{F}_Q \to F_Q$. In view of Lemma 3.3.2, we say that the $\mathbb{A}^1_*$-fibration f is *unmixed* if $\widetilde{F}_Q$ for every $Q \in Y$ is $\mathbb{A}^1_*$, $\mathbb{A}^1 + \mathbb{A}^1$, or a disjoint union of $\mathbb{A}^1$s when taken with reduced structure.[8] This definition does not depend on the choice of the curve C.

Lemma 3.3.11. *Let* $f : X \to Y$ *be an unmixed* $\mathbb{A}^1_*$*-fibration such that* X *and* Y *are smooth. Then the singular locus* $\operatorname{Sing}(f)$ *is a closed subset of* Y *of pure codimension one.*

8 With the notation of Lemma 3.3.2, f being unmixed is equivalent to saying that, for each singular fiber $\widetilde{F}_Q = \Gamma + \Delta$, either $\Delta = \emptyset$ or $\Gamma = \emptyset$.

Proof. Lemma 3.3.9 or Lemma 3.3.10 implies that every irreducible component of the closure $\overline{\mathrm{Sing}(f)}$ of $\mathrm{Sing}(f)$ has codimension one.

So, we have only to show that $\mathrm{Sing}(f)$ is a closed set. It suffices to show that if the fiber F_Q over $Q \in Y$ is $\mathbb{A}^1_*$ then there exists an open neighborhood U of Q such that $F_{Q'} \cong \mathbb{A}^1_*$ for every $Q' \in U$. Let C be a curve passing through Q. If C is not a component of $\overline{\mathrm{Sing}(f)}$, then the fiber of f over the generic point of C is $\mathbb{A}^1_*$ and hence geometrically integral. Suppose that C is a component of $\overline{\mathrm{Sing}(f)}$. The unmixedness condition on f implies that for a general point Q' of C, the fiber $F_{Q'}$ is either irreducible and dominated by $\mathbb{A}^1_*$ (case (i)), or each irreducible component of $F_{Q'}$ is dominated by a contractible curve (case (ii)). Suppose that the case (ii) occurs. Then there exists a normal affine surface $\widetilde{Z}$ with an $\mathbb{A}^1$-fibration $\widetilde{f} : \widetilde{Z} \to \widetilde{C}$ satisfying the following commutative diagram:

$$\begin{array}{ccc} \widetilde{Z} & \xrightarrow{\mu} & X \times_Y C \\ \downarrow{\scriptstyle \widetilde{f}} & & \downarrow{\scriptstyle f_C} \\ \widetilde{C} & \xrightarrow{\nu} & C \end{array}$$

where μ is a quasifinite morphism and ν is the normalization morphism. Since any singular fiber of an $\mathbb{A}^1$-fibration on a normal affine surface is a disjoint union of the affine lines, it turns out that the fiber F_Q, which is a fiber of f_C in the above diagram, is dominated by the affine line, which is a connected component of the fiber of $\widetilde{f}$. This is a contradiction, and the case (ii) cannot occur. Now removing from Y all irreducible components of $\overline{\mathrm{Sing}(f)}$ for whose general points the case (ii) occurs and replacing X by the inverse image of the open set of Y thus obtained, we may assume that the case (i) occurs for general points of the irreducible components of $\overline{\mathrm{Sing}(f)}$. Let

$$df : \mathcal{T}_X \to f^*\mathcal{T}_Y$$

be the tangential homomorphism of the tangent bundles on X and Y. Let $\mathcal{C}$ be the cokernel of df. Then $\mathcal{C}$ is a coherent $\mathcal{O}_X$-Module. Hence $\mathrm{Supp}(\mathcal{C})$ is a closed set T such that $T = f^{-1}(f(T))$ and $f(T) = \overline{\mathrm{Sing}(f)}$. Hence the point Q belongs to $f(T)$. However, df is everywhere surjective on the fiber F_Q and $F_Q \not\subset \mathrm{Supp}(\mathcal{C})$. This is a contradiction. By the above argument, every irreducible curve C through the point Q has the property that the general fibers are isomorphic to $\mathbb{A}^1_*$. Hence the fiber of f over the generic point of C is geometrically integral. By Lemma 3.3.9, we know that $A \otimes_B \mathcal{O}_Q \cong \mathcal{O}_Q[x, x^{-1}]$. This implies that there exists an open neighborhood U of Q satisfying the required property. □

In the proofs of Lemmas 3.3.9, 3.3.10, and 3.3.11, the flatness condition on the morphism f should not be overlooked. Hence the assumption that Y be normal instead of being smooth does not seem to be sufficient for the conclusion. So we ask the following question.

Question 3.3.12. Let $f : X \to Y$ be an $\mathbb{A}^1_*$-fibration with a smooth affine threefold X. Is a point $Q \in Y$ smooth if the fiber $(F_Q)_{\mathrm{red}}$ is isomorphic to $\mathbb{A}^1_*$?

The following is a partial answer to this question.

Lemma 3.3.13. *Let X be a smooth affine threefold with an effective G_m-action and let $q : X \to Y$ be the quotient morphism. Suppose that the G_m-action is fixed point free. Then Y has at worst cyclic quotient singular points.*

Proof. Suppose that Q is a singular point of Y. Let P be a point of X such that $q(P) = Q$ and let G_P be the isotropy group at P which is a finite cyclic group. Then, by Luna's étale slice theorem [135], there exists an affine subvariety V of X with a G_P-action and an étale morphism $\varphi/G_P : V/G_P \to Y$ such that $P \in V$, V is smooth at P and $(\varphi/G_P)(P) = Q$. This implies that Y has at worst cyclic quotient singularity at Q. Since the ground field is $\mathbb{C}$, we can work with an analytic slice instead of an étale slice. □

Lemma 3.3.14. *With the assumptions of Lemma 3.3.13 and the notations in the proof, the following assertions hold:*
(1) *If Y is smooth at Q then the fiber F_Q has multiplicity equal to $m = |G_P|$ near the point P.*
(2) *If Y is singular at Q then the fiber F_Q is a multiple fiber.*

Proof. (1) If V is an analytic slice with coordinates x, y, the G_P-action on V is given by ${}^{\zeta}(x, y) = (\zeta x, \zeta^b y)$ with a generator ζ of G_P which is identified with an mth primitive root of unity, where $0 \le b < m$. Then Q is singular if and only if $b > 0$. Hence $b = 0$ if Q is smooth. This implies that (x^m, y) is a local system of parameters of Y at Q. Hence the fiber F_Q has multiplicity m.

(2) With the above notations, $\mathfrak{m}_{Y,Q}\widehat{\mathcal{O}}_{X,P}$ is generated by

$$\{x^m, y^m\} \cup \{x^s y^n \mid bn + s \equiv 0 \pmod{m},\ 0 < s < m,\ 0 < n < m\}.$$

Since $\widehat{\mathcal{O}}_{X,P} = \mathbb{C}[[x, y, z]]$ with a fiber coordinate z of F_Q, it follows that $\mathfrak{m}_{Y,Q}\widehat{\mathcal{O}}_{X,P}$ is not a radical ideal. Hence the fiber F_P is not reduced near the point Q. □

The singular locus $\mathrm{Sing}(f)$ may consist of a single point as shown in the following example.

Example 3.3.15. Let $f : \mathbb{A}^3 \to \mathbb{A}^2$ be the morphism defined by $(x, y, z) \mapsto (xy, x^2(xz+1))$, where (x, y, z) is a system of coordinates of $\mathbb{A}^3$. Then f is an $\mathbb{A}^1_*$-fibration and $\mathrm{Sing}(f) = \{(0, 0)\}$. In fact, set $\alpha = xy$, $\beta = x^2(xz + 1)$. Then $f^{-1}(\alpha, \beta) = \{(\alpha y^{-1}, y, \alpha^{-3}y(\beta y^2 - \alpha^2) \mid y \in \mathbb{C}^*\} \cong \mathbb{A}^1_*$ if $\alpha \ne 0$, $f^{-1}(0, \beta) = \{y = 0,\ x^2(xz + 1) = \beta\} \cong \mathbb{A}^1_*$ if $\alpha = 0$, $\beta \ne 0$ and $f^{-1}(0, 0) = \{x = 0\} \cup \{y = xz + 1 = 0\} \cong \mathbb{A}^2 \cup \mathbb{A}^1_*$ if $\alpha = \beta = 0$.

The following example shows that a disjoint union of affine lines may appear as a singular fiber of an $\mathbb{A}^1_*$-fibration.

Example 3.3.16. Let $f : \mathbb{A}^3 \to \mathbb{A}^2$ be the morphism defined by $(x, y, z) \mapsto (xy, x^2(yz+1))$. Then f is an $\mathbb{A}^1_*$-fibration such that $\mathrm{Sing}(f) \cong \mathbb{A}^1$ and the fibers are given as follows. Set $\alpha = xy$, $\beta = x^2(yz+1)$. Then $f^{-1}(\alpha, \beta) = \{(\alpha y^{-1}, y, \alpha^{-2}(\beta y - \alpha^2 y^{-1}) \mid y \in \mathbb{C}^*\} \cong \mathbb{A}^1_*$ if $\alpha \neq 0$, $f^{-1}(0, \beta) = \{y = 0,\ x^2 = \beta\} \cong \mathbb{A}^1(x = \sqrt{\beta}, y = 0) \sqcup \mathbb{A}^1(x = -\sqrt{\beta}, y = 0)$ if $\alpha = 0,\ \beta \neq 0$ and $f^{-1}(0, 0) = \{x = 0\} \cong \mathbb{A}^2$ if $\alpha = \beta = 0$.

There is an example due to Winkelmann [228] of a flat $\mathbb{A}^1_*$-fibration $p : \mathbb{A}^3 \to \mathbb{A}^2$ which is not surjective.

Proposition 3.3.17. *Let $p(x) = \prod_{i=1}^n (x - i)$ and let $f : \mathbb{A}^3 \to \mathbb{A}^2$ be the morphism defined by $(x, y, z) \mapsto (x + zx + zyp(x), x + yp(x))$. Then the following assertions hold:*
1. *A general fiber is isomorphic to $\mathbb{A}^1_{n*}$, where $\mathbb{A}^1_{n*}$ is the affine line minus n points. Hence f is an $\mathbb{A}^1_*$-fibration if $n = 1$.*
2. *$\mathrm{Im}(f) = \mathbb{A}^2 \setminus S$, where $S = \{(i, 0) \mid 1 \le i \le n\}$.*
3. *f is equidimensional. Hence f is flat.*
4. *The singular locus $\mathrm{Sing}(f)$, that is the locus of points Q of the base $\mathbb{A}^2$ over which $f^{-1}(Q) \not\cong \mathbb{A}^1_{n*}$, is the union of $\{(\alpha, 0) \mid \alpha \in \mathbb{C}\}$ and $\bigcup_{i=1}^n \{(\alpha, i) \mid \alpha \in \mathbb{C}\}$. Hence $\mathrm{Sing}(f)$ consists of disjoint $n+1$ affine lines. For any $\alpha \in \mathbb{C} - \{1, \ldots, n\}$, the fiber $p^{-1}(\alpha, 0) \cong \mathbb{A}^1$. For any $i \in \{1, \ldots, n\}$ and $\alpha \in \mathbb{C}$, the fiber $f^{-1}(\alpha, i) \cong \mathbb{A}^1 \cup \mathbb{A}^1_{(n-1)*}$.*

Proof. (1) Set $\alpha = x + z(x + yp(x))$ and $\beta = x + yp(x)$. Then $\alpha = x + z\beta$. Hence if $\beta \neq 0$, we have $z = (\alpha - x)/\beta$. Further, if $\beta \notin \{1, 2, \ldots, n\}$, then the equation $\beta = x + yp(x)$ is solved as $y = (\beta - x)/p(x)$. In fact, if (x, y) is a solution of $\beta = x + yp(x)$ and $p(x) = 0$, then $x = \beta$ and $\beta \in \{1, 2, \ldots, n\}$. Hence $f^{-1}(\alpha, \beta) \cong \mathbb{A}^1 - \{1, 2, \ldots, n\} \cong \mathbb{A}^1_{n*}$.

(2) Suppose $\beta = 0$. Then $x = \alpha$ and z is free if $\alpha = x + z\beta$ has a solution. Further, $0 = \alpha + yp(\alpha)$, which implies $\alpha = 0$ if $p(\alpha) = 0$. But this is impossible. So, $f^{-1}(i, 0) = \emptyset$ if $i \in \{1, 2, \ldots, n\}$. If $p(\alpha) \neq 0$, y is solved as $y = -\alpha/p(\alpha)$. So, $f^{-1}(\alpha, 0) = \{(\alpha, -\alpha/p(\alpha), z) \mid z \in \mathbb{C}\} \cong \mathbb{A}^1$. Suppose $\beta \in \{1, 2, \ldots, n\}$. Then $z = (\alpha - x)/\beta$ and the equation $x + yp(x) = \beta$ is written as $(x - \beta)(yq(x) + 1) = 0$, where $p(x) = q(x)(x - \beta)$. If $x = \beta$ then y is free. If $yq(x) + 1 = 0$ then $y = -1/q(x)$. This implies that $f^{-1}(\alpha, \beta) = \mathbb{A}^1 \cup \mathbb{A}^1_{(n-1)*}$ with two components meeting in a single point $(\beta, -1/q(\beta), (\alpha - \beta)/\beta)$ in the (x, y, z)-coordinates. By the above reasoning, we know that $\mathrm{Im}(f) = \mathbb{A}^2 \setminus S$ as stated above. The assertions (3) and (4) are also shown in the above argument. □

If we restrict the morphism $f : \mathbb{A}^3 \to \mathbb{A}^2$ in the above proposition to a general linear plane in $\mathbb{A}^3$, we obtain an example of a dominant endomorphism $\varphi : \mathbb{A}^2 \to \mathbb{A}^2$ which is not surjective. Such examples have been constructed by Jelonek [97].

We further study the singular fibers of $\mathbb{A}^1_*$-fibrations.

Example 3.3.18. Let $\widetilde{A} = \mathbb{C}[x, y, z, z^{-1}]$ be the Laurent polynomial ring in z over a polynomial ring $\widetilde{B} = \mathbb{C}[x, y]$. Let G_m and a cyclic group G of order n act on $\widetilde{A}$ by

$$ {}^t(x, y, z) = (x, y, tz) \quad \text{and} \quad {}^\zeta(x, y, z) = (x, \zeta y, \zeta^d z), $$

where ζ is a primitive nth root of unity, G is identified with the subgroup $\langle\zeta\rangle$ of G_m generated by ζ, and d is a positive integer with $\gcd(d,n) = 1$. Then the following assertions hold:

(1) Let A (resp., B) be the G-invariant subring of $\widetilde{A}$ (resp., $\widetilde{B}$) and let d' be a positive integer such that $dd' \equiv 1 \pmod n$. Then $A = \mathbb{C}[x, y^n, y/z^{d'}, z^n, z^{-n}] = \mathbb{C}[x, \eta, U, T, T^{-1}]/(U^n = \eta T^{-d'})$ and $B = \mathbb{C}[x, \eta]$, where $\eta = y^n$, $U = y/z^{d'}$ and $T = z^n$.
(2) Let $X = \operatorname{Spec} A$, $Y = \operatorname{Spec}(B)$ and $q : X \to Y$ be the morphism induced by the inclusion $B \hookrightarrow A$. Then q is the quotient morphism of X by the G_m-action given by ${}^t(x, \eta, U, T) = (x, \eta, t^{-d'}U, t^n T)$.
(3) The singular locus $\operatorname{Sing}(q)$ is the line $\eta = 0$ on Y, and the singular fibers over $\operatorname{Sing}(q)$ are of the form $n\mathbb{A}^1_*$.
(4) The G-action on $\widetilde{A}$ commutes with the G_m-action. Hence the G_m-action descends onto X. Write $dd' = 1 + cn$. Then ${}^t(U^d T^c) = t^{-1}(U^d T^c)$. Hence the isotropy subgroup is trivial (resp., G) if $U \neq 0$ (resp., $U = 0$).

Example 3.3.18 is a special case of a more general result.

Lemma 3.3.19. *Let $f : X \to Y$ be an $\mathbb{A}^1_*$-fibration on an affine threefold X to an affine surface Y satisfying the following conditions:*
(i) *X and Y are smooth, f is equidimensional and* $\operatorname{Sing}(f)$ *is a smooth irreducible curve, say C.*
(ii) *There exists a positive integer $n > 1$ such that the fiber F_Q over every point $Q \in C$ is of the form $n\mathbb{A}^1_*$.*
(iii) *Either Y is factorial, or there exists an invertible sheaf $\mathcal{L}$ on Y such that $\mathcal{L}^{\otimes n} \cong \mathcal{O}_Y(-C)$.*

Then there exists a cyclic covering $\mu : \widetilde{Y} \to Y$ of order n ramifying totally over C such that the normalization $\widetilde{X}$ of $X \times_Y \widetilde{Y}$ is an $\mathbb{A}^1_*$-bundle over $\widetilde{Y}$. Further, there exists a G_m-action on $\widetilde{X}$ such that the composite of the normalization morphism $\nu : \widetilde{X} \to X \times_Y \widetilde{Y}$ with the base change $f_{\widetilde{Y}}$ of f

$$\widetilde{q} : \widetilde{X} \xrightarrow{\nu} X \times_Y \widetilde{Y} \xrightarrow{f_{\widetilde{Y}}} \widetilde{Y}$$

is the quotient morphism by the G_m-action. The quotient morphism $\widetilde{q}$ commutes with the cyclic covering group G, and hence descends down to the quotient morphism $q : X \to Y$ by the induced G_m-action which coincides with the given $\mathbb{A}^1_*$-fibration f.

Proof. Write $Y = \operatorname{Spec} B$. If Y is factorial, let $p \in B$ define the curve C and let $\widetilde{Y} = \operatorname{Spec} B[\xi]/(\xi^n - p)$. If $\mathcal{O}_Y(-C) \cong \mathcal{L}^{\otimes n}$, let $\widetilde{Y} = \mathcal{S}pec_X \bigoplus_{i=0}^{n-1} \mathcal{L}^{\otimes i}$. Then the morphism $\mu : \widetilde{Y} \to Y$ is a cyclic covering of order n totally ramifying over the curve C.

Let $\mathcal{U} = \{U_i\}_{i \in I}$ be an affine open covering of $f^{-1}(C)$ and let η_i be an element of $\Gamma(U_i, \mathcal{O}_X)$ such that $\eta_i = 0$ defines $f^{-1}(C)_{\mathrm{red}}|_{U_i}$ for each i. We consider the case where

C is defined by $p = 0$. The other case can be treated in a similar fashion. We can write $p = u_i\eta_i^n$ for $u_i \in \Gamma(U_i, \mathcal{O}_X^*)$. Then $\widetilde{X}$ over U_i is defined as $\mathcal{S}pec_{U_i}\mathcal{O}_{U_i}[\xi/\eta_i]$, where $(\xi/\eta_i)^n = u_i$. This implies that the morphism $\rho : \widetilde{X} \xrightarrow{\nu} X \times_Y \widetilde{Y} \xrightarrow{p_1} X$ with the first projection p_1 is a finite étale morphism, whence $\widetilde{X}$ is smooth. It is clear that $\widetilde{q} : \widetilde{X} \to \widetilde{Y}$ is an $\mathbb{A}^1_*$-fibration. Since ρ is finite and étale, $\widetilde{q}^{-1}(\widetilde{Q})$ with $\widetilde{Q} \in \mu^{-1}(C)_{\text{red}}$ is a reduced $\mathbb{A}^1_*$ or a disjoint union of reduced $\mathbb{A}^1_*$s. Let D be a general smooth curve on $\widetilde{Y}$ passing through $\widetilde{Q}$ and let $\widetilde{Z}$ be the normalization of $\widetilde{q}^{-1}(D)$. Then the canonical morphism $\overline{q} : \widetilde{Z} \to D$ is an $\mathbb{A}^1_*$-fibration on a normal affine surface $\widetilde{Z}$ and the fiber $\overline{q}^{-1}(\widetilde{Q})$ maps surjectively onto the fiber $\widetilde{q}^{-1}(\widetilde{Q})$. By Lemma 3.3.2, $\overline{q}^{-1}(\widetilde{D})$ does not have two or more irreducible components which surject onto $\mathbb{A}^1_*$. Hence $\widetilde{q}^{-1}(\widetilde{Q})$ consists of a single reduced component isomorphic to $\mathbb{A}^1_*$. This implies that $\widetilde{X}$ is an $\mathbb{A}^1_*$-bundle over $\widetilde{Y}$.

An $\mathbb{A}^1_*$-bundle has a standard G_m-action along the fibers. Namely, if $\widetilde{q}^{-1}(V_j) \cong V_j \times \operatorname{Spec} \mathbb{C}[\tau_j, \tau_j^{-1}]$ for an open covering $\mathcal{V} = \{V_j\}_{j\in J}$ of $\widetilde{Y}$, then G_m acts as ${}^t\tau_j = t\tau_j$. This action clearly commutes with the action of the cyclic covering group G. Note that $\widetilde{X}/G \cong X$. Hence the G_m-action on $\widetilde{X}$ descends down to X, and gives rise to a G_m-action on X. The quotient morphism $q : X \to X/\!/G_m$ coincides with $f : X \to Y$ because $q = f$ on the open set $f^{-1}(Y \setminus C)$. □

A G_m-action of a smooth affine threefold X is called *equidimensional* if the quotient morphism $q : X \to Y$ is equidimensional.

Lemma 3.3.20. *Let X be a smooth affine threefold with a G_m-action and let $q : X \to Y$ be the quotient morphism with $Y = X/\!/G_m$. Suppose that q has equidimension one, q is an $\mathbb{A}^1_*$-fibration and Y is smooth. Then a singular fiber of q is either a tube or a cross.*

Proof. Let F_Q be a singular fiber F of q. Let C be a smooth irreducible curve on Y such that $Q \in C$. Let $Z = X \times_Y C$ and let $\nu : \widetilde{Z} \to Z$ be the normalization morphism. Then the projection $\rho = q_C : Z \to C$ and $\widetilde{\rho} = \rho \cdot \nu : \widetilde{Z} \to C$ are the quotient morphisms by the induced G_m-actions on Z and $\widetilde{Z}$. The fiber $\rho^{-1}(Q)$ is the fiber F_Q and is the image of $\widetilde{F}_Q = \widetilde{\rho}^{-1}(Q)$ by ν. By Lemma 3.3.2, the fiber $\widetilde{F}_Q$ is either a tube or a cross. If $\widetilde{F}_Q$ is a tube, then F_Q is also a tube because ν restricted onto $\widetilde{F}_Q$ commutes with the G_m-action. Suppose that $\widetilde{F}_Q$ is a cross. Since the intersection point $\widetilde{P}$ of two affine lines of $\widetilde{F}_Q$ is a fixed point and $\nu_{\widetilde{F}_Q}$ is surjective, either F_Q consists of two components meeting in a point $P = \nu(\widetilde{P})$ or F_Q is a contractible curve with P a fixed point and $F_Q - \{P\}$ a G_m-orbit. In the latter case, two branches of a cross are folded into a single curve. But this is impossible because the G_m-action on a cross viewed near the point $\widetilde{P}$ has weights $-a, b$, respectively, on two branches with $ab > 0$. So, F_Q consists of two branches meeting in one point P. Consider the induced representation of G_m on the tangent space $T_{X,P}$ at the fixed point P. We can write it as ${}^t(x, y, z) = (x, t^{-a}y, t^b z)$ for a suitable system of local coordinates $\{x, y, z\}$. Then the fiber F_Q is given by $x = y^b z^a = 0$ locally at P. Hence two branches of F_Q meet transversally at P, and F_Q is a cross. □

We can prove a *converse* of this result.

Theorem 3.3.21. *Let $f : X \to Y$ be an $\mathbb{A}^1_*$-fibration on a smooth affine threefold X. Suppose that Y is normal, f is equidimensional and the singular fibers are tubes or crosses over the points of* Sing(f) *except for a finite set of points. Then there exists an equidimensional G_m-action on X such that the quotient morphism $q : X \to X//G_m$ coincides with the given $\mathbb{A}^1_*$-fibration f. Hence Y has at worst cyclic quotient singularities.*

Proof. Let $Y^\circ = Y \backslash \mathrm{Sing}(Y)$. Since $\mathrm{Sing}(Y)$ is a finite set, if a G_m-action is constructed on $f^{-1}(Y^\circ)$ in such a way that the quotient morphism coincides with f restricted on $f^{-1}(Y^\circ)$, then the G_m-action extends to X by Hartog's theorem and the quotient morphism coincides with f. Since, as shown below, the construction of a G_m-action is local over Y°, we may restrict ourselves to an affine open set of Y° and assume that Y is smooth from the beginning.

By Lemmas 3.3.9 and 3.3.10, the singular locus Sing(f) has no isolated points. Let $\mathrm{Sing}(f) = C_1 \cup \cdots \cup C_r$ be the irreducible decomposition. Let

$$Y_i = Y \setminus (C_1 \cup \cdots \cup \overset{\vee}{C_i} \cup \cdots \cup C_r)$$

for $1 \le i \le r$, let $X_i = f^{-1}(Y_i)$ and let $f_i = f|_{X_i} : X_i \to Y_i$. Then X_i and Y_i are affine and the $\mathbb{A}^1_*$-fibration f_i has an irreducible singular locus, say C_i by abuse of the notation. If C_i has singular points, we replace X_i and Y_i first by $f_i^{-1}(Y \setminus \mathrm{Sing}(C_i))$ and $Y \setminus \mathrm{Sing}(C_i)$, respectively, and then by $f_i^{-1}(U_{i\lambda})$ and $U_{i\lambda}$, respectively, where $\{U_{i\lambda}\}_{\lambda \in \Lambda}$ is an affine open covering of Y_i. If there exist a family of equidimensional G_m-action $\sigma_i : G_m \times X_i \to X_i$ which induce the standard multiplication $\tau \mapsto t\tau$ for a variable on a general fiber τ, then the actions $\{\sigma_i\}_{i=1}^r$ patch together and define an equidimensional G_m-action $\sigma : G_m \times X \to X$. By Hartog's theorem, this G_m-action is also extended over the fibers lying over the singular points of Sing(f). In fact, the union of the fibers over the singular points of Sing(f) has codimension greater than one.

So, we assume that Sing(f) is an irreducible smooth curve C. By shrinking Y again to a smaller open set, we may assume that the curve C is principal, i. e., defined by a single equation $p = 0$. If the singular fibers over the points of C are tubes, the existence of a G_m-action follows from Lemma 3.3.19. Suppose that the fibers over C are crosses. Then $f^{-1}(C) = E_1 \cup E_2$ with irreducible component E_1, E_2. Then $X \backslash E_1$ and $X \backslash E_2$ are affine and the fibration f restricted to $X \setminus E_1$ and $X \setminus E_2$ have tubes over the curve C. Hence Lemma 3.3.19 implies that there exist G_m-actions on $X \setminus E_1$ and $X \setminus E_2$ and that they coincides on the general fibers of f. Hence they patch together and define a G_m-action on X.

The last assertion follows from Lemma 3.3.13. □

For examples with tubes or crosses as the singular fibers, we refer to Example 3.3.18 for tubes and Lemmas 3.3.23 and 3.3.26 below for crosses. In view of Lemma 3.3.2, we have a satisfactory description on singular fibers of $\mathbb{A}^1_*$-fibrations on normal affine surfaces. If Z is a normal affine surface with an $\mathbb{A}^1_*$-fibration $\rho : Z \to C$, then $f = \rho \times C' : Z \times C' \to C \times C'$ gives an $\mathbb{A}^1_*$-fibration on a normal affine threefold

$Z \times C'$, where C' is a smooth affine curve. Hence the same singular fibers as in the surface case appear in the product threefold case. But we can say much less in general. Let $f : X \to Y$ be an $\mathbb{A}^1_*$-fibration, where X is a smooth affine threefold and Y is a smooth affine surface. Let F_Q be a singular fiber of f lying over a point $Q \in Y$. Let C be a smooth curve on Y through Q and let Z be the normalization of $X \times_Y C$. Then the induced morphism $f_C : Z \to C$ is an $\mathbb{A}^1_*$-fibration. Hence the fiber F_Q has a finite covering $\widetilde{F}_Q \to F_Q$, where $\widetilde{F}_Q$ is a fiber of f_C and hence has the form as described in Lemma 3.3.2. We do not know exactly what the singular fiber F_Q itself looks like.

Concerning the coexistence of tubes and crosses in the quotient morphism $q : X \to Y$ by a G_m-action, we have the following result.

Lemma 3.3.22. *Let X be a homology 3-space with an effective, equidimensional G_m-action and let $q : X \to Y$ be the quotient morphism. Suppose that q is an $\mathbb{A}^1_*$-fibration, Y is smooth and* $\dim X^{G_m} = 1$. *Then there are no tubes as singular fibers of q.*

Proof. By the assertion (3) of Theorem 3.3.6, $X^{G_m} \cong \mathbb{A}^1$ and $q|_{X^{G_m}} : X^{G_m} \to Y$ is a closed embedding. By Lemma 3.3.20, any fiber through a point of X^{G_m} is a cross. Let $F = F_Q$ be a tube of multiplicity $m > 1$. Let p be a prime factor of m and let $\Gamma = \mathbb{Z}/p\mathbb{Z}$. Consider the induced Γ-action on X as Γ is a subgroup of G_m. Then F is contained in the Γ-fixed point locus X^Γ (see the proof of Lemma 3.3.14). Let P be a point on F. Then the induced tangential representation of Γ on $T_{X,P}$ is written as

$$^{\zeta}(x, y, z) = (\zeta^a x, \zeta^b y, z), \quad 0 \le a < p,\ 1 \le b < p,$$

where ζ is a primitive pth root of unity and z is a coordinate with the tangential direction of the fiber. If $a > 0$, $b > 0$, then the point $Q = q(P)$ is a singular point of Y (see the proof of Lemma 3.3.14). Hence $a = 0$. This implies that the component of X^Γ containing F has dimension two. By the Smith theory, the locus X^Γ is a connected closed set. Hence there is a point P' of X^{G_m} such that the fiber $F'_{Q'}$ through P' is contained in X^Γ, where $Q' = q(P')$. Then we can write $F' = m_1\mathbb{A}^1 + m_2\mathbb{A}^1$ with $p \mid m_i$ for $i = 1, 2$. But this is impossible. In fact, let C be a smooth curve on Y passing through the point Q' and set $Z = X \times_Y C$. Then Z has an induced G_m-action. Let $\widetilde{Z}$ be the normalization of $Z \times_C \widetilde{C}$, where $\widetilde{C} \to C$ is a finite covering of order p totally ramifying over the point Q'. Then there is an induced G_m-action on $\widetilde{Z}$ such that the induced morphism $\widetilde{q} : \widetilde{Z} \to \widetilde{C}$ is the quotient morphism. The fiber $\widetilde{q}^{-1}(\widetilde{Q}')$ with $\widetilde{Q}'$ a point of $\widetilde{C}$ lying over Q' consists of p-copies of the cross $\mathbb{A}^1 + \mathbb{A}^1$ when taken with reduced structure. But this is impossible. □

3.3.2 Contractible 3-spaces with $\mathbb{A}^1_*$-fibrations

Finally, we discuss Questions 3.3.7 and 3.3.12. In fact, we will prove a more general result.

Lemma 3.3.23. *Let Y be a homology plane of log-Kodaira dimension $-\infty$ or 1,*[9] *and let C be a curve on Y isomorphic to $\mathbb{A}^1$. Let $V = Y \times \mathbb{A}^1$ with a G_m-action induced from the standard action on the $\mathbb{A}^1$-factor, i. e., ${}^t(Q,x) = (Q, tx)$, where $Q \in Y$ and x is a coordinate of $\mathbb{A}^1$. Let $\sigma : W \to V$ be the blow-up of V with center $C \times (0)$ which is identified with C. Let $X = W \setminus \sigma'(Y \times (0))$, where $\sigma'(Y \times (0))$ is the proper transform of $Y \times (0)$. Then the following assertions hold:*

1. *X is a homology 3-space with an induced G_m-action such that the quotient morphism $q : X \to X//G_m$ has the quotient space $X//G_m \cong Y$. If Y is contractible, the threefold X is also contractible.*
2. *The singular locus* $\mathrm{Sing}(q)$ *is the curve C. For every point Q of C, the fiber $q^{-1}(Q)$ consists of two affine lines meeting in one point. Thus we have the situation treated in Theorem* 3.3.6*(3).*
3. *If $\overline{\kappa}(Y) = 1$, then $\overline{\kappa}(X) = 1$. Hence X cannot be written as $X \cong Z \times X^{G_m}$, i. e., the answer to Question* 3.3.7 *is negative.*
4. *If $\overline{\kappa}(Y) = -\infty$, i. e., $Y \cong \mathbb{A}^2$, then $X \cong Z \times X^{G_m}$ with Z a smooth affine surface with a G_m-action.*

Proof. (1) Set $D = Y \times (0)$, $L = C \times \mathbb{A}^1$, $L' = \sigma'(L)$ and $E' = \sigma^{-1}(C) \setminus \sigma'(D)$. Then $\sigma : (X, E') \to (V, D)$ is a divisorial modification with $\sigma(E') = C$ (see [104]). Write $Y = \operatorname{Spec} B$. Then $B[x]$ is the coordinate ring of V. Since B is factorial by Lemma 3.2.1, the curve C is defined by an element ξ of B. The hypersurface $D = Y \times (0)$ in V is defined by $x = 0$. Let I be the ideal of $B[x]$ generated by ξ and x. Then X has the coordinate ring $\Sigma_{I,x}(B[x])$ which is the affine modification of $B[x]$ along (x) with center I. Clearly, X is a smooth affine threefold. By [104, Proposition 3.1 and Theorem 3.1], it follows that X is a homology 3-space and is contractible provided such is Y. Let $q_0 : V \to Y$ be the projection to Y, which is in fact the quotient morphism by the G_m-action such that $V^{G_m} = D$. By the above process, the G_m-action is inherited on X and the quotient morphism $q : X \to X//G_m$ is induced by q_0.

(2) Meanwhile, in passing from V to X, the fiber over a point $Q \in Y \setminus C$ loses the point $(Q, 0)$ in D and becomes isomorphic to $\mathbb{A}^1_*$. The fiber $q^{-1}(Q)$ for $Q \in C$ is a cross $\mathbb{A}^1 + \mathbb{A}^1$ with two $\mathbb{A}^1$ meeting in the point $L' \cap E'$. Hence $E' \cap L' = X^{G_m}$ and $\mathrm{Sing}(q) = C$.

(3) Suppose that $\overline{\kappa}(Y) = 1$. In fact, there is a unique affine line lying on Y (see [75]). Since the general fibers of q are isomorphic to $\mathbb{A}^1_*$, we have an inequality $\overline{\kappa}(X) \geq \overline{\kappa}(Y) + \overline{\kappa}(F)$ by Lemma 1.2.1, where F is a general fiber of q. Hence $\overline{\kappa}(X) \geq 1$. Furthermore, Y itself has an $\mathbb{A}^1_*$-fibration $\pi : Y \to T$ such that C is a fiber of π and $\pi^{-1}(U) \cong U \times \mathbb{A}^1_*$, where U is an open set of T contained in $\mathbb{A}^1_{2*}$. In fact, since Y is factorial, the base curve T is an affine smooth rational curve and hence $T \subseteq \mathbb{A}^1$. By the above construction, $q^{-1}(\pi^{-1}(U)) \cong \pi^{-1}(U) \times \mathbb{A}^1_* \cong U \times \mathbb{A}^1_* \times \mathbb{A}^1_*$. Hence $\overline{\kappa}(X) \leq \overline{\kappa}(U \times \mathbb{A}^1_* \times \mathbb{A}^1_*) = 1$. This

9 There are no homology planes of log-Kodaira dimension zero (see [159, Theorems 4.6.1 and 4.6.5]).

implies that $\overline{\kappa}(X) = 1$. If $X \cong Z \times \mathbb{A}^1$ as inquired in Question 3.3.7, then it would follow that $\overline{\kappa}(X) = -\infty$. But this is not the case.

(4) Suppose that $\overline{\kappa}(Y) = -\infty$. Then $Y \cong \mathbb{A}^2$ by [159, Chapter 3, Theorem 4.3.1], and the affine line C is chosen to be a coordinate line by AMS theorem. Namely, there exists a system of coordinates (ξ, η) of Y such that C is defined by $\xi = 0$. The affine modification $\Sigma_{I,x}(B[x])$ is equal to $\mathbb{C}[x, \xi/x, \eta]$. Set $y = \xi/x$ and $R = \mathbb{C}[x,y]$. Then the induced G_m-action on $\Sigma_{I,x}(B[x])$ is given by ${}^t(x,y,\eta) = (tx, t^{-1}y, \eta)$. Hence the threefold X is isomorphic to $Z \times \mathbb{A}^1$, where $Z = \operatorname{Spec} R \cong \mathbb{A}^2$ and $\mathbb{A}^1 = \operatorname{Spec} \mathbb{C}[\eta]$. So, the answer to Question 3.3.7 is affirmative. □

Remark 3.3.24. For a homology 3-space X constructed in the previous lemma, the quotient morphism $q : X \to Y$ has crosses $\mathbb{A}^1 + \mathbb{A}^1$ with each multiplicity one as the singular fibers over $\operatorname{Sing}(q) \cong \mathbb{A}^1$. The locus of intersection points of crosses is the fixed point locus $\Gamma = X^{G_m}$. There are two embedded affine planes Z_1, Z_2 meeting transversally along Γ. Since X is a homology threefold, Z_1, Z_2 are defined by $f_1 = 0, f_2 = 0$. Hence Γ is defined by the ideal $I_1 = (f_1, f_2)$ of $A_1 = \Gamma(X, \mathcal{O}_X)$. The affine modification $A_2 = \Sigma_{I_1, f_1}(A)$ or $A_2' = \Sigma_{I_1, f_2}(A)$ gives rise to a smooth affine threefold $Y_2 = \operatorname{Spec} A_2$ or $Y_2' = \operatorname{Spec} A_2'$ with an equidimensional G_m-action, which gives the quotient morphism $q_2 : X_2 \to Y$ or $q_2' : X_2' \to Y$. Both X_2 and X_2' are homology 3-spaces (resp., contractible 3-spaces) if X is a homology 3-space (resp., contractible 3-space). A difference between X_2 (or X_2') and X is that the crosses have multiplicities. Namely, X_2 (resp., X_2') has crosses $2\mathbb{A}^1 + \mathbb{A}^1$ (resp., $\mathbb{A}^1 + 2\mathbb{A}^1$). We can repeat this process to produce homology 3-spaces or contractible 3-spaces which have crosses with higher multiplicities. If a cross is written as $m\mathbb{A}^1 + n\mathbb{A}^1$, then we must have $\gcd(m,n) = 1$ (see the argument in the last part of the proof of Lemma 3.3.22).

Another remark to Lemma 3.3.23 is the following

Remark 3.3.25. Take a $\mathbb{Q}$-homology plane Y instead of a homology plane in the construction of X in Lemma 3.3.23. Then we can consider such $\mathbb{Q}$-homology planes with log-Kodaira dimension $-\infty, 0$ and 1. We take an embedded line C in Y and blow up the center $C \times (0)$ in $V = Y \times \mathbb{A}^1$. By the same construction, we obtain a smooth affine threefold X with an equidimensional G_m-action. The threefold X has the same homology groups as Y. Hence X is a $\mathbb{Q}$-homology 3-space, which is not of the product type $Z \times \mathbb{A}^1$ with a $\mathbb{Q}$-homology plane Z provided $\overline{\kappa}(Y) \neq -\infty$. For the embedded lines in the case of log-Kodaira dimension 0, see [75].

In Lemma 3.3.23, the center of blow-up is $C \times (0)$, where C is an embedded line in a homology plane, and the resulting homology 3-space has log-Kodaira dimension at most one. We can apply a similar process used in Lemma 3.3.23 with a point as the center to obtain the first assertion of the following lemma. This was first constructed in [126, Example 3.7]. The following lemma depends on [104] for the statement and proof.

Lemma 3.3.26. *The following assertions hold:*

(1) *Let Y be a homology plane and let Q_0 be a point of Y. Let $X_0 = Y \times \mathbb{A}^1$ with G_m acting trivially on Y and on $\mathbb{A}^1$ with weight -1. Let $P_0 = (Q_0, 0)$. Blow up the point P_0 and remove the proper transform of $Y \times (0)$ to obtain a smooth affine threefold X. Then X is a homology 3-space with a G_m-action such that the quotient morphism $q : X \to Y$ is induced by the first projection $p_1 : X_0 \to Y$, the fixed point locus X^{G_m} consists of the unique point which is the intersection point of the proper transform of $Q_0 \times \mathbb{A}^1$ with the exceptional surface of the blow-up (whence the G_m-action on X is* hyperbolic*) and $\overline{\kappa}(X) = \overline{\kappa}(Y)$. If Y is contractible, so is X.*

(2) *Let X be a homology 3-space with a hyperbolic G_m-action. Let P_0 be the unique fixed point and let Z_0 be the two-dimensional fiber component of the quotient morphism $q : X \to Y$, where $P_0 \in Z_0$ and $Y = X//G_m$ (cf. Theorem 3.3.6). Let X' be the divisorial modification $\Sigma_{Q_0, Z_0}(X)$ which is the blow-up of X at the center P_0 with the proper transform of Z_0 deleted. Then X' is a homology 3-space. If X is contractible, then so is X'.*

The following example provided a key step in the affirmative solution by Koras–Russell [126, 127] of the *linearization problem* of G_m-actions on the affine 3-space $\mathbb{A}^3$.

Example 3.3.27. Let X be the *Koras–Russell threefold* $x + x^2y + z^2 + t^3 = 0$. Then a hyperbolic G_m-action on X is given by

$$^{\lambda}(x, y, z, t) = (\lambda^6 x, \lambda^{-6} y, \lambda^3 z, \lambda^2 t), \quad \lambda \in \mathbb{C}^*.$$

The fixed point P_0 is $(0, 0, 0, 0)$, and the two-dimensional fiber component Z_0 is defined by $y = 0$. Let A be the coordinate ring of X and let $I = (x, y, z, t)$ which is the maximal ideal of P_0. The affine modification $A' = \Sigma_{I,y}(A)$, which is the coordinate ring of X', is given by

$$A' = \mathbb{C}[x', y, z', t'], \quad x' = \frac{x}{y}, \; z' = \frac{z}{y}, \; t' = \frac{t}{y}.$$

Hence X' is a hypersurface $x' + x'^2 y^2 + z'^2 y + t'^3 y^2 = 0$ and the hyperbolic G_m-action is given by

$$^{\lambda}(x', y, z', t') = (\lambda^{12} x', \lambda^{-6} y, \lambda^9 z', \lambda^8 t'), \quad \lambda \in \mathbb{C}^*.$$

We can further repeat the affine modifications of the same kind to X', X'', etc.

The above Koras–Russell threefold and its affine modifications are examples of smooth contractible 3-space X with a hyperbolic G_m-action such that the quotient Y is isomorphic to the quotient of the tangent space T_{P_0} at the unique fixed point P_0 of X by the induced tangential representation. In [126] a description of all such threefolds is given.

In [101] the Makar-Limanov invariants ML(X) of such threefolds X are computed. To apply this result, the equation for X' has to be brought into a standard form that exhibits X' as a cyclic cover of $\mathbb{A}^3$. To this end, let $A'' = A'^{\omega} = \mathbb{C}[x', y, \zeta, t']$ with a square root ω of the unity and $\zeta = z'^2$. Put

$$x'' = -(x'^2 y + \zeta + t'^3 y).$$

Then $x' = x'' y$ and we see that $A'' = \mathbb{C}[x'', y, t']$ is a polynomial ring in three variables and

$$x'' + x''^2 y^3 + \zeta + t'^3 y = 0.$$

Hence it follows that $A' = \mathbb{C}[x'', y, z', t']$ with a defining equation

$$x'' + x''^2 y^3 + z'^2 + t'^3 y = 0.$$

It now follows from [101, Theorem 8.4] that ML(X) $= \mathbb{C}[x]$ and ML(X') $= A'$. See also Lemma 3.4.4 and Theorem 3.4.13. □

In order to show that a fixed point exists under a G_m-action on a $\mathbb{Q}$-homology 3-space X, we used the Smith theory and its variant. The following result without using the Smith theory is of some interest.

Lemma 3.3.28. *Let X be a smooth affine variety with an effective G_m-action. Suppose that there are no fixed points. Then the Euler number $e(X)$ of X is zero.*

Proof. Let $q : X \to Y$ be the quotient morphism. Since there are no fixed points, every fiber is isomorphic to $\mathbb{A}^1_*$ when taken with reduced structure. The general fibers of q are reduced $\mathbb{A}^1_*$ and special fibers are multiple $\mathbb{A}^1_*$. We work with the complex analytic topology. Considering the isotropy groups of the fibers, there exists a descending chain of closed subsets of Y

$$F_0 \supset F_1 \supset \cdots \supset F_i \supset F_{i+1} \supset \cdots$$

such that $F_0 = Y$ and the isotropy group of the fiber over a point of $F_i - F_{i+1}$ is constant, say G_i. Then $F_i - F_{i+1}$ is covered by open sets $\{U_{i\lambda}\}_{\lambda \in \Lambda_i}$ such that $q^{-1}(U_{i\lambda})_{\mathrm{red}} \cong G_m \times_{G_i} V_{i\lambda}$, where $V_{i\lambda}$ is a suitable slice (cf. [135]). Hence $q^{-1}(F_i)_{\mathrm{red}}$ is a $\mathbb{C}^*$-bundle over the open set $F_i - F_{i+1}$. This implies that the Euler number $e(X)$ is zero. □

Lemma 3.3.28 implies that any nontrivial G_m-action on a smooth affine variety X has a fixed point if $e(X) \neq 0$. Furthermore, considering the induced tangential representation at a fixed point, we know that the fixed point locus X^{G_m} is smooth. However, we do not know if the fixed point locus X^{G_m} is connected. The following example shows that the connectedness fails in general.

Example 3.3.29. Let $Y = \mathbb{A}^2$ and let $X_0 = Y \times \mathbb{A}^1$ which has a standard G_m-action on the factor $\mathbb{A}^1$ with the point (0) as a fixed point. Choose two parallel lines ℓ_1, ℓ_2 on Y, and let $Z_i = \ell_i \times \mathbb{A}^1$ for $i = 1, 2$. Let $\sigma : W \to X_0$ be the blow-ups with centers $\ell_1 \times (0)$ and $\ell_2 \times (0)$ and let X be W with the proper transform of $Y \times (0)$ removed. Then the restriction of $p_1 \cdot \sigma$ onto X gives a morphism $q : X \to Y$ which is eventually the quotient morphism by the G_m-action on X induced by the one on X_0, where p_1 is the projection $X_0 \to Y$. The singular fibers of q are crosses over $\ell_1 \cup \ell_2$. The locus of intersection points of crosses is the fixed point locus X^{G_m}. Hence X^{G_m} is not connected. The Euler number $e(X)$ of the threefold X is two.

In Theorem 3.3.6, we observed a G_m-action on a $\mathbb{Q}$-homology 3-space X such that the quotient morphism $q : X \to Y$ has relative dimension one. The following result deals with the case of q having relative dimension two.

Proposition 3.3.30. *Let X be a $\mathbb{Q}$-homology 3-space with a G_m-action. Suppose that the quotient morphism has equidimension two. Let $q : X \to C$ be the quotient morphism, where C is a smooth affine curve. Then each fiber is isomorphic to $\mathbb{A}^2$, and C is isomorphic to $\mathbb{A}^1$. Hence X is isomorphic to $\mathbb{A}^3$.*

Proof. Let F be a fiber of q. Since $\dim F = 2$, there is a unique fixed point P such that the closure of every orbit passes through P. The locus Γ of points P is the fixed point locus X^{G_m} and $q|_\Gamma : \Gamma \to C$ is a bijection. Hence it is an isomorphism. Since X is smooth and the local intersection multiplicity $i(F, \Gamma; P) = 1$, it follows that F is smooth near P. Hence F itself is smooth. Then it is easy to show that F is isomorphic to $\mathbb{A}^2$ with an elliptic G_m-action. By Theorem 2.1.16, $q : X \to C$ is an $\mathbb{A}^2$-bundle over C. Since X is then contractible to C, the curve C is $\mathbb{Q}$-acyclic. Hence $C \cong \mathbb{A}^1$ and q is necessarily trivial. □

This result inspires the following question.

Question 3.3.31. Let X be a $\mathbb{Q}$-homology n-space with a G_m-action, where $n \geq 2$. Suppose that the quotient morphism has equidimension $n-1$. Is X isomorphic to the affine space $\mathbb{A}^n$?

In fact, one can show that each fiber of the quotient morphism $q : X \to C$ is isomorphic to $\mathbb{A}^{n-1}$ with the induced elliptic G_m-action.[10] If q is locally trivial in the Zariski topology, C is $\mathbb{Q}$-acyclic. Hence $C \cong \mathbb{A}^1$ and $X \cong \mathbb{A}^n$. If X is a homology n-space, Question 3.3.31 has an affirmative answer by a theorem of Kraft–Shwarz [133].

In [125], the following general result has been proved.

10 In fact, each fiber F of q has a unique G_m-fixed point, say P. Then the G_m-action at (X, P) is locally analytically equivalent to the tangential representation of G_m on $T_{X,P}$ (see the footnote to the proof of Theorem 3.3.6). Hence all the G_m-orbits in F pass through P and P is a smooth point of F. This implies that F is isomorphic to $\mathbb{A}^{n-1}$.

Proposition 3.3.32. *Let X be a homology n-space. Suppose that X is dominated by an affine space and X is endowed with an effective action of $T = G_m^{n-2}$ such that $\dim X^T > 0$. Then X is T-equivariantly isomorphic to the affine space $\mathbb{A}^n$ with a linear action of T on $\mathbb{A}^n$. Hence a contractible 3-space X having a nonhyperbolic G_m-action is G_m-equivariantly isomorphic to A^3 provided X is dominated by an affine space.*

The following result deals with more general cases of the quotient morphism of equidimension one which has been treated in Theorem 3.3.6.

Proposition 3.3.33. *Let X be a smooth affine threefold with a G_m-action. Suppose that the quotient morphism $q : X \to Y$ has equidimension one. Then the following assertions hold:*

(1) *If $\dim X^{G_m} = 2$, then X^{G_m} is isomorphic to Y and hence X^{G_m} is connected. The morphism q defines a line bundle over Y.*
(2) *If $\dim X^{G_m} = 1$ and $e(X) > 0$, then X^{G_m} is smooth and consists of connected components $\Gamma_1, \dots, \Gamma_r$, one of which is isomorphic to $\mathbb{A}^1$. Further, if X^{G_m} is connected, then $e(X) = 1$ and the quotient surface Y is a normal affine surface with an embedded line and at worst cyclic quotient singularities.*

Proof. (1) Since q does not contain a fiber component of dimension two, X^{G_m} lies horizontally to the fibration q. Hence each fiber contains a unique fixed point. This implies that each fiber is isomorphic to $\mathbb{A}^1$. Considering the tangent space $T_{X,P}$ and the induced tangential representation for each $P \in X^{G_m}$, we know that X^{G_m} is smooth and isomorphic to Y. Namely, q is an $\mathbb{A}^1$-fibration with all reduced fibers isomorphic to $\mathbb{A}^1$ and has two cross-sections X^{G_m} and a section at infinity. Hence q is in fact a line bundle.

(2) The morphism q is then an $\mathbb{A}^1_*$-fibration and X^{G_m} is a smooth curve (see the proof of the assertion (3) in Theorem 3.3.6). Let $X^{G_m} = \Gamma_1 \sqcup \cdots \sqcup \Gamma_r$ be the decomposition into connected components. Let $\overline{\Gamma}_i$ be the smooth completion of Γ_i. Let g_i be the genus of $\overline{\Gamma_i}$ and let n_i be the number of points in $\overline{\Gamma}_i \setminus \Gamma_i$. Note that G_m acts on $q^{-1}(Y - q(X^{G_m}))$ without fixed points. Hence, by Lemma 3.3.28, the Euler number of $q^{-1}(X - q(Y^{G_m}))$ is zero. Note that $q^{-1}(q(\Gamma_i))$, taken with reduced structure, is a union of two $\mathbb{A}^1$-bundles over $q(\Gamma_i)$ meeting transversally along the section Γ_i. In fact, since the fiber $q^{-1}(Q)$ over a point $Q \in q(\Gamma_i)$ is a cross with each branch meeting Γ_i transversally in one point, $q^{-1}(q(\Gamma_i))$ consists of two irreducible components $W_i^{(1)}$ and $W_i^{(2)}$, each of which has an $\mathbb{A}^1$-fibration with a cross-section Γ_i. Hence $W_i^{(1)}$ and $W_2^{(2)}$ are $\mathbb{A}^1$-bundles over $q(\Gamma_i)$ meeting transversally along Γ_i. Hence the Euler number of $q^{-1}(q(\Gamma_i))$ is equal to $2 - 2g_i - n_i$. This observation yields a relation

$$0 < e(X) = \sum_{i=1}^{r} (2 - 2g_i - n_i).$$

Since Γ_i is an affine curve, we have $n_i \geq 1$. If none of $\Gamma_1, \dots, \Gamma_r$ is isomorphic to $\mathbb{A}^1$, then the right-hand side of the above equality is less than or equal to zero, which is a contradiction. Hence one of them is isomorphic to $\mathbb{A}^1$. If X^{G_m} is connected, then $X^{G_m} = \Gamma_1 \cong \mathbb{A}^1$. It follows from the footnotes in the proof of Theorem 3.3.6 that q induces a closed embedding of X^{G_m} into Y and that Y is smooth near $q(X^{G_m})$. By Lemma 3.3.13, Y has at worst cyclic quotient singularities in the open set $Y \setminus (q(X^{G_m}))$. □

Remark 3.3.34. Given a smooth affine surface Y_0, we can produce a smooth affine surface Y by a half-point attachment [159] which contains Y_0 as an open set and has an embedded affine line C. By the same procedure as in Lemma 3.3.23, we take a product $Y \times \mathbb{A}^1$ and apply the divisorial modification of $Y \times \mathbb{A}^1$ with center $C \times (0)$ in $Y \times (0)$. The resulting threefold is a smooth affine threefold X with a G_m-action such that the quotient morphism $q : X \to X//G_m$, where $Y \cong X//G_m$ and $X^{G_m} \cong C$. Further, $\overline{\kappa}(X) = \overline{\kappa}(Y)$.

3.4 Algebraic varieties with $\mathbb{A}^n$-fibrations

An affine variety of dimension $n+1$ containing an $\mathbb{A}^n$-cylinder is studied. The existence of $\mathbb{A}^n$-cylinder is almost equivalent to the existence of mutually commuting, independent G_a-actions $\sigma_1, \ldots, \sigma_n$. A typical example of such an affine variety is a hypersurface $x^m y = f(x, z_1, \ldots, z_n)$, and we specialize such a hypersurface in dimension three to define an affine pseudo-3-space.

3.4.1 Varieties of dimension $n + 1$ with $\mathbb{A}^n$-cylinders

An $\mathbb{A}^n$-cylinder in an algebraic variety X of dimension $n + 1$ is a Zariski open set U such that $U \cong \mathbb{A}^n \times C$, where C is an affine curve. If $n = 1$ and X is a smooth affine surface, the existence of $\mathbb{A}^1$-cylinder is equivalent to $\overline{\kappa}(X) = -\infty$. If $n > 1$, the existence of $\mathbb{A}^n$-cylinder cannot be detected by the condition $\overline{\kappa}(X) = -\infty$. But the existence of an $\mathbb{A}^n$-cylinder implies some interesting structure of the variety in the case X is affine and factorial. We shall begin with the following general result.

Lemma 3.4.1. *Let X be an affine algebraic variety of dimension $n+1$. Write $X = \operatorname{Spec} A$. Then the following assertions hold:*

(1) *Let $\sigma_1, \ldots, \sigma_n$ be G_a-actions on X and let $\delta_1, \ldots, \delta_n$ be the corresponding lnds on A. Assume that $\delta_i\delta_j = \delta_j\delta_i$ for all i, j and that we have the proper containments* (we then say that they are *independent*)

$$\operatorname{Ker}\delta_1 \supsetneqq \operatorname{Ker}\delta_1 \cap \operatorname{Ker}\delta_2 \supsetneqq \cdots \supsetneqq \bigcap_{i=1}^{n} \operatorname{Ker}\delta_i \supsetneqq k.$$

Let $B = \bigcap_{i=1}^{n} \operatorname{Ker}\delta_i$. Then B is a normal affine domain of dimension one, and there exists an element $b \in B$ such that $A[b^{-1}] = B[b^{-1}][t_1, \ldots, t_n]$, where $t_i \in (\bigcap_{j\neq i} \operatorname{Ker}\delta_j) \setminus B$ and $t_1, \ldots, t_n$ are algebraically independent over B.

(2) *Assume that A is factorial. Let $U = \mathbb{A}^n \times C$ be an $\mathbb{A}^n$-cylinder in X, where $C = \operatorname{Spec} B'$ is an affine curve. Let $B = B' \cap A$. Then B is an affine domain of dimension one, and there exists G_a-actions $\sigma_1, \dots, \sigma_n$ on X such that $B = \cap_{i=1}^n \operatorname{Ker} \delta_i$ and $U = \operatorname{Spec} A[b^{-1}]$ with $b \in B$, where $\delta_1, \dots, \delta_n$ are the lnds corresponding to $\sigma_1, \dots, \sigma_n$.*

Proof. (1) Let $A_n = \operatorname{Ker} \delta_n$ and $B = \bigcap_{i=1}^n \operatorname{Ker} \delta_i$. Since we have by hypothesis

$$\left(\bigcap_{i=1}^{n-1} \operatorname{Ker} \delta_i\right) \cap \operatorname{Ker} \delta_n = B \subsetneq \bigcap_{i=1}^{n-1} \operatorname{Ker} \delta_i,$$

there exists an element t_n of $\bigcap_{i=1}^{n-1} \operatorname{Ker} \delta_i$ such that $t_n \notin \operatorname{Ker} \delta_n$. We may assume that $\delta_n(t_n) = b_n$ is a nonzero element of $\operatorname{Ker} \delta_n$, whence $b_n \in B$. Since t_n is thus a local slice, we have $A[b_n^{-1}] = A_n[b_n^{-1}][t_n]$. Then $A_n[b_n^{-1}]$ is an affine domain over k since so is $A[b_n^{-1}]$. Now the lnds $\delta_1, \dots, \delta_{n-1}$ restrict on $A_n[b_n^{-1}]$ as mutually commuting lnds δ_i' $(1 \le i \le n-1)$ such that they are independent, where $\operatorname{Ker} \delta_i' = (\operatorname{Ker} \delta_i)[b_n^{-1}] \cap A_n[b_n^{-1}]$ and $\bigcap_{i=1}^{n-1} \operatorname{Ker} \delta_i' = B[b_n^{-1}]$. Hence, by induction on n, there exists an element $b_n' \in B$ such that $A_n[b_n^{-1}][b'^{-1}_n] = B[b_n^{-1}][b'^{-1}_n][t_1, \dots, t_{n-1}]$, where we take $t_1, \dots, t_{n-1}$ as elements of A. Let $b = b_n b_n'$. Then we have

$$\begin{aligned} A[b^{-1}] &= A[b_n^{-1}, b'^{-1}_n] = A[b_n^{-1}][b'^{-1}_n] = A_n[b_n^{-1}][t_n][b'^{-1}_n] \\ &= A_n[b_n^{-1}, b'^{-1}_n][t_n] = B[b_n^{-1}, b'^{-1}_n][t_1, \dots, t_{n-1}][t_n] \\ &= B[b^{-1}][t_1, \dots, t_n]\,. \end{aligned}$$

It is clear that $t_1, \dots, t_n$ are algebraically independent over B.

(2) Since A is factorial, $X \setminus U$ is a closed set of pure codimension one. Hence there exists an element $b \in A$ such that $U = D(a) = \operatorname{Spec} A[b^{-1}]$. Then we can write $A[b^{-1}] = B'[t_1, \dots, t_n]$, where $t_1, \dots, t_n$ are elements of A such that they are algebraically independent over B'. Let $B = A \cap B'$. Let $Q(B')$ be the quotient field of B'. Then $Q(B')$ has transcendence degree one over k. Since we have

$$\begin{aligned} A \cap B' &= A \cap A[a^{-1}] \cap B' = A \cap (B'[t_1, \dots, t_n] \cap B') \\ &= A \cap (B'[t_1, \dots, t_n] \cap Q(B')) = A \cap Q(B'), \end{aligned}$$

the subalgebra B is an affine domain of dimension one over k by Zariski's finiteness theorem (see Theorem 2.2.4). Since $b, b^{-1} \in B'[t_1, \dots, t_n]$, it follows that $b \in B'$. Hence $b \in A \cap B' = B$. Then $B[b^{-1}] \subseteq B'$ because $b^{-1} \in B'$. Furthermore, every element of B' multiplied by a high power of b becomes an element of A. Hence $B' \subseteq B[b^{-1}]$. So, $B' = B[b^{-1}]$. Let $D_i = \partial/\partial t_i$ for $1 \le i \le n$. Then $D_1, \dots, D_n$ are mutually commuting lnds on $A[b^{-1}] = B[b^{-1}][t_1, \dots, t_n]$. Write $A = k[a_1, \dots, a_r]$. Then there exists a positive integer N such that $b^N D_i(a_j) \in A$ for every i and j. Let $\delta_i = b^N D_i$. Then $\delta_i(A) \subseteq A$. Hence it follows that $\delta_1, \dots, \delta_n$ are mutually commuting lnds on A. Since

$$\bigcap_{i=1}^{j} \operatorname{Ker} D_i \supsetneq \bigcap_{i=1}^{j+1} \operatorname{Ker} D_i$$

for $1 \le j < n$ and $(\bigcap_{i=1}^{j} \operatorname{Ker}\delta_i)[b^{-1}] = \bigcap_{i=1}^{j} \operatorname{Ker} D_i$, it follows that $\delta_1,\dots,\delta_n$ are independent. The lnds $\delta_1,\dots,\delta_n$ define mutually commuting G_a-actions $\sigma_1,\dots,\sigma_n$. □

Lemma 3.4.2. *The independence of $\delta_1,\dots,\delta_n$ in Lemma 3.4.1 does not depend on the ordering of $\delta_1,\dots,\delta_n$. Namely, if $\delta_{i_1},\dots,\delta_{i_n}$ is another ordering, then $\delta_{i_1},\dots,\delta_{i_n}$ are independent.*

Proof. Let A be an integral domain over k such that $Q(A)$ is finitely generated over k, and let δ be a nontrivial lnd of A. Let $\widetilde{\delta}$ be the canonical extension of δ as a derivation on $Q(A)$. Then $\operatorname{Ker}\widetilde{\delta} = Q(\operatorname{Ker}\delta)$ and $\operatorname{tr.deg}_k Q(\operatorname{Ker}\delta) = \operatorname{tr.deg}_k Q(R) - 1$ (see [150]). With mutually commuting lnds $\delta_1,\dots,\delta_n$ on A, let $K_i = Q(\operatorname{Ker}\delta_i)$. Then $Q(\bigcap_{j=1}^{r} \operatorname{Ker}\delta_{i_j}) = \bigcap_{j=1}^{r} K_{i_j}$ for $\{i_1,\dots,i_r\} \subseteq \{1,\dots,n\}$. This needs a proof. Since $Q(\bigcap_{j=1}^{r} \operatorname{Ker}\delta_{i_j}) \subseteq \bigcap_{j=1}^{r} K_{i_j}$ is clear, we show that the other inclusion holds. With the notations in Lemma 3.4.1, let $K_0 = Q(B)$. Then $A \otimes_B K_0 = K_0[t_1,\dots,t_n]$ which is a polynomial ring over K_0. After extending $\delta_1,\dots,\delta_n$ to the lnds on $A \otimes_B K_0$ and replacing t_i by t_i/b_i with $b_i = \delta(t_i) \in B$, we may assume that $\delta_i(t_j) = \delta_{ij}$ (Kronecker's delta). Write an element $\xi \in K = Q(A)$ as $\xi = \beta(t_1,\dots,t_n)\alpha(t_1,\dots,t_n)^{-1}$ with $\alpha,\beta \in K_0[t_1,\dots,t_n]$ and $\gcd(\alpha,\beta) = 1$. Then $\xi \in \bigcap_{j=1}^{r} K_{i_j}$ if and only if $\delta_{i_j}(\alpha) = \delta_{i_j}(\beta) = 0$ for $1 \le j \le r$. In fact, $\xi \in K_{i_j}$ implies that $\xi = ba^{-1}$ with $a, b \in \operatorname{Ker}\delta_{i_j}$. Then $b\alpha = a\beta$ in $K_0[t_1,\dots,t_n]$. Since $\gcd(\alpha,\beta) = 1$, α (resp., β) divides a (resp., b). Hence $\alpha,\beta \in \operatorname{Ker}\delta_{i_j}$. So, α and β are polynomials in variables $t_i \in \{t_1,\dots,t_n\} \setminus \{t_{i_1},\dots,t_{i_r}\}$. This implies $\xi \in Q(\bigcap_{j=1}^{r} \operatorname{Ker}\delta_{i_j})$. Furthermore, the proper containments

$$\operatorname{Ker}\delta_1 \supsetneq \cap_{i=1}^{2} \operatorname{Ker}\delta_i \supsetneq \cdots \supsetneq \bigcap_{i=1}^{n} \operatorname{Ker}\delta_i$$

imply that $\operatorname{tr.deg}_k(\bigcap_{j=1}^{r} K_{i_j}) = \operatorname{tr.deg}_k K - r$ for $1 \le r \le n$. From these observations, it is now easy to show that $\delta_{i_1},\dots,\delta_{i_n}$ are independent. □

Let $X = \operatorname{Spec} A$ be a smooth affine variety of dimension n and let δ be an lnd of the affine domain A. As a k-derivation of A, δ is an element of $\operatorname{Der}_k(A,A) \cong \operatorname{Hom}_A(\Omega_{A/k}, A)$. In fact, $\operatorname{Der}_k(A,A)$ is isomorphic to $T_{X,k} := \Gamma(X, \mathcal{T}_{X/k})$, where $\mathcal{T}_{X,k}$ is the tangent bundle of X. We call $T_{X,k}$ is the *tangent space* of X. If X is a hypersurface of $\mathbb{A}^{n+1} = \operatorname{Spec} k[X_1,\dots,X_{n+1}]$ defined by $f = 0$, we have $A = k[X_1,\dots,X_{n+1}]/(f)$. We denote the residue class of X_i by x_i. An element $D = \sum_{i=1}^{n+1} A_i(\partial/\partial X_i)$ of the tangent space $T_{\mathbb{A}^{n+1},k}$ induces an element d of $T_{X,k}$ as defined by the residue class $d(a) = D(A)+(f)$ for $a = A+(f)$ if $D(f) = \sum_{i=1}^{n+1} A_i(\partial f/\partial X_i) \in (f)$, which we denote by $\sum_{i=1}^{n+1} a_i(\partial/\partial x_i)$ where $a_i = A_i + (f)$. An element of $T_{X,k}$ is also called a *vector field* of X. For a point $P \in X$, let $\{t_1,\dots,t_n\}$ be a regular system of parameters of X at P. Then δ is written as $\delta = \sum_{i=1}^{n} a_i(\partial/\partial t_i)$ with $a_i \in \mathcal{O}_{X,P}$, and $\delta_P = \sum_{i=1}^{n} a_i(P)(\partial/\partial t_i)$ is the *direction* of the vector field δ. If σ is the G_a-action associated with an lnd δ and $\sigma(P)$ is the orbit through P, then δ_P is the tangential direction of the orbit curve $\sigma(P) := \sigma(G_a)(P)$ at P.

We shall prove the following algebraic characterization of the affine 3-space.

Theorem 3.4.3. *Let $X = \operatorname{Spec} A$ be a smooth affine variety of dimension three. Then X is isomorphic to $\mathbb{A}^3$ if and only if the following conditions are satisfied:*
(1) *A is factorial and $A^* = k^*$.*
(2) *There exist mutually commuting lnds δ_1, δ_2 of A such that δ_1, δ_2 are independent and $(\delta_1)_P, (\delta_2)_P$ span a two-dimensional subspace of the tangent space $T_{X,P}$ for every point P of X.*

Proof. The "only if" part is clear. In fact, if $X = \mathbb{A}^3$, write $A = k[x,y,z]$ and take $\delta_1 = \partial/\partial y$ and $\delta_2 = \partial/\partial z$. We prove the "if" part. Let $B = \operatorname{Ker}\delta_1 \cap \operatorname{Ker}\delta_2$. Since A is factorial and $A^* = k^*$, B is a factorial affine domain of dimension one with $B^* = k^*$. Hence B is a polynomial ring in one variable. Write $B = k[x]$. Let $T = \operatorname{Spec} B$ and let $f : X \to T$ be the $\mathbb{A}^2$-fibration associated with the inclusion $B \hookrightarrow A$ (see the argument in the proof of Lemma 3.4.1). Since B is factorially closed in A and every fiber component of f has dimension two, every fiber of f is integral, i. e., reduced and irreducible. Let X_0 be a fiber outside the $\mathbb{A}^2$-cylinder $U = \mathbb{A}^2 \times C$, where C is an open set of T.

We claim that X_0 is a smooth surface and isomorphic to $\mathbb{A}^2$. By Problem 11 in Section 3.6, the fiber X_0 has an $\mathbb{A}^1$-fibration $\rho_0 : X_0 \to Z_0$ inheriting the quotient morphism by the G_a-action σ_1 associated with δ_1. Let P be an arbitrary point of X_0. By the condition (2), the vectors $(\delta_1)_P, (\delta_2)_P$ are linearly independent in $T_{X,P}$ and they are contained in $T_{X_0,P}$. Furthermore, the induced G_a-actions σ_1, σ_2 (with σ_2 being associated with δ_2) are along the fiber X_0. Then the P-orbit $\sigma_1(\sigma_2(P)) := \{\sigma_1(g_1)\cdot\sigma_2(g_2)(P) \mid g_1, g_2 \in G_a\}$ has dimension two because the orbit directions at P are transversal. Hence it meets the smooth locus of X_0. Namely, $\sigma_1(g_1)\cdot\sigma_2(g_2)(P) = Q$ is a smooth point of X_0 for some $g_1, g_2 \in G_a$. Then, reversing the actions, P is also a smooth point. Hence X_0 is smooth.

Now the $\mathbb{A}^1$-fibration $\rho_0 : X_0 \to Z_0$ may have a singular fiber, which is then a disjoint union of the affine lines, say $C_1, \dots, C_r$. We assumed that one of σ_1, σ_2, say σ_1, acts along the fibers of ρ_0 and the other, say σ_2, acts transversally along the fibration ρ_0. Note that σ_2 preserves the $\mathbb{A}^1$-fibration ρ_0. In fact, since X_0 is dominated by $\sigma_1(\sigma_2(P)) := \{\sigma_1(g_1)\cdot\sigma_2(g_2)(P) \mid g_1, g_2 \in G_a\}$ which is isomorphic to $\mathbb{A}^2$, the curve Z_0 is isomorphic to $\mathbb{A}^1$. Write $Z_0 = \operatorname{Spec} k[u]$. Hence $\sigma_2(g)$ with $g \in G_a$ sends ρ_0 to the $\mathbb{A}^1$-fibration defined by $\sigma_2(g)^*(u)$. However, since σ_1, σ_2 commute, the element $\sigma_2(g)^*(u)$ is σ_1-invariant. Hence $k[u] = k[\sigma_2(g)^*(u)]$. This implies that σ_2 preserves the $\mathbb{A}^1$-fibration ρ_0. Then the translates $\sigma_2(g_1)(C_1)$ and $\sigma_2(g_2)(C_2)$ for some $g_1, g_2 \in G_a$ coincide with a smooth fiber of ρ_0. Then $\sigma_2(g_2^{-1}\cdot g_1)(C_1) = C_2$. Hence G_a acts transitively via σ_2 on the set $\{C_1, \dots, C_r\}$. This is a contradiction if $r \geq 2$ because G_a is connected. This argument also shows that the fiber consisting of a single component C_1 is reduced. So, $X_0 \cong \mathbb{A}^2$, and every fiber of ρ_0 is isomorphic to $\mathbb{A}^2$. Then X is isomorphic to $\mathbb{A}^3$ by Theorem 2.1.16. □

In the condition (2) of Theorem 3.4.3, the assumption that $(\delta_1)_P$ and $(\delta_2)_P$ span a two-dimensional subspace of $T_{X,P}$ is indispensable to conclude $X \cong \mathbb{A}^3$ as shown in the following result.

Lemma 3.4.4. *Let X be a hypersurface in $\mathbb{A}^4 = \operatorname{Spec} k[x,y,z,t]$ defined by $x^m y = f(x,z,t)$, where $m \geq 1$. Then the following assertions hold:*

(1) *X is factorial if the curve C_0 defined by $f(0,z,t) = 0$ in $\mathbb{A}^2 = \operatorname{Spec} k[z,t]$ is an integral curve.*
(2) *$A^* = k^*$ if $X_0 := \{x = 0\}$ is a nonempty closed set of X, where A is the coordinate ring of X.*
(3) *If $m > 1$, X is smooth if and only if there are no singular points of C_0 which satisfies $f_x(0,z,t) = 0$, where $f_x = \partial f/\partial x$. If $m = 1$, X is smooth if and only if C_0 is smooth.*
(4) *A has two different, mutually commuting and independent lnds δ_1, δ_2 which are defined by*

$$\delta_1(x) = 0,\ \delta_1(t) = 0,\ \delta_1(y) = f_z,\ \delta_1(z) = x^m,$$
$$\delta_2(x) = 0,\ \delta_2(z) = 0,\ \delta_2(y) = f_t,\ \delta_2(t) = x^m.$$

The vector fields associated with these two lnds have the proportional directions on the fiber X_0.

Proof. (1) We consider the fibration $p : X \to T$ defined by $(x,y,z,t) \mapsto x$, where $T = \operatorname{Spec} k[x] \cong \mathbb{A}^1$. Then the fiber $X_0 = p^{-1}(0)$ is isomorphic to the product $C_0 \times \mathbb{A}^1$. The result then follows as a direct application of Nagata's lemma (see Lemma 2.4.22).

(2) The localization $A[x^{-1}]$ of A is isomorphic to a polynomial ring $k[x,x^{-1}][z,t]$ over $k[x,x^{-1}]$. Hence if u is a unit of A, then u is an element of $k[x,x^{-1}]$, and $u = cx^r$ for $c \in k^*$ and $r \in \mathbb{Z}$. If $x \in A^*$, then $X_0 = \emptyset$. So, by the hypothesis $X_0 \neq \emptyset$, $u \in k^*$.

(3) Suppose $m > 1$. By the Jacobian criterion of smoothness, X has a singularity if and only if the following equations have common roots

$$f(0,z,t) = f_x(0,z,t) = f_z(0,z,t) = f_t(0,z,t) = 0.$$

This is equivalent to saying that the curve C_0 has a singular point satisfying $f_x(0, z,t) = 0$. Suppose $m = 1$. Then a similar argument shows that X has a singularity if and only if the curve C_0 has a singular point.

(4) Note that $\operatorname{Ker}\delta_1 = k[x,t]$ and $\operatorname{Ker}\delta_2 = k[x,z]$. Hence $\operatorname{Ker}\delta_1 \cap \operatorname{Ker}\delta_2 = k[x]$. Thus δ_1, δ_2 are independent. The rest of the assertion follows from a straightforward computation. □

We need the following lemma in the subsequent argument.

Lemma 3.4.5. *Let $X_0 = \operatorname{Spec} \overline{A}$ be an affine algebraic surface. Assume that there exist two independent, mutually commuting lnds on $\overline{A}$. Then X_0 is isomorphic to $\mathbb{A}^2$.*

Proof. Let $\overline{\delta}_1, \overline{\delta}_2$ be two independent lnds on $\overline{A}$ such that $\overline{\delta}_1\overline{\delta}_2 = \overline{\delta}_2\overline{\delta}_1$. Let $\overline{\sigma}_i$ $(i = 1,2)$ be the G_a-action on X_0 associated to $\overline{\delta}_i$. Let $\overline{A}_i = \operatorname{Ker}\overline{\delta}_i$, let $Z_i = \operatorname{Spec}\overline{A}_i$ and $q_i : X_0 \to Z_i$ be the morphism defined by the injection $\overline{A}_i \hookrightarrow \overline{A}$. Then q_i is the quotient morphism by the action $\overline{\sigma}_i$. If $\overline{A}_1 \subseteq \overline{A}_2$, then the G_a-orbits through a given general point of P with respect to $\overline{\sigma}_1$ and $\overline{\sigma}_2$ coincide with each other. This is a contradiction to the assumption that $\overline{\delta}_1$ and $\overline{\delta}_2$ are independent. Hence $\overline{A}_1 \not\subseteq \overline{A}_2$. This implies that $\overline{\delta}_2$ induces a nonzero lnd on $\overline{A}_1$. Since $\dim \overline{A}_1 = 1$, there exists an element $u_2 \in \overline{A}_1$ such that $\overline{\delta}_2(u_2) = 1$ and $\overline{A}_1 = k[u_2]$. Similarly, there exists an element $u_1 \in \overline{A}_2$ such that $\overline{\delta}_1(u_1) = 1$ and $\overline{A}_2 = k[u_1]$. Since u_1 is a slice of $\overline{\delta}_1$, it follows that $\overline{A} = \overline{A}_1[u_1] = k[u_1, u_2]$. Hence $X_0 \cong \mathbb{A}^2$. □

3.4.2 Hypersurfaces $x^m y = f(x, z_1, \ldots, z_{n-1})$

In this subsection, we take the base field to be the complex number field $\mathbb{C}$. We prove the following result.

Theorem 3.4.6. *Let X be a homology n-space and let $p : X \to \mathbb{A}^1 = \operatorname{Spec} \mathbb{C}[t]$ be a faithfully flat morphism such that p restricted to $p^{-1}(\mathbb{A}^1_*)$ is an $\mathbb{A}^{n-1}$-bundle over $\mathbb{A}^1_*$ in the sense of Zariski topology, where $\mathbb{A}^1_* = \mathbb{A}^1 \setminus \{t = 0\}$ and $X_0 := p^{-1}(0)$ is a smooth irreducible variety. Then X_0 is a homology $(n-1)$-space.*

Proof. The long exact sequence of integral cohomology groups with compact support for a pair (X, X_0) looks like

$$\begin{aligned} H_c^{2i}(X) \to H_c^{2i}(X_0) \to H_c^{2i+1}(X, X_0) \\ \to H_c^{2i+1}(X) \to H_c^{2i+1}(X_0) \to H_c^{2i+2}(X, X_0) \end{aligned}$$

where, by the Poincaré duality, $H_c^{2i}(X) \cong H_{2n-2i}(X) = 0$ and $H_c^{2i+1}(X) \cong H_{2n-2i-1}(X) = 0$ if $i \le n-1$. Furthermore, by the Lefschetz duality, we have $H_c^{2i+1}(X, X_0) \cong H_{2n-2i-1}(X \setminus X_0) \cong H_{2n-2i-1}(\mathbb{A}^1_*) = 0$ and $H^{2i+2}(X, X_0) \cong H_{2n-2i-2}(X \setminus X_0) \cong H_{2n-2i-2}(\mathbb{A}^1_*) = 0$ if $i \le n-2$. Here $X \setminus X_0$ is an $\mathbb{A}^{n-1}$-bundle over $\mathbb{A}^1_*$ by the hypothesis, whence $X \setminus X_0$ is contractible to $\mathbb{A}^1_*$. Hence $H_c^{2i}(X_0) \cong H_{2n-2-2i}(X_0) = 0$ and $H_c^{2i+1}(X_0) \cong H_{2n-2i-3}(X_0) = 0$ if $i \le n-2$. This implies that $H_j(X_0) = 0$ if $j \ge 1$. So, X_0 is a homology $(n-1)$-space. □

Remark 3.4.7.

(1) If $n = 3$, the hypothesis in the above theorem that $X \setminus X_0$ is an $\mathbb{A}^2$-bundle over $\mathbb{A}^1_*$ in the sense of Zariski topology can be replaced by the hypothesis that $p^{-1}(t) \cong \mathbb{A}^2$ for $t \ne 0$. Then one concludes by Theorem 2.1.16 that such an $\mathbb{A}^2$-fibration is an $\mathbb{A}^2$-bundle in the sense of Zariski topology.

(2) With the notations and assumptions of the above theorem, we have by Lemma 1.1.6 an exact sequence of the fundamental groups

$$\pi_1(X_t) \to \pi_1(X) \to \pi_1(\mathbb{A}^1) \to (1)$$

where X_t is a general fiber of p. Then $\pi_1(X) = (1)$ and hence X is a contractible n-space.

We consider an affine hypersurface $x^m y = f(x, z_1, \dots, z_{n-1})$ in

$$\mathbb{A}^{n+1} = \operatorname{Spec} \mathbb{C}[x, y, z_1, \dots, z_{n-1}]$$

and apply the above result.

Theorem 3.4.8. *Let X be a smooth affine hypersurface $x^m y = f(x, z_1, \dots, z_{n-1})$, where $m \geq 1$. Let $p : X \to \mathbb{A}^1$ be the projection $(x, y, z_1, \dots, z_{n-1}) \mapsto x$. Assume that the fiber X_0 is smooth. Then the following conditions are equivalent:*
(1) *X is a contractible variety of dimension n.*
(2) *X is a homology n-space.*
(3) *Let Y be the affine hypersurface $f(0, z_1, \dots, z_{n-1}) = 0$ in*

$$\mathbb{A}^{n-1} = \operatorname{Spec} \mathbb{C}[z_1, \dots, z_{n-1}].$$

Then Y is a homology $(n-2)$-space.

If $n = 3$, these three conditions are equivalent to the condition that $X \cong \mathbb{A}^3$.

Proof. The implication (1) ⇒ (2) is clear. We have (2) ⇒ (1) by Remark 3.4.7. By Theorem 3.4.6, the condition (2) implies that X_0 is a homology $(n-1)$-space. Since $X_0 \cong Y \times \mathbb{A}^1$, it follows that Y is a homology $(n-2)$-space. Hence (2) ⇒ (3). Consider the implication (3) ⇒ (2). The condition (3) implies that X_0 is a homology $(n-1)$-space since $X_0 \cong Y \times \mathbb{A}^1$. The long exact sequence in the proof of Theorem 3.4.6 shows that $H_j(X) = 0$ if $j \geq 3$. Writing the exact sequence for $i = n-1$, we obtain a long exact sequence

$$\begin{aligned} H_c^{2n-2}(X, X_0) \to H_c^{2n-2}(X) \to H_c^{2n-2}(X_0) \\ \to H_c^{2n-1}(X, X_0) \to H_c^{2n-1}(X) \to H_c^{2n-1}(X_0), \end{aligned}$$

which, by the Poincaré and Lefschetz dualities, becomes the following exact sequence

$$H_2(\mathbb{A}^1_*) \to H_2(X) \to H_0(X_0) \to H_1(\mathbb{A}^1_*) \to H_1(X) \to 0.$$

Here $H_2(\mathbb{A}^1_*) = 0$ and $H_0(X_0) \xrightarrow{\sim} H_1(\mathbb{A}^1_*)$ because both are isomorphic to $\mathbb{Z}$ and X is simply connected by Lemma 1.1.6. Hence $H_2(X) = H_1(X) = 0$. This shows that (3) ⇒ (2).

As for the last assertion, note that a homology 1-space is necessarily the affine line. Hence $X_0 \cong \mathbb{A}^2$ and $X \cong \mathbb{A}^3$ by Theorem 2.1.16. It is clear that the condition (1) follows if $X \cong \mathbb{A}^3$. □

For $n = 4$, the surface Y in the condition (3) above is a homology plane which is not necessarily the affine plane. For example, let a, d be positive integers such that $1 < a < d$ and $\gcd(a, d) = 1$. Set

$$f_{d,a}(z_1, z_2, z_3) = \frac{(z_1 z_3 + 1)^d - (z_2 z_3 + 1)^a}{z_3}.$$

Let Y be a hypersurface $f_{d,a} = 1$. By [221], the surface Y is a smooth contractible surface with $\bar{\kappa}(Y) = 1$. Let $f(x, z_1, z_2, z_3) = x + f_{d,a}(z_1, z_2, z_3) - 1$. Then X defined by $x^m y = f(x, z_1, z_2, z_3)$ gives such an example.

Problem 3.1. Let X be a smooth affine variety of dimension $n \geq 3$ and let $p : X \to \mathbb{A}^1 = \operatorname{Spec} k[x]$ be a faithfully flat $\mathbb{A}^{n-1}$-fibration. Suppose that $p^{-1}(\mathbb{A}^1_*) \cong \mathbb{A}^1_* \times \mathbb{A}^{n-1}$ and the fiber $X_0 = p^{-1}(0)$ is integral, i. e., irreducible and reduced. What are the additional conditions with which X becomes isomorphic to an affine hypersurface $x^m y = f(x, z_1, \ldots, z_{n-1})$?

We consider the above problem when $n = 3$.

Definition 3.4.9. Let $X = \operatorname{Spec} A$ be an affine threefold with an $\mathbb{A}^2$-fibration $p : X \to \mathbb{A}^1 = \operatorname{Spec} k[x]$. Assume that $p^{-1}(\mathbb{A}^1_*) \cong \mathbb{A}^1_* \times \mathbb{A}^2$ and $X_0 = p^{-1}(0)$ is an integral surface, where $\mathbb{A}^1_* = \mathbb{A}^1 \setminus \{0\}$. Hence we can write $A[x^{-1}] = k[x, x^{-1}, z, t]$. By [155, Lemma 2.3] (see Problem 11 in Section 3.6), X_0 has an $\mathbb{A}^1$-fibration $\rho_0 : X_0 \to C$, where C is an affine integral curve. We call such an affine threefold X *an affine pseudo-3-space*. Clearly, $\rho_0^{-1}(\operatorname{Sing} C) \subseteq \operatorname{Sing} X_0$.

We do not assume that an affine pseudo-3-space X is smooth. If it is singular, we say that X is *singular*. An affine pseudo-3-space X is isomorphic to $\mathbb{A}^3$ if X_0 is smooth. This follows from Theorems 3.4.6 and 2.1.16 since a homology plane with $\bar{\kappa}(X_0) = -\infty$ is isomorphic to $\mathbb{A}^2$. The converse is true if Abhyankar–Sathaye conjecture (see Conjecture 3.6.1) is true for dimension three.

We have the following result.

Lemma 3.4.10. *Let $X = \operatorname{Spec} A$ be an affine pseudo-3-space. Then the following assertions hold:*

(1) *X is factorial.*

(2) *Write $A[x^{-1}] = k[x, x^{-1}, z, t]$, where we may assume that $z, t \in A$ and $z - \alpha \notin xA$ and $t - \beta \notin xA$ for $\alpha, \beta \in k$. Write $X_0 = \operatorname{Spec} \bar{A}$, where $\bar{A} = A/xA$ is an integral domain. Define a k-algebra homomorphism $\varphi : k[z, t] \to \bar{A}$ as the composite of the inclusion $k[z, t] \hookrightarrow A$ and the residue homomorphism $A \to \bar{A}$. If φ is injective, i. e., $\operatorname{Ker} \varphi = 0$, then $A \cong k[x, z, t]$. It is not possible that $\operatorname{Ker} \varphi$ is a maximal ideal of $k[z, t]$.*

(3) *Suppose that $\operatorname{Ker} \varphi$ is a prime ideal of height one. Write $\operatorname{Ker} \varphi = (f_0(z, t))$, where $f_0(z, t)$ is an irreducible polynomial. Let $T = \operatorname{Im} \varphi = k[z, t]/(f_0(z, t))$. Then the morphism $q : X_0 \to D := \operatorname{Spec} T$ induced by the inclusion $T \hookrightarrow \bar{A}$ is factored by the projection $\rho_0 : X_0 \to C$, i. e., $q = \nu \cdot \rho_0$ for a morphism $\nu : C \to D$ (see Definition 3.4.9 for the notations).*

Proof. (1) The assertion holds by Lemma 2.4.22.

(2) If $z - \alpha \in xA$ for $\alpha \in k$, then $z - \alpha = xz_1$ with $z_1 \in A$ and $k[x, x^{-1}, z, t] = k[x, x^{-1}, z_1, t]$. Hence we can replace z, t by z_1, t. After all possible replacements of this type, we may assume that $z - \alpha \notin xA$ and $t - \beta \notin xA$ for $\alpha, \beta \in k$. Now assume

that φ is injective. Let $w \in A$. Since $A[x^{-1}] = k[x, x^{-1}, z, t]$, there exists an integer $r \geq 0$ such that $x^r w = g(x, z, t) \in k[x, z, t]$. We choose r to be the smallest. If $r = 0$ then $w \in k[x, z, t]$. If $r > 0$, then $g(0, \overline{z}, \overline{t}) = 0$, where $\overline{z} = \varphi(z)$ and $\overline{t} = \varphi(t)$. Write $g(x, z, t) = g_s(z, t)x^s + \cdots + g_1(z, t)x + g_0(z, t)$. Since φ is injective, $g(0, \overline{z}, \overline{t})$ is identified with $g_0(z, t)$. Hence $g_0(z, t) = 0$ and the exponent r can be reduced by 1. This contradicts the choice of r. Hence it follows that every element w of R belongs to $k[x, z, t]$. So, $A = k[x, z, t]$. Suppose that $\operatorname{Ker} \varphi$ is a maximal ideal of $k[z, t]$. Then $z - \alpha \in xA$ and $t - \beta \in xA$ for some $\alpha, \beta \in k$. But this possibility is already excluded.

(3) By Lemma 3.4.1, there exist mutually commuting lnds δ_1 and δ_2 of A such that δ_1 and δ_2 are $k[x]$-trivial and induce the nonzero lnds $\overline{\delta}_1$ and $\overline{\delta}_2$ on $\overline{A}$. In fact, we may take $\delta_1 = x^\alpha(\partial/\partial z)$ and $\delta_2 = x^\beta(\partial/\partial t)$ for suitable integers $\alpha, \beta \geq 0$. If $\overline{\delta}_1 = 0$ for example, then $\delta_1(A) \subseteq xA$. Hence we can replace δ_1 by $x^{-1}\delta_1$. Then the lnds $\overline{\delta}_1$ and $\overline{\delta}_2$ define mutually commuting G_a-actions $\overline{\sigma}_1$ and $\overline{\sigma}_2$ on $X_0 = \operatorname{Spec} \overline{A}$. If $\overline{\delta}_1$ and $\overline{\delta}_2$ are independent, Lemma 3.4.5 implies that $X_0 \cong \mathbb{A}^2$ and hence $X \cong \mathbb{A}^3$. In view of the assertion (2) above, we can exclude this case. Hence we may assume that $\operatorname{Ker} \overline{\delta}_1 = \operatorname{Ker} \overline{\delta}_2$. In fact, since $\overline{\delta}_1$ and $\overline{\delta}_2$ are dependent, $\operatorname{Ker} \overline{\delta}_1 = \operatorname{Ker} \overline{\delta}_1 \cap \operatorname{Ker} \overline{\delta}_2$, whence $\operatorname{Ker} \overline{\delta}_1 \subseteq \operatorname{Ker} \overline{\delta}_2$. Similarly, $\operatorname{Ker} \overline{\delta}_2 = \operatorname{Ker} \overline{\delta}_1 \cap \operatorname{Ker} \overline{\delta}_2$ by Lemma 3.4.2, whence $\operatorname{Ker} \overline{\delta}_2 \subseteq \operatorname{Ker} \overline{\delta}_1$. This implies that $\operatorname{Ker} \overline{\delta}_1$ is the coordinate ring S of the curve C and all the G_a-orbits of $\overline{\sigma}_1$ and $\overline{\sigma}_2$ are the fibers of ρ_0. This implies that $\overline{a}_1\overline{\delta}_1 = \overline{a}_2\overline{\delta}_2$ (see Problem 17 of Section 2.10), where $\overline{a}_1, \overline{a}_2$ are nonzero elements of $\overline{A}$. It is clear that $\overline{\delta}_1(\varphi(z)) = 0$ if $\alpha > 0$ because $\delta_1 = x^\alpha(\partial/\partial z)$. Suppose that $\alpha = 0$. Then $\delta_1 = \partial/\partial z$ and z is a slice of the lnd δ_1 of R. Hence $A = B[z]$ with $B = \operatorname{Ker} \delta_1$ and $k[x, t] \subseteq B$. Since A is a factorial domain with $A^* = k^*$, it follows that B is a factorial domain with $B^* = k^*$. Furthermore, δ_2 induces a nontrivial lnd on B. Then B is a polynomial ring in two variables, and A is hence a polynomial ring in three variables. We can exclude this case, and assume that $\alpha > 0$. By a similar argument, we may assume that $\beta > 0$. Hence $\varphi(z), \varphi(t) \in S$ and $\operatorname{Im} \varphi \subseteq S$. This implies that $q : X_0 \to T = \operatorname{Spec} S$ is factored by $\rho_0 : X_0 \to C = \operatorname{Spec} S$. □

The following result describes a concrete construction of affine pseudo-3-spaces. We employ freely the notations in Lemma 3.4.10 and its proof.

Theorem 3.4.11. *Let X be an affine pseudo-3-space. Then X is either isomorphic to the affine 3-space $\mathbb{A}^3$ or obtained from a hypersurface $xy = f_0(z, t)$ by successive affine modifications, where $f_0(z, t)$ is an irreducible polynomial. Furthermore, if the morphism $\nu : C \to D$ is birational, then X is a hypersurface $x^m y = f(x, z, t)$ for some integer $m > 0$ and a polynomial $f(x, z, t)$ of the form*

$$f(x, z, t) = f_{m-1}(z, t)x^{m-1} + \cdots + f_1(z, t)x + f_0(z, t),$$

where m is the number of affine modifications. Conversely, if X is a hypersurface of this form, it is obtained from the hypersurface $xy = f_0(z, t)$ by affine modifications and the polynomials $f_i(z, t)$ $(1 \leq i \leq m - 1)$ are related to the sections of the $\mathbb{A}^1$-fibrations appearing in the course of the construction. In the last case, the morphism ν is an isomorphism, *and the singular locus of the hypersurface is $E \times \mathbb{A}^1$ if $m > 1$ and $\operatorname{Sing} X \cong \operatorname{Sing} D$ if $m = 1$, where $E = \{P = (\alpha, \beta) \mid P \in (\operatorname{Sing} D) \cap \{f_1(\alpha, \beta) = 0\}\}$.*

Proof. We consider the case in the assertion (3) of Lemma 3.4.10. Since $f_0(\overline{z}, \overline{t}) = 0$, there exists $y_0 \in A$ such that $xy_0 = f_0(z,t)$. Let $Y(0)$ be the hypersurface $xy = f_0(z,t)$ in $\mathbb{A}^4 = \operatorname{Spec} k[x,y,z,t]$. Then the homomorphism $k[x,y,z,t] \to A$ which sends (x,y,z,t) to (x, y_0, z, t) defines an $\mathbb{A}^1$-morphism $\theta(0) : X \to Y(0)$, i. e., $\theta(0)$ is a morphism over $\mathbb{A}^1 = \operatorname{Spec} k[x]$. It is clear that $\theta(0)$ is biregular over $\mathbb{A}^1_* = \mathbb{A}^1 \setminus \{0\}$. Let $\theta(0)_0 = \theta(0)|_{X_0} : X_0 \to Y(0)_0$, where $Y(0)_0$ is a trivial $\mathbb{A}^1$-bundle over the curve D. Let $\pi(0)$ be the projection $Y(0)_0 \to D$. Since $q : X_0 \to D$ is factored by ρ_0 by Lemma 3.4.10, (3), it follows that the diagram

$$\begin{array}{ccc} X_0 & \xrightarrow{\theta(0)_0} & Y(0)_0 \\ {\scriptstyle \rho_0}\downarrow & & \downarrow{\scriptstyle \pi(0)} \\ C & \xrightarrow{\nu} & D \end{array}$$

is commutative. If general fibers of ρ_0 are mapped dominantly onto the fibers of $\pi(0)$, then Zariski's main theorem shows that $\theta(0)$ is an open immersion, whence $\theta(0)$ is an isomorphism because both X and $Y(0)$ are affine and normal. Otherwise, every general fiber of ρ_0 collapses to a point on a fiber of $\pi(0)$. Let D_1 be the closure of the images of general fibers of ρ_0 by $\theta(0)_0$. Then D_1 is a multisection of $\pi(0)$, i. e., a curve lying horizontally on the fibers of $\pi(0)$. Let $Y(1)$ be an affine threefold obtained from $Y(0)$ by blowing up the multisection D_1 and throwing the proper transform of the fiber $Y(0)_0$. This is an affine modification of $Y(0)$ with respect to the element x and the defining ideal of D_1 in $Y(0)$ (see [104]); $Y(1)$ is biregular to $Y(0)$ over $\mathbb{A}^1_* = \mathbb{A}^1 \setminus (0)$, and the fiber $Y(1)_0$ over the point (0) has an $\mathbb{A}^1$-fibration $\pi(1)$ over D_1. Since $Y(1)_0$ over the smooth locus of D_1 is smooth, $Y(1)$ has singular points possibly over $\pi(1)^{-1}(\operatorname{Sing} D_1)$ which has codimension ≥ 2 in $Y(1)$. This implies by Serre's criterion of normality [63, Théorème 5.8.6] that $Y(1)$ is normal. Then we have an $\mathbb{A}^1$-morphism $\theta(1) : X \to Y(1)$ which is biregular over $\mathbb{A}^1_*$ and induces a commutative diagram

$$\begin{array}{ccc} X_0 & \xrightarrow{\theta(1)_0} & Y(1)_0 \\ {\scriptstyle \rho_0}\downarrow & & \downarrow{\scriptstyle \pi(1)} \\ C & \xrightarrow{\nu_1} & D_1 \end{array}$$

where $\pi(1)$ is the $\mathbb{A}^1$-fibration on the fiber $Y(1)_0$ and ν is factored as $\nu : C \xrightarrow{\nu_1} D_1 \longrightarrow D$. In fact, $Y(1)_0$ has an $\mathbb{A}^1$-fibration $\pi(1)$ over the curve D_1. We repeat this argument inductively as long as the closure D_i of the images of general fibers of ρ_0 by $\theta(i-1)_0$ is a multisection of the $\mathbb{A}^1$-fibration on $Y(i-1)_0$. By the affine modification of $Y(i-1)$ with respect to the element x and the defining ideal of the curve D_i in $Y(i-1)$, we obtain an affine threefold $Y(i)$ and an $\mathbb{A}^1$-morphism $\theta(i) : X \to Y(i)$ which is biregular over $\mathbb{A}^1_*$ and induces a morphism $\theta(i)_0 : X_0 \to Y(i)_0$ satisfying the commutative diagram as above. The affine threefolds $Y(i)$ thus constructed are normal by the same reason as for $Y(1)$. Eventually the morphism $\theta(i)$ maps general fibers of ρ_0 dominantly to the fibers of $\pi(i)$. Then, by Zariski's main theorem, the morphism $\theta(i)$ becomes an isomorphism and C is isomorphic to D_i. So, we have a description of X as obtained by successive affine modifications of the hypersurface $Y(0)$.

Now we assume that $\nu : C \to D$ is birational. Then D_1 is a cross-section of the trivial $\mathbb{A}^1$-bundle $\pi(0) : Y(0)_0 \to D$. Hence it is defined by an equation $y = -f_1(\overline{z}, \overline{t})$. Then, as an element of A, $y_0 + f_1(z,t) = xy_1$ and the equation $xy_0 = f_0(z,t)$ is written as $x^2y_1 = xf_1(z,t) + f_0(z,t)$. Let $Y(1)$ be the hypersurface defined by the equation $x^2y = xf_1(z,t) + f_0(z,t)$. Then there exists an $\mathbb{A}^1$-morphism $\theta(1) : X \to Y(1)$ such that $\theta(1)$ is biregular over $\mathbb{A}^1_*$ and we have a commutative diagram as above. In fact, $Y(1)$ is the affine modification of $Y(0)$ with respect to x and the ideal $(x, y + f_1(z,t))$. Suppose that $\theta(1)_0$ is not dominant. Then the birational morphism $\nu_1 : C \to D_1$ is factored by D_2 which is the image of $\theta(1)_0$. So, D_2 is a cross-section of the $\mathbb{A}^1$-bundle $\pi_1(1) : Y(1)_0 \to D_1$, and it is defined by $y_1 = -f_2(\overline{z}, \overline{t})$. Hence we find an element $y_2 \in R$ such that $y_1 + f_2(z,t) = xy_2$. Arguing inductively, we obtain a defining equation of X as in the statement which is thus a hypersurface. Now the remaining assertions are easy to verify. □

The following example due to S. M. Bhatwadekar gives an affine pseudo-3-space which is not a hypersurface. Hence the morphism $\nu : C \to D$ is not birational. The fibers of ρ_0 are mapped to points on the space curve C in $Y(0)_0 = \operatorname{Spec} k[y,z,t]$.

Example 3.4.12. Let $R = k[x,y,z,t,u]/(xy - z^2 + t^3, xu - y^2 + zt)$ and let $X = \operatorname{Spec} R$. Let $p : X \to \mathbb{A}^1 = \operatorname{Spec} k[x]$ be the projection. Then X is an affine pseudo-3-space such that, with the notations of Lemma 3.4.10, $C = \operatorname{Spec} k[y,z,t]/(z^2 - t^3, y^2 - zt)$ and $D = \operatorname{Spec} k[z,t]/(z^2 - t^3)$. In fact, $X_0 \cong \mathbb{A}^1 \times C$. Hence the morphism $\nu : C \to D$ is not birational and X is not a hypersurface with an equation of the type given in Theorem 3.4.11. The threefold X is obtained from $Y(0) = \operatorname{Spec} k[x,y,z,t]/(xy - z^2 + t^3)$ by applying an affine modification with respect to the element x and the defining ideal of the curve C in $Y(0)$. The curve C is in fact a space rational curve parametrized by $y = v^5, z = v^6$, and $t = v^4$.

3.4.3 The ML-invariant of affine pseudo-3-spaces

Following Hedén [92] and Gupta [66], we can give a rather transparent geometric proof of the following result. We note that a more general result showing the nontriviality of $\mathrm{ML}(X)$ which can be applied to the present case is obtained in [102, Theorem 9.1].

Theorem 3.4.13. *Let $X = \operatorname{Spec} A$ be an affine pseudo-3-space which is a hypersurface in $\mathbb{A}^4$,*

$$x^m y = f(x,z,t) = f_{m-1}(z,t)x^{m-1} + \cdots + f_1(z,t)x + f_0(z,t), \tag{3.9}$$

where $m \geq 2$ and $f_0(z,t)$ is an irreducible polynomial. Assume that the curve C on $\mathbb{A}^2 = \operatorname{Spec} k[z,t]$ defined by $f_0(z,t) = 0$ is not transformed by an automorphism of $\mathbb{A}^2$ to the affine line $\mathbb{A}^1$ defined by $z = 0$ or a smooth rational curve defined by $g(z)t + h(z) = 0$ with $g(z), h(z) \in k[z]$ and $\gcd(g(z), h(z)) = 1$. Then x is a G_a-invariant element of A with respect to any G_a-action on X. Hence the Makar-Limanov invariant $\mathrm{ML}(X)$ contains $k[x]$.

Proof. The proof is based on a modification of an argument in [92]. For the convenience of the readers, we outline the argument rather than only indicate the changes in the notations and modifications.

(1) Let $\pi : X \to \mathbb{A}^3$ be the projection $(x,y,z,t) \mapsto (x,z,t)$. Then the singular locus of X is $\pi^{-1}(V(f_1) \cap \operatorname{Sing}(C))$ which is a curve unless it is the empty set. Hence X is a normal variety of dimension three. Let v be the normalized valuation of the discrete valuation ring $\mathcal{O}_{X,\pi^{-1}(C)}$. Since $\pi^{-1}(C)$ is defined by $x = 0$, we have $v(x) = 1, v(y) = -m$ and $v(z) = v(t) = 0$. We then define a filtration in A by the opposite value $v^o = -v$. Hence $v^o(x) = -1, v^o(y) = m$ and $v^o(z) = v^o(t) = 0$. Let $A_{\leq n} = \{a \in A \mid v^o(a) \leq n\}$ for every $n \in \mathbb{Z}$. Since every element $a \in A \setminus \{0\}$ is written in the form

$$a = \sum_{i=0}^{\ell} y^i p_i(x,z,t), \quad p_i(x,z,t) \in k[x,z,t], \tag{3.10}$$

where $p_i(x,z,t)$ has x-degree at most $m-1$ if $i > 0$, we can write A as a direct sum $\bigoplus_{\ell=-\infty}^{\infty} A_\ell$ of free $k[z,t]$-modules A_ℓ of rank 1 defined by

$$A_\ell = \begin{cases} k[z,t]x^{|\ell|} & \text{if } \ell \leq 0, \\ k[z,t]x^j y^r & \text{if } \ell = mr - j > 0,\ 0 \leq j < m. \end{cases}$$

In fact, by making use of the relation (3.9), we can rewrite $a \in A$ in the form

$$a = p_0(x,z,t) + \sum_{0 \leq j < m,\ r > 0} p_{jr}(z,t)x^j y^r,$$

where $p_0(x,z,t) \in k[x,z,t]$ and $p_{jr}(z,t) \in k[z,t]$. Then $v^o(p_{jr}(z,t)x^j y^r) = -j + mr$ if $p_{jr}(z,t) \neq 0$, and $v^o(x^j y^r) \geq v^o(x^{j'} y^{r'})$ if and only if either $r > r'$ or $r = r'$ and $j' \geq j$ with the equality holding if and only if $r = r'$ and $j = j'$. Hence follows the above assertion. Then the family $\{A_{\leq n}\}_{n \in \mathbb{Z}}$ satisfies the following properties:

(1) $B := A_{\leq 0} = k[x,z,t]$, $A_{\leq n} \subseteq A_{\leq n'}$ if $n \leq n'$ and $A = \bigcup_{n \in \mathbb{Z}} A_{\leq n}$,

(2) $(A_{\leq n}) \cdot (A_{\leq n'}) \subseteq A_{\leq (n+n')}$.

Thus the family $\{A_{\le n}\}_{n\in\mathbb{Z}}$ defines a filtration in A. We set $R = \mathrm{gr}(A) = \bigoplus_{n\in\mathbb{Z}} R_n$ with $R_n := A_{\le n}/A_{\le(n-1)}$. By the construction, $R_n \ne (0)$ for any $n \in \mathbb{Z}$. The graded ring R is then generated by $\mathrm{gr}(x) \in R_{-1}$, $\mathrm{gr}(y) \in R_m$ and $\mathrm{gr}(z)$, $\mathrm{gr}(t) \in R_0$. Set $W = \operatorname{Spec} R$. Then W is isomorphic to the hypersurface $x^m y = f_0(z,t)$ in $\mathbb{A}^4 = \operatorname{Spec} k[x,y,z,t]$.

(2) We set $B = k[x,z,t]$. It suffices to show that every G_a-action on X preserves the subring B of A. In fact, if so, $\pi|_X : X \to \mathbb{A}^3$ is a G_a-equivariant morphism, and hence $\pi(X)$ is G_a-stable, where $\pi(X) = (\mathbb{A}^1_* \times \mathbb{A}^2) \cup ((0) \times C)$. Since $\mathbb{A}^1_* \times \mathbb{A}^2$ is the unique irreducible component of dimension 3 in $\pi(X)$, it follows that $\mathbb{A}^1_* \times \mathbb{A}^2$ is G_a-stable. This implies that an invertible element x of $k[x,x^{-1},z,t]$ is G_a-invariant.

Let ∂ be the lnd on A which is associated to the given G_a-action. Let ℓ be anew the smallest integer such that $\partial(A_{\le n}) \subseteq A_{\le n+\ell}$ for every $n \in \mathbb{Z}$. The existence of ℓ follows from the fact that $B = A_{\le 0}$ and A are finitely generated over k. If $\partial \ne 0$ then $\ell \ne -\infty$, and ∂ induces a nonzero homogeneous lnd $\delta = \mathrm{gr}(\partial)$ on R of degree ℓ.

(3) The next step is to show that $\ell < 0$. This entails $\partial(B) \subseteq B$ because $B = A_{\le 0}$. First of all, note that $\mathrm{Sing}(W) = (0) \times \mathbb{A}^1 \times \mathrm{Sing}(C)$. Hence W is a normal variety of dimension 3. Further, W has the G_a-action $(\tau, w) \mapsto \tau(w)$ induced by δ and the G_m-action induced by the grading

$$(\lambda, (x,y,z,t)) \mapsto {}^{\lambda}w = (\lambda^{-1}x, \lambda^m y, z, t),$$

if $w = (x,y,z,t)$. The G_m-quotient morphism $p : W \to \mathbb{A}^2 = \operatorname{Spec} k[z,t]$ is given by $(x,y,z,t) \mapsto (z,t)$. It is trivial over $\mathbb{A}^2 \setminus C$. In fact, the morphism

$$(\mathbb{A}^2 \setminus C) \times G_m \xrightarrow{\sim} p^{-1}(\mathbb{A}^2 \setminus C), \quad ((z,t),\lambda) \mapsto (\lambda^{-1}, f_0(z,t)\lambda^m, z, t)$$

is a G_m-equivariant isomorphism, and the inverse is given by

$$p^{-1}(\mathbb{A}^2 \setminus C) \xrightarrow{\sim} (\mathbb{A}^2 \setminus C) \times G_m, \quad (x,y,z,t) \mapsto ((z,t), x^{-1}).$$

The inverse image $p^{-1}(C)$ is the union $F_- \cup F_+$ with $F_- = \{y = 0\}$ and $F_+ = \{x = 0\}$, where

$$F_- = \left\{w \in W \mid \lim_{\lambda\to\infty} {}^{\lambda}w \text{ exists}\right\},$$

$$F_+ = \left\{w \in W \mid \lim_{\lambda\to 0} {}^{\lambda}w \text{ exists}\right\}.$$

Furthermore, the above trivialization $(\mathbb{A}^2 \setminus C) \times G_m \xrightarrow{\sim} p^{-1}(\mathbb{A}^2 \setminus C)$ extends to a trivialization $\mathbb{A}^2 \times G_m \xrightarrow{\sim} W \setminus F_+$. Meanwhile, there is no such trivialization between $\mathbb{A}^2 \times G_m$ and $W \setminus F_-$ because the isotropy subgroup of a point of $F_+ \setminus F_-$ is a cyclic group of order m.

(4) Consider the G_a-action on W induced by δ. Since δ is homogeneous of degree ℓ, we have

$${}^{\lambda}(\varphi_\tau(s)) = \varphi_{\lambda^{\ell}\tau}({}^{\lambda}s), \quad s \in R,\ \lambda \in G_m,\ \tau \in G_a,$$

where $\varphi_\tau : R \to R$ is given by $\varphi_\tau = \sum_{n\geq 0} \frac{1}{n!}\tau^n\delta^n$. In fact, for $s \in R_m$, we have

$$^\lambda(\varphi_\tau(s)) = {}^\lambda\Big(\sum_{n\geq 0} \frac{1}{n!}\delta^n(s)\tau^n\Big) = \sum_{n\geq 0} \frac{1}{n!}\lambda^{m+n\ell}\delta^n(s)\tau^n$$

and

$$\varphi_{\lambda^\ell\tau}({}^\lambda s) = \sum_{n\geq 0} \frac{1}{n!}\delta^n(\lambda^m s)(\lambda^\ell\tau)^n = \sum_{n\geq 0} \frac{1}{n!}\lambda^{m+n\ell}\delta^n(s)\tau^n .$$

We have the following lemma, whose statement and proof are the same as in [92]. But the argument to show that $\ell \neq 0$ is slightly different.

Lemma 3.4.14. *One of the following cases takes place:*
(1) *$\ell < 0$ and F_+ is G_a-stable, but F_- is not.*
(2) *$\ell > 0$ and F_- is G_a-stable, but F_+ is not.*

Proof. Let $R^\delta = \operatorname{Ker}\delta$. Then $R^\delta = \bigoplus_{n\in\mathbb{Z}} R_n^\delta$, where $R_n^\delta = R_n \cap R^\delta$. For $s \in R_i \setminus \{0\}$, there exists some $\nu \geq 0$ such that $\delta^\nu s \in R^\delta_{i+\nu\ell} \setminus \{0\}$. Then the following assertions hold:
(1) If $\ell = 0$, we have $R_n^\delta \neq (0)$ for all n.
(2) If $\ell > 0$, we have $R_n^\delta \neq (0)$ for some $n > 0$.
(3) If $\ell < 0$, we have $R_n^\delta \neq (0)$ for some $n < 0$.

In fact, if $R_n = R_n^\delta$ then we are done because $R_n \neq (0)$. Otherwise, $\delta^\nu(s) \in R_{n+\nu\ell} \setminus \{0\}$ for $s \in R_n$ and some $\nu > 0$. If $\ell = 0$, we are done. If $\ell > 0$, we find $n > 0$ for which $R_n^\delta \neq (0)$. If $\ell < 0$, a similar argument works.

Suppose $\ell > 0$. Then, for $s \in R_n^\delta \setminus \{0\}$ with $n > 0$, we have $s({}^\lambda w) = \lambda^n s(w)$. Since $\lim_{n\to\infty} {}^\lambda w$ exists for $w \in F_-$, we have $s(w) = 0$. So, $F_- \subseteq V(s) := \{w \in W \mid s(w) = 0\}$. Since F_- is an irreducible subvariety of dimension 2 in the G_a-stable hypersurface $V(s)$, F_- is G_a-stable. Similarly, if $\ell < 0$, the set F_+ is G_a-stable. Suppose $\ell = 0$. Then both F_+ and F_- are G_a-stable. Hence $p^{-1}(C) = F_+ \cup F_-$ is G_a-stable as well as so is $W \setminus p^{-1}(C)$. For any nontrivial G_a-orbit $O = G_a(w) \subset W \setminus p^{-1}(C)$, the morphism $f_0(z,t)\cdot p|_O$ has no zeroes. Hence it is a constant $\alpha \in k^*$. Note that $p : W \setminus p^{-1}(C) \to \mathbb{A}^2 \setminus C$ is dominant. Hence we may assume that α is a general element of k^*, i. e., an element of k^* except for a finite set. Then $p|_O$ induces a dominant morphism $\rho : \mathbb{A}^1 \to C_\alpha$, where C_α is the curve $f_0(z,t) = \alpha$ in $\mathbb{A}^2$. Then C_α is isomorphic to $\mathbb{A}^1$ because C_α is smooth for a general $\alpha \in k$, whence $C \cong \mathbb{A}^1$ by the AMS theorem (see Section 1.3.6). This contradicts the beginning hypothesis of the theorem. The proof in the case $\ell = 0$ also shows that F_+ (resp., F_-) is not G_a-stable if $\ell > 0$ (resp., $\ell < 0$). □

(5) By the above lemma, it suffices to show that $F_- = V(y)$ in W is not G_a-stable. Suppose that F_- is G_a-stable. Then $W \setminus F_- = D(y)$, the open set where $y \neq 0$, is G_a-stable. Let O be a G_a-orbit in $W \setminus F_-$. Then the morphism $y : W \setminus F_- \to \mathbb{A}^1_*, w \mapsto y(w)$ restricted on O is a constant morphism because there is no dominant morphism $\mathbb{A}^1 \to \mathbb{A}^1_*$. Hence

any level surface $V_c = V(x^m c - f_0(z,t))$ in W is G_a-stable. In particular, so is $V_1 := V(x^m - f_0(z,t))$. After a change of coordinate x, we may assume that the G_a-action on V_1 is nontrivial. Now we make use of the following result of Maharana [136, Theorem 5.28] for $V = V_1$.

Lemma 3.4.15. *Let V be an affine surface in $\mathbb{A}^3$ defined by $x^m = f_0(z,t)$, where the curve C defined by $f_0(z,t) = 0$ is an irreducible curve. Assume that V has a nontrivial G_a-action. Then the polynomial $f_0(z,t)$ has one of the following forms after a suitable change of coordinates z, t:*
(i) $f_0(z,t) = z$.
(ii) $f_0(z,t) = g(z)t + h(z)$, *where* $g(z,h(z)) \in k[z]$ *and* $\gcd(g(z), h(z)) = 1$.
(iii) $m = 2$ *and* $f_0(z,t) = t^2 + \varphi(z)$, *where* $\varphi(z) \in k[z]$.

The lemma will be proved after the proof of Theorem 3.4.13 is completed. We have excluded the cases (i) and (ii) by the hypothesis. We show below that the case (iii) does not take place under the above set-up in the step (5). In this case, the affine threefold W is defined by the equation $x^2 y = t^2 + \varphi(z)$. By the hypothesis that the level surfaces V_c are G_a-stable, it follows that y is G_a-invariant. Let δ be the lnd corresponding to the given G_a-action on W. Let $K = k(y)(\xi)$ with $\xi^2 = y$. We consider the surface $W \otimes_{k[y]} K$. Let $X = x\xi - t$ and $T = x\xi + t$. Then we have $XT = \varphi(z)$. The lnd δ extends to an lnd of $\overline{K}[X,z,T]/(XT - \varphi(z))$. By [137], either $\delta(X) = 0$ or $\delta(T) = 0$. Suppose that $\delta(X) = 0$ as the other case is treated similarly. Then we have $\delta(T) = \varphi'(z)\rho(X)$ and $\delta(z) = X\rho(X)$ with $\rho(X) \in K[X]$. It follows then that $\delta(t) = \frac{1}{2}\varphi'(z)\rho(X)$, $\delta(x) = \frac{1}{2\xi}\varphi'(z)\rho(X)$, $\delta(z) = X\rho(X)$, and $\delta(y) = 0$. Since $\delta(t)$ and $\varphi'(x)$ are invariant under the involution $\iota : \xi \mapsto -\xi$, $\rho(X)$ is ι-invariant. Then $\delta(x)$ is not ι-invariant. Since the lnd δ is defined over $k[y]$, this is a contradiction. Thus the case (iii) cannot take place.

So, the case $\ell > 0$ is impossible. This completes the proof of Theorem 3.4.13. □

Proof of Lemma 3.4.15. Note that the surface V is normal. Let $q : V \to B$ be an $\mathbb{A}^1$-fibration, where B is a smooth affine curve. This q exists since we can take B as the quotient curve $V/\!/G_a$. Let H be the Galois group of the cyclic covering $V \to \mathbb{A}^2$, $(x,z,t) \mapsto (z,t)$. Then $H \cong \mathbb{Z}/m\mathbb{Z}$ with H acting on V via ${}^{\zeta}(x,z,t) = (\zeta x, z, t)$, where ζ is a primitive mth root of unity. We consider two cases separately.

Suppose that H permutes the fibers of q. Then the q induces an $\mathbb{A}^1$-fibration q_0 on $\mathbb{A}^2$. We analyze how the irreducible curve C is situated with respect to q_0. If the curve C is contained in a fiber of q_0, then C is isomorphic to $\mathbb{A}^1$ and hence defined by $z = 0$ after a change of coordinates z, t by Abhyankar–Moh–Suzuki theorem. Suppose that C is horizontal to the fibration q_0. Since the covering map $V \to \mathbb{A}^2$ is totally ramified over C, the curve C is embedded into V as a closed irreducible curve which we denote by $\widetilde{C}$ in order to avoid confusion. Then $\widetilde{C}$ lies horizontally to the fibration q. If $\widetilde{C}$ meets a general fiber F of q in more than one point, then the fiber F is stable under H and the action of H has at least two fixed points. Since the point at infinity of F is also

H-stable, H fixes at least three points of $\mathbb{P}^1$. So, the action of H is trivial on F. This is a contradiction. Thus, $\widetilde{C}$ is a cross-section of q. This implies that the curve C is a cross-section of the fibration q_0. Assuming that q_0 is given by $(z,t) \mapsto z$, the cross-section C is defined by an equation $g(z)t + h(z) = 0$, where $g(z), h(z) \in k[z]$ with $\gcd(g(z), h(z)) = 1$.

Suppose that H does not permute the fibers of q. Hence V has two independent G_a-actions. Using the fact that V is a hypersurface it is proved by Daigle–Kolhatkar [31] (see also [142] for the relative case) that V is defined by an equation of the form $XT = P(z)$ for some polynomial $P(z) \in k[z]$. By [137], V then has only involution $\iota(X,T,z) = (T,X,z)$ as a finite automorphism which interchanges two G_a-actions on V. Hence $m = 2$. Let $x = \frac{1}{2}(X-T), t = \frac{1}{2}(X+T)$, and $\varphi(z) = -P(z)$. Then $\iota(x) = -x, \iota(t) = t$ and $x^2 = t^2 + \varphi(z)$. □

The significance of affine pseudo-3-space is observed in the following result.

Theorem 3.4.16. *Let δ be a nonzero lnd on a polynomial ring $A := k[x_1, x_2, x_3, x_4]$ and let $B := \operatorname{Ker}\delta$. Assume that x_1 belongs to the plinth ideal of δ, i. e., $x_1 \in \operatorname{Ker}\delta \cap \delta(A)$. Let $X = \operatorname{Spec} B$. Then X is an affine pseudo-3-space.*

Proof. The proof consists of several steps.

(1) We set $x = x_1$ if there is no fear of confusion. The inclusion $k[x] \hookrightarrow B$ defines a morphism $p : X \to \mathbb{A}^1$. By Theorem 2.2.5, the ring B is finitely generated over k since x_1 is in the plinth ideal of δ, hence in B.

(2) We show that p is surjective and $p^{-1}(\mathbb{A}^1_*) \cong \mathbb{A}^1_* \times \mathbb{A}^2$, where $\mathbb{A}^1_* = \mathbb{A}^1 \setminus \{x = 0\}$. Since x is in the plinth ideal of δ, we have $x = \delta(u)$ for $u \in A$. Then $A[x^{-1}] = B[x^{-1}][u]$. Hence the restriction of p to $p^{-1}(\mathbb{A}^1_*)$ is surjective. If $p^{-1}(0) = \emptyset$ then $A = xA$. This implies that x is a unit of A, which is a contradiction because x is a variable in $A \supset B$. Hence p is surjective. For $x = c$ with $c \in k^*$, the fiber $p^{-1}(c)$ is isomorphic to $\mathbb{A}^2$ because $B \otimes_{k[x]} (k[x]/(x-c)) = \operatorname{Ker}\delta_c$ with δ_c induced on $A/(x-c)A = k[x_2, x_3, x_4]$ and $\operatorname{Ker}\delta_c$ is a polynomial ring in two variables by Theorem 3.2.5. Note that the equality $B \otimes_{k[x]} (k[x]/(x-c)) = \operatorname{Ker}\delta_c$ holds if the class of u is a slice of δ_c in $A \otimes_{k[x]} k[x]/(x-c)$. If the class of u is not a slice, the equality does not necessarily hold. By Theorem 2.9.15, the generic fiber of p is also isomorphic to $\mathbb{A}^2$ over $k(x)$. By Theorem 2.1.16, $X \setminus X_0$ is an $\mathbb{A}^2$-bundle over $\mathbb{A}^1_*$, where $X_0 = p^{-1}(0)$. Hence $p^{-1}(\mathbb{A}^1_*) \cong \mathbb{A}^1_* \times \mathbb{A}^2$ by Lemma 2.1.7 as vector bundles over $\mathbb{A}^1_*$ are trivial.

(3) Since δ is an lnd of A, $B = \operatorname{Ker}\delta$ and $x \in B$, the natural homomorphism $B/xB \to A/xA$ is injective. Since $A/xA = k[x_2, x_3, x_4]$ is an integral domain, so is B/xB. Hence $X_0 = \operatorname{Spec} B/xB$ is integral. Furthermore, X_0 has an $\mathbb{A}^1$-fibration by Problem 11 in Section 3.6. Hence X is an affine pseudo-3-space. □

There are two remarks to the proof of Theorem 3.4.16.

Remark 3.4.17.

(1) As in the proof of Lemma 3.4.10, Step (3), we assume that $X \not\cong \mathbb{A}^3$ and construct an $\mathbb{A}^1$-fibration $\rho_0 : X_0 \to C$, where $C = \operatorname{Spec}\operatorname{Ker}\overline{\delta}_1 = \operatorname{Spec}\operatorname{Ker}\overline{\delta}_2$ with the notations

there. The curve C is then a rational curve with one place at infinity. In fact, we may assume that the derivation δ induces a nonzero lnd $\overline{\delta}$ on A/xA. If $\overline{\delta} = 0$ then $\delta(A) \subseteq xA$ and replace δ by $x^{-1}\delta$. By repeating this replacement, we may assume that $\overline{\delta} \neq 0$. Then $B/xB \subseteq \operatorname{Ker}\overline{\delta}$. Since $A/xA = k[x_2, x_3, x_4]$, $\operatorname{Ker}\overline{\delta}$ is a polynomial ring in two variables by Theorem 3.2.5. Hence there exists a dominant morphism $\mathbb{A}^2 \to X_0 \xrightarrow{\rho_0} C$. This implies our assertion.

(2) By [30], there exists an example of a lnd δ on $k[x_1, x_2, x_3, x_4]$ such that x_1 is in the plinth ideal of δ and $R = \operatorname{Ker}\delta$ needs as many generators as N to generate B over k, where N tends to ∞. Hence $X = \operatorname{Spec} B$ is not necessarily a hypersurface as in Theorem 3.4.13.

Example 3.4.18. Let δ be the *Weitzenböck derivation* of $k[x_1, x_2, x_3, x_4]$, i. e., $\delta = \sum_{i=2}^4 x_{i-1}(\partial/\partial x_i)$. Then $x_1 \in \operatorname{pl}(\delta)$, $\operatorname{Ker}\delta$ is finitely generated over k and is the coordinate ring of an affine hypersurface $x^2y = z^2 + t^3$ (see Problem 11 in Section 3.6). This is a singular, contractible, factorial affine threefold and also a singular affine pseudo-3-space. Note that the equation misses the term x in the equation of the Koras–Russell threefold.

3.5 More on G_a-actions on affine threefolds

In Chapters 2 and 3, we considered $\mathbb{A}^1$-fibrations and the quotient morphisms of G_a-actions and observed that they provide us with rich results in the surface and threefold cases. In the present section, we elaborate some of previously treated results and obtain more refined structure theorems on affine threefolds.

3.5.1 Supplements on $\mathbb{P}^1$-fibrations

Let $f : X \to Y$ be a surjective morphism between smooth algebraic varieties such that a general fiber of f is irreducible. Let $F = f^*(Q)$ be a fiber of f. As in the surface case (Section 2.5), we write $F = \sum_{i=1}^r m_i F_i$ and call it the irreducible decomposition of F, where the F_i are irreducible components of F and $m_i = \operatorname{length}\mathcal{O}_{F,\xi_i}$ with the generic point ξ_i of F_i is the *multiplicity* of F_i. The component F_i being reduced is equivalent to the condition that $m_i = 1$. The greatest common divisor $m = \gcd(m_1, \dots, m_r)$ is called the *multiplicity* of F. If $m > 1$ we say that F is a *multiple* fiber of f.

Theorem 3.5.1. *Let $\overline{f} : V \to Y$ be a proper morphism from a smooth algebraic threefold V to a normal affine surface Y such that a general fiber is isomorphic to $\mathbb{P}^1$. Then the set of multiple fibers of $\overline{f}$ is a finite set.*

Proof. For the proof, we may assume that Y is smooth because $\operatorname{Sing}(Y)$ is a finite set. Let C_0 be an irreducible component of the closure of $\operatorname{Sing}(f)$ of codimension one (if it exists) in Y such that the (scheme-theoretic) fiber $F = \overline{f}^*(Q)$ of $\overline{f}$ over a general point Q of C_0 is a singular fiber of multiplicity m. Let $F = \sum_{i=1}^r m_i F_i$ be the irreducible decomposition. We note that the multiplicity $m = \gcd(m_1, \dots, m_r)$ is constant on an open set of C_0. Let H be a general hypersurface section of Y meeting the curve C_0. Namely, we find projective completions $V \hookrightarrow \overline{V}$ and $Y \hookrightarrow \overline{Y}$ such that $\overline{V}$ and $\overline{Y}$ are smooth and the morphism $\overline{f}$ extends to a projective morphism $\tilde{f} : \overline{V} \to \overline{Y}$. Let $\mathcal{L} = \mathcal{O}_{\overline{Y}}(d)$ the sheaf of hypersurface sections of degree $d \gg 0$. Then the linear system $|\tilde{f}^* \mathcal{L}|$ has dimension ≥ 2 and has no base points. Hence, by Bertini's theorem [90, Corollary 10.9], there is a general member of $|\mathcal{O}_{\overline{Y}}(d)|$ whose intersection $\overline{H}$ on $\overline{Y}$ as well as the inverse image $\tilde{f}^{-1}(\overline{H})$ is smooth. In particular, $H = \overline{H} \cap Y$ and $\overline{f}^{-1}(H)$ are smooth. Furthermore, H intersects the curve C_0 in smooth points transversally. Let Q be one of the intersection points. Then the fiber $F = \overline{f}^*(Q)$ is viewed as a fiber of the $\mathbb{P}^1$-fibration $\overline{f}|_{\overline{f}^{-1}(H)} : \overline{f}^{-1}(H) \to H$. The multiplicities of the irreducible components are unchanged. Namely, with F viewed as a singular fiber of the $\mathbb{P}^1$-fibration $\overline{f}|_{\overline{f}^{-1}(H)}$, the multiplicity of each irreducible component F_i is the same m_i. It is now well-known that a singular fiber of a $\mathbb{P}^1$-fibration on a smooth surface over a smooth curve has a reduced component. This implies that $m = 1$. □

The following result is also worth mentioning.

Lemma 3.5.2. *Let a finite group G act algebraically on a simply-connected complete reduced curve Γ. Then Γ/G is simply-connected.*

Proof. Using Van Kampen's theorem, we see easily that each component of Γ is a cuspidal rational curve and the dual graph of Γ is a tree. We will prove the result by induction on $|G|$ and the number of components of Γ. We call a component C_1 of Γ a tip of Γ if C_1 meets the union of the remaining components of Γ in a single point.

First assume that all the components of Γ meet in a single point P. Then every component of Γ is a tip of Γ. For any component C_1 of Γ the union of its translates by G, say Γ_1 is stable under G. It is easy to see that Γ_1/G is an irreducible rational curve. Since any two components of Γ meet only at P, we see that the various components of Γ/G meet only at the image of P in Γ/G. This implies, again by Van Kampen's theorem, that Γ/G is simply-connected.

Now we consider the case when not every component of Γ is a tip of Γ. Since the dual graph of Γ is a tree, there is an irreducible component C_1 of Γ which is a tip of Γ. Then each translate of C_1 by G is also a tip of Γ. Let Γ_1 be the union of translates of C_1. The union of the components of Γ which are not translates of C_1, say Γ_2, is also G-stable and easily seen to be connected. It is also simply-connected, being a subcurve of Γ. By induction, both Γ_1/G and Γ_2/G are simply-connected. They meet in a single point in Γ/G, which is the image of $\Gamma_1 \cap \Gamma_2$. This shows that Γ/G is simply-connected. □

Next we will prove the following closely related result.

Lemma 3.5.3. *Let $f : X \to Y$ be a proper morphism from a smooth threefold X onto a normal surface such that a general fiber of f is $\mathbb{P}^1$. If a fiber F of f (taken with reduced structure) is simply-connected, then the exceptional divisor in any resolution of singularity of Y at $Q := f(F)$ is simply-connected. Conversely, if the exceptional divisor in a resolution of a singular point Q of Y is simply-connected, then the corresponding fiber $f^{-1}(Q)$ is simply-connected.*

Proof. Assume that F_{red} is simply-connected. We can assume that Y is a small Stein neighborhood of Q. Since F_{red} is a strong deformation retract of X and F_{red} is simply-connected as a degenerate fiber of a $\mathbb{P}^1$-fibration, it follows that X is simply-connected. Let $\widetilde{Y} \to Y$ be a resolution of singularities. We can find a sequence of blow-ups with smooth centers of X, say $\widetilde{X} \to X$, such that f extends to a proper morphism $\widetilde{f} : \widetilde{X} \to \widetilde{Y}$. Since $\widetilde{X} \to X$ is a proper birational morphism and X is smooth, we know that $\widetilde{X}$ is also simply-connected. Then we have a surjection $\pi_1(\widetilde{X}) \to \pi_1(\widetilde{Y})$ by Lemma 3.2.4. This implies that $\widetilde{Y}$ is simply-connected. Since the exceptional divisor E is a strong deformation retract of $\widetilde{Y}$, it follows that E is simply-connected.

Assume that the exceptional divisor in the resolution $\widetilde{Y} \to Y$ of a singular point Q of Y is simply-connected, where Y is a small Stein neighborhood of Q. Then $\widetilde{Y}$ is simply-connected. As in the above argument, we consider the proper morphism $\widetilde{f} : \widetilde{X} \to \widetilde{Y}$. By Theorem 3.5.1, there are finitely many multiple fibers. Let $\widetilde{F}_1 = \widetilde{f}^{-1}(Q_1)$ be a singular fiber of $\widetilde{f}$ which is not a multiple fiber. Then $\widetilde{F}_1$ has an irreducible component of multiplicity one. This can be verified by replacing $\widetilde{Y}$ and $\widetilde{X}$ by a general hyperplane section H passing through Q_1 and $\widetilde{f}^{-1}(H)$ respectively and by making use of the corresponding result in the surface case. For a choice of H, refer to the argument in the proof of Theorem 3.5.1. Then we have $\pi_1(\widetilde{X}) \cong \pi_1(\widetilde{Y})$ by Lemma 1.1.6. Since $\pi_1(\widetilde{X}) \cong \pi_1(X)$ as above, we infer that $\pi_1(X) = (1)$. Since $f^{-1}(Q)$ is a strong deformation retract of X, we know that $f^{-1}(Q)$ is simply-connected. □

Lemma 3.5.4. *Let $f : X \to Y$ be a proper morphism from a smooth threefold onto a normal surface Y. Let F be a one-dimensional fiber of f. Then Y has at worst a quotient singularity at the point $f(F)$.*

Proof. Let $Q := f(F)$. Let C be a component of F and let $P \in C$ be a general point. For a general transverse hyperplane section H of X at P, the point P is isolated in the inverse image of Q for the morphism $f|_H$. It follows that the completion of the local ring of H at P is integral over the completion of the local ring of Y at Q. This implies that Y has at worst a quotient singularity at Q. In fact, by Mumford's theorem (see Theorem 1.3.17 and its remark), if a small punctured neighborhood of a normal surface singular point P is simply connected, then P is a smooth point of the surface. This implies by an easy covering space argument that if $\varphi : (\mathbb{C}^2, O) \to (S, Q)$ is a finite surjective complex analytic map of germs with S normal, then the fundamental group G of the germ $S \setminus \{Q\}$ is a finite group and the germ (S, Q) is isomorphic to the germ $(\mathbb{C}^2, O)/G$, where G acts effectively on the germ $(\mathbb{C}^2, O)$. This observation applied to a finite analytic map of the germs $f : (H, P) \to (Y, Q)$ implies that Q is a quotient singular point. See Brieskorn [20, Satz 2.8]. □

Combining Lemmas 3.5.3 and 3.5.4, we obtain the following result.

Theorem 3.5.5. *Let $f : X \to Y$ and F be the same as in the statement of Lemma 3.5.4. Assume that a general fiber of f is $\mathbb{P}^1$. Then the fiber F is simply-connected. Even if F has no one-dimensional component, the conclusion holds if Y is smooth at the point $f(F)$.*

Proof. This follows since the exceptional divisor of a resolution of a quotient singularity of Y at $f(F)$ is a tree of nonsingular rational curves. □

The morphism $f : X \to Y$ as in Lemma 3.5.3 or Theorem 3.5.5 has a general fiber isomorphic to $\mathbb{P}^1$. But it is not necessarily a generically trivial $\mathbb{P}^1$-fibration. For f to be a generically trivial $\mathbb{P}^1$-fibration, the generic fiber of f, i. e., $X \times_Y \operatorname{Spec} k(Y)$, is isomorphic to $\mathbb{P}^1$, where $k(Y)$ is the function field of Y over k (see the assertion (3) of Lemma 2.1.12).

3.5.2 The quotient surface of a smooth affine threefolds by G_a-actions

We are interested in a smooth, factorial, affine threefold X with G_a-actions and its algebraic quotient surface Y by one of the G_a-actions. Main problems here are if Y is smooth or what kind of singularity Y admits provided it is singular. Recall that two G_a-actions on X are *independent* if two respective G_a-orbits passing some point P of X have independent tangential directions (see Subsection 2.4.3). If X admits two independent G_a-actions, we have the following result.

Theorem 3.5.6. *Let X be a smooth affine factorial threefold with two independent G_a-actions, with groups G_1, G_2, say. Assume that $\Gamma(X, \mathcal{O}_X)^* = k^*$. Then $X//G_1$ is isomorphic to either $\mathbb{A}^2$ or an affine hypersurface $x^2 + y^3 + z^5 = 0$. Furthermore, X is simply-connected if and only if $X//G_1 \cong \mathbb{A}^2$. Finally, if the two G_a-actions commute, then $X//G_1 \cong \mathbb{A}^2$.*

Proof. Let $q_1 : X \to Z := X//G_1$ be the quotient morphism. For a general point $p \in Z$, let $C_p = q_1^{-1}(p)$ and Y_p be the closure of $\bigcup_{P \in C_p} G_2 P$. Then $C_p \cong \mathbb{A}^1$ and Y_p is an affine surface equipped with an $\mathbb{A}^1$-fibration $\{G_2 P \mid P \in C_p\}$. In fact, for two orbits $G_2 P$ and $G_2 P'$, either $G_2 P = G_2 P'$ or $G_2 P \cap G_2 P' = \emptyset$. Consider the following two cases (i) and (ii) separately. Write $Y = Y_p$.

(i) Suppose that $q_1|_Y : Y \to Z$ is dominant. Note here that the coordinate ring $\Gamma(Z, \mathcal{O}_Z)$ is factorial as X is factorial. Hence q_1 has no fibers of dimension 2. Let Z° be the smooth part of the normal surface Z. Then $q_1^{-1}(Z \setminus Z^\circ) \cap Y$ is a finite set and lies on a union of finitely many orbits $G_2 P$ with $P \in C_p$. Further, by eliminating finitely many G_2-orbits, we have a dominant morphism from a smooth affine surface with an $\mathbb{A}^1$-fibration to Z°. Hence $\overline{\kappa}(Z^\circ) = -\infty$. If Z^0 has an $\mathbb{A}^1$-fibration, it extends to Z. In this case, $Z \cong \mathbb{A}^2$ by Theorem 1.2.7 since Z is factorial and $\Gamma(Z, \mathcal{O}_Z^*) = \Gamma(Y, \mathcal{O}_Y^*) = k^*$. Otherwise, Z° contains an open set U of the form $(\mathbb{A}^2/\Gamma)^\circ$, where Γ is a finite group, $(\mathbb{A}^2/\Gamma)^\circ$ is the smooth part of $\mathbb{A}^2/\Gamma$ and $Z^\circ \setminus U$ is a disjoint union of curves isomorphic to $\mathbb{A}^1$ which are called *half-point attachments* (see Section 2.6). Hence the curves in $Z^\circ \setminus U$ give rise to the independent classes of $\mathrm{Pic}(Z^\circ)$. Meanwhile, since Z is factorial, it follows that $\mathrm{Pic}(Z^\circ) = 0$. This implies that $Z^\circ = U \cong (\mathbb{A}^2/\Gamma)^\circ$. Then $Z = \mathbb{A}^2/\Gamma$ because $Z \setminus Z^\circ$ is a finite set. Since Z is factorial, Γ must be the binary icosahedral group of $\mathrm{SL}(2, k)$. Hence Z is an affine hypersurface $x^2 + y^3 + z^5 = 0$ (see [159, Chapter 3, Theorem 2.6.1]).

Suppose that X is simply-connected. Let $\rho : \mathbb{A}^2 \to \mathbb{A}^2/\Gamma$ be the quotient morphism by the Γ-action on $\mathbb{A}^2$. Then $\rho^\circ : \mathbb{A}^2_* := \mathbb{A}^2 \setminus \{0\} \to (\mathbb{A}^2/\Gamma)^\circ$ is a universal covering with Galois group Γ. Let $X^\circ := X \setminus q_1^{-1}(\mathrm{Sing}(\mathbb{A}^2/\Gamma)) = q_1^{-1}((\mathbb{A}^2/\Gamma)^\circ)$. Note that $q_1^{-1}(\mathrm{Sing}(\mathbb{A}^2/\Gamma))$ is a closed set of codimension ≥ 2 in X. Hence X° is simply-connected. Then $q_1^\circ := q_1|_{X^\circ} : X^\circ \to Z^\circ$ is factored by $\mathbb{A}^2_*$ as $q_1^\circ : X^0 \xrightarrow{\pi^\circ} \mathbb{A}^2_* \xrightarrow{\rho^\circ} Z^0$. Then a general fiber of q_1 (and hence q_1°) is a disjoint union of as many affine lines as the order of Γ. This is a contradiction. Hence the case $Z \cong \mathbb{A}^2/\Gamma$ does not occur.

(ii) Suppose that $q_1|_Y : Y \to Z$ is not dominant. Then the image of $q_1|_Y$ is a rational curve B_p with one place at infinity which passes through the point p. The assumption implies that for any point $P \in C_p$ the orbit G_2P is mapped surjectively onto B_p. If $B_p \cap B_{p'} \neq \emptyset$ for distinct points p, p' of Z, then $G_2P \cap G_2P' \neq \emptyset$ for $P \in C_p$ and $P' \in C_{p'}$. Since G_2P, G_2P' are G_2-orbits, it follows that $G_2P = G_2P'$ and hence $B_p = B_{p'}$.[11] Thus the family $\{B_p\}_{p \in Z}$ has no base points. This implies that a general member B_p is smooth, hence isomorphic to $\mathbb{A}^1$. So, Z has an $\mathbb{A}^1$-fibration. Since Z is factorial and $\Gamma(Z, \mathcal{O}_Z)^* = k^*$, it follows that $Z \cong \mathbb{A}^2$.

By the above argument, we have shown that Z is isomorphic to either $\mathbb{A}^2$ or $\mathbb{A}^2/\Gamma$. Finally, we prove that if $Z \cong \mathbb{A}^2$ then X is simply-connected. Let S be the closed curve on Z such that for each general point $p \in S$ the fiber $q_1^{-1}(p)$ is not isomorphic to $\mathbb{A}^1$ in the scheme-theoretic sense. Let S_1 be an irreducible component of S and let f be a prime element of $\Gamma(Z, \mathcal{O}_Z) \cong k^{[2]}$ such that $S_1 = V(f)$. Since f is a prime element of $\Gamma(X, \mathcal{O}_X)$, the surface $T_1 := q_1^{-1}(S_1)$ is an irreducible surface. Considering the Stein factorization of $q_1|_{T_1} : T_1 \to S_1$, we know that the general fibers consist of reduced irreducible components. Now let Z° be the open set of Z such that for every point $p \in Z^\circ$ the fiber $q_1^{-1}(p)$ is reduced. Then $Z \setminus Z^\circ$ is a finite set. Let $X^\circ = q_1^{-1}(Z^\circ)$ and $q_1^\circ = q_1|_{X^\circ}$. We apply Nori's exact sequence of the fundamental groups (see Lemma 1.1.6)

$$\pi_1(F) \to \pi_1(X^\circ) \to \pi_1(Z^\circ) \to (1),$$

11 To be more accurate, we have to consider the case where the surfaces Y_p and $Y_{p'}$ contain separately disjoint fiber components of the same singular (degenerate) fiber of q_1 resulting the nonempty intersection of the images B_p and $B_{p'}$ of Y_p and $Y_{p'}$ although $Y_p \cap Y_{p'} = \emptyset$. By Theorem 3.1.8, q_1 has equidimension one since X is factorial. Since one singular fiber contains finitely many irreducible components, this situation does not occur if we take general points p, p' in Z. Namely, if $B_p \cap B_{p'} \neq \emptyset$, then $Y_p \cap Y_{p'} \neq \emptyset$ and hence $G_2P \cap G_2P' \neq \emptyset$ for $P \in C_p$ and $P' \in C_{p'}$.

where $F \cong \mathbb{A}^1$ as a general fiber of q_1°. Since $X \setminus X^\circ$ and $Z \setminus Z^\circ$ have codimension larger than one, we have $\pi_1(X^\circ) = \pi_1(X)$ and $\pi_1(Z^\circ) = \pi_1(Z) = (1)$. Hence $\pi_1(X) = (1)$.

Finally, if the two G_a-actions commute, $Z = X//G_1$ has a G_a-action induced by the G_2-action. Hence $Z \cong \mathbb{A}^2$ by Theorem 1.2.7. □

3.5.3 Singularities of the quotient surface $X//G_a$

Recall that an isolated normal surface singularity (S, P) is called *quasirational* if a desingularization of P has the exceptional locus consisting of a tree of smooth rational curves (see the definition after Remark 3.1.32).

Lemma 3.5.7. *Let X be a smooth affine threefold such that $H_i(X; \mathbb{Z}) = 0$ for $i = 1, 2$. Then X is factorial. If X has a nontrivial G_a-action then $Y := X//G_a$ has only quasirational singular points with unimodular intersection form, and $H_i(Y^\circ; \mathbb{Z}) = 0$ for $i = 1, 2$, where Y° is the smooth part of Y.*

Proof. Let $X \subset V$ be an embedding in a smooth projective threefold such that $D := V - X$ is a divisor with simple normal crossings. Consider the relative cohomology exact sequence of the pair (V, D) with integer coefficients

$$\cdots \to H^4(V, D) \to H^4(V) \to H^4(D) \to H^5(V, D) \to H^5(V) \to \cdots.$$

By the duality and the assumption $H^4(V, D) \cong H_2(X) = (0)$ and $H^5(V, D) \cong H_1(X) = (0)$. Hence we have $H^4(V) \xrightarrow{\sim} H^4(D)$, and it follows that all the topological 4-cycles on V are algebraic and generated freely (over the integers) by the cohomology classes of the irreducible components of D. This means that X is factorial. Assume that X has a nontrivial G_a-action and let $Y := X//G_a$. Then Y is also factorial. Then the quotient morphism $q : X \to Y$ has no two-dimensional components in any fiber of q. Since Y is normal the inverse image of the singular locus of Y in X is at most one-dimensional. This implies that there is a surjective homomorphism $H_1(X) \xrightarrow{\sim} H_1(q^{-1}(Y^\circ)) \twoheadrightarrow H_1(Y^\circ)$. This means that $H_1(Y^\circ) = 0$.

To show that $H_2(Y^\circ) = 0$ we use an argument from [103, Lemma 2.2]. First note that any prime element in the coordinate ring of Y remains a prime element in the coordinate ring of X. If Γ is any irreducible component of the curve in Y over which q has singular fibers, then for a general point $y \in \Gamma$ there is a point $x \in X$ lying over y such that q is smooth at x. This observation is used in the proof of Lemma 2.2 in [103]. Hence $T := \{P \in Y \mid q^{-1}(P) \text{ is a singular fiber}\} \cup \mathrm{Sing}(Y)$ is a finite set of Y. By a result of R. Thom [220], since Y has real dimension 4, any homology 2-class in Y° is the image of the fundamental class of a compact orientable surface S without boundary by a continuous map $S \to Y^\circ$. The argument in [103, Lemma 2.2] shows that there is a lift of this map $S \to X - q^{-1}(T)$. Hence the homomorphism $H_2(X - q^{-1}(T)) \to H_2(Y^\circ)$ is surjective. Since $\dim q^{-1}(T) \le 1$, the long cohomology exact sequence for the pair $(X, q^{-1}(T))$ with compact supports shows that $H_2(X - q^{-1}(T)) \cong H_2(X) = (0)$. Now $H_2(Y^\circ) = 0$.

We can embed Y° in a smooth projective surface W such that $W \setminus Y^\circ$ is a divisor Δ with simple normal crossings which contains the exceptional locus of the singular points of Y. Note that $H_1(W) = 0$ since $H_1(Y^\circ) = 0$. Now the long exact cohomology sequence of the pair (W, Δ) shows that $H_1(\Delta) = 0$. This proves that Y has only quasirational singular points. In order to prove the unimodularity of the intersection form, note that $H^2(W) \xrightarrow{\sim} H^2(\Delta)$ since $H^2(W, \Delta) \cong H_2(Y^\circ) = 0$ and $H^3(W, \Delta) \cong H_1(Y^\circ) = 0$ as shown above. Then the intersection form of Δ is viewed as the Poincaré pairing $H^2(W) \times H_2(W) \to \mathbb{Z}$ which is a perfect pairing. Hence the intersection form (and that of the exceptional locus of the singular points of Y as disjoint components of Δ) is unimodular. □

The next result is a generalization of a result of Kaliman–Saveliev [103, Proposition 2.5] which is stated for a smooth *contractible* affine threefold, i. e., a contractible 3-space.

Theorem 3.5.8. *Let X be a homology 3-space with a nontrivial G_a-action. Let $Y := X//G_a$. Then there is a natural isomorphism $\pi_1(X) \xrightarrow{\sim} \pi_1(Y^\circ)$. Further, either Y is smooth or Y is isomorphic to an affine hypersurface $\{x^2 + y^3 + z^5 = 0\}$.*

Proof. Since X is a homology 3-space, X is factorial by Lemma 3.5.7. Hence Y is also factorial as the quotient surface $X//G_a$. The quotient morphism $q : X \to Y$ is surjective by [103, Lemma 2.1] and [99, Remark 3.3] (also by Bonnet [18] if $X \cong \mathbb{C}^3$), and all the fibers of q are purely one-dimensional since X is factorial. This implies that Y has at most quotient singular points by Lemma 3.1.30. It is proved in [103, Lemma 2.4] that Y is a logarithmic homology plane.[12] In fact, it is proved in Lemma 3.5.7 that $H_1(Y^\circ) = H_2(Y^\circ) = 0$, where Y° is the smooth part of Y. Since $Y \setminus Y^\circ$ is a finite set, it implies that $H_1(Y) = (0)$ and $H_2(Y) = 0$. Again, the factoriality of X implies that at most finitely many fibers of q have multiplicity > 1. Since the inverse image in X of a finite set of points in Y has codimension 2, by Lemma 1.1.6, we obtain an isomorphism $\pi_1(X) \xrightarrow{\sim} \pi_1(Y^\circ)$.

We will now assume that Y is not smooth and show that $Y \cong \{x^2 + y^3 + z^5 = 0\}$. The proof uses some results from the classification theory of incomplete algebraic surfaces (see [159]).

Claim. *The logarithmic Kodaira dimension $\overline{\kappa}(Y^\circ) < 2$.*

If this is not true, then Y has at most one cyclic quotient singular point by [86] since Y is a logarithmic homology plane. Meanwhile, Y has only singular points of

12 A normal algebraic surface Y with at worst quotient singularities is called a *logarithmic surface*. If $H_i(Y; \mathbb{Z}) = 0$ for $i = 1, 2$, it is called *logarithmic homology plane*.

E_8-type by [103, Proposition 2.5(1)], where the proof uses only the condition that $H_1(Y) = 0$. This contradiction proves the claim.

Claim. *The logarithmic Kodaira dimension* $\overline{\kappa}(Y^\circ) \neq 1$.

Assume that $\overline{\kappa}(Y^\circ) = 1$. Then Y° has an $\mathbb{A}^1_*$-fibration, say φ. Suppose that φ extends to a morphism on Y. By the result about singularities of a logarithmic affine surface with a $\mathbb{A}^1_*$-fibration [166, Lemma 2.9], Y cannot have an E_8 singular point. In fact, by the same result [166, Lemma 2.9], the singularities allowed on a logarithmic affine surface with an $\mathbb{A}^1_*$-fibration are cyclic singularities $C_{n,q}$ or the one with the graph (1) of Section 5.3.4 (see the list in [159, pages 54 and 55]). Hence φ has a point of indeterminacy at a singular point of Y. After resolving the indeterminacy we get an $\mathbb{A}^1$-fibration on the resolution of singular points of Y. But since the singular point of type E_8 is a rational double point, it does not affect the logarithmic Kodaira dimension by the resolution.[13] Hence the resolved surface also has $\overline{\kappa} = 1$ and contains an $\mathbb{A}^1$-cylinder. This contradiction proves the second claim.

Claim. *The logarithmic Kodaira dimension* $\overline{\kappa}(Y^\circ) \neq 0$.

This is a stronger form of T. Fujita's result that there is no homology plane with $\overline{\kappa} = 0$. If Y is singular, K. Palka [185] showed that either Y° has an $\mathbb{A}^1_*$-fibration, or Y has a unique cyclic quotient singular point of type A_1 or A_2 (exceptional case). In the first case, we will be done by the same argument as before. Our surface Y does not fall in the second case because Y has a singular point of type E_8.

Now $\overline{\kappa}(Y^\circ) = -\infty$. Then the proof is almost the same as in the step (i) in the proof of Theorem 3.5.6, where Z° is replaced by Y°. Since Y is factorial and $\Gamma(Y, \mathcal{O}_Y^*) = k^*$, we conclude that in this case $Y \cong \{x^2 + y^3 + z^5 = 0\}$. □

Concerning the smoothness of the quotient surface $X//G_a$, the following problem is interesting, where the assumption that X is simply-connected and rational is new.

Problem 3.2. Let X be a smooth, factorial, simply-connected, rational, affine threefold with a nontrivial G_a-action. Is $Y := X//G_a$ then smooth?

---

13 Let $(V, D + \Gamma)$ be a pair of a smooth projective surface and an effective reduced divisor $D + \Gamma$ with simple normal crossings. Assume that $D \cap \Gamma = \emptyset$ and Γ is the exceptional locus of the minimal resolution of a rational double point. Hence each irreducible component of Γ is a smooth rational curve with self-intersection -2, and the intersection form of Γ is negative definite. If $|n(K_V + D + \Gamma)| \neq \emptyset$ for $n > 0$, then $n\Gamma$ is contained in the fixed part. In fact, let Γ_1 be the maximal effective divisor such that Γ_1 is supported by the irreducible components of Γ and Γ_1 is contained in the fixed part of $|n(K_V + D + \Gamma)|$. Let A be a general member of $|n(K_V + D + \Gamma)|$. Then $A - \Gamma_1$ has no component of Γ. Suppose that $n\Gamma > \Gamma_1$, and write $n\Gamma = \Gamma_1 + \Gamma_2$. Then $(A - \Gamma_1)\cdot\Gamma_2 \geq 0$. Meanwhile, $(A - \Gamma_1)\cdot\Gamma_2 = (n(K_V + D + \Gamma) - \Gamma_1)\cdot\Gamma_2 = (n(K_V + D) + \Gamma_2)\cdot\Gamma_2 = \Gamma_2^2 < 0$, which is a contradiction. So, $\Gamma_1 \geq n\Gamma$. This implies that $\overline{\kappa}(V \setminus (D + \Gamma)) = \overline{\kappa}(V \setminus D)$. The case of nonminimal resolution is reduced to the case of minimal resolution.

Proposition 3.5.9. *With the notations and assumptions in Problem 3.2, Y is a factorial, rational, affine surface such that:*
(1) $Y^\circ := Y \setminus \mathrm{Sing}(Y)$ *is simply-connected, and*
(2) *a point $P \in Y$ has quotient singularity at worst of E_8-type, provided $q^{-1}(P) \neq \emptyset$.*

Proof. Since X is factorial, the quotient morphism $q : X \to Y$ does not contain fiber components of codimension one. Hence $q^{-1}(\mathrm{Sing}(Y))$ has codimension larger than one in X. So, $\pi_1(X - q^{-1}(\mathrm{Sing}(Y))) = (1)$. Let $p : Z \to Y^\circ$ be the universal covering. Then the restriction of q onto $X - q^{-1}(\mathrm{Sing}(Y))$ is factored by the mapping p. This implies that p is a finite covering and p is the identity since the general fibers of q are connected. So, $\pi_1(Y^\circ) = (1)$. This proves the assertion (1). If $q^{-1}(P) \neq \emptyset$, then the singularity of Y at P is at most quotient singularity by Lemma 3.1.30. Since $\mathcal{O}_{Y,P}$ is factorial, the singularity is at worst of E_8-type. This proves the assertion (2). □

When we ask if a factorial, rational, affine surface Y is smooth provided $\pi_1(Y^\circ) = (1)$, we have the following counterexample. Note that if we assume additionally that Y is contractible, then Y is smooth by an affine Mumford theorem (see Theorem 3.2.7). We also note that a factorial affine surface does not necessarily have quotient singularities.

Example 3.5.10.
(1) Let V be the affine surface constructed in [85, Proposition 3.8]. Then V is a factorial, rational, affine surface with an E_8-singularity and $\pi_1(V^\circ) = (1)$.
(2) Let a, b, c be mutually coprime positive integers. Then the affine hypersurface $x^a + y^b + z^c = 0$ is factorial, though it has a nonquotient singularity for a suitable choice of a, b, c. We refer to [169].

3.5.4 Makar-Limanov invariant

As a corollary of Theorem 3.5.6, we can prove the following result which is essentially the algebraic proof of the cancelation theorem due to Crachiola and Makar-Limanov [29] (see also Bandman and Makar-Limanov [11]).

Theorem 3.5.11. *Let R be a regular factorial affine domain of dimension two and let $R[x]$ be a polynomial ring in one variable x over R. Let $X = \mathrm{Spec}\, R[x]$, $Y = \mathrm{Spec}\, R$ and $p : X \to Y$ be the projection. Then the following three conditions are equivalent:*
(1) *The Makar-Limanov invariant* $\mathrm{ML}(X)$ *is equal to k. Namely, there are three independent G_a-actions on X.*
(2) *Y is isomorphic to $\mathbb{A}^2$.*
(3) *X is isomorphic to $\mathbb{A}^3$.*

Proof. It suffices to show that the condition (1) implies (2). The existence of three independent G_a-actions on X implies that there is a dominant morphism from $G_a \times G_a \times G_a$ to X. Hence the unit group of $R[x]$, which is equal to R^*, is k^*. Consider the lnd δ of $R[x]$ such that $\delta(R) = 0$ and $\delta(x) = 1$. Then δ gives rise to a G_a-action along the fibers of the projection p. Namely, p is the quotient morphism under the G_a-action. Note that $R[x]$ is factorial. By Theorem 3.5.6, the surface Y is isomorphic to the affine plane $\mathbb{A}^2$ or the hypersurface $x^2 + y^3 + z^5 = 0$ in $\mathbb{A}^3$. Suppose that Y is isomorphic to the hypersurface $x^2+y^3+z^5 = 0$. Since the singularity of Y is not a cyclic singularity, $\mathrm{ML}(Y) = R$ by [153]. By Crachiola and Makar-Limanov [29, Theorem 3.1], we then have $\mathrm{ML}(X) = \mathrm{ML}(Y) = R$. This contradicts the assumption. Hence Y is isomorphic to $\mathbb{A}^2$. □

Generalizing Theorem 3.5.11, we raise the following problem.

Problem 3.3. Let R be an affine domain of dimension n over k and let $R[x]$ be a polynomial ring in a variable x over R. Suppose that $\mathrm{ML}(R[x]) = k$. Is $\mathrm{ML}(R)$ then equal to k?

But the problem has negative answers. We give two counterexamples. In the first example, $\dim R = 2$ but R is not factorial. In the second example, R is factorial but $\dim R = 3$. This question was treated, for example, by Bandman–Makar-Limanov [11].

Example 3.5.12. For an integer $n \geq 1$, let S_n be the Danielewski surface $x^n y = z^2 - 1$. Then $\mathrm{ML}(S_n) = k$ if $n = 1$ and $\mathrm{ML}(S_n) = k[x]$ if $n \geq 2$ [138]. It holds that $S_1 \times \mathbb{A}^1 \cong S_n \times \mathbb{A}^1$ for $n > 1$. Hence $\mathrm{ML}(S_1 \times \mathbb{A}^1) = \mathrm{ML}(S_n \times \mathbb{A}^1) = k$, but $\mathrm{ML}(S_n) = k[x]$ if $n > 1$.

The following example is due to Dubouloz [41]. It is also a counterexample to the conjecture in [11, p. 209].

Example 3.5.13. Let X be Koras–Russell cubic threefold $x^2 y = x + z^2 + t^3$, which is a hypersurface in $\mathbb{A}^4$. Let R be the coordinate ring of X. Then R is factorial, $\mathrm{ML}(X \times \mathbb{A}^1) = k$ and $\mathrm{ML}(X) = k[x]$.

3.5.5 Singular fibers of the G_a-quotient morphism

In Lemma 3.1.31, we proved that if X is a smooth factorial affine threefold with a G_a-action, a singular fiber of the quotient morphism $q : X \to Y := X//G_a$ is a disjoint union of contractible curves,[14] and asked if the singular fiber is indeed a disjoint union of affine lines. In the present subsection, we give a positive answer to this question. A fiber component means always an irreducible fiber component with reduced structure.

14 The hypothesis $\dim X^{G_a} \leq 1$ is automatic if X is factorial. In fact, if X^{G_a} has an irreducible component F of dimension two, it is defined by $f = 0$ with $f \in A = \Gamma(X, \mathcal{O}_X)$ such that $\delta(A) \subset fA$. Then $f \in \operatorname{Ker} \delta$. So, we can replace δ by $f^{-1}\delta$ and exclude F from X^{G_a}.

Theorem 3.5.14. *Let $f : V \to Y$ be a projective morphism from a normal variety V of dimension n onto a smooth variety Y of dimension $n-1$, where $n = 2$ or 3. Assume that a general fiber of f is isomorphic to $\mathbb{P}^1$. For a point $P \in Y$, let C be a one-dimensional component of the fiber $F_P := f^{-1}(P)$. Then C is isomorphic to $\mathbb{P}^1$.*

Proof. It is proved in Kollàr [120, p. 107, (2.8.6.2)] that $R^i f_*(\mathcal{O}_V) = 0$ for $i > 0$ with f, V, Y as above. We can assume that Y is affine. Then, by the spectral sequence $E_2^{p,q} = H^p(Y, R^q f_*\mathcal{O}_V) \Longrightarrow H^{p+q}(V, \mathcal{O}_V)$, we obtain $H^i(V, \mathcal{O}_V) = 0$ for every $i > 0$.

Consider first the case $n = 2$. Since V is not a complete surface, its cohomological dimension is less than 2 by Lichtenbaum's theorem [89, Corollary 3.2]. So, if $\mathcal{I}$ is the defining ideal sheaf of the fiber component C of F_P in V, then $H^2(V, \mathcal{I}) = 0$. Hence the natural homomorphism $H^1(V, \mathcal{O}_V) \to H^1(C, \mathcal{O}_C)$ is a surjection. This implies that $H^1(C, \mathcal{O}_C) = 0$, whence $C \cong \mathbb{P}^1$.

We assume that $n = 3$. Now we will use the formal functions' theorem [90, Chapter III, Theorem 11.1]. Let $\mathcal{O}$ be the local ring of Y at P and let $\mathfrak{m}$ be the maximal ideal of $\mathcal{O}$. As usual, let $V_n := V \times_Y \operatorname{Spec}(\mathcal{O}/\mathfrak{m}^n)$ and let $\mathcal{O}_n := \mathcal{O}_V/\mathfrak{m}^n\mathcal{O}_V$ considered as the structure sheaf of V_n. For each $n \geq 0$, we have natural morphisms $R^1 f_*(\mathcal{O}_V) \otimes \mathcal{O}/\mathfrak{m}^n \to H^1(V_n, \mathcal{O}_n)$. As n varies, we have two inverse systems, inducing a natural morphism, which is an isomorphism by the formal functions' theorem,

$$R^i f_*(\mathcal{O})^\wedge_P \xrightarrow{\sim} \varprojlim H^i(V_n, \mathcal{O}_n).$$

For $i > 0$, the left term is zero, hence we get $\varprojlim H^i(V_n, \mathcal{O}_n) = 0$.

Note that the underlying reduced complex space of V_n has dimension ≤ 2. We can write $V_n = U_{1n} \cup U_{2n}$, where U_{1n} is a suitable open neighborhood in V_n of the union of all one-dimensional components of V_n and U_{2n} is an open neighborhood in V_n of the union of all the irreducible components of dimension > 1. We can assume that $U_{1n} \cap U_{2n}$ is a disjoint union of connected (nonreduced) Stein spaces. We denote the sheaves of abelian groups $\mathcal{O}_{1n}, \mathcal{O}_{2n}$ on U_{1n}, U_{2n}, respectively, which are just restrictions of $\mathcal{O}_n$ to U_{1n}, U_{2n}, respectively. In this situation there is a Mayer–Vietoris sequence [5, p. 236],

$$\begin{aligned} \cdots \to H^1(V_n, \mathcal{O}_n) &\to H^1(U_{1n}, \mathcal{O}_{1n}) \oplus H^1(U_{2n}, \mathcal{O}_{2n}) \\ &\to H^1(U_{1n} \cap U_{2n}, \mathcal{O}_n) \to \cdots. \end{aligned}$$

Since $U_{1n} \cap U_{2n}$ is a disjoint union of finitely many connected Stein spaces, the last cohomology group is trivial. As n varies, we get Mayer–Vietoris sequences with maps from the groups in the $(n+1)$th sequence to the corresponding groups in the nth sequence making all the diagrams commute. Since $\varprojlim H^1(V_n, \mathcal{O}_n) = 0$, we deduce that $\varprojlim H^1(U_{1n}, \mathcal{O}_{1n}) = 0$.

Let I_n be the ideal sheaf of U_{1n} in $U_{1(n+1)}$. From the exact sequence $0 \to I_n \to \mathcal{O}_{1(n+1)} \to \mathcal{O}_{1n} \to 0$, and using the fact that if $U_{1n} \cap U_{2n} \neq \emptyset$ then U_{1n} is a noncompact 2-dimensional complex space without 2-dimensional compact components (if $U_{1n} \cap U_{2n} = \emptyset$, U_{1n} is one-dimensional) so that $H^2(U_{1(n+1)}, I_n) = 0$, we know that the natural maps $H^1(U_{1(n+1)}, \mathcal{O}_{n+1}) \to H^1(U_{1n}, \mathcal{O}_n)$ are surjections for each n. Since $\varprojlim H^1(U_{1n}, \mathcal{O}_{1n}) = 0$ we deduce that $H^1(U_{1n}, \mathcal{O}_n) = 0$ for $n > 0$.

The reduced curve $C \subseteq F_P$ is a closed subscheme of U_{1n} for each n. By the same argument as above, we deduce that $H^1(U_{1n}, \mathcal{O}_n) \to H^1(C, \mathcal{O}_C)$ is a surjection. This means that $H^1(C, \mathcal{O}_C) = 0$, proving that $C \cong \mathbb{P}^1$. □

Kollàr [120, p. 107, (2.8.6.3)] states the following result.

Proposition 3.5.15. *Let $f : X \to Y$ be a proper morphism with X normal, Y smooth and a general fiber $\mathbb{P}^1$. Let F be a one-dimensional fiber of f. Then $H^1(F_{\mathrm{red}}, \mathcal{O}) = (0)$.*

This result shows that if C is an irreducible component of the one-dimensional fiber F_{red} then $C \cong \mathbb{P}^1$. In fact, we have a surjection

$$H^1(F_{\mathrm{red}}, \mathcal{O}) \twoheadrightarrow H^1(C, \mathcal{O}_C).$$

Hence $H^1(C, \mathcal{O}_C) = (0)$ and $C \cong \mathbb{P}^1$. But this argument cannot be applied in the case F has irreducible components of dimension greater than one.

As a corollary of Theorem 3.5.14, we obtain the following result.

Theorem 3.5.16. *Let G_a act on a smooth affine variety X of dimension n and let $q : X \to Y$ be the quotient morphism where Y is the quotient affine variety of dimension $n-1$. We assume that $n \le 4$ and the ring of G_a-invariants of $\Gamma(X, \mathcal{O}_X)$ is an affine domain if $n = 4$. Let C_0 be a one-dimensional component of a fiber of q. If $n = 4$, we assume that Y is smooth at $P = q(C_0)$, the fiber $q^{-1}(P)$ has no components of dimension > 1 and C_0 is a reduced component. Then $C_0 \cong \mathbb{A}^1$.*

Proof. Since the case $n = 2$ is already treated in Theorem 2.5.6, we will first consider the case $n = 3$. Since $A := \Gamma(X, \mathcal{O}_X)$ is regular, $\operatorname{Ker} \delta$ is normal. Hence Y is a normal affine surface and Y has only isolated singular points. Let $P = q(C_0)$. As in the proof of Lemmas 3.1.30 and 3.5.4, Y has at worst quotient singularity at P. Assume first that Y is smooth. We can embed X as a Zariski-open set into a smooth quasiprojective threefold V in such a way that q extends to a projective morphism $f : V \to Y$ whose general fibers are isomorphic to $\mathbb{P}^1$. Then the closure C of C_0 in V is a one-dimensional component of the fiber of f containing C_0. As already mentioned in Lemma 3.1.31, C_0 is contractible. By Theorem 3.5.14, we have $C \cong \mathbb{P}^1$. It then follows that $C_0 \cong \mathbb{A}^1$ as C_0 is contractible.

We consider the general case $n = 3$. If the point P is smooth, we replace Y by a smooth affine open neighborhood U of P and replace X by $q^{-1}(U)$. Since q is an affine morphism, the restriction $q|_{q^{-1}(U)} : q^{-1}(U) \to U$ satisfies the conditions of the theorem with Y smooth. If P is a singular point of Y, it is a quotient singular point, which is locally analytically isomorphic to $\mathbb{C}^2/G$ with a finite subgroup G of $\mathrm{GL}(2, \mathbb{C})$. Take a small open neighborhood U of the singular point P and the quasiuniversal covering $\widetilde{U}$

of U,[15] which is a covering smooth and unramified outside the point P. Removing all two-dimensional components from the fiber $q^{-1}(P)$ if they exist, we may assume that the fiber $q^{-1}(P)$ is purely one-dimensional. By this operation, X stays affine because X is assumed to be smooth. Let $\widetilde{X} = (X \times_Y \widetilde{U})$ be the pull-back of f by the covering morphism $\widetilde{U} \to U$. Then, by the purity of branch loci (see [175, p. 158]) applied to the projection $\widetilde{X} \to q^{-1}(U)$ which is unramified outside $q^{-1}(P)$, $\widetilde{X}$ is smooth, $\widetilde{X} \to q^{-1}(U)$ is a unramified Galois extension and the natural projection $\widetilde{f} : \widetilde{X} \to \widetilde{U}$ fits to the situation where the quotient surface is smooth. Hence any one-dimensional component of the *new* fiber of $\widetilde{f}$ is $\mathbb{A}^1$ as shown above in the case where P is a smooth point. But the one-dimensional component of the *old* fiber is a quotient of this new $\mathbb{A}^1$ by a finite group. Hence it is smooth and simply-connected by Lemma 3.5.2. This implies that $C \cong \mathbb{A}^1$. This completes the proof in the case $n = 3$.

We consider the case $n = 4$. Since Y is smooth at the point $P = q(C_0)$ by the assumption, we may replace Y by a small affine open neighborhood of P. Now Y is a smooth affine threefold. Let H be a general hyperplane section of Y passing through P. The inverse image $q^{-1}(H)$ has a unique irreducible component X_H of dimension 3 which dominates H. As H varies over all the hyperplane sections of Y passing through P we obtain a linear system of irreducible threefolds X_H with induced $\mathbb{A}^1$-fibration onto H. Now, Bertini's second theorem says that, for a general H, the threefold X_H is smooth off the fiber $q^{-1}(P)$. Note that $q^{-1}(P)$ has codimension 2 in X_H by the assumption. By making use of Serre normality criterion [90, Theorem 8.22A, p.185], we show that X_H is a normal variety. The R_1-condition is satisfied because the singular set of X_H is contained in the fiber $q^{-1}(P)$. Since X is smooth and X_H is a locally principal divisor in X, X_H is Cohen–Macaulay. Hence the S_2-condition is satisfied. Hence $q_H : X_H \to H$ is an $\mathbb{A}^1$-fibration from an affine normal threefold to a smooth affine surface H. Finally, we can embed X_H as a closed Zariski-dense open subset of a normal threefold V_H and extend q_H to a projective morphism $f_H : V_H \to H$ which is a $\mathbb{P}^1$-fibration. By Theorem 3.5.14, any 1-dimensional irreducible component of a fiber of f_H is isomorphic to $\mathbb{P}^1$. By Lemma 3.5.17 below, the component C_0 has only one place point missing. Hence $C_0 \cong \mathbb{A}^1$. □

Lemma 3.5.17. *Let $q : X \to Y$ be an $\mathbb{A}^1$-fibration from a smooth affine fourfold X to a smooth affine threefold Y. Let C_0 be a one-dimensional, reduced component of a fiber q, which satisfies the same conditions as in Theorem* 3.5.16 *in the case $n = 4$. Then C_0 has only one place at infinity and is disjoint from other components of the fiber.*

Proof. Embed X into a smooth fourfold V as a Zariski-open set so that q extends to a projective morphism $f : V \to Y$ and $D := V - X$ is the support of an effective divisor.

15 The set $U \setminus \{P\}$ has a finite fundamental group, hence its universal covering $\widetilde{U}^\circ$ is a finite covering of $U \setminus \{P\}$. Then the covering morphism is uniquely extended to a finite covering $\widetilde{U} \to U$, where $\widetilde{U}$ is normal. It is called the *quasiuniversal covering* of U. If P is a quotient singularity then $\widetilde{U}$ is smooth.

Then f defines a $\mathbb{P}^1$-fibration. Since X is affine, D is connected and supports a relatively ample divisor. Then there is a unique horizontal irreducible component D_0 of D which meets a general fiber of f transversally in one point. Hence D_0 is a birational cross-section of f. Furthermore, since $f|_{D_0} : D_0 \to Y$ is a proper morphism, its fiber is connected. In particular, $D_0 \cap F$ is a connected, non-empty set, where $F = f^{-1}(P)$ and $P = q(C_0)$. Let C be the closure of C_0 in V and let F_0 be the union of all the irreducible components of F except for C.

Claim. *F_0 is connected.*

In fact, every point of V is *path-connected* to a point of D_0 within the fiber to which the given point belongs. Since F and D_0 are connected, this implies that F_0 is connected.

Claim. *The fiber F is simply connected. Hence the integral homology group $H_1(F)$ vanishes.*

In fact, let U be a small Euclidean open neighborhood of the point P. Then we apply Lemma 1.1.6 to obtain an exact sequence

$$\pi_1(\mathbb{P}^1) \longrightarrow \pi_1(f^{-1}(U)) \longrightarrow \pi_1(U) \longrightarrow (1).$$

This application is possible because the component C is reduced in F and we can take U so small that every fiber of f contains a reduced component. Hence $\pi_1(f^{-1}(U)) \cong \pi_1(U) \cong (1)$. Since the fiber $f^{-1}(P)$ is a strong deformation retract of $f^{-1}(U)$, we infer that $F = f^{-1}(P)$ is simply-connected.

Since we have $F = F_0 \cup C$, we have the Mayer–Vietoris exact sequence for integral homology groups

$$H_1(F) \to H_0(F_0 \cap C) \to H_0(F_0) \oplus H_0(C) \to H_0(F) \to 0.$$

Since $H_1(F) = 0$ and $H_0(F_0) \cong H_0(C) \cong H_0(F) \cong \mathbb{Z}$, we have $H_0(F_0 \cap C) = \mathbb{Z}$. This implies that $F_0 \cap C$ consists of a single point. Since $C \cong \mathbb{P}^1$ by the proof of Theorem 3.5.16, this implies that C_0 has only one place at infinity. If C_0 meets the other components of $q^{-1}(P)$ then a complete curve C is contained in the affine variety X. This is a contradiction. □

Remark 3.5.18. The assumption in Theorem 3.5.16 that $q^{-1}(P)$ has no components of dimension > 1 is unnecessary. Indeed, by Lemma 3.5.17, C_0 is a connected component of the fiber. Thus we can remove the other components from X_H appearing in the proof of Theorem 3.5.16. Then the fiber over P of the resulting threefold X'_H is C_0 and it is of codimension 2 which implies the normality by the Serre criterion. Though X'_H is not necessarily affine, the argument in the proof of Theorem 3.5.14 applies to a suitable projective completion of X'_H with q_H extended to a $\mathbb{P}^1$-fibration, and shows that C_0 is smooth. Since C_0 is also contractible as a component of $q^{-1}(P)$, C_0 is isomorphic to $\mathbb{A}^1$.

3.5.6 Locus of singular fibers

Let $f : X \to Y$ be an $\mathbb{A}^1$-fibration of normal affine varieties. We defined $\mathrm{Sing}(f)$ as the set of points $Q \in Y$ such that either the fiber $F_Q = f^*(Q)$ is a singular fiber or $f^{-1}(Q) = \emptyset$. For an $\mathbb{A}^1$-fibration, see Definition 2.1.1 and also a definition in Subsection 3.1.2 before Lemma 3.1.12. For an $\mathbb{A}^1_*$-fibration, see a definition after Corollary 3.3.8. As for the closedness of $\mathrm{Sing}(f)$, we mentioned it in Lemma 3.2.11 in the case where f is the equidimensional G_a-quotient morphism $q : X \to Y$ with a smooth affine threefold X. Here we add the following result which is slightly generalized than Lemma 3.2.11.

Theorem 3.5.19. *Let $f : X \to Y$ be an equidimensional $\mathbb{A}^1$-fibration on a smooth affine threefold to a normal affine surface Y. Then* $\mathrm{Sing}(f)$ *is a closed set.*

Proof. It is proved in Lemma 3.1.2 that f can be factored as $X \xrightarrow{q} \widetilde{Y} \xrightarrow{\nu} Y$ for a suitable G_a-action on X and $\widetilde{Y} = X//G_a$. Here, since $\dim X = 3$, the quotient $\widetilde{Y}$ exists as a normal affine surface by Theorem 2.2.4. Further the morphism $\nu : \widetilde{Y} \to Y$ is birational. Since $f : X \to Y$ is equidimensional, so is $q : X \to \widetilde{Y}$. Then the complement of the image of q in $\widetilde{Y}$ is at most a finite set. In fact, suppose that there exists an irreducible codimension one subvariety Z of $\widetilde{Y}$ such that $\overline{q(X) \cap Z} \neq Z$. Let $\mathfrak{p}$ be a prime ideal of $\widetilde{B} = A^{G_a}$ with $A = \Gamma(X, \mathcal{O}_X)$ such that $V(\mathfrak{p}) = Z$. Let $\mathcal{O} = \widetilde{B}_{\mathfrak{p}}$, which is a discrete valuation ring of the function field $k(\widetilde{Y}) = Q(\widetilde{B})$. Let t be a uniformizant of $\mathcal{O}$, which we may assume to be an element of $\mathfrak{p}$. Then the hypothesis implies that $t(A \otimes_{\widetilde{B}} \mathcal{O}) = A \otimes_{\widetilde{B}} \mathcal{O}$. Hence there exist elements $b \in \widetilde{B} \setminus \mathfrak{p}$ and $a \in A$ such that $ta = b$. Meanwhile, a is then a G_a-invariant element, and hence $a \in \widetilde{B}$. This is a contradiction. Now we infer that ν is quasifinite. By Zariski's main theorem, $\widetilde{Y}$ is embedded as an open subset of Y. If some fiber $f^{-1}(Q)$ of $f : X \to Y$ is a regular fiber, then Q is in the image of $\widetilde{Y} \to Y$. Since $\widetilde{Y} \to Y$ is an open immersion, the corresponding fiber of $q : X \to \widetilde{Y}$ is a regular fiber. Since $\mathrm{Sing}(q)$ is a closed set by Lemma 3.2.11, all the fibers of $q : X \to \widetilde{Y}$ for points close to $\widetilde{Q}$ are regular, where $\widetilde{Q}$ is the unique point of $\widetilde{Y}$ lying over Q. Since $\widetilde{Y}$ is a Zariski-open subset of Y the same is true for fibers of f lying over points close to Q. This implies that $Y \setminus \mathrm{Sing}(f) \cong \widetilde{Y} \setminus \mathrm{Sing}(q)$ and $\mathrm{Sing}(f) = (Y \setminus \nu(\widetilde{Y})) \cup \mathrm{Sing}(q)$, which is a closed set. □

3.6 Problems for Chapter 3

Let $A = k[x, y, z]$ be a polynomial ring in three variables and let f be a nonconstant element of A. Then the *Abhyankar–Sathaye conjecture* in dimension three asserts the following.

Conjecture 3.6.1. *If the affine hypersurface $X_0 = \{f = 0\}$ in $X = \mathrm{Spec}\, A \cong \mathbb{A}^3$ is isomorphic to the affine plane $\mathbb{A}^2$ then so is the hypersurface $X_c = \{f = c\}$ for every $c \in k$.*

Suppose that the conjecture holds. Then, since $f : X \to \mathbb{A}^1$ defined by $P \in X \mapsto f(P) \in \mathbb{A}^1$ is a trivial $\mathbb{A}^1$-bundle by Theorem 2.1.16, it follows that $A = k[f, g, h]$ with certain elements $g, h \in A$. Namely, f becomes a coordinate of $\mathbb{A}^3$.

We consider the conjecture under the assumptions that *the additive group* G_a *acts nontrivially on* $X = \operatorname{Spec} k[x, y, z]$ *and* f *is* G_a*-invariant*. Let δ be a locally nilpotent derivation on A which corresponds to the G_a-action. Let $B = \operatorname{Ker} \delta$. Then B is a polynomial ring in two variables by Theorem 3.2.5. Let $Y = \operatorname{Spec} B$ and let $q : X \to Y$ be the quotient morphism defined by the inclusion $B \hookrightarrow A$, which is equidimensional and surjective by [18] (see Subsection 1.3.2). Since f is G_a-invariant by the assumption, $f \in B$. Since $q : X \to Y$ is an $\mathbb{A}^1$-fibration, we define the *singular locus* of q by $\operatorname{Sing}(q) = \{Q \in Y \mid q^{-1}(Q) \not\cong \mathbb{A}^1\}$, where $q^{-1}(Q)$ is the scheme-theoretic fiber over a point $Q \in Y$. Then $\operatorname{Sing}(q)$ is a closed set of Y by Lemma 3.2.11, and $\operatorname{Sing}(q)$ has pure dimension one by Lemma 2.1.6.

We define the *plinth ideal* of δ by $\operatorname{pl}(\delta) = B \cap \delta(A)$ which is an ideal of B, and the *plinth locus* of q by $\operatorname{pl}(q) = V(\operatorname{pl}(\delta))$. Problem 6 is our final result.

1. Let A be a regular factorial affine domain of dimension 3 and let δ be a nonzero lnd on A. Prove that with the above notations, $\operatorname{Sing}(q)$ is the codimension one part of $\operatorname{pl}(q)$.

   Answer. (1) Let S be an irreducible component of $\operatorname{Sing}(q)$. Then it is defined by a prime element p of B because B is factorial. Suppose that $S \not\subset \operatorname{pl}(q)$. Then there exists a maximal ideal $\mathfrak{m}$ of B such that $p \in \mathfrak{m}$ but $\operatorname{pl}(\delta) \not\subset \mathfrak{m}$. Then there exist elements $b \in B \setminus \{0\}$ and $u \in A$ such that $b = \delta(u) \notin \mathfrak{m}$ and hence $A[b^{-1}] = B[b^{-1}][u]$. Since $b^{-1} \in B_{\mathfrak{m}}$, it follows that $A_{\mathfrak{m}} := A \otimes_B B_{\mathfrak{m}} = B_{\mathfrak{m}}[u]$. Namely, the point $Q \in Y$ defined by $\mathfrak{m}$ is not a point of $\operatorname{Sing}(q)$. This is a contradiction because $Q \in V(pB) \subset \operatorname{Sing}(q)$. Hence $\operatorname{Sing}(q) \subseteq \operatorname{pl}(q)$.
   (2) Take an irreducible component $V(pB)$ of $\operatorname{pl}(q)$, where p is a prime element of B. Suppose that $V(pB) \not\subset \operatorname{Sing}(q)$. Take a closed point Q of $V(pB)$ such that $Q \notin \operatorname{Sing}(q)$. Let $\mathfrak{m}$ be the maximal ideal of B corresponding to Q. Then $A_{\mathfrak{m}} = B_{\mathfrak{m}}[u]$ since the quotient morphism $q : X \to Y$ is an $\mathbb{A}^1$-bundle in a small open neighborhood of Q. We can take u to be an element of A such that $\delta(u) = b \in B \setminus \mathfrak{m}$. Hence $b \in \operatorname{pl}(\delta)$, and $\operatorname{pl}(\delta) \not\subset \mathfrak{m}$. This is a contradiction to the choice of $\mathfrak{m}$. Hence every irreducible component of $\operatorname{pl}(q)$ of codimension one is contained in $\operatorname{Sing}(q)$. □

2. Let $f \in B$ be an element such that the hypersurface Y_0 in $X = \mathbb{A}^3$ defined by the ideal fA is isomorphic to $\mathbb{A}^2$. Let $\varphi : X \to \mathbb{A}^1 = \operatorname{Spec} k[f]$ be the morphism defined by $P \mapsto f(P)$. The hypersurface X_c is the scheme-theoretic fiber $\operatorname{Spec} A/(f - c)A$ of the morphism φ over the point of $\mathbb{A}^1$ defined by $f = c$. The morphism φ is decomposed as

$$\varphi : X \xrightarrow{q} Y \xrightarrow{p} \mathbb{A}^1,$$

where p is induced by the inclusion $k[f] \hookrightarrow B$. Prove that the G_a-equivariant Abhyankar–Sathaye conjecture is proved if the following two assertions hold true for every $c \in k$:
(1) The curve $Y_c = \operatorname{Spec} B/(f-c)B$ in Y is the affine line in the affine plane Y.
(2) The restriction of the quotient morphism $q|_{X_c} : X_c \to Y_c$ is an $\mathbb{A}^1$-bundle.

Answer. If these two assertions are proved, then X_c is an $\mathbb{A}^1$-bundle over $\mathbb{A}^1$, which is trivial. Thus, $X_c \cong \mathbb{A}^2$ for every $c \in k$. Then X is an $\mathbb{A}^2$-bundle over $\mathbb{A}^1$ and $X \cong Y \times \mathbb{A}^2$ by Theorem 2.1.16. This implies that f is a coordinate of A. □

3. We consider the fibration $p : Y \to \mathbb{A}^1$. We may assume that the derivation δ is *irreducible* in the sense that $\delta(A) \subset uA$ with $u \in A$ implies $u \in A^*$. Then the induced G_a-action on X_c is nontrivial. Hence $q|_{X_c}$ is decomposed as

$$q|_{X_c} : X_c \xrightarrow{q_c} X_c//G_a \xrightarrow{p_c} Y_c,$$

where q_c is the quotient morphism and $X_c//G_a$ is the algebraic quotient. Prove that the following assertions hold:
(1) For every $c \in k$, the element $f - c$ is irreducible in B.
(2) The curve Y_c is isomorphic to $\mathbb{A}^1$ for every $c \in k$.

Answer. (1) Consider $q|_{X_0} : X_0 \to Y_0$. Since $X_0 \cong \mathbb{A}^2$ by the assumption, it follows that $X_0//G_a \cong \mathbb{A}^1$. Then the morphism $p_0 : X_0//G_a \to Y_0$ is a finite morphism (Stein factorization of $q|_{X_0}$). Hence Y_0 is an irreducible curve in $Y \cong \mathbb{A}^2$ with only one place at infinity. Then by the irreducibility theorem [150, p. 89], the curve $f - c = 0$ is an irreducible curve with only one place at infinity. Hence $f - c$ is an irreducible element of B.
(2) We show that Y_0 is a smooth curve. By the initial assumption that $X_0 \cong \mathbb{A}^2$, Y_0 is a rational affine curve with one place at infinity. Hence if it is smooth, then $Y_0 \cong \mathbb{A}^1$. By Abhyankar–Moh–Suzuki theorem, it follows that $Y_c \cong \mathbb{A}^1$ for all $c \in k$. Now note that $B = k[v, w]$ a polynomial ring over k. Then we have

$$\frac{\partial f}{\partial x} = \frac{\partial f}{\partial v}\frac{\partial v}{\partial x} + \frac{\partial f}{\partial w}\frac{\partial w}{\partial x},$$
$$\frac{\partial f}{\partial y} = \frac{\partial f}{\partial v}\frac{\partial v}{\partial y} + \frac{\partial f}{\partial w}\frac{\partial w}{\partial y},$$
$$\frac{\partial f}{\partial z} = \frac{\partial f}{\partial v}\frac{\partial v}{\partial z} + \frac{\partial f}{\partial w}\frac{\partial w}{\partial z}.$$

If Q is a singular point of Y_0 then $\frac{\partial f}{\partial v}(Q) = \frac{\partial f}{\partial w}(Q) = f(v(Q), w(Q)) = f(P) = 0$, where P is a point of X such that $Q = q(P)$. Hence P is a singular point of X_0 which contradicts the smoothness of X_0. Hence Y_0 is smooth. □

4. Prove that the following assertions hold:

(1) The condition $(f-c)B \supset \mathrm{pl}(\delta)$ for $c \in k$ is equivalent to the condition that Y_c is an irreducible component of $\mathrm{Sing}(q)$.
(2) Suppose that Y_c is not an irreducible component of $\mathrm{Sing}(q)$ for some $c \in k$. Then the morphism $p_c : X_c/\!/G_a \to Y_c$ is an isomorphism.

Answer. (1) The condition is equivalent to the condition that Y_c is an irreducible component of $\mathrm{pl}(q)$. By Problem 1, the assertion follows.
(2) To simplify the notations, we consider the case $c = 0$. The proof is the same in the general case. The morphism $q|_{X_0} : X_0 \to Y_0$ is induced by the natural inclusion $B/fB \hookrightarrow A/fA$. Let δ_0 be the lnd on A/fA induced by δ. Then we have an inclusion $B/fB \hookrightarrow \mathrm{Ker}\,\delta_0$, which gives the morphism p_0 in the above decomposition of $q|_{X_0}$. In order to show that p_0 is an isomorphism, it suffices to show that if $\delta(a) \in fA$ then there exists an element $b \in B$ such that $a - b \in fA$. Suppose first that δ has a slice u, i. e., $\delta(u) = 1$. Then $A = B[u]$, and we can write

$$a = b_0 + b_1u + \cdots + b_nu^n, \quad b_0, \ldots, b_n \in B.$$

Since $\delta(a) = b_1 + 2b_2u + \cdots + nb_nu^{n-1}$ is divisible by f in A, the coefficients $b_1, \ldots, b_n$ are divisible by f in B. Write

$$\delta(a) = f(b_1' + 2b_2'u + \cdots + nb_n'u^{n-1}).$$

Set

$$a' = b_1'u + b_2'u^2 + b_3'u^3 + \cdots + b_n'u^n.$$

Then $\delta(a) = f\delta(a')$, and hence $a - fa' \in B$. Set $b = a - fa'$. Then $a - b = fa' \in fA$. Suppose next that u is a local slice. Namely $\alpha = \delta(u)$ is a nonzero element of B. Then δ extends to an lnd δ' of $A[\alpha^{-1}]$ and $B[\alpha^{-1}] = \mathrm{Ker}\,\delta'$. Since $\delta'(u/\alpha) = 1$, by the previous result, we have $a - (b/\alpha^r) \in fA[\alpha^{-1}]$. Hence, changing b by a multiple of b by an element of the form α^s, we have $\alpha^r a - b \in fA$. Let $\mathfrak{p}$ be a prime ideal of B such that $f \in \mathfrak{p}$ and $\mathfrak{p} \notin V(\mathrm{pl}(\delta))$, i. e., $\mathfrak{p} \in Y_0 \setminus V(\mathrm{pl}(\delta))$. Such a prime ideal $\mathfrak{p}$ exists by the assumption. Let $\delta_{\mathfrak{p}}$ be the extension of δ to $A_{\mathfrak{p}} = A \otimes_B B_{\mathfrak{p}}$. Then we can find an element $\alpha \in \mathrm{pl}(\delta)$ such that $\alpha \notin \mathfrak{p}$. Let $(\delta_{\mathfrak{p}})_0$ be the restriction of $\delta_{\mathfrak{p}}$ onto $A_{\mathfrak{p}}/fA_{\mathfrak{p}}$. Then we have $\mathrm{Ker}(\delta_{\mathfrak{p}})_0 = B_{\mathfrak{p}}/fB_{\mathfrak{p}}$ because u/α is a slice of $\delta_{\mathfrak{p}}$ if $\alpha = \delta(u)$. This implies that p_c is birational.
Since $Y_c \cong \mathbb{A}^1$ by Problem 3, the birational morphism p_c is an isomorphism. □

5. Assume that $(f-c)B \not\supset \mathrm{pl}(\delta)$ for every $c \in k$ and that $Y_0 \cap \mathrm{Sing}(q) = \emptyset$. Prove that f is a variable of A, i. e., $A = k[f, g, h]$ for some $g, h \in A$.

Answer. Since $Y_0 \cap \mathrm{Sing}(q) = \emptyset$ by the assumption, $\mathrm{Sing}(q)$ is either the empty set or a finite disjoint union $\coprod_i Y_{c_i}$ with $c_i \in k$. Meanwhile, the condition that $(f-c)B \not\supset \mathrm{pl}(\delta)$ for every $c \in k$ implies that the last case does not occur. Hence $\mathrm{Sing}(q) = \emptyset$. Hence the conditions (1) and (2) in Problem 2 are fulfilled, so the conclusion follows. □

6. Assume that $Y_c \cap \mathrm{Sing}(q) = \emptyset$ for some $c \in k$. Then f is a variable of A.

   Answer. Since $\mathrm{Sing}(q) \cap Y_c = \emptyset$, $q|_{X_c} : X_c \to Y_c$ is an $\mathbb{A}^1$-fibration which has no singular fibers. It implies, in particular, that $p_c : X_c//G_a \to Y_c$ is an isomorphism by Problem 4(2) and $q_c : X_c \to Y_c$ is an $\mathbb{A}^1$-bundle. Furthermore, since $Y_c \cong \mathbb{A}^1$ as a parallel line by Problem 3, it follows that $X_c \cong \mathbb{A}^2$, and $\mathrm{Sing}(q)$ is a disjoint sum of Y_{c_i} for $c_1, \dots, c_r$ unless $\mathrm{Sing}(q) = \emptyset$. This implies that almost all fibers of $\varphi : X \to \mathbb{A}^1 = \mathrm{Spec}\, k[f]$ are isomorphic to $\mathbb{A}^2$. Then f is a variable by Kaliman [98]. □

7. By following arguments below, show that the space of isomorphism classes $H^1(\mathbb{A}^3_*, \mathcal{O}_{\mathbb{A}^3})$ of G_a-torsors over the punctured affine 3-space $X := \mathbb{A}^3_* := \mathbb{A}^3 \setminus \{(0,0,0)\}$ is isomorphic to the k-vector space consisting of Laurent polynomials

$$\sum_{i_1>0, i_2>0, i_3>0} \frac{c_{i_1 i_2 i_3}}{x_1^{i_1} x_2^{i_2} x_3^{i_3}}, \quad c_{i_1 i_2 i_3} \in k.$$

   (1) Let $T = \mathrm{Spec}\, R$ be an affine scheme over k and let $\mathbb{A}^2_T = \mathrm{Spec}\, R[x_1, x_2]$. Let $U_{1,R} = D(x_1)$ and $U_{2,R} = D(x_2)$ and let $\mathbb{A}^2_{T,*} = U_{1,R} \cup U_{2,R}$. Then $H^1(\mathbb{A}^2_{T,*}, \mathcal{O}_{\mathbb{A}^2_T})$ is an R-module of Laurent polynomials with coefficients in R,

$$\sum_{i_1>0, i_2>0} \frac{c_{i_1 i_2}}{x_1^{i_1} x_2^{i_2}}, \quad c_{i_1 i_2} \in R.$$

   (2) In the case of $\mathbb{A}^3_* = U_1 \cup U_2 \cup U_3$, set $U_3' = U_1 \cup U_2$. Then $U_3 \cap U_3' = U_{1,R} \cup U_{2,R}$ with $R = k[x_1, x_1^{-1}, x_2, x_3]$. Hence, by (1), $H^1(U_3 \cap U_3', \mathcal{O}_{\mathbb{A}^3})$ consists of Laurent polynomials of the type

$$\sum_{i_1>0, i_2>0} \frac{c_{i_1 i_2}}{x_1^{i_1} x_2^{i_2}} \quad c_{i_1 i_2} \in k[x_3, x_3^{-1}].$$

   (3) By the Mayer–Vietoris exact sequence, we have a long exact sequence

$$\begin{aligned} 0 &\to H^0(X, \mathcal{O}_X) \to H^0(U_3, \mathcal{O}_X) \oplus H^0(U_3', \mathcal{O}_X) \to H^0(U_3 \cap U_3', \mathcal{O}_X) \\ &\to H^1(X, \mathcal{O}_X) \to H^1(U_3, \mathcal{O}_X) \oplus H^1(U_3', \mathcal{O}_X) \to H^1(U_3 \cap U_3', \mathcal{O}_X), \end{aligned}$$

   where $H^1(U_3, \mathcal{O}_X) = 0$ and $H^1(U_3', \mathcal{O}_X)$ consists of Laurent polynomials of the same type as in (2) with coefficients $c_{i_1 i_2} \in k[x_3]$.

   Answer. (1) Use the argument in the proof of Theorem 3.1.21.
   (2) Straightforward.
   (3) In the exact sequence, note that the last homomorphism

$$H^1(U_3', \mathcal{O}_X) \to H^1(U_3 \cap U_3', \mathcal{O}_X)$$

   is injective. Hence, as in the case $n = 2$, we obtain an expression of elements in $H^1(X, \mathcal{O}_X)$ as given in the statement. □

8. Let ℓ, m, n be integers greater than 1 such that $n > m$ and $\gcd(m, n) = 1$. Let X be an affine hypersurface in $\mathbb{A}^4$ defined over $\mathbb{C}$ by an equation

$$x + x^\ell y + z^m + t^n = 0.$$

Prove the following assertions:
(1) X is a smooth rational threefold which is contractible.
(2) Let s be the least common multiple of m, n. Then X has a hyperbolic G_m-action, i. e., X^{G_m} consists of one point, which is defined by

$$^\lambda(x, y, z, t) = (\lambda^s x, \lambda^{-(\ell-1)s} y, \lambda^{s/m} z, \lambda^{s/n} t), \quad \lambda \in \mathbb{C}^*.$$

(3) X has two independent G_a-actions whose associated lnds are given by

$$\delta_1(x) = 0, \quad \delta_1(y) = -mz^{m-1}, \quad \delta_1(z) = x^\ell, \quad \delta_1(t) = 0,$$
$$\delta_2(x) = 0, \quad \delta_2(y) = -nt^{n-1}, \quad \delta_2(z) = 0, \quad \delta_2(t) = x^\ell.$$

Answer. (1) Let A be the coordinate ring of X. Then $A[x^{-1}] = \mathbb{C}[x, x^{-1}, z, t] = \mathbb{C}[x, x^{-1}] \otimes_{\mathbb{C}} \mathbb{C}[y, z]$. Hence the open set $D(x)$ is isomorphic to $\mathbb{A}^1_* \times \mathbb{A}^2$. This implies that X is retracted to the fiber $F_0 = V(x) = \{x = 0\}$ which is a direct product $\mathbb{A}^1 \times C$, where C is an affine plane curve $z^m + t^n = 0$. The curve C has only cusps as singularities and one place point at infinity. Hence the fiber F_0 is contracted to a point. So, X is a contractible threefold. The other conditions are easily verified. (2)–(3) Straightforward. □

9. Prove the following assertions:
(1) Let $\sigma : G_m \times X \to X$ be an algebraic G_m-action on an affine k-scheme $X = \operatorname{Spec} A$. Let $\varphi : A \to A \otimes_k k[t, t^{-1}]$ be the k-algebra homomorphism corresponding to σ. For an element $a \in A$, write

$$\varphi(a) = \sum_{i \in \mathbb{Z}} \delta_i(a) t^i.$$

Then $\{\delta_i\}_{i \in \mathbb{Z}}$ is a set of k-module endomorphisms of A such that $\delta_j \cdot \delta_i$ is equal to δ_i if $i = j$ and 0 if $i \neq j$ and that $\sum_{i \in \mathbb{Z}} \delta_i = \mathrm{id}_A$. Hence the last equality gives an idempotent decomposition of id_A. Set $A_i = \delta_i(A)$. Then $A = \bigoplus_{i \in \mathbb{Z}} A_i$. In particular, $A_0 = A^{G_m}$ and A_i is an A_0-module.
(2) Conversely, suppose that $A = \bigoplus_{i \in \mathbb{Z}} A_i$ is a graded ring indexed by $\mathbb{Z}$. For an element a of A, let $a = \sum_{i \in \mathbb{Z}} a_i$, where $a_i = 0$ for all but finitely many i. Define δ_i by $\delta_i(a) = a_i$ and $\varphi : A \to A \otimes_k k[t, t^{-1}]$ by

$$\varphi(a) = \sum_{i \in \mathbb{Z}} \delta_i(a) t^i.$$

Then φ defines a G_m-action $\sigma : G_m \times X \to X$.

Answer. (1) The multiplicative groups scheme $G_m = \operatorname{Spec} k[x, x^{-1}]$ has the comultiplication (counit, or coinverse, resp.) defined by

$$\Delta(t) = t \otimes t \quad (\varepsilon(t) = 1, \text{ or } \iota(t) = t^{-1}, \text{ resp.}).$$

The associativity law $\sigma \cdot (\mathrm{id}_{G_m} \times \sigma) = \sigma \cdot (\mu \times \mathrm{id}_X) : G_m \times G_m \times X \to X$ gives a commutative diagram

$$\begin{array}{ccc} A & \xrightarrow{\varphi} & A[t,t^{-1}] \\ {\scriptstyle\varphi}\downarrow & & \downarrow{\scriptstyle \mathrm{id}_{k[t,t^{-1}]}\otimes\varphi} \\ A[t,t^{-1}] & \xrightarrow[\mathrm{id}_A\otimes\Delta]{} & A \otimes_k k[t,t^{-1}] \otimes_k k[t,t^{-1}] \end{array}$$

Hence we have

$$(\mathrm{id}_A \otimes \Delta) \cdot \varphi(a) = \sum_{i\in\mathbb{Z}} \delta_i(a) t^i \otimes t^i,$$
$$(\mathrm{id}_{k[t,t^{-1}]} \otimes \varphi) \cdot \varphi(a) = \sum_{i,j\in\mathbb{Z}} \delta_j \delta_i(a) t^j \otimes t^i,$$

whence we have $\delta_i^2 = \delta_i$ and $\delta_j \cdot \delta_i = 0$ if $i \neq j$.
On the other hand, the equality of morphisms $\sigma \cdot (e \times \mathrm{id}_X) = \mathrm{id}_X : \operatorname{Spec} k \times X \to G_m \times X \xrightarrow{\sigma} X$ gives the equality

$$\varphi(a)|_{t=1} = a, \quad a \in A.$$

This implies that $\sum_{i\in\mathbb{Z}} \delta_i = \mathrm{id}_A$, which is the idempotent decomposition of id_A in the ring of k-endomorphisms of A. If we put $A_i = \delta_i(A)$, it follows that $A_i = \{a \in A \mid \delta_i(a) = a\} = \{a \in A \mid \delta_j(a) = 0 \text{ for all } j \neq i\}$. If we write

$$^{\lambda}a = \varphi(a)|_{t=\lambda},$$

we have $A^{G_m} = \{a \in A \mid {}^{\lambda}a = a \text{ for all } \lambda \in k\} = \{a \in A \mid \delta_i(a) = 0 \text{ for all } i \neq 0\}$. Since φ is a k-algebra homomorphism, A_0 is a k-subalgebra of A.
(2) Left to the readers. □

10. Let $A_n = k[x_0, x_1, x_2, \ldots, x_{n-1}]$ be a polynomial ring in n variables. Consider a k-derivation

$$D_n = \sum_{i=1}^{n} x_{i-1} \frac{\partial}{\partial x_i}$$

which we call the *Weitzenböck derivation* in dimension n. Verify the following assertions:

(1) D_n is an lnd of A_n. Let $B_n = \operatorname{Ker} D_n$.

(2) Set

$$\varphi_1 = x_0, \quad \varphi_2 = x_2x_0 - \frac{1}{2}x_1^2, \quad \varphi_3 = x_3x_0^2 - x_2x_1x_0 + \frac{1}{3}x_1^3,$$
$$\psi_4 = x_0^{-2}[((2\varphi_2)^3 + (3\varphi_3)^2]$$
$$= 9x_3^2x_0^2 + (8x_2^3 - 18x_3x_2x_1)x_0 + 3x_1^2(2x_3x_1 - x_2^2).$$

Show that $B_2 = k[\varphi_1]$, $B_3 = k[\varphi_1, \varphi_2]$, and $B_3 = k[\varphi_1, \varphi_2, \varphi_3, \psi_4]$.

(3) $\operatorname{Spec} B_3$ is a normal hypersurface $x^2w = y^3 + z^2$, which has the singular locus $\{x = y = z = 0\} \cong \mathbb{A}^1$.

Answer. D_n defines a G_a-action on $\mathbb{A}^n$, and the action is linear. Namely, the G_a-action is determined by a matrix representation on the k-vector space $\sum_{i=0}^{n-1} kx_i$. In such a case, B_n is an affine domain by *Weitzenböck theorem* (see Seshadri [209]). Hence the algebraic quotient $\mathbb{A}^n//G_a$ exists in this case. The quotient is isomorphic to $\mathbb{A}^{n-1}$ only if $n = 2, 3$. If $n = 4$, the quotient has singularities. The above proof is left to the readers, though the computation is not easy. □

11. Let $(\mathcal{O}, t\mathcal{O})$ be a discrete valuation ring with uniformizant t. Let K and k be the quotient field and the residue field of $\mathcal{O}$, respectively. Assume that k has characteristic zero. Let A be an integral domain which is finitely generated over $\mathcal{O}$ such that A is $\mathcal{O}$-flat. Assume that $A_K = A \otimes_{\mathcal{O}} K$ and $A_k = A \otimes_{\mathcal{O}} k$ are integral domains. Prove that if A_K is affine 1-ruled, i. e., $\operatorname{Spec} A_K$ has an $\mathbb{A}^1$-fibration, so is A_k.

Answer. Write $A = \mathcal{O}[a_1, \ldots, a_r]$. Then A_K has a nonzero K-trivial lnd Δ. Then there exists an integer m such that $t^m\Delta(a_i) \in A$ for every $1 \le i \le r$. Choose m so that $D := t^m\Delta$ is an $\mathcal{O}$-trivial lnd of A such that $D(tA) \subset tA$ and $D(A) \not\subset tA$. Then D induces a nontrivial k-trivial derivation δ on A_k by setting $\delta(\overline{a}) = D(a) \pmod{tA}$ for $\overline{a} = a + tA$. Since $\delta^n(\overline{a}) = D^n(a) \pmod{tA}$, δ is a nonzero lnd on A_k. Hence $\operatorname{Spec} A_k$ has an $\mathbb{A}^1$-fibration, i. e., A_k is affine 1-ruled. □

12. Let X be a hypersurface in $\mathbb{A}^{n+1} = \operatorname{Spec} k[x, y, z_1, \ldots, z_{n-1}]$ defined by $x^my = f(x, z_1, \ldots, z_{n-1})$, where $n \ge 3$ and $m \ge 1$. Prove the following assertions which are generalizations of the assertions in Lemma 3.4.4:

(1) X is factorial if the hypersurface Z_0 defined by $f(0, z_1, \ldots, z_{n-1}) = 0$ in $\mathbb{A}^{n-1} = \operatorname{Spec} k[z_1, \ldots, z_{n-1}]$ is an irreducible and reduced variety.

(2) $A^* = k^*$ if $X_0 := \{x = 0\}$ is a nonempty closed set of X, where A is the coordinate ring of X.

(3) If $m > 1$, X is smooth if and only if there are no singular points of Z_0 which satisfy $f_x(0, z_1, \ldots, z_{n-1}) = 0$, where $f_x = \partial f/\partial x$. If $m = 1$, X is smooth if and only if Z_0 is smooth.

(4) A has mutually commuting and independent lnds $\delta_1, \dots, \delta_{n-1}$ which are defined by

$$\delta_i(x) = 0, \quad \delta_i(y) = f_{z_i}(x, z_1, \dots, z_{n-1}),$$
$$\delta_i(z_j) = 0 \ (j \neq i), \quad \delta_i(z_i) = x^m.$$

The vector fields associated with these lnds $\delta_1, \dots, \delta_{n-1}$ have the proportional directions on the fiber X_0, i. e., $\dim(\sum_{i=1}^{n-1} k\delta_{i,P}) \leq 1$ for every $P \in X_0$.

Answer. Follow the proof of Lemma 3.4.4 verbatim. □

13. Let X be the Koras–Russell threefold $x + x^2y + z^2 + t^3 = 0$. Since every G_a-action on X has x as an invariant element, there are two independent G_a-actions which correspond to the following lnds:

$$\delta_1 = -2z\frac{\partial}{\partial y} + x^2\frac{\partial}{\partial z}, \quad \delta_2 = -3t^2\frac{\partial}{\partial y} + x^2\frac{\partial}{\partial t}.$$

Prove the following assertions:

(1) The quotient morphism is given by the projection $q_1 : X \to \mathbb{A}^2 = \operatorname{Spec} k[x, t]$ for δ_1 and $q_2 : X \to \mathbb{A}^2 = \operatorname{Spec} k[x, z]$ for δ_2.

(2) Singular fibers for q_1 and q_2 are given as follows:

$$q_1^{-1}(\alpha, \beta) = \begin{cases} \mathbb{A}^1 & \text{if } \alpha \neq 0, \\ \mathbb{A}^1 + \mathbb{A}^1 & \text{if } \alpha = 0,\ \beta \neq 0, \\ 2\mathbb{A}^1 & \text{if } \alpha = \beta = 0, \end{cases}$$

$$q_2^{-1}(\alpha, \beta) = \begin{cases} \mathbb{A}^1 & \text{if } \alpha \neq 0, \\ \mathbb{A}^1 + \mathbb{A}^1 + \mathbb{A}^1 & \text{if } \alpha = 0,\ \beta \neq 0, \\ 3\mathbb{A}^1 & \text{if } \alpha = \beta = 0. \end{cases}$$

Answer. Straightforward. □

3.7 Open problems

Besides the problems given in the three chapters, we list up some of open problems.

1. Let X be a homology n-space. Is X then rational?
   If $n = 2$, X is a homology plane. There is a theorem of Gurjar–Pradeep–Shastri [78].

   Theorem 3.7.1. *Let X be a log $\mathbb{Q}$-homology plane. Namely, X is a complex normal affine surface with at worst quotient singularities and $H_i(X; \mathbb{Q}) = 0$ for every $i > 0$. Then X is rational.*

This is a very important result in developing the theory of homology planes. In fact, the vanishing of reduced homology groups works as a condition to restrict the structures of algebraic surfaces. So, one can expect the same effects in the case of dimension three or higher.

2. If X is a smooth algebraic surface defined over $\mathbb{C}$. Then the condition that $H_i(X;\mathbb{Q}) = 0$ for every $i > 0$ implies that X is affine (see [56]). If X is a smooth algebraic variety of dimension $n > 2$, does the vanishing of $H_i(X;\mathbb{Q})$ for every $i > 0$ imply that X is affine? Is this the case if $n = 3$?

Remark 3.7.2. Winkelmann [226] gave an example of a unipotent group $G \cong G_a \times G_a$ acting on $\mathbb{A}^6$ freely by affine linear transformations such that the quotient $X := \mathbb{A}^6/\!/G$ is diffeomorphic to $\mathbb{C}^4$, but not biholomorphic to $\mathbb{A}^4$. In fact, X is neither affine nor Stein. This example gives a negative answer to the question. So, the question should be modified to assume $n \geq 4$. Study a category of complex algebraic varieties X such that $H_i(X;\mathbb{Q}) = 0$ for every $i > 0$.

3. Let X, Y be homology n-spaces over $\mathbb{C}$. Suppose that $X \times \mathbb{A}^1 \cong Y \times \mathbb{A}^1$. Does it imply that $X \cong Y$? We have several results concerning this cancelation problem.
   (1) For $n > 0$, if either $\overline{\kappa}(X) \geq 0$ or $\overline{\kappa}(Y) \geq 0$, then a theorem of Fujita–Iitaka [96] implies that $X \cong Y$. Hence the case $\overline{\kappa}(X) = \overline{\kappa}(Y) = -\infty$ matters.
   (2) In this case, Lemma 3.2.1 implies that both $\Gamma(X, \mathcal{O}_X)$ and $\Gamma(Y, \mathcal{O}_Y)$ are factorial and have no nonconstant units. Hence, if $n = 2$, $X \cong Y \cong \mathbb{A}^2$ by the algebraic characterization of $\mathbb{A}^2$ (see [159, Theorems 1.3.2 and 2.2.1]).
   (3) If $n = 3$, the condition $\kappa(X) = -\infty$ does not necessarily imply that X has an $\mathbb{A}^1$-fibration. Instead, we assume that X has a nontrivial G_a-action. Then $X \times \mathbb{A}^1$ has independent and mutually commuting two G_a-actions. Hence $Y \times \mathbb{A}^1$ has also independent and mutually commuting two G_a-actions. Does this imply that Y has a nontrivial G_a-action?
4. Let $A = k[x, y, z]$ be a polynomial ring in three variables. Determine all lnds of A up to changes of coordinates. For a polynomial ring $k[x, y]$ in two variables, any lnd δ is written as $\delta(x) = 0$ and $\delta(y) = f(x) \in k[x]$ by Rentschler [193]. For an lnd δ of A, if the plinth ideal $\mathrm{pl}(\delta)$ is a principal ideal bA with $b \in \operatorname{Ker}\delta$, then $\delta' = b^{-1}\delta$ is again an lnd of A whose associated G_a-action is fixed-point free. Hence by a theorem of Kaliman [99], we have

$$\operatorname{Ker}\delta = k[x, y], \delta(z) = b \in k[x, y].$$

5. Let $f : X \to Y$ be a projective morphism from a normal variety X of dimension n to a smooth variety Y of dimension $n-1$ such that a general fiber is isomorphic to $\mathbb{P}^1$. Let C be a one-dimensional irreducible component of a fiber $F = f^{-1}(P)$. Is C isomorphic to $\mathbb{P}^1$? This is a generalization of Theorem 3.5.14 to the case $n > 3$.
6. As in section 3.1, let $\overline{U}$ be the Platonic $\mathbb{A}^1_*$-surface, i. e., $\overline{U} = (\mathbb{A}^2/\!/G) \setminus \{\overline{O}\}$, where G is a small finite subgroup of $\mathrm{GL}(2, k)$. Classify the set of isomorphism classes of

$\mathbb{A}^1$-bundles over $\overline{U}$ when G is not cyclic. In other words, determine the k-vector space $H^1(\overline{U}, G_a)$ when G is not cyclic.

7. Let $f : X \to Y$ be a dominant morphism of smooth algebraic varieties, let $Q \in Y$ and let S be an irreducible component of $f^{-1}(Q)$ with $\dim S > 1$.
   (1) Suppose that f is a projective morphism and a $\mathbb{P}^1$-fibration. Is S then rational?
   (2) Suppose that f is an affine morphism and an $\mathbb{A}^1$-fibration. Is S then rational?

Bibliography

[1] S. S. Abhyankar, Quasirational singularities, Am. J. Math. **101** (1979), no. 2, 267–300.

[2] S. S. Abhyankar and T. T. Moh, Embeddings of the line in the plane, J. Reine Angew. Math. **276** (1975), 148–166.

[3] N. A'Campo and M. Oka, Geometry of plane curves via Tschirnhausen resolution tower, Osaka J. Math. **33** (1996), no. 4, 1003–1033.

[4] A. Andreotti and T. Frankel, The Lefschetz theorem on hyperplane sections, Ann. Math. (2) **69** (1959), 713–717.

[5] A. Andreotti and H. Grauert, Théorèmes de finitude pour la cohomologie des espaces complexes, Bull. Soc. Math. Fr. **90** (1962), 193–259.

[6] A. Andreotti and Y. T. Siu, Projective embedding of pseudoconcave spaces, Ann. Sc. Norm. Super. Pisa **24** (1970), 231–278.

[7] A. Andreotti and G. Tomassini, Some Remarks on Pseudoconcave Manifolds, Essays on Topology and Related Topics, Memoire dedicated to Georges de Rham, 85–104, Springer, 1970.

[8] E. Artal-Bartolo, Une démonstration géométrique du théorème d'Abhyankar-Moh, J. Reine Angew. Math. **464** (1995), 97–108.

[9] M. Artin, Some numerical criteria for contractability of curves on algebraic surfaces, Am. J. Math. **84** (1962), 485–496.

[10] M. Artin and D. Mumford, Some elementary examples of unirational varieties which are not rational, Proc. London Math. Soc. **25** (1972), no. 3, 75–95.

[11] T. Bandman and L. Makar-Limanov, Cylinders over affine surfaces, Jpn. J. Math. (N. S.) **26** (2000), no. 1, 207–217.

[12] T. Bandman and L. Makar-Limanov, Affine surfaces with $AK(S) = \mathbb{C}$, Mich. Math. J. **49** (2001), 567–582.

[13] W. Barth, K. Hulek, C. Peters and A. Van de Ven, Compact complex surfaces, 2nd ed., Ergebnisse der Mathematik und ihrer Grenzgebiete 3. Folge., **4**, Springer-Verlag, Berlin, 2004. xii+436 pp.

[14] H. Bass, E. Connell and D. Wright, Locally polynomial algebras are symmetric algebras, Invent. Math. **38** (1976/1977), no. 3, 279–299.

[15] S. M. Bhatwadekar and A. K. Dutta, On $\mathbb{A}^*$-fibrations, J. Pure Appl. Algebra **149** (2000), 1–14.

[16] S. M. Bhatwadekar and D. Daigle, On finite generation of kernels of locally nilpotent R-derivations of $R[X, Y, Z]$, J. Algebra **322** (2009), 2915–2926.

[17] A. Białynicki-Birula, On fixed point schemes of actions of multiplicative and additive groups, Topology **12** (1973), 99–103.

[18] P. Bonnet, Surjectivity of quotient maps for algebraic $(\mathbb{C}, +)$-actions and polynomial maps with contractible fibers, Transform. Groups **7** (2002), no. 1, 3–14.

[19] G. Bredon, Topology and Geometry, Graduate Texts in Mathematics, **139**, Springer, 1993.

[20] E. Brieskorn, Ratinale Singularitäten komplexer Flächen, Invent. Math. **4** (1968), 336–358.

[21] M. Brion and D.-Q. Zhang, Log Kodaira dimension of homogeneous varieties, Algebraic varieties and automorphism groups, 1--6, Adv. Stud. Pure Math. **75**, Math. Soc. Japan, Tokyo, 2017.

[22] S. Bundagaard and J. Nielsen, On normal subgroups with finite index in F-groups, Math. Tidsskr. B (1951), 56–58.

[23] P. Cassou-Nogues and D. Daigle, Rational polynomials of simple type: a combinatorial proof, Algebraic varieties and automorphism groups, 7–28, Adv. Stud. Pure Math. **75**, Math. Soc. Japan, Tokyo, 2017.

[24] P. Cassou-Nogues and P. Russell, Birational morphismsm $\mathbb{C}^2 \to \mathbb{C}^2$ and affine ruled surfaces, Affine algebraic geometry, 57–105, Osaka Univ. Press, Osaka, 2007.

https://doi.org/10.1515/9783110577563-004

[25] S. Chakraborty, R. V. Gurjar and M. Miyanishi, Pure subrings of commutative rings, Nagoya Math. J. **221** (2016), 33–68.
[26] T. C. Chau, A note concerning Fox's paper on Fenchel's conjecture, Proc. Am. Math. Soc. **88** (1983), no. 4, 584–586.
[27] C. Chevalley, Invariants of finite groups generated by reflections, Am. J. Math. **77** (1955), 778–782.
[28] C. H. Clemens and P. A. Griffiths, The intermediate Jacobian of the cubic threefold, Ann. Math. (2) **95** (1972), 281–356.
[29] A. J. Crachiola and L. G. Makar-Limanov, An algebraic proof of a cancellation theorem for surfaces, J. Algebra **320** (2008), no. 8, 3113–3119.
[30] D. Daigle and G. Freudenburg, A note on triangular derivations of $k[X_1,X_2,X_3,X_4]$, Proc. Am. Math. Soc. **129** (2000), 657–662.
[31] D. Daigle and R. Kolhatkar, Complete intersection surfaces with trivial Makar-Limanov invariant, J. Algebra **350** (2012), 1–35.
[32] D. Daigle and P. Russell, Affine rulings of normal rational surfaces, Osaka J. Math. **38** (2001), 37–100.
[33] D. Daigle and P. Russell, On log $\mathbb{Q}$-homology planes and weighted projective planes, Can. J. Math. **56** (2004), no. 3, 1145–1189.
[34] H. Derksen, Constructive Invariant Theory and the Linearisation Problem, Ph. D. thesis, University of Basel, 1997.
[35] H. Derksen, A. van den Essen, D. R. Finston and S. Maubach, Unipotent group actions on affine varieties, J. Algebra **336** (2011), 200–208.
[36] J. K. Deveney and D. Finston, A proper G_a-action on $\mathbb{C}^5$ which is not locally trivial, Proc. Am. Math. Soc. **123** (1995), 651–654.
[37] J. K. Deveney and D. Finston, Free G_a-actions on $\mathbb{C}^3$, Proc. Am. Math. Soc. **128** (1999), 131–138.
[38] J. K. Deveney and D. Finston, Regular G_a invariants, Osaka J. Math. **39** (2002), 275–282.
[39] J. K. Deveney and D. Finston, Additive group actions with finitely generated invariants, Osaka J. Math. **44** (2007), 91–98.
[40] A. Dubouloz, Additive group actions on Danielewski varieties and the cancellation problem, Math. Z. **255** (2007), no. 1, 77–93.
[41] A. Dubouloz, The cylinder over the Koras-Russell cubic threefold has a trivial Makar-Limanov invariant, Transform. Groups **14** (2009), no. 3, 531–539.
[42] A. Dubouloz and T. Kishimoto, Log-uniruled affine varieties without cylinder-like open subsets, Bull. Soc. Math. France **143** (2015), no. 2, 383–401.
[43] A. Dubouloz and T. Kishimoto, Families of affine ruled surfaces: existence of cylinders, Nagoya Math. J. **223** (2016), no. 1, 1--20.
[44] A. Dubouloz, D. Finston and P. D. Mehta, Factorial threefolds with G_a-action. arXiv:0902.3873.
[45] E. Dufresne and H. P. Kraft, Invariants and separating morphisms for algebraic group actions, Math. Z. **280** (2015), no. 1–2, 231--255.
[46] A. K. Dutta, On $\mathbb{A}^1$-bundles of affine morphisms, J. Math. Kyoto Univ. **35** (1995), no. 3, 377–385.
[47] P. Eakin, The converse to a well-known theorem on noetherian rings, Math. Ann. **177** (1968), 278–282. Tidsskr. B (1951), 56–98.
[48] C. Ehresmann, Les connexions infinitésimales dans un espace fibré différentiable, Colloque de topologie (espaces fibrés), Bruxelles, 29–55, 1950.
[49] H. Flenner, Rationale quasihomogene Singularitöten, Arch. Math. (Basel) **36** (1981), no. 1, 35–44.
[50] H. Flenner and M. Zaidenberg, $\mathbb{Q}$-acyclic surfaces and their deformations, Classification of algebraic varieties (L'Aquila, 1992), 143–208, Contemp. Math. **162**, Amer. Math. Soc., 1994.

[51] H. Flenner, Sh. Kaliman and M. Zaidenberg, Deformation equivalence of affine ruled surfaces, 23 May 2013. arXiv:1305.5366v1.

[52] K.-H. Fieseler and L. Kaup, On the geometry of affine algebraic $\mathbb{C}^*$-surfaces, Problems in the theory of surfaces and their classification (Cortona, 1988), 111–140, Sympos. Math., XXXII, Academic Press, London, 1991.

[53] E. E. Floyd, On periodic maps and the Euler characteristics of associated spaces, Trans. Am. Math. Soc. **72** (1952), 138–147.

[54] R. H. Fox, On Fenchel's conjecture about *F*-groups, Math. Tidsskr. B (1952), 61–65.

[55] G. Freudenburg, Algebraic theory of locally nilpotent derivations, Invariant Theory and Algebraic Transformation Groups, VII, Encyclopaedia of Mathematical Sciences **136**, Springer-Verlag, Berlin, 2006. xii+261 pp.

[56] T. Fujita, On the topology of noncomplete algebraic surfaces, J. Fac. Sci., Univ. Tokyo, Sect. 1A, Math. **29** (1982), no. 3, 503–566.

[57] T. Fujita, On Zariski problem, Proc. Jpn. Acad., Ser. A, Math. Sci. **55** (1979), no. 3, 106–110.

[58] B. Giesecke, Simpliziale Zerlegung abz'ahlbarer analytische R'aume, Math. Z. **83** (1964), 173–213.

[59] M. H. Gizatullin, Affine surfaces that can be augmented by a nonsingular rational curve, Izv. Akad. Nauk SSSR, Ser. Mat. **34** (1970), 778–802. Math. USSR, Izv. **4** (1970), no. 4, 787–810.

[60] M. H. Gizatullin, Quasihomogeneous affine surfaces, Izv. Akad. Nauk SSSR, Ser. Mat. **35** (1971), 1047–1071.

[61] J. E. Goodman, Affine open subsets of algebraic varieties and ample divisors, Ann. Math. **89** (1969), 160–183.

[62] H. Grauert and R. Remmert, Theory of Stein spaces. Translated from the German by Alan Huckleberry, Grundlehren der Mathematischen Wissenschaften **236**, Springer, Berlin-New York, 1979. xxi+249 pp.

[63] A. Grothendieck and J. Dieudonné, Éléments de Géométrie Algébrique, Inst. Hautes Etudes Sci. Publ. Math. **24** (1965). 231 pp.

[64] A. Grothendieck, Cohomologie locale des faisceaux cohérents et théorèmes de Lefschetz locaux et globaux, Séminaire de Géométrie Algb́rique du Bois-Marie, 1962, Advanced Studies in Pure Mathematics **2**, North-Holland Publishing Co., Amsterdam, Masson Cie, 1968. vii+287 pp.

[65] A. Grothendieck, Techniques de construction et théorèmes d'existence en géométrie algébrique IV: Les schémas de Hilbert, Séminaire Bourbaki, 13e année **221** (1960/1961), 1–28.

[66] N. Gupta, On the cancellation problem for the affine space $\mathbb{A}^3$ in characteristic p, Invent. Math. **195** (2014), 279–288.

[67] R. V. Gurjar, Affine varieties dominated by $\mathbb{C}^2$, Comment. Math. Helv. **55** (1980), no. 3, 378–389.

[68] R. V. Gurjar, A new proof of the Abhyankar-Moh-Suzuki theorem, Transform. Groups **7** (2002), no. 1, 61–66.

[69] R. V. Gurjar, A new proof of Suzuki's formula, Proc. Indian Acad. Sci. Math. Sci. **107** (1997), no. 3, 237–242.

[70] R. V. Gurjar and S. R. Gurjar, Topological characterization of $\mathbb{C}^2$ via open algebraic surface theory, Int. J. Math. **31**, 2050055 (2020), no. 7, 8 pp.

[71] R. V. Gurjar and M. Miyanishi, Affine surfaces with $\overline{\kappa} \leq 1$, Algebraic geometry and commutative algebra, Vol. I, 99–124, Kinokuniya, Tokyo, 1988.

[72] R. V. Gurjar and M. Miyanishi, Affine lines on logarithmic $\mathbb{Q}$-homology planes, Math. Ann. **294** (1992), no. 3, 463–482.

[73] R. V. Gurjar and M. Miyanishi, On contractible curves in the complex affine plane, Tohoku Math. J. (2) **48** (1996), no. 3, 459–469.

[74] R. V. Gurjar and M. Miyanishi, Automorphisms of affine surfaces with $\mathbb{A}^1$-fibrations, Mich. Math. J. **53** (2005), no. 1, 33–55.
[75] R. V. Gurjar and A. J. Parameswaran, Affine lines on $\mathbb{Q}$-homology planes, J. Math. Kyoto Univ. **35** (1995), no. 1, 63–77.
[76] R. V. Gurjar and A. R. Shastri, A topological characterization of $\mathbb{C}^2/G$, J. Math. Kyoto Univ. **25** (1985), no. 4, 767–773.
[77] R. V. Gurjar and A. R. Shastri, The fundamental group at infinity of affine surfaces, Comment. Math. Helv. **59** (1984), no. 3, 459–484.
[78] R. V. Gurjar, C. R. Pradeep and A. R. Shastri, On rationality of logarithmic $\mathbb{Q}$-homology planes. II, Osaka J. Math. **34** (1997), no. 3, 725–743.
[79] R. V. Gurjar, K. Masuda and M. Miyanishi, $\mathbb{A}^1$-fibrations on affine threefolds, J. Pure Appl. Algebra **216** (2012), no. 2, 296–313.
[80] R. V. Gurjar, K. Masuda and M. Miyanishi, Surjective derivations in small dimensions, J. Ramanujan Math. Soc. 28A (2013), 221--246.
[81] R. V. Gurjar, K. Masuda and M. Miyanishi, Deformations of $\mathbb{A}^1$-fibrations, Automorphisms in birational and affine geometry, 327–361, Springer Proc. Math. Stat. **79**, Springer, Cham, 2014.
[82] R. V. Gurjar, K. Masuda and M. Miyanishi, Affine threefolds with $\mathbb{A}^2$-fibrations, Transform. Groups **22** (2017), no. 1, 187–205.
[83] R. V. Gurjar, K. Masuda, M. Miyanishi and P. Russell, Affine lines on affine surfaces and the Makar-Limanov invariant, Can. J. Math. **60** (2008), no. 1, 109–139.
[84] R. V. Gurjar, M. Koras, K. Masuda, M. Miyanishiand and P. Russell, $\mathbb{A}^1_*$-fibrations on affine threefolds, Affine algebraic geometry, 62–102, World Sci. Publ., Hackensack, NJ, 2013.
[85] R. V. Gurjar, M. Koras, M. Miyanishi and P. Russell, Affine normal surfaces with simply-connected smooth locus, Math. Ann. **353** (2012), no. 1, 127–144.
[86] R. V. Gurjar, M. Koras, M. Miyanishi and P. Russell, A homology plane of general type can have at most a cyclic quotient singularity, J. Algebraic Geom. **23** (2014), 1–62.
[87] R. V. Gurjar, K. Masuda, M. Miyanishi and P. Russell, Affine lines on affine surfaces and the Makar-Limanov invariant, Can. J. Math. **60** (2008), no. 1, 109–139.
[88] H. A. Hamm, Zur Homotopietyp Steinscher Räume, J. Reine Angew. Math. **338** (1983), 121–135.
[89] R. Hartshorne, Cohomological dimension of algebraic varieties, Ann. Math. **88** (1968), 403–450.
[90] R. Hartshorne, Algebraic Geometry, Graduate Texts in Mathematics **52**, Springer, New York-Heidelberg, 1977.
[91] I. Hedén, Affine extensions of principal additive bundles over a punctured surface, Transform. Groups **21** (2016), no. 2, 427–449.
[92] I. Hedèn, Russell's hypersurface from a geometric point of view, Osaka J. Math. **53** (2016), no. 3, 637–644.
[93] G. Horrocks, Fixed point schemes of additive group actions, Topology **8** (1969), 233–242.
[94] S. Iitaka, Deformations of compact complex surfaces, Global Analysis (Papers in Honor of K. Kodaira), 267–272, Univ. Tokyo Press, Tokyo, 1969.
[95] S. Iitaka, Algebraic geometry. An introduction to birational geometry of algebraic varieties, Graduate Texts in Mathematics, **76**, North-Holland Mathematical Library, 24 Springer, New York-Berlin, 1982. x+357 pp.
[96] S. Iitaka and T. Fujita, Cancellation theorem for algebraic varieties, J. Fac. Sci., Univ. Tokyo, Sect. 1A, Math. **24** (1977), no. 1, 123–127.
[97] Z. Jelonek, A number of points in the set $\mathbb{C}^2 \setminus F(\mathbb{C}^2)$, Bull. Pol. Acad. Sci., Math. **47** (1999), no. 3, 257–261.
[98] Sh. Kaliman, Polynomials with general $\mathbb{C}^2$-fibers are variables, Pac. J. Math. **203** (2002), no. 1, 161–190.
[99] Sh. Kaliman, Free $\mathbb{C}^+$-actions on $\mathbb{C}^3$ are translations, Invent. Math. **156** (2004), no. 1, 163–173.

[100] Sh. Kaliman, One more proof of the Abhyankar-Moh-Suzuki theorem, Algebraic varieties and automorphism groups, 183–188, Adv. Stud. Pure Math. **75**, Math. Soc. Japan, Tokyo, 2017.
[101] Sh. Kaliman and L. Makar-Limanov, On the Russell-Koras contractible threefolds, J. Algebraic Geom. **6** (1997), no. 2, 247–268.
[102] Sh. Kaliman and L. Makar-Limanov, *AK*-invariant of affine domains, Affine algebraic geometry, 231–255, Osaka Univ. Press, Osaka, 2007.
[103] Sh. Kaliman and N. Saveliev, $\mathbb{C}^+$-actions on contractible threefolds, Mich. Math. J. **52** (2004), no. 3, 619–625.
[104] Sh. Kaliman and M. Zaidenberg, Affine modifications and affine hypersurfaces with a very transitive automorphism group, Transform. Groups **4** (1999), no. 1, 53–95.
[105] Sh. Kaliman and M. Zaidenberg, Families of affine planes: the existence of a cylinder, Mich. Math. J. **49** (2001), no. 2, 353–367.
[106] T. Kambayashi, On the absence of nontrivial separable forms of the affine plane, J. Algebra **35** (1975), 449–456.
[107] T. Kambayashi and M. Miyanishi, On flat fibrations by the affine line, Ill. J. Math. **22** (1978), no. 4, 662–671.
[108] T. Kambayashi and P. Russell, On linearizing algebraic torus actions, J. Pure Appl. Algebra **23** (1982), 243–250.
[109] T. Kambayashi and D. Wright, Flat families of affine lines are affine-line bundles, Ill. J. Math. **29** (1985), no. 4, 672–681.
[110] M.-C. Kang, On Abhyankar-Moh's epimorphism theorem, Am. J. Math. **113** (1991), no. 3, 399–421.
[111] T. Kishimoto, A new proof of a theorem of Ramanujam-Morrow, J. Math. Kyoto Univ. **42**, no. 1, 117–139 (2002).
[112] L. Kaup, Eine topologische Eigenschaft Steinscher Räume, Nachr. Akad. Wiss. Göttingen Math.-Phys. Kl. II **1966** (1966), 213–224.
[113] Y. Kawamata, Addition formula of logarithmic Kodaira dimensions for morphisms of relative dimension one, Proc. of the Internat. Symposium on Algebraic Geometry, Kyoto Univ., Kyoto, 1977, 207–217, Kinokuniya Book Store, Tokyo, 1978.
[114] Y. Kawamata, On deformations of compactifiable complex manifolds, Math. Ann. **235** (1978), no. 3, 247–265.
[115] Y. Kawamata, On the classification of non-complete algebraic surfaces, Algebraic Geometry, 215–232, Lecture Notes in Math., **732**, Springer, Berlin, 1979.
[116] F. Knop, H.-P. Kraft, D. Luna and T. Vust, Local properties of algebraic group actions, Algebraische Transformationsgruppen und Invariantentheorie, 63–75, DMV Sem. **13**, Birkhauser, Basel, 1989.
[117] R. Kobayashi, Uniformization of complex surfaces, Kähler metric and moduli spaces, 313–394, Adv. Stud. Pure Math. **18-II**, Academic Press, Boston, MA, 1990.
[118] K. Kodaira, On the stability of compact submanifolds of complex mnaifolds, Am. J. Math. **85** (1963), 79–94.
[119] K. Kodaira, Complex manifolds and deformation of complex structures, Classics in Mathematics, Springer, 1986.
[120] J. Kollár, Rational Curves on Algebraic Varieties, Ergebnisse der Mathematik und ihrer Grenzgebiete, 3 Folge **32**, Springer, 1996.
[121] R. Kobayashi, S. Nakamura and F. Sakai, A numerical characterization of ball quotients for normal surfaces with branch loci, Proc. Jpn. Acad. Ser. A **65** (1989), 238–241.
[122] J. Kollar and S. Mori, Birational geometry of algebraic varieties, Cambridge Tracts in Mathematics **134**, Cambridge University Press, Cambridge, 1998.
[123] M. Koras, Linearization of reductive group actions, Group Actions and Vector Fields, 92–98, Lecture Notes in Math. **956**, 1982.

[124] M. Koras, On contractible plane curves, Affine algebraic geometry, 275–288, Osaka Univ. Press, Osaka, 2007.

[125] M. Koras and P. Russell, Codimension 2 torus actions on affine n-space, Group actions and invariant theory, Montreal, PQ, 1988, 103–110, CMS Conf. Proc. **10**, Amer. Math. Soc., Providence, RI, 1989.

[126] M. Koras and P. Russell, Contractible threefolds and $\mathbb{C}^*$-actions on $\mathbb{C}^3$, J. Algebraic Geom. **6** (1997), no. 4, 671–695.

[127] M. Koras and P. Russell, $\mathbb{C}^*$-actions on $\mathbb{C}^3$: the smooth locus of the quotient is not of hyperbolic type, J. Algebraic Geom. **8** (1999), no. 4, 603–694.

[128] H. P. Kraft, Geometrische Methoden in der Invariantentheorie, Aspects of Mathematics, D1, Friedr. Vieweg & Sohn, Braunschweig, 1984.

[129] H.-P. Kraft, Free $\mathbb{C}^+$-actions on affine threefolds, Contemp. Math. **369** (2005), 165–175.

[130] H.-P. Kraft and V. L. Popov, Semisimple group actions on the three-dimensional affine space are linear, Comment. Math. Helv. **60** (1985), no. 3, 466–479.

[131] H. P. Kraft, T. Petrie and J. D. Randall, Quotient varieties, Adv. Math. **74** (1989), no. 2, 145–162.

[132] H.-P. Kraft and P. Russell, Families of group actions, generic isotriviality, and linearization, Transform. Groups **19** (2014), no. 3, 779–792.

[133] H. P. Kraft and G. W. Schwarz, Reductive group actions with one-dimensional quotient, Publ. Math. IHÉS **76** (1992), 1–97.

[134] S. Kuroda, A counterexample to the fourteenth problem of Hilbert in dimension three, Mich. Math. J. **53** (2005), no. 1, 123–132.

[135] D. Luna, Slices étales, Sur les groupes algébriques, 81–105, Bull. Soc. Math. France, Paris, Mémoire **33**, Soc. Math. France, Paris, 1973.

[136] A. Maharana, $\mathbb{Q}$-homology planes as cyclic covers of $\mathbb{A}^2$, J. Math. Soc. Jpn. **61** (2009), no. 2, 393–425.

[137] L. Makar-Limanov, On groups of automorphisms of a class of surfaces, Isr. J. Math. **69** (1990), 250–256.

[138] L. Makar-Limanov, On the group of automorphisms of a surface $x^n y = P(z)$, Isr. J. Math. **121** (2001), 113–123.

[139] L. Makar-Limanov, A new proof of the Abhyankar-Moh-Suzuki theorem via a new algorithm for the polynomial dependence, J. Algebra Appl. **14** (2015), no. 9, 1540001.

[140] M. Maruyama, On classification of ruled surfaces, Lectures in Mathematics, Department of Mathematics, Kyoto University **3**, Kinokuniya Book-Store Co., Ltd., Tokyo, 1970, iv+75 pp.

[141] K. Masuda, G-endomorphisms of affine G-varieties which induce automorphisms of the invariant subrings of the coordinate rings, J. Algebra **307** (2007), no. 1, 97–105.

[142] K. Masuda, Characterizations of hypersurfaces of a Danielewski type, J. Pure Appl. Algebra **218** (2014), no. 4, 624–633.

[143] K. Masuda and M. Miyanishi, The additive group actions on $\mathbb{Q}$-homology planes, Ann. Inst. Fourier (Grenoble) **53** (2003), no. 2, 429–464.

[144] K. Masuda and M. Miyanishi, Affine pseudo-planes and cancellation problem, Trans. Am. Math. Soc. **357** (2005), no. 12, 4867–4883.

[145] K. Masuda and M. Miyanishi, Lifting of the additive group scheme actions, Tohoku Math. J. (2) **61** (2009), no. 2, 267–286.

[146] H. Matsumura, Commutative ring theory, 2nd ed., Cambridge Studies in Advanced Mathematics **8**, Cambridge University Press, Cambridge, 1989. xiv+320 pp. .

[147] Y. Matsushima, Éspaces homogènes de Stein des groupes de Lie complexes, Nagoya Math. J. **16** (1960), 205–218.

[148] M. Miyanishi, An algebraic characterization of the affine plane, J. Math. Kyoto Univ. **15** (1975), 169–184.

[149] M. Miyanishi, Analytic irreducibility of certain curves on a nonsingular affine rational surface, Proceedings of the International Symposium on Algebraic Geometry, Kyoto Univ., Kyoto, 1977, 575–587, Kinokuniya Book Store, Tokyo, 1978.
[150] M. Miyanishi, Curves on rational and unirational surfaces, Tata Institute of Fundamental Research Lectures on Mathematics and Physics, **60**. Tata Institute of Fundamental Research, Bombay; by the Narosa Publishing House, New Delhi, 1978. ii+302 pp.
[151] M. Miyanishi, Regular subrings of a polynomial ring, Osaka J. Math. **17** (1980), no. 2, 329–338.
[152] M. Miyanishi, Non-complete algebraic surfaces, Lecture Notes in Mathematics **857**, Springer-Verlag, Berlin-New York, 1981. xviii+244 pp.
[153] M. Miyanishi, Singularities of normal affine surfaces containing cylinderlike open sets, J. Algebra **68** (1981), no. 2, 268–275.
[154] M. Miyanishi, Normal affine subalgebras of a polynomial ring, Algebraic and topological theories, Kinosaki, 1984, 37–51, Kinokuniya, Tokyo, 1986.
[155] M. Miyanishi, An algebro-topological characterization of the affine space of dimension three, Am. J. Math. **106** (1984), no. 6, 1469–1485.
[156] M. Miyanishi, Algebraic characterizations of the affine 3-space, Algebraic Geometry Seminar, Singapore, 1987, 53–67, World Sci. Publishing, Singapore, 1988.
[157] M. Miyanishi, Recent topics in open algebraic surfaces, Sūgaku **46** (1994), no. 3, 243–257. Amer. Math. Soc. Transl. (2), **172** (1996), 61–76.
[158] M. Miyanishi, Algebraic geometry, Translated from the 1990 Japanese original by the author, Translations of Mathematical Monographs **136**, Amer. Math. Soc., Providence, RI, 1994. xii+246 pp.
[159] M. Miyanishi, Open algebraic surfaces, CRM Monograph Series **12**, Amer. Math. Soc., Providence, RI, 2001. viii+259 pp.
[160] M. Miyanishi, Completely parametrized $\mathbb{A}^1_*$-fibrations on the affine plane, Computational commutative algebra and combinatorics, Osaka, 1999, 185–201, Adv. Stud. Pure Math. **33**, Math. Soc. Japan, Tokyo, 2002.
[161] M. Miyanishi, Recent developments in affine algebraic geometry, Affine Algebraic Geometry, edited by T. Hibi, 307–378, Osaka University Press, 2007.
[162] M. Miyanishi, G_a-actions and completions, J. Algebra **319** (2008), no. 7, 2845–2854.
[163] M. Miyanishi and K. Nakamura, On the structure of minimal surfaces of general type with $2p_g = (K^2) + 2$, J. Math. Kyoto Univ. **18** (1978), no. 1, 137–171.
[164] M. Miyanishi and T. Sugie, Affine surfaces containing cylinderlike open sets, J. Math. Kyoto Univ. **20** (1980), no. 1, 11–42.
[165] M. Miyanishi and T. Sugie, Generically rational polynomials, Osaka Math. J. **17** (1980), no. 2, 339–362.
[166] M. Miyanishi and T. Sugie, Homology planes with quotient singularities, J. Math. Kyoto Univ. **31** (1991), no. 3, 755–788.
[167] M. Miyanishi and S. Tsunoda, Noncomplete algebraic surfaces with logarithmic Kodaira dimension $-\infty$ and with nonconnected boundaries at infinity, Jpn. J. Math. New Ser. **10** (1984), no. 2, 195–242.
[168] M. Miyanishi and S. Tsunoda, Logarithmic del Pezzo surfaces of rank one with noncontractible boundaries, Jpn. J. Math. New Ser. **10** (1984), no. 2, 271–319.
[169] S. Mori, Graded factorial domains, Jpn. J. Math. New Ser. **3** (1977), no. 2, 223–238.
[170] J. A. Morrow, Minimal normal compactifications of $\mathbb{C}^2$, Complex Analysis, (Proc. Conf. Rice Univ., Houston, Tex., 1972), Vol. I: Geometry of singularities, 97–112, no. 1, Rice Univ. Studies **59**, 1973.
[171] D. Mumford, The topology of normal singularities of an algebraic surface and a criterion for simplicity, Publ. Math. IHES **9** (1961), 229–246.

[172] D. Mumford, J. Fogarty and F. Kirwan, Geometric invariant theory, 3rd ed., Ergebnisse der Mathematik und ihrer Grenzgebiete (2) **34**, Springer-Verlag, Berlin, 1994. xiv+292 pp.
[173] M. Nagata, A treatise on the fourteenth problem of Hilbert, Mem. Sci. Kyoto Univ. **30** (1956), 57–70.
[174] M. Nagata, A remark on the unique factorization theorem, J. Math. Soc. Jpn. **9** (1957), 143–145.
[175] M. Nagata, Local rings, Interscience Tracts in Pure and Applied Mathematics **13**, John Wiley & Sons, New York-London, 1962.
[176] M. Nagata, A generalization of the imbedding problem of an abstract variety in a complete variety, J. Math. Kyoto Univ. **3** (1963), 89–102.
[177] M. Nagata, A type of subrings of a noetherian ring, J. Math. Kyoto Univ. **8** (1968), 465–467.
[178] R. Narasimhan, On the homology groups of Stein spaces, Invent. Math. **2** (1967), 377–385.
[179] W. D. Neumann, Complex algebraic plane curves via their links at infinity, Invent. Math. **98** (1989), no. 3, 445–489.
[180] W. D. Neumann, On the topology of curves in complex surfaces, Topological methods in algebraic transformation groups, New Brunswick, NJ, 1988, 117–133, Progr. Math. **80**, Birkhauser Boston, Boston, MA, 1989.
[181] W. D. Neumann and P. Norbury, Vanishing cycles and monodromy of complex polynomials, Duke Math. J. **101** (2000), no. 3, 487–497.
[182] M. Nori, Zariski's conjecture and related problems, Ann. Scit. École Norm. Sup. 4 Ser. **16** (1983), 305–344.
[183] A. Nowicki and M. Nagata, Rings of constants for k-derivations in $k[x_1, \dots, x_n]$, J. Math. Kyoto Univ. **28** (1988), no. 1, 111–118.
[184] S. Y. Orevkov and M. G. Zaidenberg, On rigid rational cuspidal plane curves, Russ. Math. Surv. **51** (1996), no. 1, 179–180.
[185] K. Palka, Exceptional singular $\mathbb{Q}$-homology planes, Ann. Inst. Fourier (Grenoble) **61** (2011), no. 2, 745–774.
[186] K. Palka, A new proof of the theorems of Lin-Zaidenberg and Abhyankar-Moh-Suzuki, J. Algebra Appl. **14**, no. 9, 1540012, 15 pp. (2015).
[187] V. L. Popov, Classification of affine algebraic surfaces that are quasihomogeneous with respect to an algebraic group, Izv. Akad. Nauk SSSR, Ser. Mat. **37** (1973), 1038–1055.
[188] V. L. Popov, Picard groups of homogeneous spaces of linear algebraic groups and one-dimensional homogeneous vector fiberings, Izv. Akad. Nauk SSSR, Ser. Mat. **38** (1974), 294–322.
[189] V. L. Popov, On the Makar-Limanov, Derksen invariants, and finite automorphism groups of algebraic varieties, Affine Algebraic Geometry, the Russell Festschrift, 289–311, CRM Proceedings and Lecture Notes **54**, Amer. Math. Soc., 2011.
[190] V. L. Popov and E. B. Vinberg, Invariant theory, Encyclopaedia of Mathematical Science **55**, 123–278, Springer, 1994.
[191] S. Ramanan, A note on C. P. Ramanujam, C. P. Ramanujam—a tribute, 11–13, Tata Inst. Fund. Res. Studies in Math. **8**, Springer, Berlin-New York, 1978.
[192] C. P. Ramanujam, A topological characterization of the affine plane as an algebraic variety, Ann. Math. **94** (1971), 69–88.
[193] R. Rentschler, Opérations du groupe additif sur le plan affine, C. R. Acad. Sci. Paris Sér. A-B **267** (1968), A384–A387.
[194] D. R. Richman, On the computation of minimal polynomials, J. Algebra **103** (1986), no. 1, 1–17.
[195] P. Roberts, An infinitely generated symbolic blow-up in a power series ring and a new counterexample to Hilbert's fourteenth problem, J. Algebra **132** (1990), 461–473.
[196] M. Rosenlicht, Some basic theorems on algebraic groups, Am. J. Math. **78** (1956), 401–443.
[197] R. Lee, Embeddings of the line in the plane, J. Reine Angew. Math. **337** (1982), 113–118.

[198] P. Russell, Gradings of polynomial rings, Algebraic geometry and its applications, edited by C. L. Bajaj, Springer, 1994.
[199] P. Russell, Embeddings problems in affine algebraic geometry, Polynomial automorphisms and related topics, 113–135, Publishing House for Science and Technology, Hanoi, 2007.
[200] J. Rynes, Nonsingular affine k^*-surfaces, Trans. Am. Math. Soc. **332** (1992), no. 2, 889–921.
[201] H. Saito, Fonctions entières qui se réduisent à certains polynomes I, Osaka Math. J. **9** (1972), 293–332. II, Osaka Math. J. **14** (1977), 649–674.
[202] F. Sakai, Classification of normal surfaces, Algebraic geometry, Bowdoin, 1985, 451–465, Proc. Sympos. Pure Math., **46**, Part 1, Amer. Math. Soc., Providence, RI, 1987.
[203] I. Sasao, Generically rational polynomials of quasi-simple type, J. Algebra **298** (2006), no. 1, 58–104.
[204] A. Sathaye, Polynomial ring in two variables over a DVR: a criterion, Invent. Math. **74** (1983), no. 1, 159–168.
[205] G. Scheja and U. Storch, Fortsetzung von Derivationen, J. Algebra **54** (1978), no. 2, 353–365.
[206] A. Seidenberg, The hyperplane sections of normal varieties, Trans. Am. Math. Soc. **69** (1950), 357–386.
[207] A. Seidenberg, Derivations and integral closure, Pac. J. Math. **16** (1966), 167–173.
[208] F. Serrano, Multiple fibres of a morphism, Comment. Math. Helv. **65** (1990), no. 2, 287–298.
[209] C. S. Seshadri, On a theorem of Weitzenböck in invariant theory, J. Math. Kyoto Univ. **1** (1961/62), 403–409.
[210] C. S. Seshadri, Quotient spaces modulo reductive algebraic groups, Ann. Math. **95** (1972), 511–556.
[211] A. R. Shastri, Divisors with finite local fundamental group on a surface, Proc. Sympos. Pure Math. **46**, Part 1, 467–481, Amer. Math. Soc., Providence, RI, 1987.
[212] G. C. Shephard and J. A. Todd, Finite unitary reflection groups, Can. J. Math. **6** (1954), 274–304.
[213] E. H. Spanier, Algebraic topology, Springer, 1966.
[214] R. Steinberg, Conjugacy classes in algebraic groups, Lect. notes in math. **366**, Springer, 1974.
[215] T. Sugie, On T. Petrie's problem concerning homology planes, J. Math. Kyoto Univ. **30** (1990), no. 2, 317–342.
[216] H. Sumihiro, Equivariant completion, J. Math. Kyoto Univ. **14** (1974), 1–28. II, J. Math. Kyoto Univ. **15** (1975), 573–605.
[217] M. Suzuki, Propriétés topologiques des polynômes de deux variables complexes, et automorphismes algébriques de l'espace $\mathbb{C}^2$, J. Math. Soc. Jpn. **26** (1974), 241–257.
[218] M. Suzuki, Sur les opérations holomorphes du groupe additif sur l'espace de deux variables complexes, Ann. Sci. Éc. Norm. Sup. (4) **10** (1977), 517–546.
[219] R. Tanimoto, Algebraic torus actions on affine algebraic surfaces, J. Algebra **285** (2005), no. 1, 73–97.
[220] R. Thom, Quelques propriétés globales des variétés différentiables, Comment. Math. Helv. **28** (1954), 17–86.
[221] T. tom Dieck and T. Petrie, Contractible affine surfaces of Kodaira dimension one, Jpn. J. Math. **16** (1990), 147–169.
[222] B. Totaro, Affine varieties and Stein manifolds, Int. J. Math. **2** (1991), no. 5, 563–566.
[223] W. V. Vasconcelos, Derivations of commutative noetherian rings, Math. Z. **112** (1969), 229–233.
[224] R. Weitzenböck, Über die Invarianten von linearen Gruppen, Acta Math. **58** (1932), no. 1, 231–293.
[225] R. O. Wells, Differential Analysis on Complex Manifolds, Graduate Texts in Mathematics, **65**, Springer, 2008.

[226] J. Winkelmann, On free holomorphic $\mathbb{C}$-actions on $\mathbb{C}^n$ and homogeneous Stein manifolds, Math. Ann. **266** (1990), 593–612.
[227] J. Winkelmann, Invariant rings and quasi-affine quotients, Math. Z. **244** (2003), no. 1, 163–194.
[228] J. Winkelmann, A correspondence to P. Russell, August 23, 2011.
[229] G. Xiao, Surfaces fibrées en courbes de genre deux, Lecture Notes in Mathematics **1137**, Springer-Verlag, Berlin, 1985. x+103 pp.
[230] G. Xiao, π_1 of elliptic and hyperelliptic surfaces, Int. J. Math. **2** (1991), 599–615.
[231] H. Yoshihara, On plane rational curves, Proc. Jpn. Acad. Ser. A **55** (1979), 152–551.
[232] M. Zaidenberg, Isotrivial families of curves on affine surfaces and characterizations of the affine plane, Math. USSR, Izv. **30** (1988), 503–532.
[233] M. G. Zaıdenberg and V. Ya. Lin, An irreducible, simply connected algebraic curve in $\mathbb{C}^2$ is equivalent to a quasihomogeneous curve, Dokl. Akad. Nauk SSSR **271** (1983), no. 5, 1048–1052.
[234] O. Zariski, Interprétations algébrico-géométriques du quatorzième probléme de Hilbert, Bull. Sci. Math. (2) **78** (1954), 155–168.
[235] O. Zariski, Studies in equisingularity I. Equivalent singularities of plane algebroid curves, Am. J. Math. **87** (1965), 507–536.
[236] H. Żołądek, A new topological proof of the Abhyankar-Moh theorem, Math. Z. **244** (2003), no. 4, 689–695.

Index

De Gruyter Studies in Mathematics

Volume 78
Kai Liu, Ilpo Laine, Lianzhong Yang
Complex Delay-Differential Equations, 2021
ISBN 978-3-11-056016-9, e-ISBN 978-3-11-056056-5,
e-ISBN (ePUB) 978-3-11-056040-4

Volume 77
Laura Costa, Rosa Maria Miró-Roig, Joan Pons-Llopis
Ulrich Bundles. From Commutative Algebra to Algebraic Geometry, 2021
ISBN 978-3-11-064540-8, e-ISBN 978-3-11-064768-6,
e-ISBN (ePUB) 978-3-11-064580-4

Volume 76
Anton Zettl
Recent Developments in Sturm-Liouville Theory, 2021
ISBN 978-3-11-071884-3, e-ISBN 978-3-11-071900-0,
e-ISBN (ePUB) 978-3-11-071938-3

Volume 75
Elena Cordero, Luigi Rodino
Time-Frequency Analysis of Operators, 2020
ISBN 978-3-11-053035-3, e-ISBN 978-3-11-053245-6,
e-ISBN (ePUB) 978-3-11-053060-5

Volume 25
Karl H. Hofmann, Sidney A. Morris
The Structure of Compact Groups, 4th Edition, 2020
ISBN 978-3-11-069595-3, e-ISBN 978-3-11-069599-1,
e-ISBN (ePUB) 978-3-11-069601-1

Volume 74
Alessandro Carbotti, Serena Dipierro, Enrico Valdinoci
Local Density of Solutions to Fractional Equations, 2019
ISBN 978-3-11-066069-2, e-ISBN 978-3-11-066435-5,
e-ISBN (ePUB) 978-3-11-066131-6

www.ingramcontent.com/pod-product-compliance
Ingram Content Group UK Ltd.
Pitfield, Milton Keynes, MK11 3LW, UK
UKHW050954290926
15470UKWH00006B/1808

9 783110 577365